Advances in Intelligent Systems and Computing

Volume 691

Series editor

Janusz Kacprzyk, Polish Academy of Sciences, Warsaw, Poland
e-mail: kacprzyk@ibspan.waw.pl

About this Series

The series "Advances in Intelligent Systems and Computing" contains publications on theory, applications, and design methods of Intelligent Systems and Intelligent Computing. Virtually all disciplines such as engineering, natural sciences, computer and information science, ICT, economics, business, e-commerce, environment, healthcare, life science are covered. The list of topics spans all the areas of modern intelligent systems and computing.

The publications within "Advances in Intelligent Systems and Computing" are primarily textbooks and proceedings of important conferences, symposia and congresses. They cover significant recent developments in the field, both of a foundational and applicable character. An important characteristic feature of the series is the short publication time and world-wide distribution. This permits a rapid and broad dissemination of research results.

Advisory Board

Chairman

Nikhil R. Pal, Indian Statistical Institute, Kolkata, India
e-mail: nikhil@isical.ac.in

Members

Rafael Bello Perez, Universidad Central "Marta Abreu" de Las Villas, Santa Clara, Cuba
e-mail: rbellop@uclv.edu.cu

Emilio S. Corchado, University of Salamanca, Salamanca, Spain
e-mail: escorchado@usal.es

Hani Hagras, University of Essex, Colchester, UK
e-mail: hani@essex.ac.uk

László T. Kóczy, Széchenyi István University, Győr, Hungary
e-mail: koczy@sze.hu

Vladik Kreinovich, University of Texas at El Paso, El Paso, USA
e-mail: vladik@utep.edu

Chin-Teng Lin, National Chiao Tung University, Hsinchu, Taiwan
e-mail: ctlin@mail.nctu.edu.tw

Jie Lu, University of Technology, Sydney, Australia
e-mail: Jie.Lu@uts.edu.au

Patricia Melin, Tijuana Institute of Technology, Tijuana, Mexico
e-mail: epmelin@hafsamx.org

Nadia Nedjah, State University of Rio de Janeiro, Rio de Janeiro, Brazil
e-mail: nadia@eng.uerj.br

Ngoc Thanh Nguyen, Wroclaw University of Technology, Wroclaw, Poland
e-mail: Ngoc-Thanh.Nguyen@pwr.edu.pl

Jun Wang, The Chinese University of Hong Kong, Shatin, Hong Kong
e-mail: jwang@mae.cuhk.edu.hk

More information about this series at http://www.springer.com/series/11156

Feng Qiao · Srikanta Patnaik
John Wang
Editors

Recent Developments in Mechatronics and Intelligent Robotics

Proceedings of the International Conference on Mechatronics and Intelligent Robotics (ICMIR2017) - Volume 2

 Springer

Editors
Feng Qiao
Shenyang Jianzhu University
Shenyang
China

John Wang
Montclair State University
Montclair, NJ
USA

Srikanta Patnaik
SOA University
Bhubaneswar
India

ISSN 2194-5357   ISSN 2194-5365   (electronic)
Advances in Intelligent Systems and Computing
ISBN 978-3-319-70989-5   ISBN 978-3-319-70990-1   (eBook)
DOI 10.1007/978-3-319-70990-1

Library of Congress Control Number: 2017958551

Printed on acid-free paper

This Springer imprint is published by Springer Nature
The registered company is Springer International Publishing AG
The registered company address is: Gewerbestrasse 11, 6330 Cham, Switzerland

Preface

On behalf of the Organizing Committee I welcome all the delegates to the International Conference on Mechatronics and Intelligent Robotics (ICMIR2017) held at Kunming, China, during May 20–21, 2017. This annual conference is being organized each year by Interscience Research Network, an international professional body, in association with International Journal of Computational Vision and Robotics and International Journal of Simulation and Process Modelling, published by Inderscience Publishing House. I must welcome this year's General Chair Prof. Feng Qiao, Shenyang JianZhu University, Shenyang, China, for his generous contribution to ICMIR-2017. He has also contributed an issue of his journal International Journal of Simulation and Process Modelling to the selected papers of the conference.

Like every edition, this edition of ICMIR2017 was academically very rich and we had three eminent professors as keynote speakers namely Prof. John Wang, Dept. of Information Management & Business Analytics, School of Business Montclair State University, USA, Prof. Kevin Deng, Distinguished Professor and Executive Director of Automotive Research Institute, Jilin University, and Dr. Nilanjan Dey, Department of Information Technology, Techno India College of Technology, Kolkata, India.

There has been a rapid developments during last five years in various directions such as: robotic-assisted manufacturing; advanced mechanisms and robotics; systems modelling and analysis; instrumentation and device control; automation systems; intelligent sensing and control; medical robotics; and autonomous and complex systems. New technologies are constantly emerging, which are enabling applications in various domains and services. Intelligent Mechatronics and Robotics is no longer a functional area within the department of mechanical or electronics, but is an integral part of the manufacturing function of any organization. In the recent time, Intelligent Mechatronics and Robotics is probably the single most important facilitator of the manufacturing process. The result of research in this domain is now influencing the process of globalization, particularly in the productive, manufacturing and commercial spheres. Creating economic opportunities and contributing to monotony reduction is another thrust area for the emerging epoch of Intelligent Mechatronics and Robotics.

This edition of ICIMR covered the following areas but not limited to intelligent mechatronics, robotics and biomimetics, novel and unconventional mechatronic systems, modelling and control of mechatronics systems, elements, structures, mechanisms of micro- and nano-systems, sensors, wireless sensor networks and multi-sensor data fusion, biomedical and rehabilitation engineering, prosthetics and artificial organs, AI, neural networks and fuzzy logic in mechatronics and robotics, industrial automation, process control and networked control systems, telerobotics, human–computer interaction, human–robot interaction, artificial intelligence, bio-inspired robotics, control algorithms and control systems, design theories and principles, evolutionary robotics, field robotics, force sensors, accelerometers, and other measuring devices, healthcare robotics, human–robot interaction, kinematics and dynamics analysis, manufacturing robotics, mathematical and computational methodologies in robotics, medical robotics, parallel robots and manipulators, robotic cognition and emotion, robotic perception and decision, sensor integration, fusion, and perception. This volume covers various articles covering the recent developments in the area of Intelligent Mechatronics and Robotics categorized into seven (7) tracks, such as:

1. Intelligent Systems
2. Intelligent Sensor & Actuator
3. Robotics
4. Mechatronics
5. Modelling & Simulation
6. Automation & Control and
7. Robot Vision

Srikanta Patnaik
Programme Chair: ICMIR-2017

Conference Organizing Committee

General Chair

Feng Qiao   Shenyang JianZhu University, Shenyang, China

Programme Chair

Srikanta Patnaik   SOA University, Bhubaneswar, Odisha, India

Organizing Chair

Zhengtao Yu   Kunming University of Science and Technology, Kunming, China

Technical Programme Committee

Xilong Qu	Hunan University of Finance and Economics, China
Yong Ma	Electronic Information School Wuhan University, China
Xiaokun Yang	University of Houston Clear Lake, China
Hao Wang	State Key Laboratory of Rolling and Automation, Northeastern University, China
Vladicescu Popentiu	Florin, City University, UK
Imran Memon	Zhejiang University, China
Guangzhi Qu	Oakland University, USA
V.S.S. Yadavalli	University of Pretoria, South Africa
Bruno Apolloni	Università degli Studi di Milano, Italy

Harry Bouwman	Delft University of Technology, Netherlands
Shyi-Ming Chen	National Taiwan University of Science and Technology, Taiwan
Yahaya Coulibaly	University Technology Malaysia, Malaysia
B.K. Das	Government of India, India
Joseph Davis	The University of Sydney, Australia
Arturo De La Escalera Hueso	Universidad Carlos III de Madrid, Spain
Ali Hessami	Vega Systems, UK
Yen-Tseng Hsu	National Taiwan University of Science and Technology, Taiwan
Lakhmi C. Jain	Bournemouth University, UK
Sanjay Jain	National University of Singapore, Singapore
Chidananda Khatua	Intel Corporation Inc., USA
Ayse Kiper	Middle East Technical University, Turkey
Ladislav J. Kohout	Florida State University, USA
Reza Langari	Texas A&M University, USA
Maode Ma	Nanyang Technological University, Singapore
N.P. Mahalik	California State University, Fresno, USA
Rabi N. Mahapatra	Texas A&M University, USA

Invited Speakers

John Wang	Department of Information Management & Business Analytics, School of Business Montclair State University, USA
Kevin Deng	Automotive Research Institute, Jilin University
Nilanjan Dey	Department of Information Technology, Techno India College of Technology, Kolkata, India

Acknowledgement

Like every year, this edition of ICMIR-2017 was also attended by more than 150 participants and 172 papers were shortlisted and published in this proceeding. The papers covered in this proceeding are the result of the efforts of the researchers working in various domains of Mechatronics and Intelligent Robotics. We are thankful to the authors and paper contributors of this volume.

We are thankful to the editor in chief and the anonymous review committee members of the Springer series on *Advances in Intelligent Systems and Computing* for their support to bring out the proceedings of 2017 International Conference on Mechatronics and Intelligent Robotics. It is noteworthy to mention here that this was really a big boost for us to continue this conference series.

We are thankful to our friends namely Prof. John Wang, from School of Business Montclair State University, USA, Prof. Kevin Deng, from Automotive Research Institute, Jilin University, and Dr. Nilanjan Dey, from Techno India College of Technology, Kolkata, India, for their keynote address. We are also thankful to the experts and reviewers who have worked for this volume despite the veil of their anonymity.

We are happy to announce here that next edition of the International Conference on Mechatronics and Intelligent Robotics (ICMIR2017) will be held at Kunming, China in association with Kunming University of Science and Technology, Kunming, China during last week of April 2018.

It was really a nice experience to interact with the scholars and researchers who came from various parts of China and outside China to participate the ICMIR-2017 conference. In addition to the academic participation and presentation, the participants must have enjoyed their stay, during the conference and sightseeing trip at Kunming.

I am sure that the readers shall get immense ideas and knowledge from this volume of AISC series volume on "Recent Developments in Mechatronics and Intelligent Robotics".

Contents

Robot Vision

Mechatronics

Research on the Application of Beidou System in Aviation Ammunition Support

Wei Chen[1(✉)], Di Zhang[1], Haikun Zhang[2], and Henhao Zhu[3]

[1] Air Force Logistics College, Xuzhou, 210000, Jiangsu, China
`chenweijiangyiwa@126.com`
[2] 94608 Troops, Nanjing, 221000, Jiangsu, China
[3] 94647 Troops, Fuzhou, 350026, Fujian, China

Abstract. This paper introduced the structure of the Beidou navigation and positioning system, analyzed the Beidou positioning solution principle, elaborated the application of Beidou system in aviation ammunition support command, could improve the efficiency of aviation ammunition support, and visual positioning of various support materials in the process of aviation ammunition support.

Keywords: Beidou system · Aviation ammunition · Support command

1 Introduction

In the processing of research of aviation ammunition visualization based on Beidou satellite, used positioning function of Beidou satellite to achieve precise positioning of the personnel, vehicles, ammunition [1, 2], and used the real-time transmission of positioning information to achieve the communication function of its command prompt feedback timely and ensure the implementation of present situation.

2 Beidou Navigation and Positioning System Structure

Beidou Satellite navigation and positioning system is an all-weather, all day long, high precision, regional satellite navigation and positioning system, can achieve 3 functions, including fast positioning, two-way short message communication, the two last functions are which global positioning system (GPS) couldn't provide, and the positioning accuracy in the areas of our country and the positioning accuracy of GPS, covering all areas of China and the surrounding area. The whole system consists of two geostationary satellites (respectively in the 80 o east longitude and 140o east longitude, 36000 km above the equator), the central control system, calibration system and user machine is composed of 4 parts, each part of the outbound link (i.e., the center of the ground to satellite to user link) and inbound links (i.e. the user machine to the satellite link is connected to the central station). As shown in Fig. 1.

© Springer International Publishing AG 2018
F. Qiao et al. (eds.), *Recent Developments in Mechatronics and Intelligent Robotics*,
Advances in Intelligent Systems and Computing 691, DOI 10.1007/978-3-319-70990-1_1

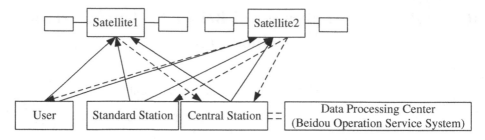

Fig. 1. Beidou navigation and positioning system structure

3   Beidou Positioning Principle

The central station at a specific frequency signal with a continuous wave emission of super frame, divided the frame structure of the pseudo code spread spectrum, the signal by two geostationary satellites to their respective antennas covering all user broadcast regions. When the user needs to position/communication service, location specific process is: first by the center signal, respectively by the two satellite reflection to the user terminal receiving part, and then reflected back by the user terminal two satellites were back to the central station, the central station is calculated in two ways the time required for T1, T2, and after calculation, you can measure the user to two satellite distance R1, R2, complete the positioning (as shown in Fig. 2).

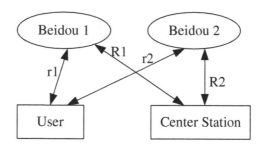

Fig. 2. Beidou positioning principle

The calculation formula is as follows:

$$c \times T1 = 2(r1 + R1) \tag{1}$$

$$c \times T2 = 2(r2 + R2) \tag{2}$$

In the formula (1) and (2), R1, R2 is known, T1, T2 is the light passes through two different outbound links in the link station added time, C is the speed of light, leaving two unknown r1, r2, the two equation solution. The location for geographic coordinate calculation of the user, because the two satellite locations are known, respectively two and two of the center of the sphere, spherical radius of r13, r2, the other is the spherical earth ellipsoid basic parameters have been determined (in the center station store digital ground height data), the intersection point of three balls for the user position measurement (due to the two center in the equatorial plane of the earth, so the earth reference ellipsoid is two, the intersection point of one point in the northern hemisphere, as the position of the user, another point is in the southern hemisphere).

Positioning by the user terminal sends a request to the central station, the central station to locate its position information broadcast will get by the user, you can also take the initiative to locate the user specified by the central station, located after the location information will be sent to the user, and stored by the central station. The location information of the user is placed in the data section of a frame of the outbound signal and transmitted to the user via satellite 1 or satellite 2. The application of communication, the communication contents in the same way to the user. The first frame in each period data segment send standard time (day, ground station, signal and time correction data) and the position of the satellite information, the user receives the signal is compared with the local clock, and calculate the user's local clock and standard time signal of the difference, and then adjust the local clock with the standard time alignment (one-way timing); or comparing the results through inbound links via satellite to the ground by the forwarding center, the center of the ground accurately calculate the difference between the local clock and standard time, and then through the outbound signal via satellite or satellite 1 2 forwarded to the user, the user at this time to adjust the local clock and standard time signal align (two-way timing).

Signal frequency in the vicinity of the L band, the band of radio signals by the atmosphere, rain, fog, the impact is very small, so the positioning and communication process is very reliable, especially suitable for all-weather protection.

Different from the GPS system, the calculation of all user terminal positions is done in the ground control center. Therefore, the control center can retain all the Beidou terminal user location and time information. At the same time, the ground control center station is responsible for monitoring and management of the entire system.

The calibration station has a total of 20, to ensure the positioning accuracy, it is known as the point position compared with the actual measured position, the difference value were adjusted for position measurement data near the user's location, and the calculation process are completed in the ground station control center, is transparent to the user.

The user terminal is a device which is directly used by the user to receive the ranging signal transmitted by the satellite to the ground station. According to the different tasks, the user terminal is divided into the positioning communication terminal, the group user management terminal, the differential terminal, the terminal, etc.

4   Beidou System Application Solution

The general application of Beidou system by the Beidou terminal (including Beidou mobile command user machine and common user machine, Beidou Beidou) service center system composed of four Beidou navigation positioning satellite Beidou satellite navigation system, the ground control center. Beidou satellite navigation system for real-time access to the user access, and the use of the Beidou satellite communication functions for positioning information and other resources to ensure real-time transmission of information. All Beidou users are subject to the Beidou satellite navigation system ground control center station management, as shown in Fig. 3.

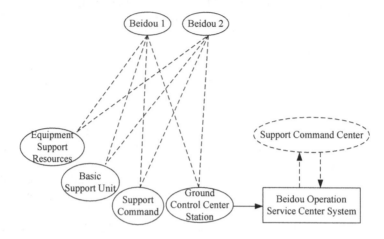

Fig. 3. Beidou system application solution

The service center system is designed for the security command center set up, user location information and short message by Beidou satellite transmitted to the ground station, and real-time push to Beidou service center, security command center through DDN, VSAT, etc. a variety of ways to access the Beidou dial-up service center, Beidou operation the service center sends information to security command center, security command center to realize the monitoring and management of its affiliated users.

There are two kinds of system structure modes of the Beidou missile ammunition support command system:

4.1 Security Command Center + Beidou User Machine (Including the Beidou Mobile Command Users and Beidou Basic User Machine)

The application mode of command system in large scale, through the ground command automation network center, interconnected to the service center to complete the Beidou mobile target monitoring and management of large-scale cross regional command and dispatch.

4.2 Mobile Command User + Beidou Basic User

Because of the limited communication capacity constraints, commanding user command capacity, this application mode is suitable for a small group of users, couldn't transmit through the terrestrial network, a communication device command type subscriber machine as the command and dispatch center use and rapid construction of practical management system.

References

1. Wang, Y., Zhou, L.: Research on the characteristics of ground to air missile equipment support command system. Compl. Syst. Sci. **10**(4), 41–48 (2013)
2. Shao, Y., Liu, X.: Research on the equipment support command based on informanization. J. Mil. Traffic Coll. **13**(4), 22–24 (2011)
3. Wang, X.: Research on comparison of Beidou 2 navigation and positioning system and GPS sole point positioning capacity. Geomat. Spatial Inf. Technol. **39**(2), 97–99 (2016)

Research on the Aviation Ammunition Support Technology of American Air Force

Di Zhang[1(✉)], Wei Che[1], Jie Xu[2], and Jiejun Dai[3]

[1] Air Force Logistics College, Xuzhou 210000, Jiangsu, China
chenweijiangyiwa@126.com
[2] 94865 Troops, Hangzhou 310021, Zhejiang, China
[3] Taiji Factory, Dongchen, Beijing 100005, China

Abstract. This paper analyzed the current situation of American Air Force ammunition support technology, introduced the application plans of American GPS, and the RFID technology in military ammunition transport application plans in detail, with reference to the construction of our military aviation ammunition support system.

Keywords: American air force · Aviation ammunition · Support technology

1 Introduction

In order to adapt to the characteristics of modern warfare, to achieve the rapid tracking of the rapid process of aviation ammunition support, the U.S. department of defense used the visual technology as an important issue of Aviation Ammunition command. In this paper, the technology of aviation ammunition support was analyzed, which provided the reference for the application of aviation ammunition support technology.

2 The Present Situation of American Aviation Ammunition Support Technology

Currently the U.S. military is speeding up the development and equipment of the whole material visual system (TAV: Total Asset Visibility) and in the transport of materials visible (IAV: Intransit Asset Visibility) system [1, 2]. The former can be sent to the factory to the front from the rear of a foxhole material tracking, including the location, status of materials and equipment and carrier. The latter will use radio frequency apparatus, search apparatus and satellite network computer system, material database, track transportation supplies. These two systems will greatly improve the speed, accuracy and economy of the U.S. According to the U.S. Department of defense report, the U.S. military has built four sub systems: the Army (Army Total Asset Visibility visual materials); Navy materials (Navy Total Asset visual Visibility); air supplies (Air Force Total Asset visual Visibility); the Marines (Marine Corps Total visual material Asset Visibility [3]). "The whole material visual system enables the commander to timely and accurately

© Springer International Publishing AG 2018
F. Qiao et al. (eds.), *Recent Developments in Mechatronics and Intelligent Robotics*,
Advances in Intelligent Systems and Computing 691, DOI 10.1007/978-3-319-70990-1_2

understand the location, quantity, status, trend and status of the materials of the Department of defense, and control the information". "For ammunition storage, processing, transportation and demand information have accurate knowledge", "rapidly reduce the processing time distribution, the main benefits include the storage of material list and receipt, and because of the material tracking and less demand list". U.S. military plans to build a joint asset visualization management system based on RFID technology in 2030.

Because the Americans use such as RFID technology and advanced automatic identification technology in ammunition supply chain, through the RFID card reader, etc., the data can be easily and quickly entered into a shared information system. At the same time, it uses cellular communication, satellite communication, GPS positioning and other wireless tracking methods. Figure 1 is the American global tracking network based on RFID technology, satellite positioning technology, satellite communications technology. Built by the U.S. Savi-Technology for the department of defense, the world's largest real-time tracking system, able to track about 250000 pieces of moving objects from toilet paper to ammunition. In more than 480 locations in 40 countries have built automatic identification device with radio frequency identification as the main body, such as by satellite, military communication network connected, as long as users have access permissions, you can have the exclusive right to conduct real time management of goods and materials.

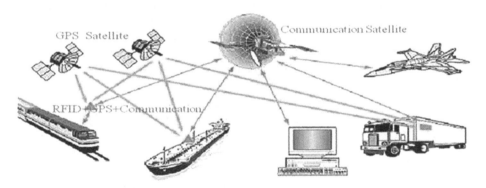

Fig. 1. The American global tracking network

Aviation ammunition support process started from the defense warehouse to the pallet. A automatic radio frequency identification card list of aviation ammunition support materials and a Savi-Technology card could be attached to the tray, the tray of the aviation ammunition support materials in the information contained above, including the name of goods, quantity, number, storage and transport control number, document number. The information is then automatically sent to a central information system. Only in almost all of the supply chain, such as aircraft, ships, truck loading ramps are mobile, manual or automatic Savi-Technology reader, used to track goods to where. To achieve the whole process of tracking the whole process.

3 GPS Application

GPS (Global Positioning System global positioning system) is a global, all-weather, real-time, automation and high efficiency advantages of navigation, positioning and timing, speed measuring system, which can provide seven dimensional high precision for the user (three-dimensional position, three-dimensional velocity and time). GPS consists of three parts: satellite, ground control center and user. GPS uses the satellite as a reference point to calculate the location, positioning accuracy to meter level. There are 24 satellites in the air, located in the orbit of the surface of about 6 km, about 201 million 830 thousand km, about 55° orbit. Each star has 4 high-precision cesium atomic clock and a real-time update of the database, the satellite is constantly sending data to the earth's surface. GPS receiver on the ground whenever and wherever the data can be received at the same time 4 satellites, and then according to these data automatically calculate the latitude and longitude of the receiver and the direction of movement speed, time and other data.

The GPS receiver uses a three edge measurement mechanism to calculate the position of a point in a three-dimensional space. So at least three satellites are needed. GPS satellites broadcast their location information and the exact time at this time. GPS receiver through the reception of the signal, to know the location of the satellite and the signal transmission time, you can measure the distance to the satellite, and then calculate their position, as shown in Fig. 2. Suppose we measured a satellite distance of 11 000 km, that we are in a satellite as the center, 11 000 km radius of the sphere. As of two satellites in addition, also in their center, the corresponding distance as the radius of the sphere. The three balls intersect at two points, one of which is the position of the receiver, and the other can be easily eliminated by judgement.

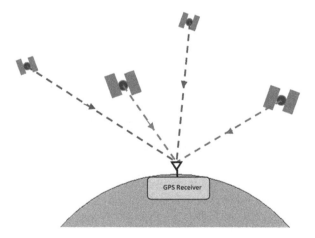

Fig. 2. GPS positioning principle

The information transmission time between the satellite and the ground receiver is measured, which is the basis of the GPS distance measurement. For satellites, the use of very accurate atomic clock timing, very accurate. But the GPS receiver atomic clock on the ground is too expensive, so the use of fourth satellites for the timing, time synchronization and GPS receiver time, accurate time synchronization to ensure accurate positioning.

There are two methods of GPS positioning: one is the single positioning method, by a receiver to receive the data of the 4 stars, the positioning accuracy is low, the error has tens of meters. The other is a relative positioning method, at the same time from the same set of satellite receiving data from multiple receivers, and at least one receiver at a known point (known longitude, height), this point as the reference point for the other point to the reference point of the distance. This approach can achieve very high accuracy, also known as DGPS (wide area GPS differential correction), which requires a large number of reference points composed of infrastructure to support. GPS positioning data through the GSM (global mobile communication network), CDMA SMS (short message) or cluster communications and other means to the ground monitoring center, displayed on the mobile phone with GIS or large screen projector. The center can monitor the current position, the route, the direction and the speed of the target for 24 h.

4   RFID Application Plan of U.S

American Air Force established the visual logistics supply network coverage of important regions of the world based on RFID standard EPC, extensive communication network, efficient information processing, strong security against ability. There are 480 monitoring point of 38 countries in the world, real-time tracking of 270000 container including military and commercial purposes and important goods.

Figure 3 is the RFID application plan of U.S.

1. air ammunition support material can only be read through the passage, in the case of the destruction of the database, the site can not achieve the automatic statistics;

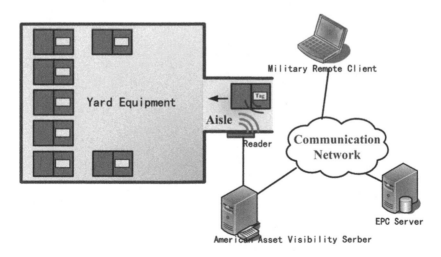

Fig. 3. RFID application plan of U.S.

2. using UHF passive tags to identify the distance of less than 5 meters, and has a directional;
3. tags can only store a EPC code, equipment information must be resolved through the EPC server to obtain.

References

1. Ma, H., Zhang, Z.: Apocalypse of GPS Used in the American FBCB2. 8, Apr. Vol. 8 (2010)
2. Li, Y., Huang, H.: The Analysis of <<21st Century Troops Combat Command System Efficiency >> of American Army. Foreign Military Academics (2009)
3. PU Linke. American army combat command system. Modern Def. Technol. 06, 30–34 (2002)

The Assembly Redesign of Cage Rotor Separator

Li Kunshan and Li Yang[✉]

School of Mechanical Engineering, University of Jinan, Jinan 250022, People's Republic of China
742674@qq.com

Abstract. The method of using three-dimensional design of the main components cage rotor separator were optimized, according to the green design theory, to meet the minimum number of parts of the design method and the direction of the minimum assembly direction; according to the part interlock feature, to achieve the main assembly process in the equipment outside the design method and structure of the principle of symmetrical design. The numerical simulation method is used to simulate the flow field of the separator, especially the influencing factors of the flow field of the rotor system. The optimal performance of the redesign is verified by the experimental data.

Keywords: Separator · Assembly · Design · Numerical simulation · Green environmental friendly

1 Preface

The purpose of the cage-type rotor separator assembly is to assemble the rotor system, housing, power and drive system of the separator as quickly as possible and to ensure the assembly quality of the separator. Therefore, the quantitative analysis of the application to the cage rotor separator design process is essential, especially in the cage rotor separator design research early, the program demonstration and selection stage is particularly important, the purpose of optimizing the design of the separator in order to reduce the equipment assembly time, reduce assembly costs.

2 Three-Dimensional Design and Assembly Design Analysis of Cage Rotor Separator

Three-dimensional solid modeling software has a powerful parametric design and surface modeling capabilities, high-performance assembly and finite element structure analysis functions. The design of the cage-type rotor separator is due to the different specifications and performance specifications. At the same time, due to technical reasons, it is not possible to form a standardized design. The two-dimensional design can not meet the requirements of its assembly design. The three-dimensional design can be used to solve the redesign some of the problems in the process.

© Springer International Publishing AG 2018
F. Qiao et al. (eds.), *Recent Developments in Mechatronics and Intelligent Robotics*,
Advances in Intelligent Systems and Computing 691, DOI 10.1007/978-3-319-70990-1_3

In the re-design of the cage-type rotor separator, the use of the least part design method to minimize the number of parts and optimize the structure through the three-dimensional design analysis, such as the eddy current wear protection plate of the cage type rotor separator design and fixed method, the use of hanging structure and the combination of pin connection method. To avoid the cylinder and wear protection board between the use of a large number of bolts and easy to disassemble the problem.

Figure 1 shows the three-dimensional design of the shaft system of the separator. The advantage of the three-dimensional design method is to solve the possible interdependence between parts and parts. At the same time, the three-process is the least, so it will fit the assembly direction of the least design concept, from the figure can be seen, the miner shaft on all the parts are assembled in the same direction, so greatly improved assembly efficiency.

 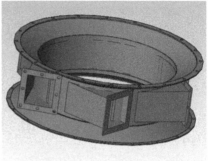

Fig. 1. Assembly direction of separator   **Fig. 2.** Integrated modular design

The three-dimensional design of outlet of cage rotor is as shown in Fig. 2. It utilized overall welded structure technique which enhances the ease upgrade and services.

The cage rotor separator' parts design allows them to be installed externally therefore the observations become handy, and errors are easily to be detected. As shown in Fig. 3: oil duct, electric motor belt, pedestal spigot and platform fixations.

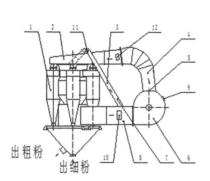 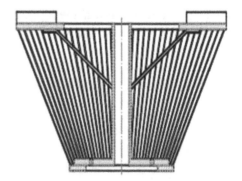

Fig. 3. Typical external installations   **Fig. 4.** Symmetric and grabbing mechanism optimized design

Parts on cage rotor separator are mainly symmetric, as shown in Fig. 4. The primary parts include rotor, barrel body, etc. (Fig. 5).

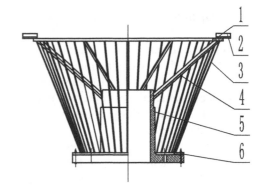

Fig. 5. Spigot design   **Fig. 6.** The cage-type rotor

3   Redevelopment of Fan Blades and Auxiliary Fan Blades of Cage Type Rotor

According to the working principle and structural characteristics of cage type rotor separator, compared with the ordinary centrifugal separator, install the small fan of the cyclone-type separator and the third generation SEPAX separator, comprehensive analysis, cage rotor separator of the cone-shaped grading structure is simple and practical, high precision classification. In order to further improve the grading performance of the classification rotor of the cage type separator, this paper uses the numerical simulation method and the three-dimensional design method to discuss the structure, distribution and installation of the blade of the cage-type graded rotor. As shown in Fig. 6, the structure of the cage rotor separator is composed of the upper cover plate, the lower cover plate, the strip rotor blade and the upper eddy current auxiliary wear-resistant blade. Considering the specific law of the airflow movement in the cage type rotor separator, the characteristics of the flow field vortex, the cage-like rotor bar structure with large and small design to ensure that the classifier in the classifier within the particle size of the various parts of the separator on the various sections of the grading force is basically the same to achieve the purpose of accurate classification, high precision grading, adaptability, significantly improve the efficiency of powder. A sufficient number of vortex-assisted blades are provided on the outside of the cover on the rotor of the cage-type rotor separator. The positive pressure generated by the vortex-assisted fan enhances the ability of the particles to fix the blades through the graded rotor, thus avoiding the occurrence of running roughness phenomenon.

Cage rotor separator: the rotor blades are mainly used in the welding structure, between the parts and components used to facilitate the assembly of the lap structure and bolt connection method, of course, is also easier to disassemble, so that a better solution rotor replacement problems, to avoid the overall removal of the rotor scrapped, to achieve re-use, to achieve the purpose of green design.

4   Application as an Example, Production and Evaluation

The numerical simulation of the flow field is the mathematical method to simulate the flow field of the separator, from the basic theory of hydrodynamics, heat and mass transfer, thermodynamics. Establishment of mass, momentum, energy, components, turbulence characteristics conservation equations, the equation of continuity, equation diffusion, turbulent energy equation, etc., followed by numerical solution of to solve these equations [9].

Conventional simulation of flow field between blades, distribution of isobars in flow, effect of blade length on flow field, influence of rotational speed on flow field. The simulation results show that with the increase of air flow, the radial velocity of gas flow increases, the tangential velocity decreases, and the area of non-working blade return area decreases, and the flow field in the graded area tends to be consistent. Point of view, is conducive to the classification of particles.

However, due to the increase of the radial velocity, the drag force of the particles increases, and the particle size of the fine powder of the separator increases, failing to achieve the fine classification. But the air volume of the fan is too small and not conducive to produce enough negative pressure in the classification area, is not conducive to the rapid discharge of fine powder. So the classifier air volume to select a reasonable value, and with the rotor system with good speed to achieve a good grading effect. The actual process from the classification point of view, it is best in the air volume and speed are higher in the case of classification, because the wind is too small is not conducive to the transport of materials.

5   Application as an Example, Production and Evaluation

The use of three-dimensional design methods, re-design theory, numerical simulation processing technology on the existing cage rotor separator part of the engineering time transformation, and achieved a little success, the focus of the project group in China's cement developed areas and cement province of Shandong Province, Hebei Province and other parts of the implementation of the transformation of the cement plant, and achieved good results, accumulated a lot of data and a certain experience, but also verified our design concept.

In order to evaluate the comprehensive index and performance of the separator in a comprehensive and accurate manner, the main technical parameters such as feeding, returning powder, finished particle size, grinding speed and ball ratio are measured. At the same time, consumption changes are also compared. The specific results are shown in Tables 1 and 2.

Table 1. Feeding particle size composition

Particle size (μm)	<5	5–10	10–20	20–30	30–40	40–50	50–60	60–70	70–80	>80
Percentage (%)	7.10	9.14	16.29	13.25	9.11	5.12	6.70	6.17	7.16	19.46

Table 2. Contrast result of classifier performance between foreign and homeland.

Type of separator	FLS–REC	O–SEPA	SEPAX	LAROX	HES-I	Improved classifier
Separator efficiency (%)	76	69	71	86	76	84.1
Newtonian efficiency (%)	55	56	56	33	65	67.5
Cycle load	1.9	2.7	2.5	1.3	2.1	0.58
Energy efficiency (%)	66	74	71	35	82	62
Coarse powder recovery rate (%)	79	87	85	47	89	83.2
Energy saving value	23	31	27	9	33	33.7
Coefficient of definition	0.51	0.53	0.57	0.37	0.41	0.52
Circulation load	1.7	2.6	2.4	1.4	1.5	2.1
Excision particle size	45	28	37	65	35	42
By-pass value	9.5	10	9.5	11	9	10.8

6   Conclusions and Improvements

The separator is the main auxiliary equipment of the circle grinding system in the production process of the cement industry. In the optimization design of the separator, the overall design of the separator is taken into consideration, taking into account the design of the parts. Fully consider the machine assembly performance and the actual assembly requirements, the use of green assembly design concept, optimize the traditional design. Through the assembly design analysis of the separator, we have redesigned the structure of the rotor system, the outlet, the spindle and the cage rotor of the cage type rotor separator, thus improving the convenient performance of the installation and maintenance of the separator, grading efficiency is also significantly improved, to achieve through the green design optimization of the principle of structural design.

References

1. Kunshan, L.: The design and development of the non-ball mill. Miner. Mt. Mach. **7**, 24–27 (2002)
2. Kunshan, L.: The characteristic and theory of LKS non-ball mill. Symp. Refract. **5**, 316 (2002)
3. Kunshan, L., Mo, Y.: The design and development of non-ball
4. Boger, Z.: Application of neural networks to water and wastewater treatment plant operation. ISA Trans. **31**(1), 25–33 (1992)

5. Kunshan, L.: Rare earth superfine powers fabrication and characterization by novel non-ball miller. Mechatron. Mater. Process. **11**, 648–653 (2011)
6. Kunshan, L.: Simulation of failure detection based on neural network for no-ball mill. Adv. Manuf. Syst. **4**, 627–631 (2011)
7. Kunshan, L.: The re-design of centrifugal classifier. Adv. Metall. Min. Eng. **12**, 820–823 (2011)
8. Kunshan, L.: Classifers' environmental friendly(green)re-design. Autom. Equip. Syst. **3**, 658–662 (2012)
9. Kunshan, L.: The use of solid edge in the classifiers design. Adv. Manuf. Technol. **3**, 2080–2083 (2012)
10. Kunshan, L.: The assembly design of non-ball mills. Adv. Des. Res. Manuf. **8**, 65–68 (2013)
11. Kunshan, L.: Design of non-ball mill based on the green conception. Adv. Des. Res. Manuf. **8**, 73–76 (2013)
12. Leschinski, K.: Classification of particles in the submicron range in an impeller wheel air classifier. KONA Powder Part. **14**, 52–60 (1996)

Design and Analysis of Double-Arc Tooth Profile Harmonic Drive

Xincheng Cao[1], Shiyi Song[2], Huomu Xie[1], and Mengmeng Wang[3(✉)]

[1] School of Aerospace Engineering, Xiamen University, Xiamen 361001, China
[2] YTO Group Corporation, Luoyang 471000, China
[3] Unit of 96274 of PLA, Luoyang 471000, China
wangmmpla@sina.com

Abstract. Harmonic drive show an extraordinary transmission characteristic with high with high efficiency and attract considerable attention. In this article, a double-arc tooth profile harmonic drive was designed. An explicit finite element analysis (FEA) about deformation, torsion, stress and contact force under different loads was put forward and compared with theoretical analysis and experimental results. The effect of different load to redistribution of stress on the flexspline (FS) and the change of the contact force distribution are also researched. The result shows that the hysteresis behavior in the torsion of FS cup due to energy storage and release lead to the change of distribution. Besides, the hysteresis behavior in FEA is also identified by the agreement with experimental results. Furthermore, the torsion of FS cup of double-arc tooth profile HD independent on kinematic errors is elaborated based on FEM. The article shows a meaningful attempt for the double-arc tooth profile harmonic drive improvement.

Keywords: Harmonic drive (HD) · Double-arc · Flexspline (FS) · Contact force · Torsion

1 Introduction

Harmonic drive has become popular for high transmission ratio, near-zero backlash and precise operation with high efficiency [1]. Due to the complexity of flexible deformation in the structure, many studies have focused on its meshing theory, tooth profile design and optimization, kinematic and dynamic characteristics and so on [2]. Compared with the involute profile HD, the double-arc tooth profile HD has the advantages of moderating the stress concentration on the dedendum and improving the quality of meshing, carrying capacity and torsion stiffness [3]. However, a comprehensive study of double-arc tooth profile HD including deformation, stress, contact force under different loads and torsion of FS cup is rarely seen. Xin et al. proved the rationality of double-arc tooth profile HD for precise transmission. Moreover, he did some experiments to clarify the advantage of double-arc tooth profile HD, including the higher transmission precision and higher torsion stiffness than the involute profile HD [4]. Dong proposed the kinematic model under no load and offers a rigorous analysis of the kinematics of harmonic drives. The unique feature of cup in transmission and its effect

© Springer International Publishing AG 2018
F. Qiao et al. (eds.), *Recent Developments in Mechatronics and Intelligent Robotics*,
Advances in Intelligent Systems and Computing 691, DOI 10.1007/978-3-319-70990-1_4

on tooth conjugation was discussed [5, 6]. Cui and Shen has carried out research on the contact force under different loads based on flexibility matrix [7]. Available literature focuses on the mechanism responsible for the kinematic error a lot. But, as part of kinematic error, the torsional flexibility is seldom studied independently [8].

The paper designed and built a model of double-arc profile HD based on gear meshing theory, put forward the theoretical model of deformation and contact force, and then calculated the deformation, contact force and torsion of the model based on FEM. Then the comparison of FEM results with the theoretical method and experimental results was elaborated. The change of the contact force distribution is studied and the torsion of FS cup of double-arc tooth profile HD independent on kinematic errors is explored based on FEM. The work would be a useful reference to study the meshing characteristics and transmission performance of double-arc tooth profile harmonic drive.

2 Design and Modeling of Double-Arc Tooth Profile HD

As the height of tooth, nominal meshing angle and diameters of arcs are important to the structure [9, 10], the parameters are settled with other main parameters, as is shown in Table 1.

Table 1. Parameters of the double-arc profile HD

Items	Value	Items	Value
Modulus, m (mm)	0.5	Height of tooth, h (mm)	1.95 m
Nominal meshing angle, α_0 (°)	25	Radius of concave arcs, ρ_f (mm)	1.307 m
Transmission ratio, i	−50	Radius of concave arcs, ρ_a (mm)	1.147 m
Radial displacement coefficient ω_0^*	1	Maximum radial displacement, ω_0 (mm)	0.5
Number of flexspline teeth, z_1	100	Thickness of flexspline cup, δ (mm)	0.4
Number of circular spline teeth, z_2	102	Gear tooth ring width, b_R (mm)	11

2.1 Design of Flexspline Tooth Profile

As is shown in Fig. 1, concave and convex arc circular splines are connected by a tangent line AB. In this way, the flexspline profile could be described by Eq. 1.

$$\begin{cases} (x_{FS} + l_a)^2 + (y_{FS} + h_f)^2 = \rho_a^2 \\ (x_{FS} - 0.5s)\cot\gamma + y_{FS} - h_f = 0 \\ (x_{FS} - l_f - t)^2 + (y_{FS} - h_f - e_f)^2 = \rho_f^2 \end{cases} \tag{1}$$

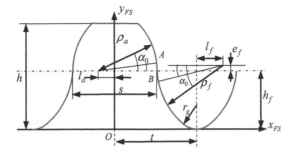

Fig. 1. Parameter of flexspline tooth profile

The Eq. 1 can be transformed into a parametric equation,

$$\begin{cases} x_{FS} = x_{FS}(u) \\ y_{FS} = y_{FS}(u) \end{cases} \tag{2}$$

Where u is the parameter of the flexspline tooth profile.

2.2 Conjugating Profile of Circular Spline (CS)

The conjugating profile of circular spline is directly associated with the profile of flexspline tooth as well as wave generator, which is a cam with cosine profile. The neutral layer line is the equidistant curve of the cam profile. The deformation of 1/4 neutral layer line is shown in Fig. 2.

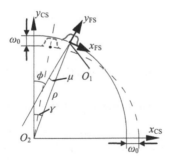

Fig. 2. Deformation of the neutral layer line of flexspline

The equation of the neutral layer line of flexspline after assembling is as below

$$\rho = r_0 + \omega_0 \cos 2\varphi + \delta/2 \tag{3}$$

where r_0 is the radius of inner cylinder of flexspline before deformation, φ is the rotational angle starting from the long axis.

According to the enveloping theory of tooth profile, the circular spline profile can be deduced as follows: [11]

$$
\begin{cases}
x_{CS}(u, \phi) = x_{FS}(u) \cos \Phi + y_{FS}(u) \sin \Phi + \rho \sin \gamma \\
y_{CS}(u, \phi) = -x_{FS}(u) \sin \Phi + y_{FS}(u) \cos \Phi + \rho \cos \gamma \\
\partial x_{CS}(u, \phi)/\partial u \bullet \partial y_{CS}(u, \phi)/\partial \phi - \partial x_{CS}(u, \phi)/\partial \phi \bullet \partial y_{CS}(u, \phi)/\partial u = 0
\end{cases}
\tag{4}
$$

Where Φ is the angle between the coordinate systems of the flexspline and circular spline.

When the flexspline rotates around the wave generator, according to the gear meshing theory, the conjugating circular spline tooth profile would be enveloped by the moving flexspline teeth [12], as is shown in Fig. 3. Scanning the envelop trace for the boundary, the circular spline tooth profile is attained. Figure 3(b) describes the theoretical profile of both flexspline tooth profile (red) and circular spline tooth space (blue) while the flexspline is deformed by wave generator and there are 34 pairs of teeth in meshing.

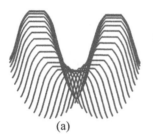 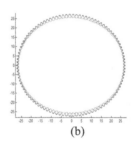

(a)   (b)

Fig. 3. (a) Circular spline tooth profile enveloped by moving FS (b) Theoretical profile of FS tooth profile (red) (Color figure online)

3   Theoretical Analysis of Deformation and Contact Force

3.1   Theoretical Analysis of Deformation of the Neutral Layer Line

As is shown in Fig. 2, O_1 is the intersection point of the neutral layer line and the normal line of the tooth after assembling. The radial and circumferential displacements of the neutral layer line as well as the rotational angle of the normal line of each tooth on the flexspline determine the position and direction of each tooth on the flexspline. According to Eq. 3, it is easy to determine the radial displacement of O_1, that is,

$$
\omega = \omega_0 \cos 2\varphi
\tag{5}
$$

Circumferential displacement can be deduced as below:

$$v = -\int \omega d\varphi = -\int_0^\varphi \omega_0 \cos 2t dt = -0.5\omega_0 \sin 2\varphi \tag{6}$$

According to theory of moment force of shell [13], rotational angle of normal line of each tooth is:

$$\mu = (1/r)(v + \partial\omega/\partial\varphi) = -\frac{5\omega_0}{2(r_0 + \delta/2)} \sin 2\varphi \tag{7}$$

3.2 Theoretical Distribution of Contact Force

The theoretical distribution of contact force has been approximated by experiment [14], as is shown in Fig. 4. Where, θ_2 and θ_3 is the meshing zone and θ_1 is the central angle between symmetric axis of contact force distribution CC and long axis of wave generator AA'.

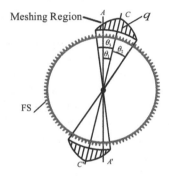

Fig. 4. Distribution of contact force

The contact force distribution of unit tooth ring width in the region of θ_2 and θ_3 can be described separately as:

$$\begin{cases} q = q_{max} \cos[\pi(\theta - \theta_1)/2\theta_2] \\ q = q_{max} \cos[\pi(\theta_1 - \theta)/2\theta_3] \end{cases} \tag{8}$$

Where θ is positive in the clockwise direction and negative anticlockwise, q is loading distribution of unit tooth ring width, q_{max} is maximum loading.

4   Simulation and Analysis of Double-Arc HD with FEM

4.1   Simulation of the Assembling and Meshing

The finite element model chose C3D8R as the element type and there have been 702537 elements in all. The whole model has two pairs of interactions: one is to set the outer face of wave generator as master surface and the inner face of flexspline as slave surface, the other one is to set the tooth surface of circular spline as master surface and the tooth surface of flexspline as slave surface. The simulation consists of two processes: the first one is the assembling simulation of wave generator and flexspline, the next is the meshing simulation of the assembly with wave generator as the input component and flexspline as the output one. The material parameters of the parts are given in Table 2.

Table 2. Material parameters of components

	Material	Young's modulus (MPa)	Poisson's ratio
Flexspline	30CrMnSiA	196000	0.3
Wave generator	45steel	206000	0.3
Circular spline	45steel	206000	0.3

After assembly of wave generator and flexspline, the radial and circumferential displacements of the neutral layer line from a section vertical to axis Z at the middle of the tooth ring on FS were extracted. In the step of meshing simulation, different torques would be exerted on the bottom of flexspline around axis Z, including 0 N m, 10 N m, 50 N m, 100 N m, 500 N m. Moreover, a comparison of stress of the neutral layer would be made to explicate the effect of load on the flexspline cup. Besides, the contact force of the tooth profile would be extracted and analyzed.

4.2   Deformation and Stress Analysis

According to theoretical calculation, the meshing pairs of teeth between flexspline and circular spline add up to 34, while the simulation shows 40 pairs in meshing as is revealed in Fig. 5. The difference between the theoretical calculation and simulation comes from the flexible deformation of the teeth, while the theoretical calculation assumes the teeth on flexspline are rigid bodies and connected by flexible tooth space and the length of deformed neutral layer line is the same as the one before assembling. However, the neutral layer line would stretch a little [15], which leads more teeth to meshing.

Figure 6 draws the comparison between the theoretical calculation and the FEA results of radial and circumferential displacement of the neutral layer line at the middle section of tooth ring. The axis X is the angle from the short axis in anticlockwise way. The positive Y value in Fig. 6(a) indicates that the neutral layer line moves in the centrifugal direction. The positive Y value in Fig. 6(b) means the neutral layer line moves in the clockwise direction. It is apparent that the circumferential displacements from the two methods are particularly the same. The maximum circumferential

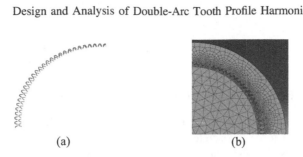

(a)   (b)

Fig. 5. (a) Meshing status in theoretical calculation. (b) Meshing status in finite element simulation

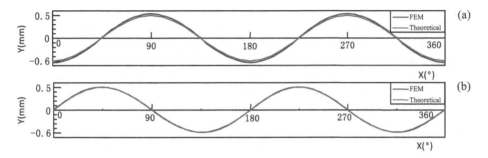

Fig. 6. (a) Radial displacement and (b) circumferential displacement

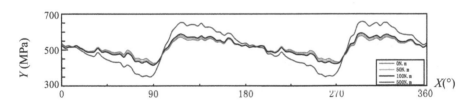

Fig. 7. Comparison of stress on the neutral layer line under different loads

displacement occurs at the middle of the long and short axis and the minimum appears at the long axis and the short axis. However, the difference from the radial displacement mainly comes from the interference between the WG and FS.

When the wave generator rotates 180°, stresses of the neutral layer line on flexspline under different loads have the similar trends. As is seen in Fig. 7, the maximum stress takes place in the region ranging from 90° to 135°, that is about 45° anticlockwise from the long axis. When the load is 0 N m, 50 N m, 100 N m, the change of stress is not significant and the amplitude is within 150 MPa. However, when the load adds up to 500 N m, the change of stress becomes quite obvious and the amplitude turns up to 400 MPa. The difference demonstrates that loads within a certain range have little effect on the stress of the flexspline cup. However, if the load increases beyond the range, the effect to redistribute stress on the flexspline cup will become obvious.

4.3   Analysis of Contact Force Distribution and FS Cup Torsion

Due to the fact that the tooth space has no contact with the meshing tooth, the distribution of contact force is discrete rather than continuous.

Figure 8 shows the increment of contact force under different loads 10 N m, 50 N m, 100 N m when WG rotates 180°. With the increase of the load, the amplitude of contact force increment increases, but the position of distribution of contact force under different load does not change.

As is shown in Fig. 9(a), and (b), when wave generator rotates anticlockwise within about 90°, $\theta_1 \neq 0°$, $\theta_2 \geq \theta_3$, but if it rotates in an approximate range of 90°–180°, $\theta_2 \leq \theta_3$. That is, the trend of distribution is contrary on the condition that WG rotates within and beyond 90°. The axis X is the central angle which starts from the short axis. The phenomenon may be caused by hysteresis behavior in the flexible torsion of FS cup.

The flexspline can be interpreted as a combination of a nonlinear torsional spring and a nonlinear viscous damping [16], as is shown in Fig. 10.

The spring stores potential energy until the compliance reversal and then the potential energy is transformed into kinetic energy of the mass block. After the energy release, kinetic energy of the mass block would store in the spring again. In this way, the structural damping and spring result in a hysteresis behavior due to the energy storage and release at compliance reversal [17].

In Fig. 10, ϕ_1 is the angular displacement of cup rim, while ϕ_2 is the angular displacement of the bottom of the cup.

$$\varphi = \varphi_1 - \varphi_2 \tag{9}$$

That is, ϕ is the torsional angle of the FS cup. The relationship between torque τ and the angular displacement ϕ can be expressed as:

$$\tau(t) = g[\dot{\varphi}(t), \ \varphi(t)] + f(\varphi(t)) \tag{10}$$

where the function $f(\cdot)$ is the stiffness curve, while the function $g(\cdot)$ is the nonlinear dynamic friction [18].

Although the structural parameters are not exactly same, the fundamental structure of the model in the paper has a lot in common with the experimental model in literature [8]. Based on the experiment data, the rotational velocity of WG in simulation is set as 120 rpm.

In order to study the torsion of FS cup, a generating line of the neutral layer is extracted and four points as Fig. 11(a) marks on the line is tracked while WG rotates. The circumferential displacements of the four points with WG rotating are compared in (b). The axis X is the angle that WG rotates. As is illustrated in (b), the rotation of the bottom of FS lags behind the rotation of tooth ring when WG rotates less than about 90°, and the condition is contrary when WG rotates over 90°.

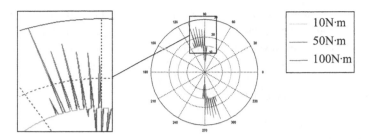

Fig. 8. Increment of contact force under different loads when WG rotates 180°

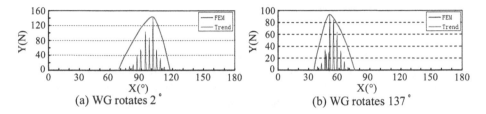

(a) WG rotates 2°

(b) WG rotates 137°

Fig. 9. Distribution of contact force on flexspline

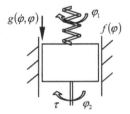

Fig. 10. Analog of hysteresis model from flexspline

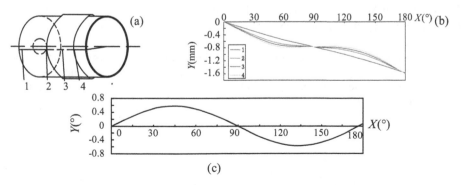

Fig. 11. (a) Definition of four measure points (b) Circumferential displacements of four points (c) Torsional angle of FS cup

Therefore, the compliance reversal took place when WG rotates about 90°, which leads to the change of the contact force distribution. In spite of the torsion of FS cup, the bottom of the cup achieves the required uniform transmission ratio at the system output. From (c), the torsional angle of FS cup changes approximately in a sinusoidal-like way as WG rotates, which reflects the flexibility of FS independently, rather than inclusive in the kinematic errors.

5 Conclusion

In this paper, the three main components of a cup-like double-arc tooth profile harmonic drive were designed and modeled according to the gear meshing theory. Then, the theoretical deformation of the neutral layer of the flexspline was put forward and the distribution of contact force on the meshing teeth was evaluated by experimental formula. After simulating the assembling and meshing processes under different loads, an explicit FEA of the model about deformation, torsion, stress and contact force was carried out and a comparison of FEA with theoretical analysis and experimental results was proposed. The contribution, findings, and unique features of the paper are summarized below:

(1) According to theoretical calculation, the meshing pairs of teeth between flexspline and circular spline add up to 34, while the simulation shows 40 pairs in contact. The difference lies in the flexible deformation of the teeth and the stretch of the neutral layer line.

(2) The load on the flexspline under certain range has little effect on redistributing stress on the flexspline cup, but the effect will be significant if the load reaches certain extent.

(3) With the increase of the load, the increment of contact force under different loads increases in amplitude, but the position of distribution does not change.

(4) The hysteresis behavior in the torsion of FS cup due to energy storage and release accounts for the change of distribution and the hysteresis behavior in FEA is identified by the agreement with experimental results.

(5) The torsional angle of FS cup changes approximately in a sinusoidal-like way as WG rotates. In spite of the torsion of FS cup, the bottom of the cup achieves the required uniform velocity at the system output.

Acknowledgements. This research was financially supported by The Ministry of Industry and Information Technology (MIIT) 2016 Comprehensive And Standardized Trial and New Model Application of Intelligent Manufacturing under Grant No. Yu Luo Industrial Manufacturing [2016]07744.

References

1. Ostapski, W.: Analysis of the stress state in the harmonic drive generator-flexspline system in relation to selected structural parameters and manufacturing deviations. Tech. Sci. **58**(4), 683–698 (2010)
2. Chen, X., Liu, Y., Xing, J.: A novel method based on mechanical analysis for the stretch of the neutral layer line of the flexspline cup of a harmonic drive. Mech. Mach. Theory **76**(1), 1–19 (2014)
3. Xin, H., He, H., Xie, J.: Proof of reasonableness of adopting circular arc profile in precision harmonic drive. J. Changchun Inst. Opt. Finemech. **23**(3), 47–50 (1997)
4. Xin, H.: Design for basic rack of harmonic drive with double-circular-arc tooth profile. China Mech. Eng. **22**(6), 656–662 (2011)
5. Dong, H., Ting, K.-L., Wang, D.: Kinematic fundamentals of planar harmonic drives. J. Mech. Des. **133**(1) (2011). 011007.1-8
6. Dong, H., Wang, D., Ting, K.-L.: Kinematic effect of the compliant cup in harmonic drives. J. Mech. Des. **133**(1) (2011). 051004.1-7
7. Cui, B., Shen, Y.: Study on load distribution in gear of harmonic drive by finite element method. Mech. Transm. **21**(1), 7–9 (1997)
8. Ghorbel, F.H., Gandhi, P.S., Alpeter, F.: On the kinematic error in harmonic drive gears. J. Mech. Des. **123**, 91–97 (1998)
9. Luo, M.: Structural design and FEA of double-arc tooth profile short cup harmonic drive. Thesis, Nanjing University of Science and Technology, 2008
10. Ostapski, W., Mukha, I.: Stress state analysis of harmonic drive elements by FEM. Tech. Sci. **55**(1), 115–123 (2007)
11. Bi, Z.: Analysis of mechanical characteristics for key parts of harmonic drive. Thesis, Jilin University, 2011
12. Ivanov, M.: Harmonic Gear Drives. Visajas Kola Press, Moscow (1981)
13. Dong, H.: Study of kinematic circular spline and meshing characteristic of harmonic gear drives based on the deformation function of the flexspline. Thesis, Dalian University of Technology, 2008
14. Chen, X., Liu, Y., Xing, J., et al.: The parametric design of double-circular-arc tooth profile and its influence on the functional backlash of harmonic drive. Mech. Mach. Theory **73**(1), 1–24 (2014)
15. Zou, C., Tao, T., Mei, X., Lv, J., Jiang, G.: Contact analysis for short harmonic reducer in robotic joints. J. Xi'an Jiaotong Univ. **47**(5), 82–87 (2013)
16. Kayabasi, O., Erzincanli, F.: Shape optimization of tooth profile of a flexspline for a harmonic drive by finite element modeling. Mater. Des. **28**, 441–447 (2007)
17. Ruderman, M., Hoffmann, F., Bertram, T.: Modeling and identification of elastic robot joints with hysteresis and backlash. Trans. Ind. Electron. **56**(10), 3840–3847 (2009)
18. Dhaouadi, R., Ghorbel, F.: Modeling and analysis of hysteresis in harmonic drive gears. Syst. Anal. Model. Simul. **1**, 1–14 (2002)

Second Order Difference Aided CRC Check Stopping Criterion for Turbo Decoding

Xuying Zhao[1(✉)], Xiaoqin Wang[2], and Donglin Wang[2]

[1] University of Chinese Academy of Sciences, Beijing, China
zhaoxuying2012@ia.ac.cn
[2] Institute of Automation, Chinese Academy of Sciences, Beijing, China

Abstract. This paper presents an enhanced CRC (Cyclic Redundancy Check) check stopping criterion to improve the throughput of turbo decoders. Turbo decoder is applied to the 3G and LTE systems. CRC check stopping criterion is normally used by the turbo decoder, and it reduces the iteration times dramatically when transmission blocks are received correctly. However, when the quality of the received signals is very low, the receiving data cannot be decoded correctly even though the turbo decoder keeps working all the time. In these scenarios, the CRC check stopping criterion has no effects any more. To reduce the number of iterations when the quality of the received signals is low, the second order difference aided CRC check stopping criterion is proposed. Based on the second order difference of soft-bit information and/or hard-bit information, some bad situations are identified before CRC check stopping condition is satisfied. Through simulations, it is proved that this method greatly reduces the decoding iteration times in bad transmission environments. On the other hand, the whole throughput of the turbo decoder is improved.

Keywords: CRC check · Second order difference · Turbo decoding

1 Introduction

Turbo codes, introduced by Claude Berrou in 1993 [1] have been widely researched and used because of good decoding performance. Typically, turbo decoding algorithms set a maximum iteration number in case of endless computing. These methods are called "Fixed-Iteration-Number" (FIN) stopping criterion. FIN has poor flexibility especially when transmission quality is good. In this situation, the decoder can correct all the errors by iterating one or two times. Combined with the FIN, many other algorithms are proposed to improve the decoding performance after some sufficient iterations.

In 1996 Joachim Hagenauer proposed the Cross-Entropy (CE) method to stop iteration before it reached the maximum iteration number [2]. CE algorithm can significantly reduce the iteration numbers with very little performance loss, but the computational complexity is relatively large. Based on CE method, some other stopping criterions were created to reduce the complexity such as SCR (Sign-Change-Ratio) algorithm and HDA (Hard-Decision-Aided) algorithm [3]. Sign-Difference-Ratio algorithm was relative to the SCR and it resolved the problem of big data storage [4]. Improved HDA

© Springer International Publishing AG 2018
F. Qiao et al. (eds.), *Recent Developments in Mechatronics and Intelligent Robotics*,
Advances in Intelligent Systems and Computing 691, DOI 10.1007/978-3-319-70990-1_5

algorithm (IHDA) reduced the storage memory without degrading the performance compared to the HDA [5]. Later on, based upon the CE method, a kind of method called Yu criterion was proposed and its decoding performance was almost near the optimal [6]. Several other early termination methods were proposed afterwards. For example, Jia Hou presented the adaptive SNR (Signal Noise Ratio) algorithm [7], Fan-Min Li proposed the MOR (Measurement of Reliability) method in 2005 [8] and Wei Jiang proposed the algorithm about the mutual information between the logarithm likelihood ratio and the data bits [9]. In recent years, some other new algorithms for early termi-nation have also been raised [10, 11].

In contrast with the above methods, the error correcting performance of CRC check stopping criterion is good and corresponds to the Genie method. Nevertheless, CRC check algorithm has one obvious drawback. When SNR is low or some outburst errors happen to the decoder, CRC check cannot pass and the decoder will iterate up to the maximum number. From simulations we can see that under this case the output of the decoder is almost the same as that with fewer iteration times, and the time consumption caused by excess iterations is invalid. A new method is proposed to solve this problem in this paper. At each iteration, we compute the correlation values of the two sub-decoders and the correlation values are called the first order difference. Then we compute the difference between two consecutive iterations and the difference is called the second order difference. The decoder stops iteration based on that the second order difference is positive or not. This method can reduce the number of iterations with almost no damage to the BLER (Block Error Rate) performance, especially in low SNR.

The remainder of the paper is organized as follows. In Sect. 2, the second order difference aided CRC check stopping criterion is presented. Simulation results and system complexity are analyzed in Sect. 3. Finally, Sect. 4 concludes the paper.

2 Second Order Difference Aided CRC Check Stopping Criterion

CRC check stopping criterion is rather reliable and early stopping hardly degrades the performance. On the basis of considerable performance, we can further decrease the average iteration times. Figure 2 shows that when SNR is low, the CRC check iteration times are high, whereas the BLER performance is still "bad". Next we will focus on this problem.

In discrete function, the difference between two continuous adjacent values is called the first order difference. If we define $x(k)$, then $y(k) = x(k + 1) - x(k)$ is the first order difference, and this value indicates the monotonicity of the discrete function. The second order difference is defined as $z(k) = y(k + 1) - y(k)$, and this indicates the speed of the change rate.

For turbo codes, we have analyzed the regularity of the internal data stream variation in the decoder. By tracking amount of intermediate process results, we discover that the difference between two sub-decoders is getting lower and lower as the iteration goes on. One case is shown in Table 1, where we compute the correlation values of the LLRs and hard bits from the two sub-decoders.

Table 1. The regularity of the data stream changes as the iteration goes on

Iteration number	First order difference	Second order difference
0	−0.47484	
1	−0.791836	0.316997
2	−0.981215	0.189379
3	−1.200226	0.219011
4	−1.375428	0.175202
5	−1.681867	0.306439
6	−1.958921	0.277054
7	−2.659726	0.700806

The above table shows that the expected hard bits of the two sub-decoders are approaching to each other as the number of iterations increases. The values of second order difference are all positive, which indicates the "approaching" tendency. Normally, when the iteration times get bigger, the first order difference will be smaller and the decreasing rate is growing. That is, the second order difference keeps positive. Once the decoder works abnormally, the first order difference becomes larger and the second order difference turns to be negative. As is shown in Table 2, at the last iteration 4, the first order difference becomes larger than −1.674556 at iteration 3, indicating that the decoder meets some errors. Correspondingly, the second order difference changes into a negative value −0.062031.

Table 2. Abnormal work of the decoder, the decoded bits of the decoder are not the same with the original information bits

Iteration number	First order difference	Second order difference
0	−0.564011	
1	−0.9835	0.419489
2	−1.244495	0.260995
3	−1.674656	0.430161
4	−1.612625	−0.062031

Based on the above analysis, we propose the second order difference aided CRC check stopping criterion. When some accidental errors happen to the decoder, the errors will pass over iteration by iteration, and the extra iterations work uselessly or even worse. The second order difference aided CRC check algorithm is as follows,

(1) Set $i = 0$.
(2) At iteration i, compute the correlation values of the two sub-decoders and save them, where the correlation values are corresponding to the first order difference. At the same time, CRC check is processing. If CRC check is passed, stop the iteration; otherwise, go to step (3).
(3) At iteration $i + 1$, compute the correlation of the two sub-decoders, then calculate the second order difference along with the previous difference values. If the second order difference is positive, go on CRC checking and save the first order difference

values; otherwise, stop the iteration. If the CRC check is passed, stop the iteration; else set $i = i + 1$ and go to step (4).

(4) If i is less than the preset maximum iteration number, go to step (3); else stop the iteration.

3   Simulation Results and System Complexity Analysis

There are several schemes to compute the first order difference. The output LLRs of each sub-decoder can be used to calculate these values. To some extent, the performance of soft information is better than the hard bits. While the arithmetic units for floating points are complex and time-consuming, especially the multiplication arithmetic. On the other hand, as the CRC check is based on the hard bits, we can take this advantage and adopt the hard bits to compute the first order difference. However, completely hard decision will damage a part of performance. Considering the system complexity and decoding performance, we can have a trade-off. A multiplier for floating point and hard bit is corresponding to a multiplexer, which is less complex than the floating-point multiplier. For simplicity, we call the soft information and hard bits assisted CRC check "CRC-SHA", and the hard bits assisted CRC check "CRC-HARD". The CRC-SHA architecture is depicted in Fig. 1.

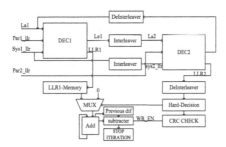

Fig. 1. CRC-SHA architecture

In this architecture, we need an extra memory to store the LLRs of the first sub-decoder. After the second sub-decoder outputs the LLRs and makes a hard decision, the computing of the first order difference begins. The process is like that the soft information is to be multiplexed and goes into an accumulator. After finishing the whole code block, the first order difference will be stored and do subtraction with the previous saved ones. Then we can get the second order difference, and this value will control the CRC check module and decide whether to stop the iteration.

There is a situation that the decoder works abnormally on occasion, but it can still repair itself to obtain the correct results. If we stopped the iteration early, the BLER performance might be a little "bad". This paper proposes a solution to reduce the decoder's accidental errors. Here a parameter "ErrFlag" is set to indicate the error detected numbers. We simulate the cases "$ErrFlag = 1$" and "$ErrFlag = 2$". The BLER performance and the average iteration times are described in Fig. 2.

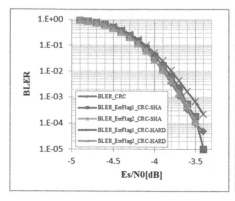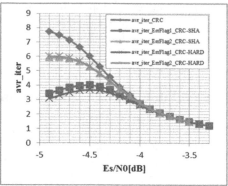

Fig. 2. The BLER and average iteration numbers of CRC-SHA and CRC-HARD

The above figures show that CRC check stopping criterion has the lowest BLER, while the average iteration times are highest. In the case *ErrFlag* = 2, the BLER performance of CRC-SHA is nearly to the CRC check, and the average iteration times are also decreased. Figure 2 shows that when SNR is in −4.9 to −4.4 dB, the average number of iterations falls in approximate two times. In the case *ErrFlag* = 1, the BLER performance of CRC-SHA has a loss of 0.1 dB, while the CRC-HARD loses nearly 0.2 dB. Both CRC-SHA and CRC-HARD can dramatically reduce the average iteration times. Whereas at a poor transmission environment, the CRC-HARD may degrade the decoding performance due to its weaker decoding ability. So considering the system complexity and error correcting performance, we take the CRC-SHA algorithm as the alternative solution.

4   Conclusion

This paper discusses a new method to overcome the CRC check drawbacks, primarily the number of iterations, which efficiently enhances the throughput of the turbo decoder. Almost all the current papers about the stopping criterion are based on the first order difference and a threshold value. We have put forward a new idea based on the second order difference. From the simulation results, we can conclude that this method can dramatically improve the decoding speed, reducing the unnecessary iterations. Actually, as the second order difference method includes the first order process, it can be used to stop the iteration alone. We can calculate a threshold about the second order difference. If the first order difference is in a certain interval and the second order difference is beyond the threshold, we can stop the iteration. Compared with the CRC-SHA, this method will reduce the power consumption and further research can be followed.

References

1. Berrou, C., Glavieux, A., Thitimajshima, P.: Near Shannon limit error-correcting coding and decoding: turbo codes. In: Proceedings IEEE International Conference on Communications, pp. 1064–1070 (1993)
2. Hagenauer, J., Offer, E., Papke, L.: Iterative decoding of binary block and convolutional codes. IEEE Trans. Inform. Theory **42**, 429–445 (1996)
3. Shao, R.Y., Lin, S., Fossorier, M.P.C.: Two simple stopping criterion for turbo decoding. IEEE Trans. Commun. **47**(8), 1117–1120 (1999)
4. Wu, Y., Woerner, B.D., Ebel, W.J.: A simple stopping criterion for turbo decoding. IEEE Commun. Lett. **4**(8), 258–260 (2000)
5. Ngatched, T.M.N., Takawira, F.: Simple stopping criterion for turbo decoding. Electron. Lett. **37**(22), 1350–1351 (2001)
6. Yu, N.Y., Kim, M.G., Chung, S.U.: Efficient stopping criterion for iterative decoding of turbo codes. Electron. Lett. **39**(1), 73–74 (2003)
7. Hou, J., Lee M.H., Park, J.Y.: Adaptive SNR turbo decoding algorithm and stop criterion. In Proceedings IEEE International Conference on PDCAT, pp. 893–895 (2003)
8. Li, F.-M., Wu, A.-Y.: A new stopping criterion for efficient early termination in turbo decoder designs. In Proceedings IEEE International Symposium on ISPACS, pp. 585–588 (2005)
9. Jiang, W., Li, D.: Two efficient stopping criteria for iterative decoding. In: First International Conference on Communications and Networking in China, 2006 , pp. 1–4 (2006)
10. Zhanji, W., Mugen, P., Wenbo, W.: A new parity-check stopping criterion for turbo decoding. IEEE Commun. Lett. **12**(4), 304–306 (2008)
11. Gazi, O.: New early termination method for turbo decoders. In: Signal Processing and Communications Applications Conference (SIU), pp. 1215–1218 (2014)

Comparative Study on Energy Efficiency Evaluation Methods of Variable Frequency Motor System

Paerhati Abudoukelimu[✉], Su'an Ge, Yanfeng Sun, Yongguang Ge, and Peng Li

Experimental Research Institute of Xinjiang Oilfield Company, Xinjiang 834000, China
pahat@petrochina.com.cn

Abstract. The evaluation of the energy consumption is the focus of the motor running condition monitoring. At present, frequency conversion motor system did not install the torque sensor. In view of this situation, first of all, this paper fully considers the basic principle of frequency conversion motor system, as well as nameplate method, slip method, current method and equivalent circuit method. Then, this paper presents a energy consumption calculation method based on the input and output data of variable frequency motor and motor speed. Therefore, non-intrusive energy efficiency on-line evaluation is realized. The method can apply to variable frequency motor system without torque sensor. In this paper, 37 kW motor is used as an example. Above-mentioned different methods were experimentally verified. The accuracy of the proposed method is verified under different operating frequencies.

Keywords: Frequency conversion motor system · Efficiency · Energy saving · Online evaluation

1 Introduction

With the popularization and large-scale application of frequency converter system, people are also very concerned about the energy consumption of variable frequency motor system. However, it is not practical to calculate power consumption of the entire motor system by measuring the output torque directly, since there is no direct mounting of the torque sensor in field operation. In this case, in order to achieve energy efficiency evaluation, *Kun Xia* use the output torque ripple and the clearance of torque noise to achieve a kind of high accuracy motor energy efficiency evaluation method [1]. The method is mostly synthesis and improvement in the method under the IEEE Standard 112 [2]. In addition to the name plate method, this method is more or less to affect the normal operation of the motor. So these methods are not suitable for direct on-site measurement. *Zhang Zhihua* who synthesizes the characteristics of ultra-high efficiency motor develops a set of energy efficiency for evaluating ultra-efficient motor [3]. Harry Li conducts a comparative study of 12 test methods in the laboratory [4]. But all of these methods fail to eliminate the effect of the measurement method on motor operation. In the paper of *Zhao hai sen*, the method of finite element analysis is used to calculate the iron loss of the induction motor [5]. The accuracy is high in this method. But it is not suitable for the accurate evaluation in the project site. In addition, some researchers

© Springer International Publishing AG 2018
F. Qiao et al. (eds.), *Recent Developments in Mechatronics and Intelligent Robotics*,
Advances in Intelligent Systems and Computing 691, DOI 10.1007/978-3-319-70990-1_6

evaluate the energy efficiency of the motor system with their own ways such as the nameplate method, slip method, current method, loss analysis method and equivalent circuit method [6–8]. At the same time, the paper chooses the appropriate algorithm to build the on-line evaluation platform for motor energy consumption [9]. But it is only suitable for the power supply. In the case of frequency conversion power supply, the calculation method of the core loss is given in the paper [10]. Although these methods have certain significance, they have certain limitations and are not suitable for the accurate evaluation for variable frequency motor system. In this paper, based on the existing methods of evaluating the energy efficiency, an energy evaluation method is proposed. This method is combined with the loss analysis method. And its advantages and disadvantages are compared by 37 kW motor experiment.

2 Introduction to Common Methods for Energy Efficiency Evaluation of Motor Systems

For variable frequency motor system, we know that the efficiency actually depends on its loss. The lower is the loss, the higher is the efficiency. In a sense, accurate measurement of motor loss is just an accurate identification to efficiency. Since the previous method can not be applied to the accurate evaluation, it is necessary to formulate a kind of energy efficiency evaluation method in accordance with the on-site frequency conversion condition. At present, there is many traditional methods such as nameplate method, slip method, current method, statistical method and equivalent circuit method.

2.1 Nameplate Method

The nameplate method is the simplest method to estimate motor efficiency. Its assumes that the motor efficiency is equal to the efficiency. The efficiency is shown on the nameplate under any load conditions. In fact, the efficiency changes with the load. Especially, for small power, its efficiency varies greatly with the load. In addition, if the motor is repaired, the nameplate data cannot represent the actual state. In this case, this method is wrong. This method generally causes at least ±1% error. And it requires the motor no-load test, load test and power failure test. If you do not do these and only obtain from nameplate parameters, you will get ±3% error. This method requires at least load test and power-off test. Many industrial occasions are uninterrupted operating. Thus, the method is difficult to achieve in the field.

2.2 Slip Method

The slip method assumes that the ratio of the output power to the rated power is equal to the ratio of the slip to the rated slip. Based on this assumption, we only need to detect the slip in the field and input power. We can obtain from the motor efficiency, as shown in equation (1).

$$\eta = \frac{slip}{slip_{rated}} \cdot \frac{P_{orated}}{P_1} \tag{1}$$

In the standard, there is a tolerance of 20% between the actual value of permissible rated slip and the value of the nameplate. So the error is large. In addition, the measured slip is achieved by detecting the motor speed. In some cases, it is difficult to get the speed.

2.3 Current Method

The current method assumes that the ratio of the output power to the rated power is equal to the ratio of the stator current to the rated current. Based on this assumption, the motor efficiency can be obtained by simply measuring the stator current and the input power in the field, as shown in equation (2).

$$\eta = \frac{I}{I_N} \cdot \frac{P_{orated}}{P_1} \tag{2}$$

I_N is the nameplate full load current. I is of the input current in motor stator. Since the motor current contains the no-load current, this part of the current component does not decrease as the load. So this method brings a greater error. Therefore, the load is usually overestimated. If it is possible to carry out a no-load test of the motor, the formula (3) can improve the current method and the detection accuracy.

$$\eta = \frac{I - I_0}{I_N - I_0} \cdot \frac{P_{orated}}{P_1} \tag{3}$$

I_0 is the no-load current. Since the relationship between the current and the load is not linear, the load is calculated by Eq. The formula (3) is usually lower than the actual load. Therefore, it is better to use the result. We obtain the average value from the Eqs. (2) and (3), which will be able to close to the actual load of the motor. This method also achieved good results, as shown in equation (4).

$$\eta = \frac{2I - I_0}{2I_N - I_0} \cdot \frac{P_{orated}}{P_1} \tag{4}$$

2.4 Equivalent Circuit Method

Motor efficiency evaluation can be obtained through the equivalent circuit calculation. Figure 1 is the equivalent circuit of asynchronous motor. There is a total of seven impedance in the Fig. 1. These parameters can be obtained through two no-load test and resistance test. A no-load test is a no-load test in the motor voltage to measure voltage, current and power data. In the no-load case, the voltage drops to the slip, at which the slip is equal to the rated load. And the relevant data is measured in this state. This equivalent circuit can obtain some parameters such as input power, output power and efficiency at different speed. This method is equivalent to the United States IEEE112-F1 standard, in which the stray loss is calculated in accordance with the IEEE prescription. It is described in the previous section. Equivalent circuit method can reach a certain accuracy, but it needs to change the voltage of the no-load test, which has some difficulties in the field condition in general.

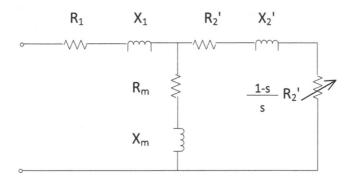

Fig. 1. Schematic diagram of equivalent circuit method

3 The Evaluation Energy Consumption of Variable Frequency Motor System Based on Loss Analysis

Based on the above analysis, a method is presented in the engineering practice. The method measures the input and output data of the inverter to evaluate the energy efficiency about variable frequency motor system. Especially, the output torque of the inverter motor is not suitable for testing. In the course of the research, we combine the actual situation of the project with the basic principle of every link loss calculation. A set of standard calculation method of the energy consumption is worked out. First of all, we get the frequency conversion motor test data without load. We combine wind friction with the method of separation of iron loss. Iron loss and wind loss of the motor are given under the power frequency. Then, combined with the results of no-load test loss calculation, copper loss can be obtained. And a comprehensive energy consumption evaluation system is summarized.

Motor efficiency according to equation (5).

$$\eta_m = \frac{P_2 - \Delta P_\Sigma}{P_2} \times 100\% \tag{5}$$

ΔP_Σ is the total loss power of the motor at a certain operating frequency in kilowatts (kW). The total loss power is calculated according to formula (6):

$$\Delta P_\Sigma = P_{cu1} + P_{fe} + P_{cu2} + P_{fw} + P_a \tag{6}$$

P_{cu1} is the operating frequency motor stator copper consumption. P_{fe} is the operating frequency of the motor iron consumption. P_{cu2} is the operating frequency of the motor rotor copper consumption. P_{fw} is the operating frequency of the motor wind friction loss. P_a is the operating frequency of the motor stray losses.

The stator copper loss is calculated according to equation (7):

$$P_{cu1} = mI_1^2 R_1 \tag{7}$$

I_1 is measured at the operating frequency of the fundamental current. M is the motor phase. R1 is the stator resistance. Iron consumption according to type (8):

$$P_{fe} = \left(\frac{f}{f_N}\right)^{1.3} P_{fe-fN} \tag{8}$$

f_N is the frequency power supply frequency. P_{fe-fN} is the frequency of the motor rated iron consumption, which provided by the manufacturer of data. Rotor copper consumption according to type (9):

$$P_{cu2} = sP_M \tag{9}$$

S is the running frequency of the motor slip. P_M is the operating frequency of the electromagnetic power of the motor.

Spurious loss is determined as follows, for the calculation of high precision requirements, the recommended method of determining the value of stray as follows:

$$P_N \leq 1kW \quad P_a = 0.025P_{in}$$
$$1kW < P_N \leq 10000kW \quad P_N \geq 10000kW$$
$$P_a = (0.025 - 0.005 \lg P_N)P_{in} \quad P_a = 0.005P_{in}$$

P_N is rated power. P_{in} is rated input power. Wind friction calculation can be carried out as follows: In accordance with the provisions of IEEE-112, the wind friction loss is 1.2% of rated power. According to the actual measured motor speed, the use of interpolation method can solve the wind friction when there is wind friction loss of the motor type test.

4   Comparison and Verification of Different Energy Efficiency Evaluation Methods for 37 kw Variable Frequency Motor System

4.1   Comparison of Measured and Evaluated Results Under 37 KW Frequency

In order to verify the accuracy of energy consumption evaluation, 37 kW variable frequency motor system test data were selected for efficiency calculation. And it is compared with other existing motor system energy efficiency evaluation methods. In order to verify the integrity of the calculation method, we have a list of three 10-load conditions at different frequencies. The nominal operating efficiency of the motor is 92%. The rated voltage is 380 V. The rated power factor is 0.85. The stator resistance R1 is measured by the bridge. The excitation resistance Rm of the reaction iron loss is obtained from the no-load test data. According to these parameters, there is for analysis and calculation:

The result of comparison is in 50 Hz under the energy consumption.

Nameplate method is taken under the nameplate value of 92%. Slip method takes the rated slip of 0.012. The current method takes the rated current 73.83 A. Equivalent circuit method uses stator resistance R1 = 0.1774 Ω, Rm = 0.69 Ω.

The method is calculated in accordance with Chapter 2, and its result is plotted as follows:

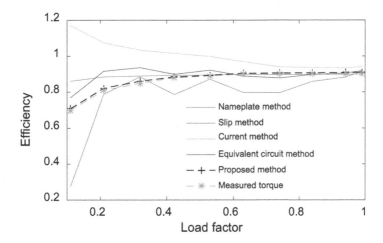

Fig. 2. Different evaluation methods comparison in 50 Hz frequency

The result of comparison is in 40 Hz under the energy consumption (Fig. 2).

The nameplate motor efficiency at any load rate is taken under the nameplate value of 92%. Slip of the rated slip takes 0.013. The current method takes the rated current

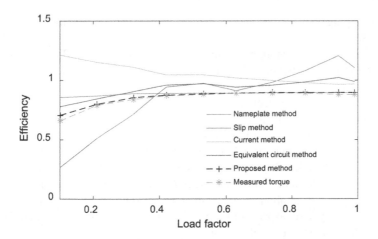

Fig. 3. Different evaluation methods comparison in 40 Hz frequency

59.06 A. Equivalent circuit method uses stator resistance R1 = 0.1774 Ω, Rm = 0.69 Ω.The method in this paper is calculated in accordance with Chapter 2, and its result is plotted as follows:

The result of comparison is in 30 Hz under the energy consumption (Fig. 3).

Nameplate method Motor efficiency at any load rate is taken under the nameplate value of 92%. Slip method takes the rated slip of 0.012. The current method takes the rated current 44.30 A. Equivalent circuit method uses stator resistance R1 = 0.1774 Ω, Rm = 0.69 Ω.

The method is calculated in accordance with Chapter 2, and its result is plotted as follows (Fig. 4):

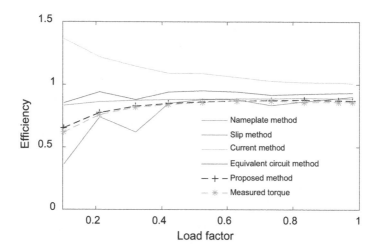

Fig. 4. Different evaluation methods comparison in 30 Hz frequency

4.2   Comparison of Experimental Results

From the mapping result of various energy consumption assessment methods at different frequencies:

Nameplate method is more stable in the entire motor system efficiency evaluation. As the corresponding efficiency is almost linear under each load rate, the nameplate method is not suitable for low-load-rate motor system for energy efficiency evaluation. The slip method has a high accuracy around the rated load factor. But the slip method can cause large errors in efficiency evaluation due to small errors in the speed test. In other words, there is no sense at low load rates.

No-load current accounts for a large part of the total current in the current method. The measurement result will obviously make the overall energy efficiency higher.

Equivalent circuit method has good precision in all frequency and load. The regularity error is not strong about energy efficiency evaluation result. It is necessary to ensure the accuracy of all aspects of the experiment in the use of the method. However, this can not be done in many field conditions.

This paper presents an energy efficiency evaluation algorithm based on loss analysis. The result is in agreement with the energy efficiency of the motor system at different frequencies and at different load rates. The most important thing is to achieve the accuracy of the error within 3% at low load rate, and reach 1% within the error at half load.

5 Conclusion

In this paper, the efficiency evaluation method is compared about common variable frequency motor system. The validity is proved by the comparison of its application range and calculation accuracy. At the same time, other evaluation methods also provide an important reference for the selection of on-site assessment methods.

The proposed method can accurately evaluate the efficiency of variable frequency motor system and ensure the accuracy at low load rate without measuring the output torque of the motor case. Therefore, it can be determined that the method is suitable for the efficiency evaluation of frequency conversion motor system in the engineering field. In most cases, this method can meet the needs of engineering evaluation.

References

1. Xia, K., Jing, L., Dong, B., et al.: A new test system for torque testing and efficiency measurement of three-phase induction motor 12(1), 103–105 (2015)
2. IEEE: IEEE Standard Test Procedure for Polyphase Induction Motors and Generators, p. 112. IEEE, New Jersey (1991)
3. Zhihua, Z., Shengyong, L., Weihua, C., et al.: Study and design of high efficiency motor energy efficiency test system based on 112b method. Chin. J. Minimicrosci. 43(12), 49–54 (2015)
4. Li, H., Curiac, R.: Motor efficiency, efficiency tolerances and the factors that influence them, IEEE Industry Applications Magazine, pp. 1–6. IEEE, New Jersey (2010)
5. Haisen, Z., Dongdong, Z., Yilong, W., et al.: Characteristics and refinement analysis of no-load iron loss of induction motor under variable frequency power supply. Proc. CSEE 8, 2260–2269 (2016)
6. Benning, W.: Method of determining the efficiency of asynchronous motors and apparatus for carrying out the method. US Patent 5659293, 8, 62–64 (1997)
7. He, Q.: Development of measurement methods for motor efficiency. Energy Effic. 12, 8–13 (2003)
8. Becnel, C.L., et al.: Determining motor efficiency by field testing. IEEE Trans. Ind. Appl. IA(23), 440–443 (1995)
9. Huaguang, Y., Ying, F., Xiaohua, D., et al.: Energy-saving motor efficiency testing method and intelligent test platform. Demand Side Manag. 3, 41–43 (2007)
10. Yalin, N., Yunkai, H., Qiansheng, H.: Research and testing of iron core loss in SPWM power supply. Motor Control Appl. 36, 10 (2009)

Integrated Optimal Speed Change Law of Hydro-Mechanical Continuously Variable Transmission Tractor

Mingzhu Zhang[1,2](✉), Xiaoyang Hao[1], Yuxin Yin[1],
Mingming Cui[1], and Tingting Wang[1]

[1] School of Mechatronics Engineering, Henan University of Science
and Technology, Luoyang, Henan Province, China
ming2000@126.com
[2] Collaborative Innovation Center of Machinery Equipment Advanced
Manufacturing of Henan Province, Luoyang, Henan Province, China

Abstract. In order to achieve the productivity-economy comprehensive best of the hydro-mechanical continuously variable transmission (HMCVT) tractor, this paper proposes the integrated productivity-economy optimum as research target. By analyzing the influence factors of productivity and economy, the comprehensive best evaluation index and variable speed change law are formulated. Taking the Dongfanghong LA4004 wheel tractor as an example establishes the mathematical mode of engine and HMCVT to optimization calculate the transmission ratio and engine speed. Optimization result shows that in the load characteristic field has one best transmission ratio and engine speed at any traction load and vehicle speed when the tractor operates in the integrated productivity - economy optimum mode. The speed change law is to adjust the hydraulic control unit and the throttle opening degree according the traction load and vehicle speed. When the tractor adopts productivity-economy integrated optimal speed change law can realize comprehensive optimum. It lays a theoretical foundation for the hydro-mechanical continuously variable transmission tractor automation developing.

Keywords: Hydro-mechanical continuously variable transmission · Productivity-economy · Comprehensive optimum · Optimum transmission ratio · Speed change law

1 Introduction

Hydro-mechanical continuously variable transmission through the power spilt and confluence mechanism can realize large power transmission and multi-range step-less speed change. It has been widely used in agricultural machinery and military vehicle [1, 2]. In order to improve the automatic operation level of CVT vehicle, the scholars have done a lot of researches on the vehicle speed regulation and control strategy [3–5].

The tractor of Hydro-mechanical continuously variable transmission as an agricultural machinery mainly hooks the plough or rake together operation. Tractor is often driven at low speed in the field because of traction requirements. Considering the

F. Qiao et al. (eds.), *Recent Developments in Mechatronics and Intelligent Robotics*,
Advances in Intelligent Systems and Computing 691, DOI 10.1007/978-3-319-70990-1_7

factors of load fluctuation and vehicle slip, the speed change law based on the power and economical performance indexes cannot reflect the tractor operation requirement. The researchers proposed the productivity and economy as the evaluation indexes [6]. In this paper analyze the factor influence the productivity and economy. Author puts forward a comprehensive evaluation index of productivity-economy and establishes the optimal speed change law.

2 Productivity-Economy Integrated Optimal Evaluation Index

The tractor ideal operation mode is high traction power to tow the agricultural machinery operation and low fuel consumption at the same time. Productivity refers to the tractor completed assignments in a certain period of time in accordance with certain quality standards. From the perspective of the energy transfer the traction power N_T directly influences the tractor's traction ability. It directly reflects the productivity of the tractor. Traction power higher suggests the productivity bigger, if the hook traction force is F_t, the vehicle speed is v, and then the traction power is obtained by the (1).

$$N_T = F_t v = (F_q - F_f)v = \frac{0.377 F_t n_e S_r \eta_\delta r_q}{i_z} = \left(\eta_z \eta_b i_z \frac{T_e}{r_q S_r} - \mu mg\right) \times \frac{0.377 n_e S_r \eta_\delta r_q}{i_z} \quad (1)$$

Where, F_t is the tractive force on the hook, kN; F_q is the tractor's driving force, kN; F_f is the tractor's rolling resistance, kN; v is the tractor's velocity, m/s; η_δ is the slip rate of the driving wheel; r_q is the radius of the driving wheel, m; n_e is the engine revolution speed, r/min; S_r is the transmission speed ratio; i_z is the tractor main transmission ratio.

The formula (1) shows that the traction power is determined by engine speed, torque and HMCVT transmission ratio. When the traction power is constant adjusting the throttle and the transmission ratio a series of solutions can be obtained.

Tractor economy refers to the amount of fuel consumed per unit of output per kilowatt hour of traction. The fuel consumption ration (g_T) can be used as the economic evaluation index to reflect the economic performance of the hydraulic mechanical continuously variable tractor. If the tractor engine fuel consumption is g_e, the tractor traction efficiency is η_T, then the tractor fuel consumption ratio as shown in formula (2):

$$g_T = \frac{g_e}{\eta_T} = \frac{g_e}{\eta_b \eta_f \eta_z \eta_\delta} \quad (2)$$

Where, g_e is the engine fuel consumption ratio; η_b is transmission efficiency; η_f is rolling efficiency, η_z is main transmission efficiency, η_δ is slip efficiency.

Tractor slip efficiency is determined by soil coefficient and traction force [7]. According to the test data of tractor engine, the functional relationship between the effective fuel consumption rate and the engine speed and torque is obtained. According to the (2), it can be seen that the fuel consumption ratio of tractor is determined by the engine effective fuel consumption rate and the traction efficiency of tractor.

In order to obtain good economy and productivity performance at the same time, this paper puts forward the comprehensive evaluation index I_{EP}, which is defined as the ratio of tractor traction power and fuel consumption ratio. The Mathematical expression is (3).

$$I_{EP} = \frac{N_T}{g_T} \qquad (3)$$

According to the I_{EP} expression, when the traction power is bigger and low fuel consumption ratio the I_{EP} is bigger, which indicates that the tractor comprehensive performance is better. According to the (1) and (2) the I_{EP} can express as (4).

$$I_{EP} = \frac{N_T}{g_T} = \frac{F_t v \eta_T}{g_e} = \frac{0.377 \eta_z \eta_f r_q \eta_\delta^2 [\eta_z i_z f_b(S_r, n_e, T_e) \frac{T_e}{r_q S_r} - \mu m g] \times n_e S_r f_b(S_r, n_e, T_e)}{f_e(n_e, T_e) i_z} \qquad (4)$$

According to the (4), I_{EP} needs to consider tractor traction power, engine fuel consumption and transmission efficiency, and it is a continuous function of engine speed, torque and transmission speed ratio. A series of I_{EP} values can be obtained by adjusting the speed ratio of the transmission and the engine speed and torque in the allowable range of the maximum traction power of the tractor. Ensure the I_{EP} maximize can combine the tractor economy and productivity. Through optimization calculation gets the transmission ratio and engine speed when the I_{EP} maximum achieving the best productivity and economy.

3 The HMCVT Speed Change Law

The output power of the tractor is determined by the traction load. In the load characteristic field of tractor maximum traction power adjust transmission ratio continuously in the engine performance field to obtain a series of engine speed. In order to achieve the productivity-economy comprehensive optimum, Calculating transmission ratio and engine speed at the maximum I_{EP} is the optimal transmission ratio and engine speed. According to the optimal transmission ratio and engine speed adjusting transmission and throttle achieve variable speed transmission of tractor and productivity-economy optimal. Accordingly, the transmission control strategy is: storing the relationship by table in control unit, that between the optimal ratio and engine speed corresponding to every point in the load characteristics field. According to the tractor speed measured by the speed sensor and the tractor load measured by the force sensor, the optimal transmission ratio and the engine speed are determined by table-checking. By adjusting the HMCVT hydraulic control unit and engine throttle, changing the transmission ratio and the engine speed achieve the tractor productivity-economy comprehensive optimum control.

4 Optimization Calculation of the Optimal Productivity-Economy

The Dongfanghong LA4004 tractor which equipped hydro-mechanical continuously variable transmission is the research object. According to the speed change law, the transmission ratio and engine speed are optimized.

(1) The optimization variables include transmission ratio, engine speed and torque;

$$X = [S_r, T_e, n_e] \tag{5}$$

(2) The optimization target is maximal I_{EP};

$$\max f(X) = \max[I_{EP}] \tag{6}$$

(3) Constraint conditions: to limit the engine speed and torque range and the speed-changing ratio of the HMCVT.

$$\begin{cases} 1300 \leq n_e \leq 2300 \\ 0 \leq T_e \leq 1200 \\ 0 \leq |S_r| \leq 2.12 \end{cases} \tag{7}$$

The main steps of the optimization are as follows:

(1) Firstly, the vehicle speed and traction force is processed with the discrete way, and ensure the matrix dimension is equal.

$$F_t = \begin{bmatrix} F_{t1} & F_{t2} & F_{t3} & \cdots & F_{tm} \end{bmatrix}^T \tag{8}$$

$$v = \begin{bmatrix} v_1 & v_2 & v_3 & \cdots & v_n \end{bmatrix} \tag{9}$$

$$S_r = diag\begin{pmatrix} S_{r1} & S_{r2} & S_{r3} & \cdots & S_{rm} \end{pmatrix} \tag{10}$$

(2) When $v = v_1$, Calculating transmission efficiency and engine fuel consumption. According to the (1) and (2), traction power and fuel consumption ratio can be get as the (11) and (12).

$$N_T = F_t P_1 v_1 E = \begin{bmatrix} F_{t1}v_1 & \cdots & F_{t1}v_1 \\ F_{t2}v_1 & \cdots & F_{t2}v_1 \\ \vdots & \vdots & \vdots \\ F_{tm}v_1 & \cdots & F_{tm}v_1 \end{bmatrix}$$

$$= \begin{bmatrix} \alpha[\beta f_b(T_{e1}, n_{e1}, S_{r1})T_{e1}n_{e1} - \gamma n_{e1}S_{r1}] & \cdots & \alpha[\beta f_b(T_{en}, n_{en}, S_{rn})T_{en}n_{en} - \gamma n_{en}S_{rn}] \\ \vdots & \vdots & \vdots \\ \alpha[\beta f_b(T_{e1}, n_{e1}, S_{r1})T_{e1}n_{e1} - \gamma n_{e1}S_{r1}] & \cdots & \alpha[\beta f_b(T_{en}, n_{en}, S_{rn})T_{en}n_{en} - \gamma n_{en}S_{rn}] \end{bmatrix} \tag{11}$$

$$g_T = \frac{g_e}{\eta_T} = \frac{1}{\eta_z \eta_f \eta_\delta} g_e \eta_b^{-1} = \begin{bmatrix} \mathrm{p}f_e(T_{e1}, n_{e1}) & \cdots & \mathrm{p}f_e(T_{e1}, n_{en}) \\ \vdots & \vdots & \vdots \\ \mathrm{p}f_e(T_{en}, n_{e1}) & \cdots & \mathrm{p}f_e(T_{en}, n_{en}) \end{bmatrix} \eta_b^{-1} \tag{12}$$

(3) According to the traction power and fuel consumption ration, I_{EP} is calculated as shown in (13). Using the Matlab to calculate the maximum of I_{EP} in each row and the corresponding transmission ratio is the optimal transmission ratio of tractor at any traction load under the speed of v_1.

(4) When $v = v_2, v_3 \ldots\ldots$, Repeating the step (2) $\sim$ (3) calculate the optimal transmission ratio of each speed and traction load.

$$I_{EP} = \frac{N_T}{g_T} = F_t P_1 v_1 E \eta_b g_e^{-1}$$

$$= \begin{bmatrix} \alpha \mathrm{p}[\beta f_b(T_{e1}, n_{e1}, S_{r1})T_{e1}n_{e1} - \gamma S_{r1}n_{e1}]f_b(T_{e1}, n_{e1}, S_{r1}) & \cdots & \alpha \mathrm{p}[\beta f_b(T_{en}, n_{en}, S_{rn})T_{en}n_{en} - \gamma S_{rn}n_{en}]f_b(T_{en}, n_{en}, S_{rn}) \\ \vdots & \vdots & \vdots \\ \alpha \mathrm{p}[\beta f_b(T_{en}, n_{e1}, S_{r1})T_{en}n_{en} - \gamma S_{r1}n_{e1}]f_b(T_{en}, n_{e1}, S_{r1}) & \cdots & \alpha \mathrm{p}[\beta f_b(T_{en}, n_{em}, S_{rm})T_{en}n_{en} - \gamma S_{rm}n_{en}]f_b(T_{en}, n_{en}, S_{rm}) \end{bmatrix} g_e^{-1} \tag{13}$$

According to the above steps to optimize calculation of the optimal transmission ratio and the engine speed in the load characteristics field as shown in Fig. 1.

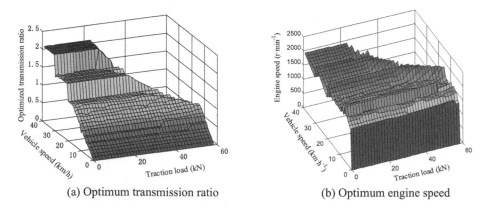

(a) Optimum transmission ratio   (b) Optimum engine speed

Fig. 1. Transmission ratio and engine speed optimization calculated results

Figure 1(a) is the optimal transmission ratio in the load characteristics field. There is one best optimal transmission ratio at any speed and traction load in the load characteristic field. The optimal speed ratio is a ladder like distribution. The platform part is the HMCVT high efficiency point. Figure 1(b) is the optimal engine speed in the load characteristic field. It can be seen from the figure that the optimal engine speed is 1600–1800 r/min. The engine has low fuel consumption in this range according to the characteristics of engine fuel consumption. The tractor traction power and fuel consumption ration for productivity-economy integrated optimization shows in Fig. 2. It can be seen the tractor has high fuel consumption ratio and low traction power in the low speed area due to the low efficiency of hydraulic transmission. Tractor main traction speed area is 5 ∼ 15 km/h. The tractor has low fuel consumption and high traction power at the main traction area. The transmission works in hydro-mechanical ranges.

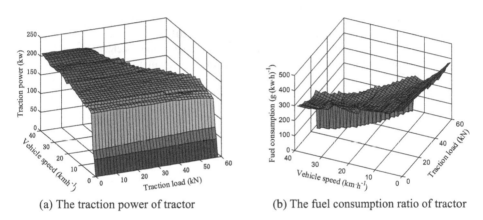

(a) The traction power of tractor   (b) The fuel consumption ratio of tractor

Fig. 2. The traction power and fuel consumption ratio

5 Conclusion

(1) The tractor operation needs to consider the requirements of productivity and economy. The productivity-economy comprehensive evaluation index I_{EP} is defined as the ratio of traction power and fuel consumption ratio. Analysis the factor affect productivity-economy comprehensive evaluation index. The speed change law is established based on the optimal transmission ratio and throttle.

(2) The I_{EP} as evaluation index considering the transmission efficiency and engine performance calculate the optimum transmission ratio and engine speed in the tractor load characteristics field. The calculation result shows that at any point in the load characteristics field there is only one optimal transmission ratio and engine speed. According to the speed change law adjust the transmission ratio and throttle to achieve the tractor productively and fuel consumption comprehensive optimal.

Acknowledgment. This paper is supported by National Natural Science Foundation of China (Grant No51375145) and the Project for Key Science and Technology Research of Henan Province(Grant No162102210205).

References

1. Antonio, R., Alarico, M.: Multi-objective optimization of hydro-mechanical power split transmission. Mech. Mach. Theor. **62**, 112–128 (2013)
2. Ni, X.D.: Design and experiment of hydro-mechanical CVT speed ratio for tracto. Trans. Chin. Soc. Agric. Mach. **44**, 15–20 (2013)
3. Ahn, S., Choi, J.: Development of an integrated engine-hydro mechanical transmission control algorithm for a tractor, pp. 34–44. Sungkyunkwan University, Seoul (2015)
4. He, R.: Method to determine target speed ratio of CVT based on driving intention. Trans. Chin. Soc. Agric. Mach. **40**, 16–19 (2009)
5. Hao, Y.Z., Sun, D.Y., Lin, Y.P.: Overall optimization control strategy of continuously variable transmission system. J. Mech. Eng. **49**, 84–91 (2013)
6. Zhang, M.Z.: Control strategy development for multi-range hydro-mechanical continuously variable transmission in tractors. Ph.D. thesis, Xi'an University of Technology, Xi'an, China, pp. 66–69 (2007)
7. Xu, R.L., Luo, Y.H., Xu, L.Y.: Analysis of wheel slip rate impact on tractor performance base on soft soil. J. Agric. Mech. Res. **1**, 230–233 (2014)

Dynamic Three Parameter Range Shifting Law of the Multi-range Hydro-Mechanical CVT

Mingzhu Zhang[1,2(✉)], Yuxin Yin[1], Xiaoyang Hao[1],
Tingting Wang[1], and Mingming Cui[1]

[1] School of Mechatronics Engineering, Henan University of Science and Technology, Luoyang, Henan Province, China
`ming2000@126.com`
[2] Collaborative Innovation Center of Machinery Equipment Advanced Manufacturing of Henan Province, Luoyang, Henan Province, China

Abstract. To improve the range shifting stability of the multi-range hydro-mechanical continuously variable transmission (HMCVT), the multi-range HMCVT assembled in Dongfanghong 400 horsepower wheeled tractor is taken as the research object. The dynamic model of range shifting is established. Combined with tractor unit dynamics, engine dynamic characteristics and the multi-range HMCVT transmission characteristics, the dynamic three parameter range shifting law is formulated. The throttle degree, vehicle speed and driving resistance are control parameters. Simulation analysis is established by applying matlab/simulink and stateflow software system. The simulation results indicate that, the jerk, sliding friction work are respectively decreased by 57.14%, 75.26% with applying the dynamic three parameter dynamic range shifting law. The dynamics, range shifting stability and durability of the multi-range HMCVT system are effectively improved. This study lays a theoretical foundation for the development and application of automatic transmission.

Keywords: Range shifting stability · The multi-range HMCVT · Dynamic three parameter range shifting law

1 Introduction

The hydro-mechanical continuously variable transmission (HMCVT) has a good application value in agricultural tractors for its high efficiency, high power and wide speed range [1]. Rang shifting as the core of the study, but there is little public information [2, 3]. Therefore, this paper takes the process of the HM2 to HM3 of the multi-range HMCVT assembled in Dongfanghong 400 horsepower wheeled tractor as the concrete research object. Combined with the dynamic characteristics of engine and tractor working characteristics, formulating multi-range HMCVT dynamic three parameter shift rule, which choose throttle opening, vehicle speed and driving resistance as control parameters. According to the established mathematical model, the proposed control strategy is verified by Matlab/simulink and stateflow platform. And it lays a theoretical

© Springer International Publishing AG 2018
F. Qiao et al. (eds.), *Recent Developments in Mechatronics and Intelligent Robotics*,
Advances in Intelligent Systems and Computing 691, DOI 10.1007/978-3-319-70990-1_8

foundation for the development and application of the tractor automatic transmission control system.

2   Analysis and Evaluation Index of Range Shifting Process

The transmission system of the multi-range HMCVT equipped with Dongfanghong 400 horsepower wheeled tractor is shown in Fig. 1.

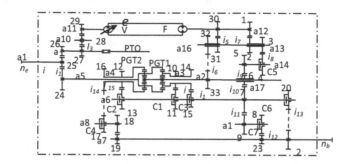

Fig. 1.   Principle diagram of the multi-range HMCVT

In Fig. 1, the number 1–33 are the gear number, a1–a17 represent the axis number, PGT1 and PGT2 refer to planet rows, i1–i18 are gear transmission ratio, C1–C7 refer to each clutch, VP, FM correspond to pump and motor.

The output speed is continuously changed by adjusting the displacement ratio (e) of pump. Five forward range and four reverse ranges of variable speed can be realized by separately controlling clutch. When switching between adjacent two segments of the multi-range HMCVT, in order to realize the continuously variable speed, the displacement ratio of two adjacent segments must be connected and continuous change, and synchronous ranges shifting must be implemented [4]. In the premise of ensuring synchronous ranges shifting to complete the shift operation, Smooth and continuous range shifting can be theoretically implemented. This paper use jerk and sliding friction work as stability evaluation indexs.

3   Mathematical Model of the Range Shifting Process

The dynamic model of HM2 to HM3 of the multi-range HMCVT is established by using the inertia damping system. For the two clutch engagement, the switching process is divided into four states. The description is as follows:

(1)   Before switching (state1): The clutch C1 is engaged, and C2 is in the state of separation, and the rotating speed and torque of C1 and C2 are both in a stable state.

$$T_{c1} = T_{a1} \cdot \left(\frac{k_1}{(1+k_1)i_1 i_{16}} + \frac{V_{pm}\eta_V^{+2}e}{(1+k_1)V_{mm}i_t i_{16}} \right)^{-1} \tag{1}$$

$$T_{C2} = 0 \tag{2}$$

Where, V_{pm}, V_{mm} respectively indicate the maximum displacement of pump and motor, mL/r; η_v represents the total volumetric efficiency of the pump and motor; $i_t = i_2 i_3 i_4 i_5 i_6$.

(2) Torque phase (state2): Clutch C1 is still engaged, but begin to drain the oil. The active disk speed n_{1Z} and driven disk speed n_{1C} still synchronized. C2 begin to fill the oil, it is from the state of separation into the sliding state, the driving wheel speed of n_{2Z} and driven disk speed n_{2C} no longer synchronized. Clutch C1 transmission torque T_{C1} is reduced, C2 start to transfer part of the torque T_{C2}.

$$T_{C1} = \frac{T_{a1}}{i_1} - \frac{2+k_2}{k_2} T_{C2} \tag{3}$$

$$T_{C2} = \mu n A r p(t) sign \left\{ \left(\frac{1+k_2}{k_2 i_1} - \frac{V_{pm}\eta_V^{-2}e}{V_{mm}k_2 i_t} \right) \cdot \frac{\omega_{a1}}{i_{13}} - \left[\frac{k_1}{(1+k_1)i_1} - \frac{V_{pm}\eta_V^{+2}e}{(1+k_1)V_{mm}i_t} \right] \cdot \frac{\omega_{a1}}{i_{13}} \right\} \tag{4}$$

Where, μ is friction coefficient of friction element; n is the number of clutch disc; A is the clutch disc area, m^2; r is the clutch disc radius, m; $sign$ () is symbolic function; $P(t)$ is variation of clutch oil pressure with time, Mpa.

(3) Inertial phase (state3): The speed of active plate and driven plate of clutch C1 enters the slipping state, and the speed of them is no longer synchronous until completely separated. C2 has not yet been fully engaged, and its speed of active plate and driven plate gradually synchronized. The friction torque of clutch C1 decreases with the decrease of oil pressure, and the torque transferred from static friction torque to dynamic friction torque. The torque of clutch C2 transmission is still dynamic friction torque.

$$T_{C1} = \mu n A r p(t) sign \left(\left[\frac{k_1}{(1+k_1)i_1} - \frac{V_{pm}\eta_V^{+2}e}{(1+k_1)V_{mm}i_t} \right] \cdot \frac{\omega_{a1}}{i_{13}} - \omega_{1c} \right) \tag{5}$$

$$T_{C2} = \mu n A r p(t) sign \left(\left[\frac{1+k_2}{k_2 i_1} - \frac{V_{pm}\eta_V^{-2}e}{V_{mm}k_2 i_t} \right] \cdot \frac{\omega_{a1}}{i_{13}} - \omega_{2c} \right) \tag{6}$$

(4) After switching (state4): The change process ends. The clutch C2 is no longer in the sliding friction state, C2 is in a joint state, Clutch C2 transmit all torque.

$$T_{C1} = 0 \tag{7}$$

$$T_{C2} = \frac{k_2}{2 + k_2} \cdot \frac{T_{a1}}{i_1} \tag{8}$$

4 Dynamic Three Parameter Ranges Shifting Law

In order to reduce the velocity fluctuation of the range shifting and improve the ranges shifting stability of the multi-range HMCVT, a kind of dynamic three parameter ranges shifting law with throttle opening, vehicle speed and driving resistance as control parameters is proposed. The dynamic three parameter dynamic ranges shifting law fully considers the characteristics of the variable operating resistance during tractor operation. The switch from HM2 to HM3 is treated as a dynamic acceleration process. Different range shifting velocity is determined under different driving resistance and throttle opening using analytical methods on the basis of the equal acceleration of the tractor before and after the segment. The speed fluctuation of tractor is reduced during the changing process. The specific calculation steps are as follows:

Speed regulation model of engine is fitted by the multivariate polynomial as follows:

$$T_e = \sum_{j=0}^{s} \sum_{i=0}^{j} \varepsilon \cdot \left[(1/2)(j+1)(j+2) - j - 1 + i \right] \cdot \alpha^i \cdot n_e^{j-i} \tag{9}$$

Where, T_e is engine torque, N·m; s is order of model, value 2; ε is the coefficient of polynomial; the variable $i \leq s$, $j \leq s$; α is throttle opening; ne is engine speed, r/min.

Second, when the throttle opening takes a fixed value, the engine torque can be fitted to the two time function curve, shown in formula 10. The driving force Fq can be expressed as formula 11. Under the constant throttle opening, the following relations are obtained from the equal acceleration of adjacent segments, as shown in formula 12.

$$T_e = a_1 n_e^2 + a_2 n_e + a_3 \tag{10}$$

$$F_q = \frac{T_e i_z \eta_b \eta_z}{r_q i_b} = \frac{i_z \eta_b \eta_z}{r_q i_b} [a_1, a_2, a_3] \left[n_e^2, n_e, 1 \right]^T \tag{11}$$

$$\left(F_{qn} - f_n \right) / \delta_{en} m = \left(F_{q(n+1)} - f_{(n+1)} \right) / \delta_{e(n+1)} m \tag{12}$$

Where, $a1$, $a2$, $a3$ represent two polynomial coefficients when throttle opening is given. ib, ηb, iz, ηz respectively indicate the transmission ratio and efficiency of the multi-range HMCVT, transmission ratio and efficiency of the tractor main drive mechanism. rq is vehicle active wheel radius, m.f is running resistance; δe is slip rate of tractor. Finally, according to the formula (10)–(12), the change section equation between section n and section n+1 can be obtained when the throttle opening is given.

$$\frac{a_1 (i_{b(n+1)}^3 - i_{bn}^3) i_z^3 (\eta_{bn} - \eta_{b(n+1)}) \eta_z}{0.377^2 i_{b(n+1)}^3 i_{bn}^3 r_q^3} v^2 + \frac{a_2 (i_{b(n+1)}^2 - i_{bn}^2) i_z^2 (\eta_{bn} - \eta_{b(n+1)}) \eta_z}{0.377 i_{b(n+1)}^2 i_{bn}^2 r_q^2} v + \frac{a_3 (i_{b(n+1)} - i_{bn}) i_z (\eta_{bn} - \eta_{b(n+1)}) \eta_z}{r_q i_{b(n+1)} i_{bn}} = 0 \tag{13}$$

According to the formula (13), change speed vn of section n and section n+1 can be calculated when the throttle opening is given. Repeated the above calculation in different throttle, a series of range shifting points are calculated under different throttle opening and driving resistance. The dynamic three parameter ranges shifting law can be obtained by categorising the range shifting points. Taking HM2 to HM3 as an example, the dynamic three parameter ranges shifting surface can be got by surface fitting, as shown in Fig. 2.

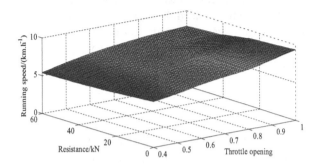

Fig. 2. Dynamic three parameter changing curve between HM2 and HM3

5   Simulation Analysis

Taking HM2 and HM3 in the multi-range HMCVT as an example to carry out the simulation analysis,the simulation conditions are as follows: the throttle opening is 1, the tractor resistance is 10 KN, the main oil pressure is 1.5 Mpa, the starting time is 0.1 s, and the clutch C1 and C2 begin to filling and draining oil at the same time, C1 uses 0.2 s to complete the draining oil, C2 uses 0.4 s to complete the oil-filled. The simulation results are shown in Fig. 3.

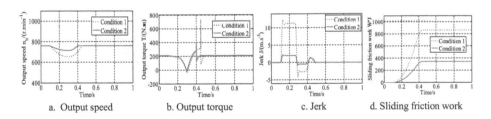

a. Output speed   b. Output torque   c. Jerk   d. Sliding friction work

Fig. 3.   Simulation result curve between HM2 and HM3

Figure 3a–d separately shows that the variation tendencies of the transmission output speed, output torque, jerk, the sliding friction work when the above-mentioned dynamic three parameter ranges shifting law (Condition 2) and the traditional static synchronous ranges shifting law (Condition 1) which is fixed throttle corresponding fixed ranges shifting speed mentioned in the literature [5] are respectively applied between HM2 and HM3.

As shown in Fig. 3a–d, by comparing the result of the dynamic three parameter ranges shift with the traditional static ranges shift, the sinkage of output speed is 46 r min^{-1} and 108 r min^{-1}, respectively, the step mutation of output torque is 399 Nm and 735 Nm, respectively. The jerk J are 11.48 m s^{-3} and 2.84 m s^{-3}, respectively. the sliding friction work W are 874 J and 390 J, respectively. So the output speed and torque fluctuation of the multi-range HMCVT are obviously reduced with appling the dynamic three parameter ranges shifting law. And by using dynamic three parameter ranges shifting law, the jerk, the sliding friction work of the multi-range HMCVT drive transmission system are significantly reduced. Therefore, the application of dynamic three parameter ranges shifting law, can reduce the range-shift impact of the multi-range HMCVT system effectively and improve the durability and dynamic performance of the multi-range HMCVT system.

6   Conclusion

Based on the range shift between HM2 and HM3 on the multi-range HMCVT of the wheel tractor, the dynamic three parameter dynamic ranges shifting law of the multi-range HMCVT is formulated by analytical method. The throttle degree, vehicle speed and driving resistance are taken as control parameters. The driving speed calculated by different throttle opening and different driving resistance is regarded as the changing speed of adjacent two ranges. The simulation results indicate that, jerk and sliding friction work are respectively decreased by 57.14% and 75.26% with applying the dynamic three parameter dynamic ranges shifting law. The dynamic performance, durability and stability of the multi-range HMCVT drive system are improved effectively.

Acknowledgment.   This paper is supported by National Natural Science Foundation of China(Grant No.51375145) and the Project for Key Science and Technology Research of Henan Province(Grant No.162102210205).

References

1. Xi, Z.Q., Zhou, Z.L.: Application status and technical analysis of automatic transmission on tractor. J. Mech. Transm. **6**, 187–195 (2015)
2. Gao, X., Zhu, Z., Zhu, Y.: Analysis of transmission characteristics of HMCVT. J. Chongqing Jiaotong Univ. **4**, 712 (2013)
3. Zhong, W.J., Gao, Q., Lu, Z.X., et al.: Test and analysis of the slip rate of tractors driven on roads. J. Huazhong Agric. Univ. **4**, 130–136 (2015)
4. Macor, A., Rossetti, A.: Optimization of hydro-mechanical power split transmissions. Mech. Mach. Theory **46**, 1901–1919 (2011)
5. Xu, L.Y., Zhou, Z.L., Peng, Q.L., et al.: Drive scheme design and characteristic analysis of multi-range hydro-mechanical CVT. China Mech. Eng. **21**, 2641–2645 (2012)

Study on Control Strategy of Microquadcopter

Hui-jun Deng, Guang-you Yang[(✉)], Fan Liu, and Jun Ma

Institute of Agricultural Machinery, Hubei University of Technology, Wuhan, Hubei, China
pekka@126.com

Abstract. The control system of quadcopter is a system with some features, such as multi-variable, nonlinear, strong coupling and under actuation, which lead to the weak anti-disturbance capability and the vulnerable control performance by parameter. Taking the Microquadcopter constructed by the hardware platform named PIXHAWK as research object, this article discusses the mainstream control method applied to the Unmanned Aerial Vehicle system and chooses PID control method and fuzzy control method as control strategy. Then, we make use of the model which is simulating by Matlab/Simulink software. The results of simulation indicate that fuzzy PID control method can satisfy the accuracy requirement of the multi-rotor Unmanned Aerial Vehicle control system.

Keywords: Quadcopter · PID control · Control method · Simulation of Matlab

1 Introduction

For the accuracy requirement of response in the multi-rotor Unmanned Aerial Vehicle control system, this requires us to select the appropriate control strategy, in which PID control, sliding mode control, fuzzy control and neural network control has been widely applied [1]. Due to the advantages such as simple structure, good reliability and high control precision, PID control has been widely adopted in the industrial process control system [2], but the gradually increased complexity of system leads to the uneasily obtained mathematical model. As a solution of the problem, fuzzy control, a major branch of intelligent control, is a relatively good method because of its low demand for the mathematical model [3].

The research object of this paper is the Microquadcopter, which is a multi-variable, nonlinear, strong coupling and under actuated control system [4]. Considering the diversity of control strategies, this paper mainly studies the PID control, fuzzy control and the control that is effectively combined of the above two control methods.

2 The Controller Design of Microquadcopter

2.1 PID Control of Microquadcopter

The PID control principle diagram is shown in Fig. 1. The diagram reveals the relationship between the control accuracy and the parameters, in which the proportion coefficient K_p is mainly used to adjust the response rate and reduce the error, but will result in

© Springer International Publishing AG 2018
F. Qiao et al. (eds.), *Recent Developments in Mechatronics and Intelligent Robotics*,
Advances in Intelligent Systems and Computing 691, DOI 10.1007/978-3-319-70990-1_9

decreased stability; the integral coefficient K_i is mainly used to eliminate steady-state error, but produces overshoot; the differential coefficient K_d will lead to a better dynamic performance of system, but the interference will be affected [5]. Therefore, the key of the PID controller is how to select appropriate parameters [6].

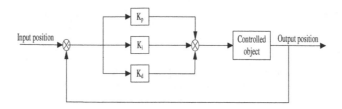

Fig. 1. PID control principle diagram

Considering the complexity of the multi-rotor Unmanned Aerial Vehicle control system, it is more suitable to take the cascade PID control (the outer ring for controlling the position of Microquadcopter, while the inner ring for controlling the attitude of Microquadcopter) as control strategy than the non-cascade PID control because of its higher control accuracy and faster response speed [7].

2.2 Fuzzy Control of Microquadcopter

The control principle diagram of fuzzy is shown in Fig. 2. The diagram reveals the importance of the design of the fuzzy controller. The fuzzy controller, composed of four function modules which are fuzzification, knowledge base, fuzzy reasoning and defuzzification, is the core of fuzzy control which is a kind of intelligent control method based on fuzzy set theory, fuzzy language variable and fuzzy logic reasoning [8].

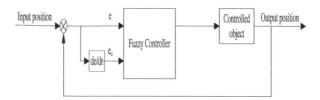

Fig. 2. Fuzzy control principle diagram

The specific design flow of fuzzy controller is: the selection of structure → the fuzzification of system variables → the algorithm design of fuzzy reasoning → the defuzzification of output variables [9].

2.2.1 The Selection of Structure

Taking error e and error change rate e_c as input variables of the controller, u as output variable.

2.2.2   The Fuzzification of System Variables

Defining the range of input variables as the domain of the fuzzy set:

$$e = \{-3, -2, -1, 0, 1, 2, 3\}$$
$$e_c = \{-3, -2, -1, 0, 1, 2, 3\}$$

Defining the range of output variable as the domain of the fuzzy set:

$$u = \{-3, -2, -1, 0, 1, 2, 3\}$$

The fuzzy subsets of the system variables are defined as {NB, NM, NS, ZO, PS, PM, PB}.

The membership function curve of input variable e is shown in Fig. 3. We can find that we mainly select the membership degree as the trigonometric function because of its advantages, such as the simple structure and the relatively good adjustment effect.

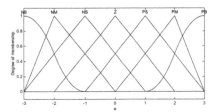

Fig. 3. Membership function plots of input variable e

Considering the weak influence of membership function on the control effect of the controlled object, we establish the membership function curve of input variable e_c and output variable u in the same way.

2.2.3   The Algorithm Design of Fuzzy Reasoning

The rule base of output variable u is shown in the Fig. 4. It is composed of 49 rules to achieve the flow of fuzzy reasoning.

u \ e_c / e	NB	NM	NS	ZO	PS	PM	PB
NB	NB	NB	NB	NB	NM	NS	ZO
NM	NB	NB	NM	NM	NS	ZO	ZO
NS	NB	NM	NM	NS	ZO	ZO	PS
ZO	NB	NM	NS	ZO	PS	PM	PB
PS	NS	ZO	ZO	PS	PM	PM	PB
PM	ZO	ZO	PS	PM	PM	PB	PB
PB	ZO	PS	PM	PB	PB	PB	PB

Fig. 4. Rule base of output variable u

2.2.4   The Defuzzification of Output Variables

The output of fuzzy controller is a fuzzy set, which is decided by the method of defuzzification. We choose centroid method as our defuzzification method.

2.3   Fuzzy PID Control of Microquadcopter

The fuzzy PID control principle diagram is shown in Fig. 5.The diagram reveals the workflow of fuzzy PID controller. The specific workflow of fuzzy PID controller is: take the current error e and error rate e_c as input variables → get the output variables of the fuzzy controller by completing the fuzzification and fuzzy reasoning of the input variables → adjust the parameters of PID controller to adapt the change of the output variables [10].

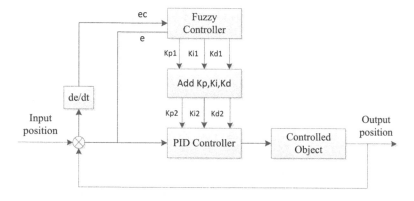

Fig. 5. Fuzzy PID control principle diagram

The design flow of fuzzy PID controller can refer to the fuzzy controller.

2.3.1   The Selection of the Structure of the Fuzzy Controller

Taking error e and error change rate e_c as input variables of the controller, K_p, Ki, K_d as output variables.

2.3.2   The Fuzzification of Input and Output Variables

The design of output variables can refer to the design of fuzzy controller, only with different output variables.

2.3.3   The Algorithm Design of Fuzzy Reasoning

The rule base of output variables are shown in the Figs. 6, 7 and 8. It is composed of 49 rules to achieve the flow of fuzzy reasoning.

2.3.4   The Defuzzification of Output Variables

We choose the same defuzzification method with the fuzzy part.

K_p \\ e_c e	NB	NM	NS	ZO	PS	PM	PB
NB	PB	PB	PM	PM	PS	ZO	ZO
NM	PB	PB	PM	PS	PS	ZO	ZO
NS	PM	PM	PM	PS	ZO	NS	NS
ZO	PM	PM	PS	ZO	NS	NM	NM
PS	PS	PS	ZO	NS	NS	NM	NM
PM	PS	ZO	NS	NM	NM	NM	NM
PB	ZO	ZO	NM	NM	NM	NB	NB

Fig. 6. Rule base of output variable K_p

K_i \\ e_c e	NB	NM	NS	ZO	PS	PM	PB
NB	PS	NS	NB	NB	NB	NM	NS
NM	PS	NS	NB	NM	NM	NS	ZO
NS	ZO	NS	NM	NM	NS	NS	ZO
ZO	ZO	NS	NS	NS	NS	NS	ZO
PS	ZO	ZO	ZO	ZO	ZO	ZO	ZO
PM	PB	NS	PS	PS	PS	PS	PB
PB	PB	PM	PM	PM	PS	PS	PB

Fig. 7. Rule base of output variable K_i

K_d \\ e_c e	NB	NM	NS	ZO	PS	PM	PB
NB	NB	NB	NM	NM	NS	ZO	ZO
NM	NB	NB	NS	NS	NS	ZO	ZO
NS	NB	NM	NS	NS	ZO	PS	PS
ZO	NM	NM	NS	ZO	PS	PM	PM
PS	NM	NS	ZO	PS	PS	PM	PB
PM	ZO	ZO	PS	PS	PM	PB	PB
PB	ZO	ZO	PS	PM	PM	PB	PB

Fig. 8. Rule base of output variable K_d

3   The Analysis of Simulation Results for Microquadcopter

3.1   Simulation of Traditional PID Control for Microquadcopter

The simulation platform of PID control is shown in Fig. 9. It reveals the workflow of PID control. We firstly select the suitable parameters of the PID controller and then analyze the simulation of PID control in Matlab/Simulink software.

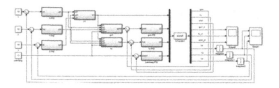

Fig. 9.   Simulation platform of PID

3.2   Simulation of Fuzzy Control for Microquadcopter

The simulation platform of fuzzy control is shown in Fig. 10. It reveals the workflow of fuzzy control. We firstly set up the suitable fuzzy controller and then analyze the simulation of fuzzy control in Matlab/Simulink software.

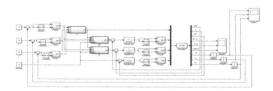

Fig. 10.   Simulation platform of Fuzzy control

3.3   Simulation of Fuzzy PID Control for Microquadcopter

The simulation platform of fuzzy PID control is shown in Fig. 11. It reveals the workflow of fuzzy PID control. We firstly set up the suitable fuzzy PID controller and then analyze the simulation of fuzzy control in Matlab/Simulink software.

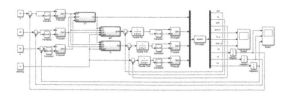

Fig. 11.   Simulation platform of Fuzzy PID control

3.4   The Analysis of Simulation Results for Microquadcopter

The comparison figure of control strategies is shown in Fig. 12. From the simulation results, we can find the difference between different control strategies in which the traditional PID control is around 15 s to make system become stable and the response time is slow, moreover, the curve is not smooth. The fuzzy control is around 43 s to make system become stable and the waveform is oscillating in a small range. The fuzzy PID control is around 7 s to make system become stable and the overshoot is smaller, moreover, the curve is smoother.

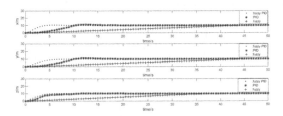

Fig. 12. Comparison of control strategies

In the multi-rotor Unmanned Aerial Vehicle control system, the most important is the stability of the aircraft position control. From Fig. 8, we can see that the control effect of single PID control or fuzzy control is not ideal, but the improved fuzzy PID controller has faster response time and the curve is smoother which make the control strategy to meet the control requirements.

4   Conclusions

In this paper, building the "X" type Microquadcopter as controlled object, we simulate for the traditional PID control, fuzzy control and fuzzy PID control in the Matlab/Simulink platform. The simulation results show that the fuzzy PID control has better dynamic performance than the other control method. In addition, due to the limitation of the fuzzy PID control method, the adaptive learning ability of the controller is bad. Therefore, we need to do further research on a more stable control method.

Acknowledgments. This paper was supported by the Key Technologies R & D Program of Wuhan Science and Technology Bureau (No. 2015020202010129) and the National Natural Science Foundation of China (No. 51174084).

References

1. Khairuddin, I.M.: Modeling and PID control of a quadrotor aerial robot. Adv. Mater. Res. **903**(1), 327–331 (2014)
2. Palunko, I., Fierro, R.: Adaptive control of a quadrotor with dynamic changes in the center of gravity. In: The 18th IFAC World Congress, Milano (2011)

3. Erginer, B.: Design and implementation of a hybrid fuzzy logic controller for a quadrotor VTOL vehicle. Int. J. Control Autom. Syst. **10**(1), 61–70 (2012)
4. Zhang, R., Quan, Q., Cai, K.Y.: Attitude control of a quadcopter aircraft subject to a class of time-varying disturbances. IET Control Theory Appl. **5**(9), 1140–1146 (2011)
5. Jin-xian, Y., Zhi-peng, L., Chao, C.: Study of a four rotor aircraft modeling. Appl. Mech. **12**(494), 293–296 (2014)
6. Mehmet onder EfeSenior Member: Neural network assisted computationally simple PID Control of a Quadrotor UAV. IEEE. Trans. Ind. Inf. **7**(2), 354–361 (2011)
7. Ming-zhi, Y., Min, W.: Design of flight control system for a four-rotor mini rotorcraft. Comput. Meas. Control **16**, 485–490 (2008)
8. Wu, Z.H., Jia, Q.L.: Several control methods of quadrotor. Modern Electron. Tech. **36**(15), 88–90, 94 (2013)
9. Duan, X.G., Li, H.X., Deng, H.: Robustness of fuzzy PID controller due to its inherent saturation. J. Process Control **22**(2), 470–476 (2012)
10. Cheng-qun, Q., Cheng-lin, L., Fa hua, S., et al.: Design of automobile cruise control system based on Matlab and Fuzzy PID. Trans. Chin. Soc. Agric. Eng. **28**(6), 197–202 (2012)

Research and Simulation Analysis on Dynamic Matching Characteristics of Electric Vehicle Transmission System

Lei Zhang[1(✉)], Zeqi Wang[1], Haixin Zhao[1], and Xiumin Yang[2]

[1] School of Mechanical Engineering and Automation, Northeastern University, Shenyang, China
leizhang@mail.neu.edu.cn
[2] Institute of Automation, Shenyang Institute of Technology, Shenyang, China

Abstract. In this paper, the vehicle simulation model of the pure electric vehicle was established by using software, AVL CRUISE. Furthermore, according to the analysis of the experimental data and characteristic curve of the electric vehicle driving motor, the motor numerical model and the power battery model were established. After above, two technical core factors, power matching and speed ratio matching, which would determine the vehicle's dynamic and economic performance when EMCVT is applied to in the power transmission system of electric vehicles, were presented. Thereby, the clamping force model of the variable speed taper disc was established, which was used to guarantee the power matching. Meanwhile, control strategies for ensuring the speed ratio matching were proposed. At last, the theoretical models established in the later experiments conformed to practical changes of the EMCVT during operations, meeting the control requirements for electric vehicles.

Keywords: Pure electric vehicles · EMCVT · Power system matching · Theoretical models

1 Introduction

Compared with traditional fuel vehicles, electric vehicles have higher requirements on the transmission. Currently, analyzed from the actual operation of some new energy passenger cars, motors relying on its own speed governing are far from meeting operating requirements of vehicles. Vehicles require great torque in starting and climbing, and the power demand when vehicles are running at high speed and their higher rate of energy conversion need to be guaranteed in this process. The variation range of the motor speed and torque is quite wide during the process of highest speed, quick starting and maximum speed climbing. Thus, the absence of transmission can easily make the temperature rise pretty high when the motor is running on the circumstance of heavy current, damaging motors or arising failure risks of the control system. When the motor is running at a lower or higher speed, the low efficiency requires connecting a transmission to the motor, making it adapt to the best working area in the motor, reducing risks of overheating and fire happening in the motor as well as the battery pack, and improving vehicles' running mileages.

© Springer International Publishing AG 2018
F. Qiao et al. (eds.), *Recent Developments in Mechatronics and Intelligent Robotics*,
Advances in Intelligent Systems and Computing 691, DOI 10.1007/978-3-319-70990-1_10

In this paper, the key technology of the CVT matching can be studied by analyzing the relation between the key components' (power battery, motor and transmission system) parameters in the electric vehicles' powertrain transmission system and the vehicle performance. Reasonable matching the parameters of electric vehicles' key components, which can play its role effectively. The matching also had a significant impact on improving vehicle power, economy, and ride comfort as well as endurance mileage.

2   Dynamic System Model and Simulation of Electric Vehicle

In this paper, the E520 type pre-propulsion is chosen as the object of study. According to the characteristics of this vehicle, using the software CRUISE in the AVL, the battery module, motor module, EMCVT module, MATLAB API module, driver module and wheel module were dragged in to the CRUISE working area. Simultaneously, brake control function, drive control function and a constant unit module were established. After selecting the subsystem modules, mechanical connection and the electrical connection have been connected between the modules according to the configuration scheme of the vehicle and the connection relation between the components, as is shown in Fig. 1.

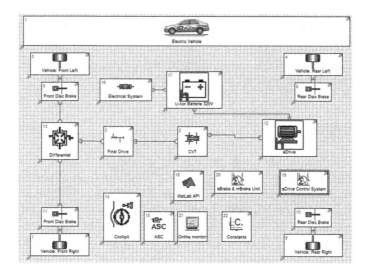

Fig. 1. Total drive system model of battery electronic vehicle

2.1   Motor Characteristics Experimental Data Processing and Modeling

The motor test is an indispensable means of detecting power, economy and reliability of the motor, as well as inspecting the manufacturing quality, reliability and wear resistance of the whole machine and parts, and detecting the accuracy, stability and reliability of the control system. Meanwhile, it is also a key link of researching, designing and manufacturing the new type of motor or corresponding parts. By testing the rated working condition and peak working condition under the rated voltage as well as peak working condition under low voltage working condition for the motor, the efficiency of the motor and the controller according to the test results can be achieved. According to the data tested from the motor test bench under the rated voltage and rated working condition, the external characteristics curve and the efficiency characteristics diagram of the motor and the controller by using MATLAB can be obtained. From the test results, the best working area of the motor can be achieved, thereby providing the experimental data for the formulation of the power system control strategy Fig. 2.

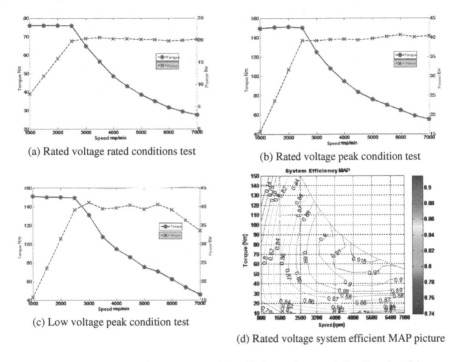

(a) Rated voltage rated conditions test

(b) Rated voltage peak condition test

(c) Low voltage peak condition test

(d) Rated voltage system efficient MAP picture

Fig. 2. The external characteristics curve and the efficiency characteristics diagram of the motor. (a) Rated voltage rated conditions test. (b) Rated voltage peak condition test. (c) Low voltage peak condition test. (d) Rated voltage system efficient MAP picture

2.2 Battery Model

In CRUISE, a resistive-capacitance model was used. The basic model of the battery consisted of a voltage source and an ohmic resistance, and the resistance model contained two internal RC components is used to describe the concentration overvoltage and transmission overvoltage respectively, as is shown in Fig. 3. Meanwhile, the battery thermal model in CRUISE takes into account of the losses caused by the heat generated by the battery and the convection caused by cooling. A single battery or its combination can be simulated to build any desired module through internal calculations of CRUISE.

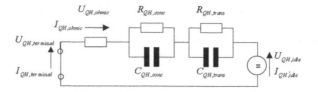

Fig. 3. Single battery model

The power battery capacity is setted according to the requirements of the vehicle driving range, and the maximum output power of the power battery is setted according to the demand power of the vehicle acceleration and climbing. In CRUISE, the model of the power battery is built based on the relation between the electric charge state and the voltage at the time of charge-discharge. According to the charge-discharge curve obtained by the experiment and the energy balance method, the relation between the batteries transient SOC and the voltage can be achieved, as is shown in Fig. 4.

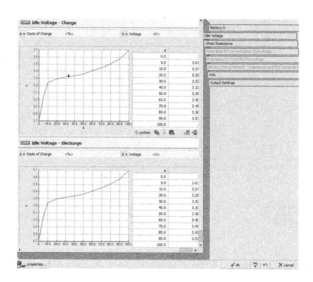

Fig. 4. Relations between voltage and SOC when charge and discharge

3 Matching Model of Continuously Variable Transmission and Motor

The power system performance of the electric vehicle loaded with EMCVT (Electromechanical Continuously Variable Transmission) is mainly depending on two factors.

(1) The power matching between the motor and the CVT.
(2) The speed ratio matching between the motor and the CVT.

3.1 Mathematical Model of Axial Clamping Force of Main and Driven Pulley

The relation between the maximum torque can be transmitted by the metal belt type continuously variable transmission and the axial clamping force of the driven pulley [1] is shown as below,

$$Q_{DN} = \frac{T_{max} \cdot \cos \alpha}{2\mu \cdot R_{DR}} \tag{1}$$

In this equation, α is pulley angle, μ is the friction coefficient between the belt and the pulley, R_{DR} is the drive wheel working radius. The relation among the ratio of the drive wheel and the driven pulley $k = Q_{DR}/Q_{DN}$, the speed ratio i and the torque ratio $\lambda = T_e/T_{emax}$ is shown as Eq. (2) [2],

$$k = f(i, \lambda) \tag{2}$$

When the clamping force of the driven pulley is determined, the clamping force converted to the driving pulley by the metal belt can be determined by the clamping force ratio k.

3.2 EMCVT Speed Ratio Control Strategy

The speed ratio control strategy study of the CVT was based on the characteristics of the motor and the power battery, and the performance parameters of the vehicle. A speed ratio control method can be chosen compromising to the economy and the dynamic performance, making full use of the torque margin of the motor. The vehicle driving ability of the vehicle is improved. Simultaneously, the battery energy consumption can be reduced.

As is shown in Fig. 5, the motor speed is adjusted by controlling the torque of the motor and the speed ratio of the CVT, making it vary along the equal power curve of the target power, until it intersect with the optimum power line, which is the target operating point P_2. The operating point of the control motor is moved along the maximum torque line until it reached the target power curve. In the EMCVT, the displacement sensor of the taper disc is installed on the position of the movable taper disc of the drive shaft, which can detect axial displacement of the movable taper disc. The corresponding transmission ratio can achieved through this function and the displacement sensor of the taper disc, the formula is [3]

$$\begin{cases} R_{DR} = R_{DRmin} + \dfrac{S_1}{2\tan\alpha} \\ i = \dfrac{-(\pi - \dfrac{2R_{DR}}{d}) + \sqrt{(\pi - \dfrac{2R_{DR}}{d})^2 - 4\dfrac{R_{DR}}{d}(\pi + \dfrac{R_{DR}}{d} + \dfrac{2d}{R_{DR}} - \dfrac{L}{R_{DR}})}}{2\dfrac{R_{DR}}{d}} \end{cases} \tag{3}$$

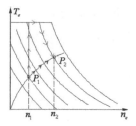

Fig. 5. CVT drive system control model

In the equation, where R_{DR} is the driving plate pitch circle radius, S_1 is the driving plate axial displacement, α is the angle between the plate generatrix and the vertical surface of plate axis, d is the center distance of driving shaft, L is the length of the metal strip.

4   Conclusion

Electric vehicles have a higher demand for the transmission, and power matching and speed ratio matching are the two core technology issues of electric vehicle power system which introduces CVT, determining the vehicle's dynamic performance and economic performance. The later experimental results showed that the theoretical model established in this paper can reflect the actual operation of EMCVT and meet the control requirements.

Acknowledgement. This research was supported by the science and technology funds from Liaoning Education Department (Serial Number: L2015375).

References

1. Fujii, T., Kurokawa, T., Kanehara, S.: A study on a metal pushing V-belt type CVT-part 2: compression force between metal blocks and ring tension. SAE Technical Paper 930666 (1993)
2. Boos, M., Mozer, H.: Electronic-the continuously variable ZF transmission(CVT). SAE Technical Paper 970685 (1997)
3. Zhang, L., Xiaomei, C., Pan, H., Zuge, C., Xiumin, Y.: The control system modeling and the mechanical structure analysis for EMCVT. TELKOMNIKA **11**(7), 4159–4167 (2013)

The Clutch Control Strategy of EMCVT in AC Power Generation System

Xiumin Yang[1(✉)], Yangyang Zhao[1], Song Zhang[1], and Lei Zhang[2]

[1] Shenyang Institute of Technology, Shenyang, China
yxiumin@tom.com
[2] School of Mechanical Engineering and Automation, Northeastern University, Shenyang, China

Abstract. An original control model based on electric-mechanic continuously variable transmission (EMCVT) and the corresponding control mechanism were put forward to solve the frequency conversion problem of alternating-current generator, and the major research focused on the control strategy of external clutch adopting mechanical variable frequency technology. By building the model of dynamic clutch, the control objectives and control parameters were determined. The control method of smooth access or delay disconnect when the clutch running was studied, which realizes the switching control between micro power grid interconnection and isolated island operation in the alternating-current generator. A control experiment of EMCVT external clutch was designed in the paper. Through the experiment it testify that adopting EMCVT in AC generating set, it can product without harmonic pollution and the control of high efficiency constant frequency to come true. Besides, the clutch model and the fuzzy control strategy were present and stable access or disconnect of the output power were realized according to certain rule, achieving the phase consistency of grid frequency and follow-up control of output power.

Keywords: EMCVT · AC power frequency control · Clutch · Fuzzy control strategy

1 Introduction

With the development of continuously variable transmission (CVT) technology and the increased strength research, the ability to transfer power of CVT is continuously improved and the application space also will be continuously expanded. This paper tries to apply new-pattern EMCVT to the frequency conversion control of the output micro power grid interconnection in the alternator and its corresponding clutch control strategies was studied.

2 The Control Requirements of EMCVT Clutch in the Alternator

As for adopting mechanical variable frequency technology, almost the frequency of the output power is determined by the generator rotor speed, where excitation system (AVR) controls the voltage of the generator's export. No matter before no-load or

© Springer International Publishing AG 2018
F. Qiao et al. (eds.), *Recent Developments in Mechatronics and Intelligent Robotics*,
Advances in Intelligent Systems and Computing 691, DOI 10.1007/978-3-319-70990-1_11

after interconnection (access to the load), the speed of the source motive power generation system are decided by the carrying capacity of the unit, that is to say, decided by the governor and TCU control where the carrying capacity increases while speed increases before the no-load, and magnitude is controlled when the rotating speed is constant after interconnection. It need control the AVR to maintain terminal voltage of the generator set.

Speed control system, CVT and excitation system control generator set control together, including diesel engine throttle, fan adjustable paddle institutions, the yaw mechanism and brake mechanism, where the three sides cooperate with each other to make sure grid frequency and voltage constant. When the user load changed violently, the external friction clutch of EMCVT was controlled timely to achieve the output power smooth access or delay disconnect, consistency of voltage, frequency, phase of grid, follow-up control of output power, switching control between micro-grid interconnection and isolated island operation.

When the micro grid from the normal operation turning to the isolated island operation on account of some reason resulting in splitting of grid, or the inverse operation, this important point is how to control the micro grid switch smoothly and keep the system stable. So to make sure the control strategy of the CVT transmission of external clutch, by controlling the generator input shaft angle, the output voltage phase consistency with grid systems is guaranteed, avoiding the impact in interconnection.

3   The Establishment of the Clutch Model

3.1   Dynamic Model of the Clutch

Clutch combination process can be simplified as shown in Fig. 1 [1]:

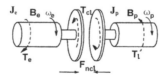

Fig. 1. Schematic drawing of clutch combining process

By the above model, the dynamics equations of driving disc and driven disc of clutch are as shown in formula (1) and formula (2).

$$T_e = T_{cl} + J_e \omega_e + B_e \omega_e \tag{1}$$

$$\eta T_{cl} = T_l + J_p \omega_p + B_p \omega_p \tag{2}$$

In the formula, ω_e, ω_p are the angular velocity of driving disc and driven disc of clutch. B_e, B_p are the equivalent damping coefficient of driving disc and driven disc of clutch. J_e is the rotational inertia of driving disc including source motive and the flywheel. J_p is the rotational inertia of clutch driven disc. T_e is the source motive output torque. T_{cl} is the clutch torque and $T_{cl} = K \cdot F_{ncl}$, K is transfer coefficient. F_{ncl} is the positive pressure. T_l is the drag torque from output load converting to clutch output shaft. η is the transmission efficiency from 0.88 to 0.93 approximately. When the clutch occludes completely, the dynamics equation of the transmission system is as shown in formula (3).

$$T_e = \frac{T_l}{\eta} + (\frac{J_p}{\eta} + J_e)\dot{\omega}_p + B_e \omega_p \qquad (3)$$

3.2 Control Objectives and Control Parameters

The combination process of external clutch should be smooth, soft and small friction losses to ensure no shake and shock in interconnection. Usually, j represents the shock degree and L represents friction loss.

(1) **Shock degree j**. The shock degree refers to the reciprocal of rotational speed on time. If ignoring the influence of equivalent damping and regarding the drag force as a fixed value in interconnection, the shock degree can be expressed as following [2].

$$j = \frac{r_w}{i_0 i_t} \cdot \frac{d^2\omega_p}{dt^2} = \frac{r_w}{i_0 i_t J_p} \frac{d(T_{cl})}{dt} \qquad (4)$$

In the formula, r_w is the tachometer disc radius of output shaft. $\sum J_w$ is rotational inertia of front wheel and back wheel. i_0, i_t is transmission ratio of main reducing gear and transmission ratio of transmission.
From the above formula, the impact degree is mainly related to the torque rate of the clutch in dynamic access. From the above analysis, torque is related to the positive pressure F_{ncl}, so by controlling the positive pressure rate F_{ncl} reasonably, the starting will be smooth and soft.

(2) **Friction loss L**. Friction loss refers to the sliding friction work from clutch driving friction plate and clutch driven friction plate. Friction loss can be divide two parts, the stage from friction plate just contacting to torque overcoming the friction drag and the stage driving disc speed being equal with the driven disc speed. The mathematical expression is [3]:

$$L = L_1 + L_2 = \int_0^{t_1} T_{cl}\omega_e dt + \int_{t_1}^{t_2} T_{cl}(\omega_e - \omega_p)dt \qquad (5)$$

In the formula, t_1 is the time from friction plate contacting to torque overcoming the friction drag. t_2 is the time from driving disc speed being equal with the driven disc speed.

Because the time from friction plate contacting to torque overcoming the friction drag is very short, the previous item can be ignored, so the formula (5) can be transformed to:

$$L = \int_0^{t_2} T_{cl}(\omega_e - \omega_p)dt \qquad (6)$$

From the above formula, the source motive speed has a great influence on friction loss in interconnection.

4   The Fuzzy Controller Design of EMCVT

4.1   Control Strategy of Clutch Engaging Fuzzy

The engaging process of external clutch includes two closely related links. The actuation speed is decided by output power demand and grid condition and the actuator is controlled to make the clutch pull in regularly. The clutch intelligent control scheme is shown in Fig. 2.

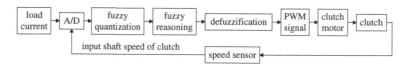

Fig. 2. The method of starting clutch intelligent control

4.2   Selection of Input and Output Variable in Fuzzy Control

According to the control requirements, choose power grid load current i, source machine speed ne and EMCVT disc position variation ΔL as the inputs and clutch motor control signal as the output.

4.3   Fuzzification of the Input Variables

In the research, the prime mover of alternator source of micro-grid uses a gasoline engine. The fuzzy language of input and output can be described as {very small, small, smaller, medium, bigger, big, very big}, expressed as {VS, S, LS, M, LB, B, VB}. The PWM control signal u of clutch motor can be described as {very low, low, lower, medium, higher, high, very high}, expressed as {VS, S, LS, M, LB, B,VB}. The set corresponds to the 13.quantitative levels [4].The membership degree between fuzzy subset and quantitative level is determined by experience.

4.4 Establishment of the Fuzzy Control Table

The purpose of establishing the fuzzy control table is to find out a proper accurate value representing the fuzzy quantity. The gravity method is applied in this paper [1].

$$z_0 = [\sum_{j=1}^{n} \mu_x(x_j) \cdot x_j]/ \sum_{j=1}^{n} \mu_x(x_j) \tag{7}$$

In the formula, n is output quantitative series. x_j is domain element value of domain. $\mu_x(x_j)$ is the membership degree of domain.

4.5 Solve the Fuzzy Process

The basic domain of controlled quantity z_0 can be determined by the total control table of n_e and i. The control voltage signal of clutch motor can be calculate by the following formula.

$$u = kz_0 \tag{8}$$

Coefficient k can be determined by trial and error and the experience of the technical personnel. Fuzzy control table can be represented as a two-dimensional matrix $n \times m$.

But the controller memory is stored data according to the one-dimensional space where some internal storage location is signed by address to access data. So the matrix elements of two-dimensional matrix $n \times m$ are saved data into the ROM of single chip by row. The relative distance from the first element to u_{ij} is $i \times n + 1$. The Fig. 3 shows the flow chart of fuzzy control subroutine.

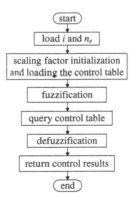

Fig. 3. The flowchart of fuzzy control sub-function

5 Experimental Research

The CVT test table was built using the modified FC2000 dynamic test system. The control test of external clutch in EMCVT was finished. The motivation of EMCVT derived from the gasoline engine. The output shaft of EMCVT was connected with the dynamometer and throttle actuator achieved an automatic control by controlling a DC motor. In the experiments, a constant output shaft speed of EMCVT was regard as a control target. By adjusting the change rate of engine throttle and dynamometer load, conversion control effect between power system interconnection of micro grid and an island operation mode was simulated while control effect of both connection and disconnection of EMCVT was observed.

The transient time of connection of clutch can be controlled within two seconds and the control accuracy of output speed of EMCVT can be controlled within four percent in the experiment.

The theory was validated on the hardware platform. Because common dc motors were applied to clutch control and throttle control and mobility of mechanical transmission mechanism was not the most optimal, the process time of clutch engagement was longer in the experiment and the control precision and stability of output speed for EMCVT were lower. If servo motors and scroll-type mechanical transmission mechanisms were applied on the system, the control precision and stability would be improve dramatically.

6 Conclusion

(1) When the EMCVT is applied into the alternating-current generator, electric power transmission, non-harmonic transformer and high efficiency, can have a function of constant frequency control.
(2) Experimental Research can draw a conclusion that on the running processing of alternating-current generator system, the external clutch of EMCVT can ensure the output power in or out stably.
(3) The fussy control strategy of external clutch of EMCVT and the clutch model were present, which achieves the satisfied control for external clutch of EMCVT.

Acknowledgement. This research was supported by the science and technology funds from Liaoning Education Department (serial number: L2015375).

References

1. Zhang, L., Zhang, X., Yang, X., Liu, Y.: The study on fuzzy control of automotive clutch. Adv. Mater. Res. **230–231**, 334–338 (2011)
2. Anlin, G., Bingzhao, G., Hui, J.: Precise speed control of AMT clutch's engagement. Chin. J. Mech. Eng. **36**(10), 68–71 (2000)
3. Guoqiang, W.A.N., Keqiang, L.I.: Power-on downshift control for clutch-to-clutch automatic transmission. J. Mech. Eng. **22**, 66–72 (2015)
4. Xiong, W.E.I., Xing-li, Z.H.O.U.: Control strategy study of electron-controlled throttle for diesel engine based on intelligent filtering. Chin. Intern. Combust. Eng. Eng. **35**(5), 77–81 (2014)

Efficient Optimization Strategy of Fuzzy Search for Asynchronous Motor Vector Control System

Xing Chen, Fang He$^{(\boxtimes)}$, and Yayun Zheng

School of Electrical Engineering, University of Jinan, Jinan, Shandong, China
cse_hf@ujn.edu.cn

Abstract. The efficiency optimization of asynchronous motor control system is meaningful for saving energy. The efficiency optimization principle of asynchronous motor is analyzed. Several efficiency optimization strategies are discussed. Efficiency optimization strategy of fuzzy search is proposed. This strategy is to realize minimum input power of motor by fuzzy searching for optimal rotor flux. The online fuzzy search controller is designed. Simulation model of motor control system is set up. By comparing and analysis of simulation results, the effectiveness of strategy proposed is validated.

Keywords: Fuzzy search · Efficiency optimization · Asynchronous motor

1 Introduction

AC asynchronous motors are widely used in advanced manufacturing system. It has great significance to optimize efficiency of asynchronous motor control system. Much relevant work has been done [1]. There are several strategies of efficiency optimization for motor. They have different optimization principles. One is online searching optimization strategy of input power [2]. The other is optimization strategy based on loss model of motor. Besides, there is an approximate optimization strategy of minimum stator current [3].

Optimal control strategy of minimum stator current is affected by variation and saturation of motor parameters, which optimization result is not accurate. The optimization strategy based on loss model requires a precise mathematical model and depends on the motor parameters. The accuracy of mathematical model and the parameter variation of motor will affect the effect of efficiency optimization. Online searching optimization strategy of input power does not request the exact parameters of motor, which can be applied for any motor. So this kind of efficient optimization method has been favored by many scholars and has been widely used.

In order to energy-saving, this paper study efficient optimization strategy for asynchronous motor vector control system. Section 2 analyzes efficiency optimization principle of asynchronous motor. Efficiency optimization strategy of fuzzy search is proposed in Sect. 3. Design of fuzzy search controller is given in Sect. 4. Simulation model of efficiency optimization control system based on fuzzy search strategy is set up, and simulation results are analyzed in Sect. 5.

© Springer International Publishing AG 2018
F. Qiao et al. (eds.), *Recent Developments in Mechatronics and Intelligent Robotics*,
Advances in Intelligent Systems and Computing 691, DOI 10.1007/978-3-319-70990-1_12

2   Efficiency Optimization Principle of Asynchronous Motor

If the speed of motor and load of motor are determined, the output power of asynchronous motor is constant. In order to improve the running efficiency of asynchronous
motor, a feasible and effective method is to reduce its input power. Because the input
power is the sum of the loss power and the output power, an important way of improving
efficiency of motor is to reduce the total loss of asynchronous motor. Figure 1 shows
the principle of efficiency optimization for asynchronous motor vector control system.

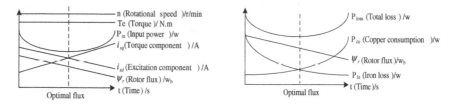

Fig. 1.   Principle of efficiency optimization for asynchronous motor vector control system

The stator current is divided into excitation component i_{sd} and torque component i_{sq}
according to the idea of rotor flux orientation vector control. The rotor flux only depends
on excitation component of stator current. The total loss P_{loss} loss of motor is sum of
copper consumption P_{cu} of motor and stator iron loss P_{Fe}. The copper consumption
includes stator copper loss and rotor copper loss.

It is assumed that the speed of motor and torque of motor are constant at certain
steady state. In Fig. 1, the total loss P_{loss} of system is smallest at the intersection of iron
loss P_{Fe} curve and copper consumption P_{cu} curve. That means that the input power P_{in}
is smallest and efficiency is highest. This corresponding flux is the optimal excitation
flux.

The efficiency optimization of asynchronous motor can realize if the motor flux is
the optimal excitation flux. Its efficiency will be best under certain speed of motor and
torque of motor.

3   Efficiency Optimization Strategies of Fuzzy Search

The advantage of efficiency optimization control strategy of minimum stator current is
easy to achieve. Its disadvantage is that the optimization results may be local optimum.
In addition, the change of motor parameters will affect the results of efficiency optimization.

The efficiency optimization control strategy based on loss model is set up for asynchronous motor vector control frequency control system, in which the flux control and
torque control are completely independent. The optimal excitation flux of minimum loss
can be derived as Formula (1) [4].

$$\psi_r^{opt} = \sqrt[4]{\frac{K_1}{K_2 + K_3(\omega_r/n_p)^2}} \sqrt{T_e} \tag{1}$$

where

$$K_1 = \left(R_s + \frac{R_m R_r}{R_m + R_r}\right)\frac{(R_m + R_r)^2}{n_p^2 R_m^2}; K_2 = \frac{R_s}{L_m^2}; K_3 = n_p^2\left(\frac{R_s}{L_m^2} + \frac{1}{R_m}\right);$$

In Formula (1), R_s and R_r is respectively stator resistance and rotor resistance. R_m is iron equivalent resistance. n_p is the number of pole pairs of asynchronous motor. L_m is rotor mutual inductance. w_r is rotor angular velocity.

When flux is optimal, the efficient of motor vector control system will be optimal. The expression of optimal efficient is Formula (2).

$$\eta^{opt} = \frac{P_{out}}{P_{in}} = \frac{P_{out}}{P_{out} + P_{loss}} = \frac{(\omega_r/n_p)T_e}{(K_2 + K_3(\omega_r/n_p)^2)\psi_r^{opt2} + K_1\frac{T_e^2}{\psi_r^{opt2}} + (K_4 + 1)(\omega_r/n_p)T_e} \tag{2}$$

where $K_4 = 2\frac{R_m + R_r}{R_m^2}\left(R_s + \frac{R_r R_m}{R_r + R_m}\right)$;

The advantage of efficiency optimization control strategy of loss model is that the excitation flux of minimum total loss can be calculated directly by formula. The implement of control is very fast. But this kind of control strategy requires a precise mathematical model of motor and depends on the motor parameters largely. Because the mathematical model of motor is very complex, it is difficult to determine its control parameters online.

Because fuzzy control is of great significance for those complex control systems which precise mathematical models are difficult to get. A kind of fuzzy search strategy of efficient optimization control is proposed bellow in this paper.

Figure 2 shows the schematic diagram of efficiency optimization strategy of fuzzy search for vector control system of asynchronous motor. The efficiency optimization strategy of fuzzy search is that an online fuzzy search controller is set to detect the input power of the motor. The optimal flux is searched according to the principle of minimum input power. Then the motor will run under the optimal flux to achieve the best efficiency.

Because the loss of frequency converter is very small, it can be ignored. The input power of DC side of inverter can be used instead of the input power of motor in Fig. 2.

Figure 3 presents the flowchart of fuzzy search method of efficiency optimization for asynchronous motor. The speed of motor and torque of the motor will affect the quantification factor and proportionality factor of fuzzy search controller. The inputs of fuzzy search controller include the input power of the DC side of inverter and the flux of the previous cycle. The optimal flux can be obtained by fuzzy reasoning. The optimal flux will determine the excitation current i_{sd} according to stator flux of motor vector control system. The stator current is synthesized by excitation current i_{sd} and torque current i_{sq}. The motor will get an optimal efficiency using this stator current.

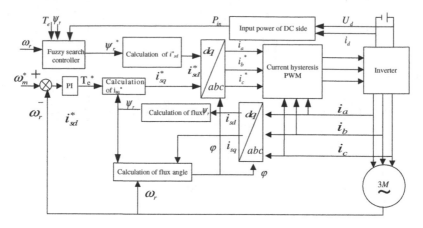

Fig. 2. Vector control system using efficiency optimization strategy of fuzzy search

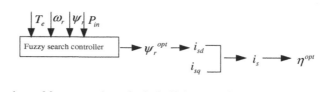

Fig. 3. Flowchart of fuzzy search method of efficiency optimization for asynchronous motor

4 Design of Fuzzy Search Controller

Figure 4 shows the block diagram of fuzzy search controller. k_a is the input quantification factor. k_b is the output proportionality factor. $P_{in(k)}$ is the input power of DC side of inverter. After the fuzzy reasoning, the flux change of this period is got. Finally, the optimal flux can be got to improve efficiency of motor control system after several steps fuzzy searching.

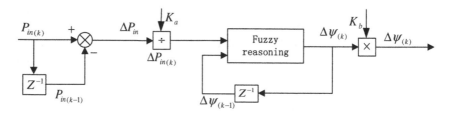

Fig. 4. Fuzzy search controller

The quantification factor k_a and proportionality factor k_b can be expressed as:

$$\begin{cases} K_a = A\omega_r + B \\ K_b = C_1\omega_r - C_2T_e + C_3 \end{cases} \tag{3}$$

Equation (3) shows that the input quantification factor and output proportionality factor are associated with the speed of motor and torque of motor. They change with speed of motor and torque of motor. The coefficients A, B, C_1, C_2, and C_3 are constants, which are determined by simulation studies.

If the input power P_{in} decreases in the previous control cycle, the change trend of flux in this cycle will be same as that of in previous cycle. The change of flux $\Delta\psi_r$ is proportional to the change of P_{in}. According to analysis above and result of simulation debugging, the correspondence fuzzy search rules can be got, which are shown in Table 1.

Table 1. $\Delta\psi(k)$ Fuzzy search rules table

$\Delta\psi_{(k-1)}$	$\Delta P_{(k)}$						
	PB	PM	PS	ZO	NS	NM	NB
N	PM	PS	PS	ZO	NS	NM	NB
P	NM	NS	NS	ZO	PS	PM	PB

5 Simulation

Simulation model can be set up using MATLAB/Simulink software. According to Fig. 2, simulation model of efficiency optimization control system based on fuzzy search method is set up as shown in Fig. 5. In Fig. 5, "Fuzzy Logical Controller" is the fuzzy search controller of efficiency optimization.

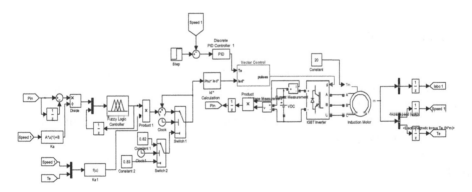

Fig. 5. Simulation model of efficiency optimization control system based on fuzzy search

In order to compare performance of the system operation, "Switch1" is set in the simulation model. It is used to switch state from constant flux supplied to optimal flux though fuzzy search. At first, the system runs under a constant flux before t = 2 s. And then, the system will run on state of fuzzy search. The simulation results are shown in Fig. 6. The Fig. 6(a) shows the curve of rotor flux with switching operation at t = 2 s. By fuzzy searching, the optimal flux of motor rotor can be got after five steps. The rotor flux is reduced from 0.83 Wb to 0.28 Wb. The Fig. 6(b) shows the change of input power of corresponding rotor flux. From this figure, it shows that the input power is gradually reduced from 3623.1 W to 2801.1 W after switching. According to the Formula (2), the efficiency of motor can be calculated. The result shows that the efficiency of motor increases from 43.4% to 56.1% after switching.

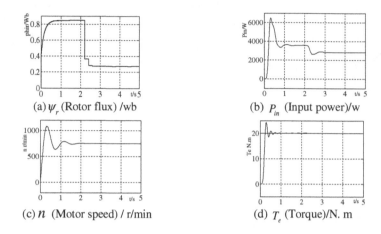

(a) ψ_r (Rotor flux) /wb   (b) P_{in} (Input power)/w

(c) n (Motor speed) / r/min   (d) T_e (Torque)/N. m

Fig. 6. Simulation results

The speed of motor is shown in Fig. 6(c). The torque of motor is shown in Fig. 6(d). They all have little change after switching. The application of fuzzy search method has no effect on the stability of system.

6   Conclusion

The simulation results show that the strategy of fuzzy search proposed above can effectively optimize efficiency of asynchronous motor control system. The effect of energy-saving is obvious. The main advantage of this control strategy is that it need not depend on exact parameters of the motor. It is applicable to efficient optimization of asynchronous motor control system.

References

1. Chen, C.: Application of fuzzy control in speed control of AC variable frequency motor system. Electr. Mach. Control Appl. **39**(08), 55–58 (2012)
2. Kioskerdis, I., Margaris, N.: Loss minimization in scalar-controlled induction motor drive with search controllers. IEEE Trans. Power Electron. **11**(12), 213–220 (1996)
3. Shao, W., Zheng, Y., Zhao, K.: Research on minimum stator current control of asynchronous motor. Electr. Switch. **6**(6), 30–34 (2007)
4. Yu, Z., Yan, W., Bao Z.: A control strategy for the optimal efficiency of induction motor based on loss model and fuzzy search. In: Proceedings of the 29th Chinese Control Conference, pp. 2493–2496 (2010)

Analysis of Anti Interference of Channel Control Measure in 2.4 GHz-Band Network

Xudong Chen[✉], Jing Fan, Kang Cheng, and Minmu Chen

The Key Lab of Wireless Sensor Networks, School of
Electrical and Information Technology, Yunnan Minzu University,
Kunming, Yunnan, People's Republic of China
596243496@qq.com

Abstract. The most recent emerging channel controls for Wi-Fi such as the Bluetooth protocol are designed to operate in the 2.4 GHz ISM band. Since Wi-Fi, Bluetooth and ZigBee are using the same frequency band, they produce mutual interference when they operate at the same time. The simultaneous operation may lead to significant performance degradation. Wireless networks can be developed based on the IEEE 802.15.4 protocol or based on another protocol which is built on IEEE 802.15.4. The ZigBee protocol standard, which is built on top of IEEE 802.15.4, provides additional benefits of a well-developed wireless networking standard. And the ZigBee is designed, built and supported by hundreds of the world's leading technology companies. In this paper, we introduced the basic concepts of the three systems that we used, based on Shannon theorem and Gaussian channel truth theory, to improve transmission performance of the systems. This improvement allocated the chaotic channel reasonably, base on Orthogonal Frequency Division Multiplexing theorem, to reduce the expenses on the channel. All of modeling and simulation result were obtained by MATLAB, and two examples that are similar with the actual application situation were exhibited respectively in main simulation section below. Finally, this paper proposed a new channel control mechanism which can make the communication performance improved. The effect of performance of the communication is obvious, especially with increasing users. Hence, during limited frequency band, more customs can exist, with higher channel utilization rate.

Keywords: Channel control · 2.4 GHz · Wi-Fi · ZigBee · Bluetooth

1 Introduction

Although the ISM (Industrial, Scientific and Medical) radio bands do not need permits, they are subject to a certain limit of transmission power (less than 1 W), and they should not cause interference to other bands. 2.4 GHz wireless technology is a radio band between 2.405–2.485 GHz, and this band is an ISM band. Wi-Fi, ZigBee, and Bluetooth, all three systems work in this band. The advantages of Wi-Fi have been gradually discovered because of the development of society. Compared to the shortcomings of the geographic limitations and long response time of wired communications, Wi-Fi has a bigger coverage area, has almost no geographical limits, and it sets

© Springer International Publishing AG 2018
F. Qiao et al. (eds.), *Recent Developments in Mechatronics and Intelligent Robotics*,
Advances in Intelligent Systems and Computing 691, DOI 10.1007/978-3-319-70990-1_13

up communication links quickly. Since the establishment of standard (IEEE 802.15.1) of Bluetooth technology, Bluetooth has been used widely in many fields such as home automation, industrial control, and toy making. Although Bluetooth has obvious advantages, it has multiple defects too. These defects include the huge power consumption, very short communication range, too small of a network size and its complexity. The emergence of ZigBee technology standards complements these advantages and disadvantages of Bluetooth. However, the three technologies have channel interference with each other, because they work in the same frequency band. Since the ZigBee is officially launched in 2004, some articles about the interference with coexistence problem have been studied.

2   Literature Survey

Reference literature indicates that WiFi and ZigBee systems need cognitive radio capability to improve the anti-interference performance of the two systems. The performance is decided by three elements: the data packet error rate, link quality and received signal strength performance metrics [1]. But it they did not propose a specific method how to improve the anti-interference performance between the two systems. In [2], they concluded that a balance between the power level of the radio frequency signal decided by bit error rate, and error vector the amplitude measurement of Wi-Fi system and ZigBee system should be considered. But low power can weaken the original transmission performance. In [3], an algorithm is proposed based on the Wi-Fi and Bluetooth systems. It reduces the expenses on the network. But the transmission time of information is extended. In [4], the authors solve the collision probability signal in a closed form after analysis of Wi-Fi system and Bluetooth systems. But there are certain limitations in it. In [5], after analysis of Wi-Fi and Bluetooth, the authors proposed a new virtual interface architecture to solve the switch problem, but it only works in these two systems. In [6], the coexistence analysis model of ZigBee and Bluetooth systems are introduced. A numerical method was proposed for solving the interference of multiple Bluetooth systems in the field of the data packets confirmation. However, this calculation method cannot be used for the number of packets that are not recognized. In [7], the interaction between ZigBee system and Bluetooth system was studied. Then the coexistence problem of equipment interference is measured in real life. But they did not propose any method to solve the coexistence problem. In [8], the authors evaluated the interaction influence among Wi-Fi, ZigBee and Bluetooth systems. But they did not take a specific research on how to reduce the interference. In [9], the packet error rate of Wi-Fi, ZigBee and Bluetooth is was evaluated. The analysis results were validated with simulation, but no reasonable methods were proposed to reduce the error rate and the collision time.

Normally, the available channels used by several systems may be overlapped in the frequency domain. A non-overlapped channel can be selected by a method of channel control to improve the capability of anti-interference. But it is very important to ensure the stability of the system when the channel is controlled. In this paper, we used the results of previous studies [1, 3, 6, 8] and proposed a simulation model of 2.4 GHz-band network. With our model, the interference in the whole system model is was simulated when the channels are were occupied by users who used a single or multiple systems. In this paper, we proposed a NCC (New Channel Control) mechanism that can be used to control channels in all of the three systems with the users of three systems in 2.4 GHz network and finally, we compared our mechanism with previous ones to decide which one is better or the best.

3   Channel for Three Kinds of Systems and Channel Fusion Model

The Wi-Fi system works in 2.4 GHz-band. And its available bandwidth of the system is 83.5 MHz. The band is divided into 13 channels and each channel bandwidth is 22 MHz in China. There are only 3 channels (channels 1, 7, 13) in the whole frequency band and the 3 channels can coexist without interference. 27 physical channels are defined in ZigBee system distributes 1 of the channels into 868 MHz-band, 10 channels into 915 MHz-band and 16 channels into 2.4 GHz. Each channel bandwidth is 2 MHz. The spacing of each channel is 5 MHz. In this paper, we only investigated 16 of them in 2.4 GHz-band. 79 physical channels are defined in Bluetooth system. There are 79 jump frequency points in the Bluetooth system. The hopping rate of each point is 1600/s. The number of Bluetooth devices that can be connected to others must not beyond 7 at the same time.

3.1   Wi-Fi System for Modulation and Demodulation of Software Simulation

The modulation and demodulation of Wi-Fi system is DQPSK (Differential Quadrature Phase Shift Keying) and its direct sequence spread spectrum is Barker. The initial signal is divided into two signals. These two signals mixed with carrier wave are processed by Barker sequence. So the output signal of DQPSK modulation is formulated [10] as

$$y_{DQPSK}(t) = \mathrm{u}(t)\cos 2\pi f_{cw}t - \mathrm{v}(t)\sin 2\pi f_{cw}t \tag{1}$$

where the parameter $\mathrm{u}(t)$, $\mathrm{v}(t)$ are two initial signals, f_{cw} is a carrier frequency, $y_{DQPSK}(t)$ is the signal mixed with the carrier wave.

With the help of channel, noise will be added to signals. And then it is dispread by Barker. Two signals are extracted from the output signals. And the output signals of DQPSK demodulation is formulated [10] as

$$\hat{u}(t) = \hat{r}_w(t) * \cos 2\pi f_{cw}t$$

$$\hat{v}(t) = -\hat{r}_w(t) * \sin 2\pi f_{cw}t \tag{2}$$

where $\hat{r}_w(t)$ is signal displayed by barker, $\hat{u}(t), \hat{v}(t)$ are two signals which are extracted from the output signals.

3.2   ZigBee System for Modulation and Demodulation of Software Simulation

Zigbee system is an OQPSK (Offset Quadrature Phase Shift Keying) modulation and demodulation. With the help of modulation, the initial signal is divided into two signals. After that, the 32 bit chip code is directly extended and mixed with signals before being sent into the channel. Demodulation uses 32 bit chip code solution to expand, and extracts the two signals from the output signals. The modulation and demodulation are similar to that of the DQPSK, the only difference being two orthogonal components staggered about half symbol period.

3.3   Bluetooth System for Modulation and Demodulation of Software Simulation

The modulation and demodulation of Bluetooth system is GFSK (Gauss Frequency Shift Keying). The initial signal accessed to the Gaussian Low Pass Filter is sent into the integrator. So the output signal of GFSK modulation is formulated [10] as

$$y_{GFSK}(t) = I(t) \cos 2\pi f_{cb} t - Q(t) \sin 2\pi f_{cb} t \qquad (3)$$

where $I(t)$ and $Q(t)$ are two initial signals, f_{cb} is a carrier frequency, $y_{GFSK}(t)$ is a signal mixed with added carrier wave.

When a signal is going through a channel, it will be mixed with the noise. The signal is extracted by LPF(Low Pass Filter) from the output signals. So the signal of GPSK demodulation is formulated [10] as

$$\hat{y}_{deGFSK}(t) = \frac{1}{2} r_b(t) r_b(t - T_b) \sin \Delta \varphi(T_b) \qquad (4)$$

where $r_b(t)$ is a signal mixed with noise, $\Delta \varphi(T_b)$ is a phase difference, $\hat{y}_{deGFSK}(t)$ is an output signal of GPSK demodulation.

4   Proposed Method

4.1   Original Model of the Channel

Based on characteristics of the modulation and demodulation used by the above three systems, we established the original model of the channel and the fusion model of the channel. The original channel model, which is consistent with the transmission of the real channel, mixes the three systems simply. In this original channel model, consumers are allowed to use the system randomly, and the system will generate information of users randomly, and finally collect statistics data of the system performance. Our fusion model is based on this original channel model and is added with the channel monitoring and distribution control mechanism, to reduce interference among the signals, caused by the overload of consumers when they access the system simultaneously.

In the original channel model, when the users access the Wi-Fi system, channels 1–13 will be distributed to the consumers randomly. Although single Wi-Fi system can accommodate that 13 users randomly accessed the systems, it will cause more channel interference when more users access to this system; when the users access to the ZigBee system, channels 11–26 will be distributed to the users randomly; when the users access to the Bluetooth system, channels 1–79 will be distributed to the users randomly. For single ZigBee or Bluetooth system, they can accommodate that 16 and 79 users randomly accessed the systems respectively. As long as the number of their users outnumber of tolerance value, the systems allows users to transfer information regularly; it will cause more channel interference when more users access to this system. When a channel has been occupied, it will not be reassigned to a new user, but other channels will be allocated to new users by supervision. In this model, multiple users access to the Wi-Fi system will cause interference. However, the ZigBee and Bluetooth technologies will not create interference when multiple users access to the systems. When there are more than two users whose channels are close or conflict, it will create higher transmission error rate. The Bluetooth system signal transmission is different from that of the other two systems. The system signal will hop to transmit randomly in 79 physical channels, while the frequency hopping transmission will also create channel interference.

4.2 Channel Control Mechanism Model

Because of the shortcomings of the original channel model, we propose a fusion model that is based on the original channel model. In the channel fusion model, when there are only Wi-Fi users, this system mechanism will monitor, control and distribute Wi-Fi channels. The channel overlap interference will not occur when multiple channels are occupied simultaneously, but the number of available channels is reduced from 13 to 3. There are only 3 channels (channels 1, 7 and 13) can exist at the same time without interference. If these three channels are occupied, no other user is allowed to use the Wi-Fi system; while there are only ZigBee or Bluetooth users, the number of available channels do not decrease as the Wi-Fi system.

No matter what kind of new users access the system, the new mechanism tech will monitor the entire 2.4 GHz band to test whether the required system's channel is occupied or not. If there are still some of them available, it will allocate the new channels to the users according to the rules of our NCC mechanism; on the contrary, the channel which controls the entire 2.4 GHz band will not allow users to access.

4.3 Channel Conflict and Control During Parallel Output of Three Kinds of Systems

The channels will be bound to be conflict when ones of parallel output with three kinds of systems are taken place, see Table 1.

In Table 1, denotes the channel without influencing other channels. For example, six users access to three kinds of systems. The ZigBee users occupied channel 1 and the Bluetooth users occupied channels 3–5. The Wi-Fi users occupied channels 1 and 2. Just these two channels of Wi-Fi system are overlapped. The channel 1 of Wi-Fi system

Table 1. Channel conflict control

No.	System	Channel		No.	System	Channel
1	Wi-Fi	1, 2, 3, 4		4	Wi-Fi	6, 7, 8, 9, 10, 11, 12, 13
	ZigBee	1, 2, 3			ZigBee	10, 11, 12
	Bluetooth	1/2, 3/4, 5, 6/7, 8/9, 10, 11/12, 13/14, 15			Bluetooth	46/47, 48/49, 50, 51/52, 53/54, 55, 56/57, 58/59, 60
2	Wi-Fi	1, 2, 3, 4, 5, 6, 7		5	Wi-Fi	9, 10, 11, 12, 13
	ZigBee	4, 5, 6			ZigBee	13, 14, 15
	Bluetooth	16/17, 18/19, 20, 21/22, 23/24, 25, 26/27, 28/29, 30			Bluetooth	61/62, 63/64, 65, 66/67, 68/69, 70, 71/72, 73/74, 75
3	Wi-Fi	3, 4, 5, 6, 7, 8, 9, 10		6	Wi-Fi	12, 13
	ZigBee	7, 8, 9			ZigBee	16
	Bluetooth	31/32, 33/34, 35, 36/37, 38/39, 40, 41/42, 43/44, 45			Bluetooth	76/77, 78/79

and channels 3–5 of Bluetooth system are overlapped. The left half of channel 5 of Bluetooth system and channel 2 of Wi-Fi system are also overlapped. The right half of channel 5 of Bluetooth system and channel 1 of ZigBee system are overlapped. So that, these overlapped channels will be interfered.

In order to eliminate the interference among them, the NCC mechanism was proposed. This mechanism will be used in Table 2. In order to eliminate the interference among three kinds of systems, we proposed the NCC mechanism. This mechanism will be used in Table 2.

Table 2.

System	Sequence of channel selection
Wi-Fi	13, 7, 1
ZigBee	5, 6, 11, 12, 1, 2, 3, 4, 7, 8, 9, 10, 13, 14, 15, 16
Bluetooth	The system randomly selected non conflicting channels with used channel of Wi-Fi system and ZigBee system.

In Table 2, the number of the channel should not be over 3 in the Wi-Fi System. Each of its channels is occupied by the order which is decided by the channel without interference and then by their names from most to least complicated. The ZigBee system will assign channels by the level of interference from high to low and then by their names from most to least complicated, with its number of channel should not beyond 16. The number of consumers which who use the Bluetooth system should not beyond 7 in the same time. And each of its channels occupied successively by the level of channel interference from low to high is shown in Table 1.

According to Tables 1 and 2, four channels (channels 26, 27, 56 and 57) of the Bluetooth users are represented by ⸬ will not be influenced by other channels. ▨

represents 4 channels (channels 5, 6, 11 and 12) and is shown in Tables 1 and 2. On the one hand, the ZigBee users have 8 channels (channels 23, 24, 28, 29, 53, 54, 58 and 59) that are shown in Tables 1 and 2. The Bluetooth users will not be affected in the case that WI-Fi users occupied all of the channel of the Wi-Fi system. On the other hand, the channels of the ZigBee system are overlapped, which is shown in Tables 1 and 2. Tables 1 and 2 show that channel 5 of ZigBee system and channels 23 and 24 of Bluetooth system are overlapped. Moreover, channel 6 of the ZigBee system and channels 28 and 29 of the Bluetooth system are overlapped. Also, channel 11 of the ZigBee system and channels 53 and 54 of the Bluetooth system are overlapped. Furthermore, the channel 12 of the ZigBee system and channels 58 and 59 of the Bluetooth system are overlapped. ☐ shows that the influence of channels will bring about channel interference when three kinds of system channels are occupied. Thus it can be seen, we propose this new model that fixes these overlap problems efficiently.

5   Experimental Results

5.1   BER of Each Kind of System with Three Users Access

When 9 users access the Wi-Fi, ZigBee and Bluetooth systems, we distribute 3 users to each system. Figure 6 shows that the No.2 Wi-Fi customer and No. 1 ZigBee customer overlap during 0.01–0.03, 0.05–0.06 and 0.09–0.1 s; No. 2 Wi-Fi customer and No. 2 Bluetooth customer overlap during 0.01–0.03, 0.05–0.06 and 0.09–0.1. No. 3 ZigBee customer and No. 3 Bluetooth customer overlap during 0–0.02 s, see Fig. 1.

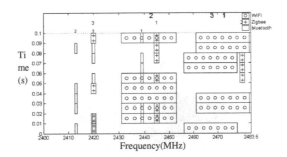

Fig. 1. Three users of Wi-Fi, ZigBee and Bluetooth before NCC

The results after we added the NCC mechanism to this model, see Fig. 2.

In Fig. 2, we added NCC mechanism to eliminate the channel overlap among the Wi-Fi, ZigBee and Bluetooth systems to reduce the interference among channels and reduce the error rate. The simulated error rates and channel data of the three systems are shown in Figs. 3, 4 and 5.

As shown in Fig. 3, the channel overlap among No. 2 Wi-Fi customer and other customers is severe because they did not use our NCC mechanism. The error rate is much higher than that of No. 1 customer and No. 3 customer. With the help of our NCC mechanism, the bit error rate is significantly reduced.

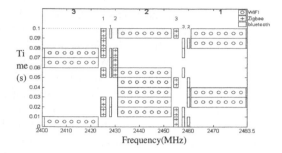

Fig. 2. Three users of Wi-Fi, ZigBee and Bluetooth after NCC

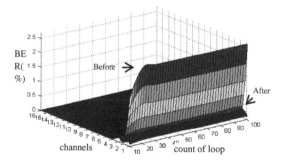

Fig. 3. Information of three users of Wi-Fi

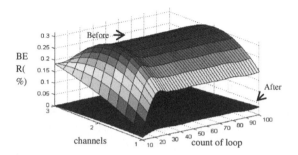

Fig. 4. Information of three users of ZigBee

See Fig. 4, the transmission power of the Wi-Fi system is stronger than that of the Bluetooth system, which uses frequency hopping technology, so the interference to the Bluetooth system channel does not persist. Although the channel overlap between No. 1 and No. 2 ZigBee customers are serious with other user channels, the interference of No. 2 Wi-Fi user to No. 3 ZigBee one is worse than that of No. 3 Bluetooth customer to No.3 ZigBee one. So the bit error rate of No. 1 ZigBee user is higher than that of No. 2 and No. 3 ZigBee users, and the bit error rates of No. 2 and No. 3 ZigBee users are almost equal.

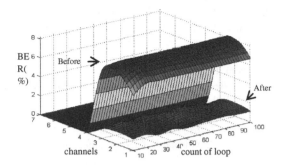

Fig. 5. Information of three users of Bluetooth

As shown in Fig. 5, the channel overlap rate between Bluetooth users and other users is higher, so that the bit error rate is higher. Bluetooth system uses frequency hopping technology, and it made frequency hopping transmission randomly in 79 channels without our NCC mechanism. So that the bit error rate is high because its channels overlapped ones of other two systems. With our NCC mechanism, the bit error rate will be low as a result of channels of Bluetooth make frequency hopping transmission according to regulations by this mechanism.

6   Conclusion

In this paper, we propose that the NCC mechanism jointly optimizes communication and controls of the systems. By using our NCC mechanism, we effectively reduced interference from co-existence of the channels and effectively reduced the BER. The improvement effect of the communication performance is obvious, especially when numerous consumers accessed to the system simultaneously.

References

1. Xhafa, A.E., Sun, Y.: Mechanism for co-existence of collocated WLAN and Bluetooth the same device. Int. Conf. Comput. **12**(9), 905–910 (2013)
2. Howitt, I.: IEEE802.11 and bluetooth co-existence analysis methodology. Veh. Technol. Conf. **2**(2), 1114–1118 (2001)
3. Rihan, M., El-Khamy, M., El-Sharkawy, M.: On ZigBee Co-existence in the ISM band. measurements and simulations. In: International Conference on Wireless Communications in Unusual & Confined Areas, vol. 7363(1), pp. 1–6 (2012)
4. Neburka, J., Tlamsa,Z., Benes, V., Polak, L., Kaller, O., Klozar, L., Blolecek, L., Zach, O., Kufa, J., Sebesta, J., Kratochvil,.T.: Study of the co-existence between ZigBee and Wi-Fi IEEE802.111b/g networks in the ISM band. In: Radioelektronika, pp. 106–109 (2015)
5. Subbu, K.P., Soman, S.: An interference mitigation scheme for IEEE 802.15.4 networks under IEEE 802.11b/g interference. In: International Conference on Computing, pp. 1–6 (2014)

6. Shin, S.Y., Kang, J. S., Park, H.S.: Packet error rate analysis of ZigBee under interference of multiple Bluetooth piconets. In: IEEE Vehicular Technology Conference, pp. 1–5 (2009)
7. Sikora, A., Groza, V.F.: Co-existence of IEEE 802.15.4 with other system in the 2.4 GHz-ISM-band. IEEE Instrum. Meas. Technol. Conf. **3**, 1786–1791 (2012)
8. Garroppo, R.G., Gazzarrini, L., Giordano, S., Tavanti, L.:. Experimental assessment of the co-existence of Wi-Fi, ZigBee and Bluetooth devices. In: World of Wireless, Mobile and Multimedia Network, pp. 1–9 (2011)
9. Shin, S., Park, H., Choi, S., Kwon, W.: Packet error rate analysis of ZigBee Under Wan and bluetooth interference. IEEE Trans. Wirele. Commun. **6**(8), 2825–2830 (2007)
10. Proakis, J.G., Salehi, M.: Fundamentals of Communication Systems, 2nd edn. Pearson Education, London (2007)

A Curve Fitting Method for Evaluating Pilot Operation Quality

Li Tong[1(✉)], Chi Ying[2], and Wu Qian[1]

[1] Civil Aviation Management Institute of China, Beijing, China
ttlitong@163.com
[2] Siemens Healthcare Technology Center, Princeton, USA

Abstract. Wind, turbulence, and other environmental interference can affect pilot flight operation quality. Especially in areas where wind is reported all year round, or where seasonal winds obviously change (like in some highlands). Serious situation will interfere with flight landing or taking off. The quality of pilot efforts in reactions to these air flow interference can be assessed. We proposed a curve fitting method that can quantify how stable the pilots control the flights. This method defines the best/perfect flight routes through controlling spline control points in different ways in curve fitting, and then computes the error score according to the difference between actual flight routes and these perfect routes curve fitting found.

Keywords: Curve fitting · B-Spline · Control points · Flight safety · Pilot operation quality · Non-linear · Regression

1 Introduction

Pilot risk assessment is a very important task for all airlines. Quick access recorder (QAR) is a powerful data source and facilitator. Currently, in airline regulations, only exceedance on very serious events has been considered to validate pilot performances [1–4], which is not sufficient. Wind and turbulence are the only environmental factors recorded by QAR, whereas other environmental interference like thunderstorm, sandstorm, etc., right now has not been written down. Influences from all remarkable environmental interference shall be evaluated although they can be anything and we don't know in detail what they are.

As China is the first country to mandate the use of QAR throughout the world, no one has ever tried to conduct the same research right now, let along using curve fitting methods. We take these as challenges.

Fitting curves to data using nonlinear regression started quite early before 1990. Harvey [5], Lancaster [6], and Guest [7] reviewed some methods like linear regression, polynomial regression, spline regression, etc. They compared methods like cubic spline versus cubic polynomial, as shown in Fig. 1. They dig into difficulties in nonlinear regression and worked out numerical implementations. However, the methods illustrated cannot fit well to our question. For instance, cubic spline goes through every data

© Springer International Publishing AG 2018
F. Qiao et al. (eds.), *Recent Developments in Mechatronics and Intelligent Robotics*,
Advances in Intelligent Systems and Computing 691, DOI 10.1007/978-3-319-70990-1_14

point, it cannot distinguish wind turbulence caused jitters; while cubic polynomial fits data points to the following equation:

$$\gamma = A + Bx + Cx^2 + Dx^3 + Ex^4 \ldots \tag{1}$$

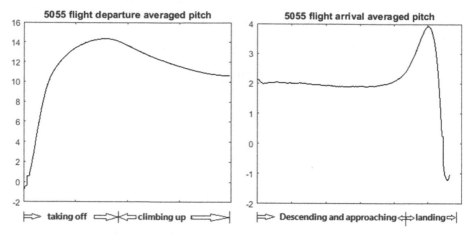

Fig. 1. An example of averaged pitch curve with the perfect performance marked in orange.

The goal of polynomial regression is to determine values for the parameters (A, B, C,...) that make the curve best fit the data points. It does not know which data points are important to keep, while which others to smooth, and to smooth to which extent. Besides, since the curves don't recognize outliers, the effects of an outlier cannot be ignored [5]. As we are the first to solve the curve fitting problem helping score the whole set continuous pilot performance at different time instances, we propose to equip eyes to regression methods which can observe and recognize pilot operation errors.

2   Proposed Method

To quantify pilot efforts in reactions to environmental interferences, and to score the whole continuous pilot performances, we need to observe carefully the continuous curves of pitch, roll, heading, and air speed of the airliners. As long as these curves are stable, we know the flight has not been affected by whatever environmental factors. However, curves have fluctuations, and these fluctuations include both correct changes following commands and unnecessary shakings, which we have to tell the difference and handle respectively.

Spline curve fitting is a suitable technology to help us solve this difficulty, whose basic idea is to use dense control points to keep the original correct curve shape, and to use sparse control points to smooth useless jitters. The equation of the B-Spline curve fitting:

$$P(t) = \sum_{i=1}^{n} B_t N_{i,k}(t) \tag{2}$$

Where the basis function $N_{i,k}(t)$ are given by

$$N_{i,k}(t) = \begin{cases} 1 & x_i \leq t < x_{1+i} \\ 0 & otherwise \end{cases} \tag{3}$$

$$N_{i,k}(t) = \frac{(t - x_1)N_{i,k-1}(t)}{x_{i+k-1}x_i} + \frac{(x_{i+k} - t)N_{i+1,k-1}(t)}{x_{i+k} - x_{i+1}} \tag{4}$$

Applications to the real scenarios:

(1) Pitch and speed

There is a common rule between pitch and speed: the averaged curve keeps the universal correct changes following commands. Therefore, we can firstly find these correct changes, and set dense control points for each airplane curve to keep them, then, set sparse control points for other time instances to smooth the remaining curve segments.

Figure 1 is an example. It is the averaged flight departure and arrival pitch curves, out of 5055 flights respectively. The main stages of the flight departure are the taking off and climbing up. During taking off, the flight raises its head, because the motion is larger than the wind's, it normally is smooth and don't have random jitters caused by the reactions to wind interference. While, the climbing stage is on the other side. During this stage, the flight nearly keeps steady pitch-angle-wise, and the small displacements affected by winds and turbulences are obvious. Thus, when the data we have are demonstrating flights taking off in different airports, the taking-off stage curve shall be kept as the perfect path (if lots of data show flight departure in the same airport, the averaged path then can be deemed as the perfect one.). And, the climbing stage curve shall be smoothed to get rid of small fluctuations to show the perfect path. Likewise in flight arrival process, during descending and approaching stage, the pitch angle nearly keeps steady, while during landing stage, a "flare" operation is performed, which creates a sharp bulge in the curve. The flare operation is very obvious around the time of touch down. And this is the only correct fluctuation in flight arrival. The remaining jitters in descending and approaching stage, although is not easy to avoid due to the complex weather situation, is actually unnecessary and can be improve to remove. Thus, the perfect, although difficult to reach, pitch operation curve at arrival time instances is the origin path of landing flare and smoothed path of the descending and approaching.

Now we know where the taking off and landing flare is. We then assign dense control points to the averaged flight pitch curve, and apply to all flights to keep the time instances consistent. Then, assign also sparse control points to the remaining parts of the averaged curve, and apply them to all flight curves, so that all control points are the same for all airplanes. Figure 2 is an example of one airplane departure pitch curve with the perfect performance marked out in orange.

Figure 3 is an example of one airplane arrival pitch curve with the perfect performance marked out in orange.

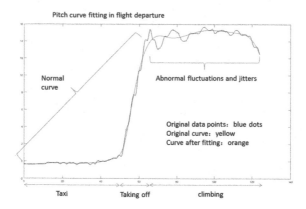

Fig. 2. An example of one airplane pitch curve during taking off and ascending, with perfect performance marked out.

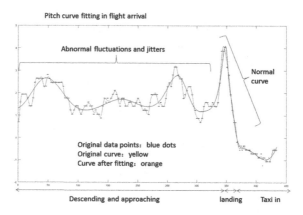

Fig. 3. An example of one airplane arrival pitch curve with perfect performance marked out in orange.

Air speed curve fitting is similar to pitch's and even simpler. Thus, we skip the explanation.

(2) Roll and heading

Roll and heading are in different situation. If flight routes are different, air planes normally do not have averaged correct rolls and heading turns in common. They turn whenever needed. It looks like no rules to follow. However, there is a common sense: whenever there is a real turn, both roll and heading are changing in large scale at the same time. Thus, we pair roll and heading curves for each airplane, and find the corresponding changes in large scale, setting dense control points to them, and setting sparse control points to the remaining of the curves to make it perfectly smooth.

Figures 4 and 5 are examples of one airplane roll curves during both taking off (and ascending) and landing (and descending) time instances. Figures 6 and 7 are examples

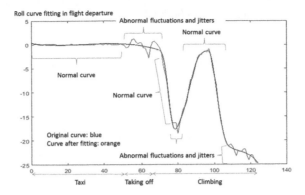

Fig. 4. Example of one airplane roll curves during both taking off (and ascending), with best performance marked out.

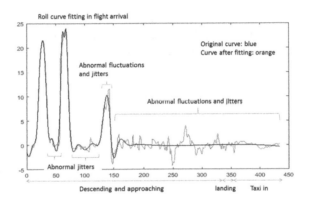

Fig. 5. Example of one airplane roll curves during landing (and descending), with perfect performance marked out.

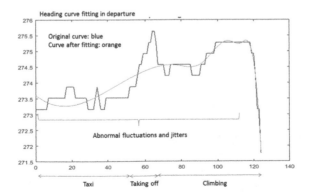

Fig. 6. Example of one airplane heading curves during both taking off (and ascending), with perfect performance marked out.

of one airplane heading curves during both taking off (and ascending) and landing (and descending) time instances.

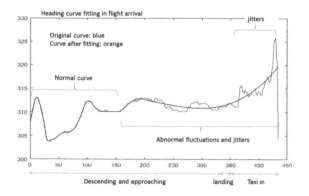

Fig. 7. Example of one airplane heading curves during landing (and descending), with perfect performance marked out.

When we pair roll and heading curves, we get Figs. 8 and 9, two examples of matching time instances for correct fluctuations and unneeded shakings. The time instances are obviously consistent.

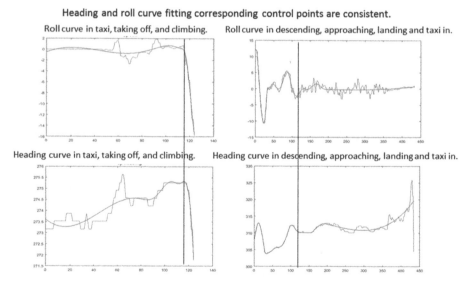

Fig. 8. Example of matching time instances for correct fluctuations in order and unneeded shakings, during taking off and ascending time instances.

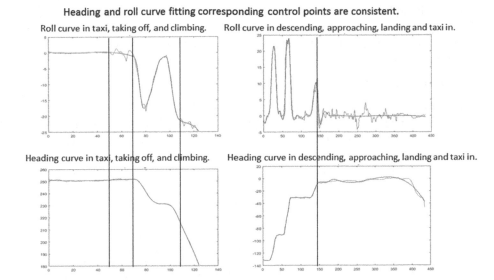

Fig. 9. Example of matching time instances for correct fluctuations in order and unneeded shakings, during descending and landing phase.

Figure 10 is a flow chart which summarizes the processes we take to find the perfect performances for all types of curves. Finding the difference between perfect performance and original curves, we can score pilot basic performances.

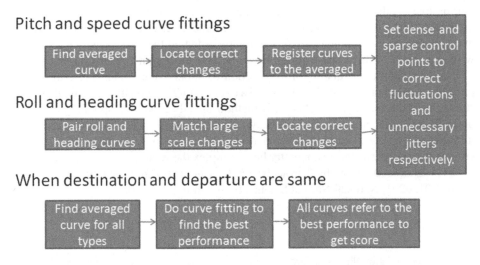

Fig. 10. Flow chart of main processes of curve fitting for all types of curves

When all airplanes are to and from two fixed airports, the whole curves, including cruise stage for all types including roll and heading, are comparable. In this case, the averaged curves can be extracted as the best performance among all individual perfect operations.

(3) **Statistical Result:**

The error score we computed equals to the sum of square errors of pitch, roll, air speed, and heading. Figure 11 shows the error scores for the departure stage of all 5055 flights. Figure 11 shows the error scores for the arrival stage of all 5055 flights.

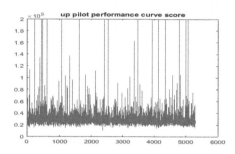 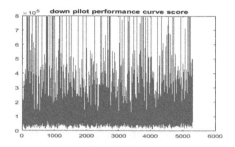

Fig. 11. Error scores for the departure & arrival stage of all 5055 flights.

3 Conclusion

During flying, pilots do not only need to monitor the aircraft, but also need to deal with the wind, turbulence, and other severe environmental interference. Especially to those extreme adverse environment where wind is recorded all the year round and changed with season, the quality of pilot efforts in reactions to these air flow can affect the safety of landing. Quantify the quality of the reaction of pilots to the threats is useful for the pilots fly skills improvement. We proposed a curve fitting method that can quantify how stable the pilots control the flights. This method unveils good corresponding relations between our proposed error scores and environmental interferences like QAR-recorded turbulence. Using this method, we can score continuous pilot performances against environmental interferences, no matter how complex the weather is.

This paper is supported by 2 projects: Civil Aviation Technology Innovation Foundation 20150201, National Natural Science Foundation U1533102.

References

1. China civil aviation flight standard office. The implementation and management of FOQA (flight operational quality assurance). AC – 121/135 – FS – 2012 – 45
2. European strategic safety & analysis unit safety regulation group civil aviation authority. CAP 739 flight data monitoring regulatory operation manual (2003). ISBN:086039. www.caa.co.uk/publications
3. FAA (American Federal aviation administration). Consultant report. AC No. 120-82 (2004)

4. International civil aviation organization. Manual on flight data analysis programmes (FDAP). Doc 10000 AN/501 (2014). ISBN:978-92-9249-416-2
5. Harvey, J.M., Lennart, A.R.: Fitting curves to data using nonlinear regression: a practical and nonmathematical review. FASEB J. **1**(5), 365–374 (1987)
6. Guest, P.G.: Numerical Methods of Curve Fitting. Cambridge University Press, Cambridge (2004)
7. Lancaster, P., Salkauskas, K.: Curve and Surface Fitting. An Introduction. Academic Press, London (1986)
8. Lu, J., Liu, H., Cao, J.: Numerical simulation of local circulation over the Cangshan Mountain-Eihai Lake area in Dali. Chin. J. Atmos. Sci. **28**(6), 1198–1210 (2014)

Modeling and Simulation

Simulation Design of Improved OPTA Thinnin Algorithm

Chenghao Zhang, Weifeng Zhong$^{(\boxtimes)}$, Chunyu Zhang, and Xi Qin

School of Automation, Harbin University of Science and Technology, Harbin, China
910381219@qq.com

Abstract. In fingerprint thinning algorithm, the thinning effect of typical serial OPTA algorithm, resulting in more burrs after thinning, and there exist many points that may affect fingerprint thinning effects. In view of above problem, a new improved algorithm is proposed, this algorithm removes the boundary points of the fingerprint image, and then it gives new removal templates and retention templates, after the first removal and retention, it again gives a remove template for remaining burrs and bifurcation points, as well as the discriminant formula of non critical points, for three different fingerprints, the thinning algorithm before improving is compared with improved thinning algorithm, the results find that the new improved thinning algorithm can greatly reduce the occurrence of burrs and other removal points in the thinning process, with good thinning effect.

Keywords: Fingerprint thinning algorithm · Improved OPTA algorithm · Burr removal · Good thinning effect

1 Introduction

After such three treatments of fingerprint image normalization, block direction finding and binarization, different fingerprint lines are gained, this is caused by different forces on the chip and the circuit noise in the acquisition process, for subsequent fingerprint image processing, the images must be refined. According to iterative partitioning, algorithms applied in fingerprint thinning include parallel algorithm and serial algorithm [1].

Simply speaking, the difference between the two algorithms is: each operation of serial thinning algorithm is based on the last operation, the iteration of the algorithm is strong; while each operation of parallel thinning algorithm is based on the same conditions, so there is no iteration between the calculations. These two have their own advantages and disadvantages, serial algorithm has the advantage of more detailed treatment on each detail point on the basis of each iteration, while the disadvantage is that the thinning effect is not ideal for each calculation can not decide whether to retain a point because of iteration of each result; the condition of each treatment of parallel thinning is same, all pixels are then calculated one by one under this condition, its advantage lies in the fast calculation speed, and the disadvantage is that different effect will be generated from different fingerprint under the premise of unified calculation template used for all fingerprints, that is to say, a certain parallel thinning algorithm must not be universal for all fingerprints, there must be unsatisfactory effect for some fingerprints after being refined by the thinning template [2–4].

© Springer International Publishing AG 2018
F. Qiao et al. (eds.), *Recent Developments in Mechatronics and Intelligent Robotics*,
Advances in Intelligent Systems and Computing 691, DOI 10.1007/978-3-319-70990-1_15

In order to solve the above mentioned problem, in this paper, the principle of OPTA (One Pass Thinning Algorithm) is firstly introduced, then, based on this algorithm, a new serial parallel hybrid algorithm based on OPTA is proposed to improve fingerprint thinning effect [5–7].

2 The Principle of OPTA Thinning Algorithm and the Improved Algorithm Design

2.1 OPTA Thinning Algorithm Principle

In this algorithm, each pixel P (non background) of each fingerprint image takes 10 neighborhood points as shown in below figure by Z character pattern.

P1	P2	P3	
P4	P	P5	P9
P6	P7	P8	
	P10		

Fig. 1. Domain extraction

The extracted neighborhoods are compared with the 8 removal templates and the 4 retention templates in literature [8], If the 10 neighborhoods are compared with the 8 removal templates, as long as there is 1 template in all 8 removal templates perfect matches with 10 neighborhoods, then the center will be compared with the 4 retention templates, if the same as one of the 4 reserved templates, then the center point is kept, otherwise, it is removed, If this point is different from any one of the removal and retention templates, the point is also kept. This method is used to remove the images after the normalization, the block direction and binarization, until it cannot be deleted, then OPTA refinement process ends.

Typical OPTA thinning algorithm often has a lot of burr, thinning effect is very bad, and it is not conducive to the latter fingerprint image processing.

0	0	0		0	x	1		1	1	1		1	x	0
x	1	x		0	1	1		x	1	x		1	1	0
1	1	1		0	x	1		0	0	0		1	x	0
	a0				b0				c0				d0	
x	0	0		0	0	x		x	0	x		x	1	x
1	1	0		0	1	1		0	1	1		1	1	0
x	1	x		x	1	x		0	1	x		x	0	0
	e0				f0				g0				h0	

Fig. 2. Removal template in document 8

x	1	x	0		x	x	0	0		x	0	x	x		x	0	x	x
0	1	1	0		0	1	1	0		x	1	x	x		x	1	x	x
x	1	x	0		x	x	1	x		1	0	0	x		0	1	0	x
x	x	x	x		x	x	x	x		x	0	0	x		0	0	0	x
	a1					b1					c1					d1		

Fig. 3. Retention templates in document 8

2.2 Serial Parallel Hybrid Algorithm Based on OPTA

In order to facilitate the latter description, the definitions of several commonly used words are given in this paper [9–12]:

Fig. 4. Schematic diagram of burr

Target, background and boundary points: Based on pixel value, define the point with a value of 1 as the target point, and point with a value of 0 is defined as a background point, fingerprint line as object point, the rest of the graph is the background, while the definition of the boundary point belongs to the target point, but where at least one background point in $N_4(P)$.

$N_4(P)$ and $N_8(P)$ points: each pixel has a neighborhood window based on it, expressed as P has $N_4(P)$ and $N_8(P)$ points, that is, the commonly known as the 4 neighborhood and the 8 neighborhood, $N_4(P)$ are the up and down, left and right four points of P, 4 diagonal points of P expressed as $N_D(P)$, $N_D(P)$ and $N_4(P)$, all consisted in $N_8(P)$.

Endpoint: Points in the boundary point range, the the restriction is that there is only one point in the $N_8(P)$ field of the point.

Bifurcation point: in this type of point, there are at least 3 and above target points in the $N_8(P)$ field.

Key points (break point): this point is located in connectivity of fingerprint lines, it plays an important role on whether fingerprint lines are connected, which can not be deleted easily.

Single pixel width: number of target points in the $N_8(P)$ field are used to limit the end point, continuous point and bifurcation point in Fingerprint line, the maximum number of target points in these three kinds of points are respective 1, 2, 3, only when fully satisfied lines in all number of target points of three kinds of points can be called as single pixel width.

P1	P2	P3
P4	P	P5
P6	P7	P8

Fig. 5. The neighborhoods of P

After the algorithm, we find all the boundary points in the fingerprint image, and based on following formula, to judge whether the boundary point is reserved, suppose a boundary point is P, and two characteristic quantities *USUM* and *VSUM* are defined on it, where:

$$USUM = \sum_{i=1}^{8} P_i \tag{1}$$

$$VSUM = \sum_{i=1}^{8} |P_{i+1} - P_i| \tag{2}$$

The deletion condition for a boundary point is established under the below formula:

$$(VSUM = 2) \delta (USUM < 6)$$
$$\delta (USUM! = 1) \tag{3}$$

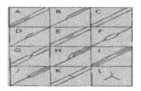

Fig. 6. Points to be removed during thinning

In addition to the boundary points, unnecessary points in fingerprint image thinning are also given in above figure, where A is bifurcation point, B is branch point, C is isolated point, D is burr, E is stub, F is eye line, G is cross point, H is double fork point, I is break line, J is triradius point, K is foot bridge, and L is trigonometrical station, while the endpoint is the point on the edge of the line.

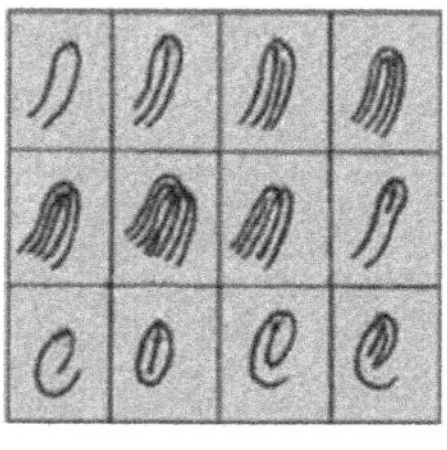

Fig. 7. Various types of core points

In addition to the above types of points, the figure gives a schematic diagram of various types of core points. The following is the introduction of a variety of unnecessary point selection and removal [12–14], before proceeding to the next step, the following definitions are made for all of the above characteristics:

The distance between p and q is $disce(p, q)$, its unit is pixel; If two points are on the same stripes $l(p, q) = 1$, otherwise $l(p, q) = 0$; Ridge orientation angle of local neighborhood in two points p and q is $\theta(p, q)$, the angle between a line and a horizontal axis of points p and q is $\theta_x(p, q)$, and $\Delta\theta$ is defined by:

$$\Delta\theta = \theta_x(p, q) - \theta(p, q) \tag{4}$$

The 8 neighborhood graph $N_8(P)$ at P is same as the used one in last section, distinguishing conditions of various types of feature points are as: Criterion of endpoint is that $N_8(P)$ shall meet the following formula:

$$N_8(P) = \frac{1}{2} \sum_{i=1}^{8} |P_{i+1} - P_i| = 1 \tag{5}$$

Criterion of bifurcation point is that $N_8(P)$ shall meet the following formula:

$$N_8(P) = \frac{1}{2} \sum_{i=1}^{8} |P_{i+1} - P_i| = 3 \tag{6}$$

The criterion of burr: for all endpoints p and bifurcation points q in streak lines, as long as any of the following two formulas is met, burr removal shall be made, the two formulas are respective criterion of short and long burrs:

$$(disce(p, q) < 4) \wedge (l(p, q) = 1) \tag{7}$$

$$(disce(p, q) < 8) \wedge (l(p, q) = 1)$$
$$\wedge (\Delta\theta < \frac{\pi}{4}) \tag{8}$$

Criterion of disconnection: for all endpoints p and bifurcation points q in streak lines, if meeting the below formula, it shall be removed;

$$(disce(p, q) < 7) \wedge (\Delta\theta \approx 0) \tag{9}$$

Criterion of foot bridge: for all endpoints p and bifurcation points q in streak lines, if meeting the below formula, it shall be removed.

$$(disce(p, q) < 5) \wedge (l(p, q) = 1)$$
$$\wedge (\Delta\theta \approx \frac{\pi}{2}) \tag{10}$$

Eye liner discriminant conditions: for all endpoints p and bifurcation points q in streak lines, if meeting the below formula, it shall be removed.

$$(disce(p, q) < 7) \tag{11}$$

Use the above formula to delete all the boundary points that can be removed and the points to be removed in the thinning process, these formulas are used to remove the boundary points and the points to be removed, while it can not remove some special boundary points and the points to be removed, then the 15 neighborhood pixels of P are defined, each pixel of the image is taken as the 15 neighboring pixels in the follow figure.

P1	P2	P3	P12
P4	P	P5	P13
P6	P7	P8	P14
P9	P10	P11	P15

Fig. 8. 15 neighborhood in P

If $N_8(P)$ of P, that is, all points in the 8 neighborhood are same with one of A ~ H removal plates (Except for 0 or 1 X points),then it shall be removed, If $N_8(P)$ of P, that is, all points in the 8 neighborhood are not matching one of A ~ H removal plates (Except for 0 or 1 X points), then it shall be retained. New removal templates A ~ H are defined as:

0	1	x		0	x	0		x	0	1		0	1	1
x	1	1		1	1	x		0	1	1		0	1	1
0	1	x		1	x	1		x	1	1		x	x	1
	A				B				C				D	
1	0	1		x	x	0		x	1	1		1	1	1
1	1	1		1	1	1		0	1	x		1	1	0
x	x	x		1	0	x		0	0	x		x	0	0
	E				F				G				H	

Fig. 9. New defined removal templates

If only using above removal templates, there must be a situation of a local area has been all deleted, which brings a great impact, then the following 4 new retention templates are given, If the 15 neighborhood of point P includes a matching between $N_8(P)$ and A ~ H, while its $N_{12}(P)$ is not matched with the following plates, then it shall be deleted, if matches, then it shall be restored and retained.

a				b				c					d			
x	0	0		1	0	0		0	0	x	0		1	x	0	1
0	0	1		1	1	0		0	0	1	1		0	x	1	1
0	1	1		x	0	x		0	1	x	0		0	0	1	1
1	x	0		0	x	0										

Fig. 10. New defined retention templates

An iteration is completed for all of the above processes, on this basis, removal of bifurcation points and burrs is carried out, in the algorithm, a total of 4 templates as I ~ K in following figure are given:

I			J			K			L		
x	x	x	x	x	x	x	1	x	x	0	x
1	1	0	0	1	1	x	1	x	x	1	x
x	x	x	x	x	x	x	0	x	x	1	x

Fig. 11. Re-matched 4 new templates

In the plates, if $N_8(P)$ of P is totally matched with one of the above 4 plates, then the algorithm is to deal with P, for the key points of $N_8(P)$ of P in the above 4 plates, formula 12 is given as judging condition:

In the formula, $i = 1, 2, \ldots, 7, 8$, if it is a key point in certain point P after judging, then the point shall be retained, if it is not the key point, then determine whether the point is the endpoint or bifurcation and burr points, remove fork and burr points and retain endpoints, using the above 4 plates to carry out iteration on images gained by OPTA algorithm from I to L, that is, I template output as input to the J template, after the completion of iteration of all the 4 templates, the final thinning image is obtained.

$$N_8(P(i)) = \begin{cases} I(i), (P_2 = 0 \wedge P_3 = 1) \vee (P_7 = 0 \wedge P_8 = 1) \\ J(i), (P_4 = 0 \wedge P_1 = 1) \vee (P_5 = 0 \wedge P_3 = 1) \\ K(i), (P_2 = 0 \wedge P_1 = 1) \vee (P_7 = 0 \wedge P_6 = 1) \\ L(i), (P_4 = 0 \wedge P_6 = 1) \vee (P_5 = 0 \wedge P_8 = 1) \end{cases} \quad (12)$$

The reason of this new algorithm called as serial parallel hybrid operation is that there is a serial relationship between two iterations, if operation of each iteration is carried put alone, then it is parallel operation.

3   Algorithm Validation

3.1   Fingerprint Image Source

In this algorithm, the fingerprint images are the images gained by Fingerprint acquisition chip FPS200 controlled by TMS320VC5509 DSP and EPM240T100C5N CPLD [15].

3.2   Fingerprint Image Simulation

The system simulation environment for the design of the algorithm is: under MATLABR2010A in Win7, image contrast between typical algorithm thinning image and improved algorithm thinning image of 3 different fingerprints is as follows:

(a) binaryzation figure of fingerprint 1 (b) OPTA thinning image of fingerprint 1(c) Improved algorithm thinning image of fingerprint 1

Fig. 12.   (a) Binarization figure of fingerprint 1. (b) OPTA thinning image of fingerprint 1. (c) Improved algorithm thinning image of fingerprint 1

(d) binaryzation image of fingerprint 2 (e) OPTA thinning image of fingerprint 2 (f) Improved thinning image of fingerprint 2

Fig. 13.   (d) Binarization image of fingerprint 2. (e) OPTA thinning image of fingerprint 2. (f) Improved thinning image of fingerprint 2

(g) binaryzation figure of fingerprint 3(k) OPTA thinning image of fingerprint 3 (h) Improved algorithm thinning image of fingerprint 3

Fig. 14.   (g) Binarization figure of fingerprint 3. (k) OPTA thinning image of fingerprint 3. (h) Improved algorithm thinning image of fingerprint 3

In Above 9 pictures:

(a), (d), (g) are binarization images of 3 fingerprints;

(b), (e), (k) are images of 3 fingerprints after typical OPTA thinning algorithm;

(c), (f), (h) are images of 3 fingerprints after OPTA serial parallel hybrid algorithm thinning in this paper.

After comparing the three fingerprints images by two different thinning algorithms, it can be seen:

Refining effect in(c), (f), (h) is obvious better than that in (b), (e), (k), its specific performance: a sharp reduction in the number of burrs, some bifurcation points, broken lines, eyeliner, bridges are removed, the thinning lines are more close to images after binarization.

4 Conclusion

In the typical OPTA thinning algorithm, there are actual a lot of thinning images, and bifurcation points, broken wires, eyeliner, and small bridges, after redefining the boundary points and the points to be removed by using formulas, if the appropriate conditions are met, the points shall be deleted; and a new removal template, retention template and re match template are defined, and based on the key point reservation condition based on 4 re matching templates, after serial parallel hybrid algorithm design, the fingerprint image thinning effect is significantly improved, so it is proved that this algorithm is superior to the typical OPTA thinning algorithm.

References

1. Golabi, S., Saadat, S., Helfroush, M.S., et al.: A novel thinning algorithm with fingerprint minutiae extraction capability. Int. J. Comput. Theory Eng. **4**(4), 504–514 (2012)
2. Xu, Y.Y., Tao, C.B., Zhu, S.W., et al.: PCNNs fingerprint thinning algorithm based on template. J. Jiangnan Univ. (NATURAL SCIENCE EDITION) **9**(2), 138–142 (2010)
3. Cai, X.M., Sun, P.: Fingerprint image thinning algorithm based on template. J. Xi'an Univ. Posts Telecommun. **21**(3), 59–63 (2016)
4. Chen, G., Chen, N., Zeng, Y.: Improved OPTA fingerprint thinning algorithm based on neighborhood search. Comput. Simul. **29**(9), 300–303 (2012)
5. Fang, B., Wen, H., Liu, R.Z., et al.: A new fingerprint thinning algorithm. In: 2010 Chinese Conference on IEEE Pattern Recognition (CCPR), pp. 1–4. IEEE (2010)
6. Wang, X.T., Xu, D.C.: Fingerprint image thinning algorithm based on improved PCNN. Comput. Eng. **36**(18), 180–181 (2010)
7. Golabi, S., Saadat, S., Helfroush, M.S., et al.: A novel thinning algorithm with fingerprint minutiae extraction capability. Int. J. Comput. Theory Eng. **4**(4), 504–514 (2012)
8. Lu, J.H., Peng, H., Wei, W.: A hybrid algorithm based on improved OPTA and parallel string refinement. Ind. Control Comput. **23**(7), 67–69 (2010)
9. Wu, G.X.: Research and simulation of missing fingerprint image recognition. Comput. Simul. **28**(8), 284–286 (2011)
10. Hu, M., Li, D.C.: Research on fingerprint image quality evaluation method. Comput. Technol. Dev. **2**(2), 125–128 (2010)

11. Song, L.L., Diao, Y.H.: Research on fingerprint image pre processing algorithm. Comput. Dev. App. **23**(10), 46–47 (2010)
12. Jain, A.K., Feng, J.: Latent fingerprint matching. IEEE Trans. Pattern Anal. Mach. Intell. **33**(1), 88–100 (2011)
13. Jorgensen, P.E.T., Pearse, E.P.J.: A discrete Gauss-Green identity for unbounded Laplace operators, and the transience of random walks. Israel J. Math. **196**(1), 113–160 (2013)
14. Deng, C.X., Wang G.B., Yang, X.R.: Image edge detection algorithm based on improved canny operator. In: IEEE 2013 International Conference on Wavelet Analysis and Pattern Recognition, pp. 168–172. IEEE (2013)
15. Luo, T., Yand, W.D., Zhang, Y.Z., et al.: Research on digital holographic imaging method for fingerprint acquisition. Photoelectron. Laser **23**(5), 966–969 (2012)

Fusion Evaluation of Ammunition Based on Bayes Method

Lei Ming and Liu Fang[✉]

Shenyang Ligong University, Shenyang, China
892339223@qq.com

Abstract. Ammunition is an important part of the weapon system, it has a "long-term storage, one-time use" features. Ammunition storage reliability test sample quantity is often small; the data source is diverse. When the data and information of the various storage tests of ammunition exist simultaneously, it has important significance to study the reliability of the comprehensive assessment method. In this paper, the Bayes method is used to evaluate the reliability of ammunition storage based on the data of the normal stress storage condition and the experimental data under the acceleration condition. The simulation results show the correctness and effectiveness of this method.

Keywords: Bayes method · Data fusion · Reliability

1 Introduction

The reliability data of ammunition storage can be divided into two aspects: First, storage reliability test data, including accelerated life test data and normal stress storage life test data; Second, field data, including ammunition factory acceptance data, military normal test data, force ammunition test data, ammunition quality survey results, acceptance records and accident reports and so on. Based on the particularity of the above ammunition storage data, when the reliability evaluation is carried out, the fusion of stored data becomes the key to solve the problem. In theory, information fusion is mainly divided into fusion method based on probability theory and non-probabilistic fusion method [1]. Bayes method, information entropy law and expert system and neural network method can be applied to solve the multi-source information fusion problem.

Due to the small number of ammunition test samples, it is difficult to obtain accurate and reasonable results by classical reliability evaluation method. Bayes method of Small sample is the model that combines the sample information, the general information, and the prior information for uncertainty reasoning [2]. In this paper, the Bayes method is used to evaluate the reliability of ammunition storage based on the data of the normal stress storage condition and the experimental data under the acceleration condition. At the same time, the Gaussian method is used to solve the Bayesian integral algorithm, which provides strong support for the development of information fusion software for ammunition storage.

© Springer International Publishing AG 2018
F. Qiao et al. (eds.), *Recent Developments in Mechatronics and Intelligent Robotics*,
Advances in Intelligent Systems and Computing 691, DOI 10.1007/978-3-319-70990-1_16

2 Fusion Evaluation Method for Ammunition Storage Reliability

Accelerated life testing can compensate for the shortcomings of traditional ammunition storage life based on long, large-scale storage life tests, which can obtain product life information in a short time, and its models and statistical methods are maturing. But by its nature, by this method to assess and predict the reliability of the product, it is a certain risk and error. In this paper, the accelerated life test data is used as pre-test information, and the storage reliability data under normal stress is used as the field data to determine the likelihood function. The Bayes method is used to combine the two test data for ammunition Evaluation of storage reliability.

2.1 Fusion Evaluation Algorithm

Assuming ammunition products have two types of data, in which the step-stress accelerated life test data is determined as: $(S_i, T_{i,j}, M_{i,j}, R_{i,j})$, S_i is the test stress, $T_{i,j}$ is the test time, $M_{i,j}$ is the amount of samples taken for each group, $R_{i,j}$ is the number of failures, $i = 1, 2, \cdots, k$ is the stress level, $j = 1, 2, \cdots, h_i$ is the number of samples at each level of stress. The normal condition storage data is: $t_i, k_i, n_i, J_i, \ i = 1, \cdots, k$. Among them, t_i is the storage time (year), k_i is the environmental factor, n_i is the test sample number, and J_i is the number of failures. The data under different environmental factors can be integrated according to the environmental factor method [3, 4], unified to a certain environment, which can be recorded as data $t_i, k_i, n_i, J_i, \ i = 1, \cdots, k$.

According to the characteristics of the test data, by calculating the likelihood of the lifetime distribution of the samples under the stress of each stage, the maximum likelihood estimation of the corresponding parameters is obtained by using the Powell extreme method [5]. Then the estimation of the storage life distribution parameters m and η of the ammunition under normal stress is obtained through the Weibull model to accelerate the model. Predict the reliability of t_i moment $R(t_i) = \exp[-(t_i/\eta)^m]$, according to the normal approximation method, the reliability lower limit t_i at the time t_i is regarded as the pre-test information, and the reliability data t_i under normal conditions is regarded as the field data. t_i Time of storage reliability is R_{si}. Its pre-test distribution is evenly distributed $U(R_L(t_i), 1)$. $R_L(t_i) = R_{idown}$, so the post-distribution of R_{si} is

$$
\pi(R_{si}/s) \begin{aligned} &= \frac{R_{si}^{s_i}(1-R_{si})^{n_i-s_i}}{\int_{R_{idown}}^{1} R_{si}^{s_i}(1-R_{si})^{n_i-s_i}\, dR} \\ &= \frac{R_{si}^{s_i}(1-R_{si})^{n_i-s_i}}{B(si+1,n_i-s_i+1)\left(1-I_{R_{idown}}(si+1,n_i-s_i+1)\right)} \end{aligned} \qquad (R_{idown} \le R_{si} \le 1) \qquad (1)
$$

The Bayes point of R_{si} is estimated as:

$$\hat{R}_{si} = \int_{R_{idown}}^{1} R\pi(R_{si}/s)dR$$

$$= \frac{(s_i+1)\left(1 - I_{R_{idown}}(si+2, n_i - s_i + 1)\right)}{(n_i+2)\left(1 - I_{R_{idown}}(si+1, n_i - s_i + 1)\right)}.$$

(2)

2.2 Integral Approximate Calculation

Numerical integration is a very important aspect of the Bayes method. Numerical integration is calculated in many ways, such as the Newton-Cotes method, the Romberg method and so on [6], but these methods are not suitable for computer programming. In this paper, the Gaussian method is used to approximate the integral calculation. Gauss integral approximate calculation pseudo code is as follows:

```
double Gaussjifen(a,b)              // a, b   represent the upper and lower points
{static const double x[] = { 0.9324695, 0.6612094, 0.2386192, −0.2386192,
−0.6612094, −0.9324695};// Gauss point initialization
static const double w[] = { 0.1713245, 0.3607616, 0.4679139, 0.4679139,
0.3607616, 0.1713245 };// Gaussian coefficient initialization
xm = 0.5*(b + a);   xr = 0.5*(b - a);
for (j = 0; j < 6; j++)
{dx = xr*x[j];
s += w[j] * func(xm + dx) ;}
return s *= xr;}
```

2.3 The Example Analysis

An ammunition product, accelerated life test data is shown in Table 1, under the standard conditions for storage reliability test data as shown in Table 2.

After the data in Table 1 above was pretreated, by comparing the test time equivalently, it was found that the life distribution parameters of the ammunition were as follows under various stress levels(The unit of characteristic life is days): $m_1 = 2.38762$, $\eta_1 = 107.11311$, $m_2 = 2.6078$, $\eta_2 = 90.49034$, $m_3 = 2.298934$, $\eta_3 = 46.46533$. And then use the Powell method to obtain the maximum likelihood estimation of the corresponding parameters, and then through the Weibull model to accelerate the model to obtain normal stress (20 °C that is 293 open) ammunition storage life distribution parameters m_0 and η_0 are estimated as: $m_0 = 2.4245$, $\eta_0 = 9982.6245$. That is, the reliability of the ammunition under normal stress conditions is:

Table 1. Some ammunition components accelerated life test results

Sampling number	Stress level (on)	Detection time (hours)	Sample capacity (piece)	Failure number (piece)
1	340	350	20	1
2	340	602	20	0
3	340	700	20	1
4	340	786	20	1
5	340	980	20	2
6	345	1150	20	3
7	345	1290	20	3
8	345	1410	20	4

Table 2. Some ammunition product under standard conditions of storage reliability test data

Stored address	Storage temperature (on)	Samples number	Failure number (piece)	Storage time (years)	Data type
A	293	10	1	5	storage
A	293	10	0	8	storage
A	293	10	1	13	storage
A	293	10	2	17	storage
A	293	10	3	23	storage

$$R(t_i) = \exp\left[-(t_i/27.3497)^{2.4245}\right].$$

The accelerated life test data and the storage reliability test data under standard conditions were fused to obtain the storage reliability data of ammunition under standard conditions $R(t_i) = \exp\left[-(t_i/30.8997)^{2.4226}\right]$., with the storage time of the law of change shown in Fig. 1.

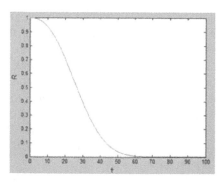

Fig. 1. Storage reliability curve of ammunition under standard conditions

3 Conclusion

When ammunition storage of a variety of data exists at the same time, storage data fusion information becomes the key to solve the problem. In this paper, Bayes method is studied in detail, and this method is proposed to realize the reliability evaluation of ammunition storage by a priori information fusion. In the process of solving the Bayes method, the Gaussian method approximation integral calculation is used to solve the problem of integral programming. At the end of this paper, an example is given. The simulation results show that the Bayes method can correctly and effectively solve the problem of ammunition storage data fusion evaluation.

References

1. Huang, Z., Yan, L.: Based on Bayesian information fusion the analysis of manufacturing execution systems reliability growth. Modern Manufact. Eng. **5**, 21–26 (2014)
2. Yang, Q., Wei, X.: Reliability prediction of travel time based on spatio-temporal Bayesian mode. J. South China Univ. Technol. **44**(4), 115–121 (2016)
3. Zhou, H.: Researches on Bayesian approach for storage reliability assessment of ammunition system in small sample circumstances and design of its application software. Graduate School of Defense Science and Technology. (2006)
4. Elsayed, E.A., Wang, II.: Bayes & classical estimation of environment factors for the binomial distribution. IEEE Trans. Reliab. **45**(4), 661–665 (1996)
5. Chen, B.: Optimization theory and algorithm. Tsinghua University Press, Beijing (1989)
6. Xiao, X.: Modern numerical calculation method. Peking University Press, Beijing (2003)

Fault Diagnosis of Analog Circuits Based on Multi Classification SVDD Aliasing Region Identification

Shuang-yan Hu[1(✉)], De-qin Shi[2], Xiao-shan Song[1], Lin-bo Fang[1],
Wei-jun Yang[1], and Qi Tong[1]

[1] The Xi'an High-Tech of Institute, Xi'an 710025, China
shuang0000shuang@163.com
[2] Aeronautics and Astronautics Engineering College,
Air Force Engineering University, Xi'an, China

Abstract. In order to solve the problem of poor diagnostic accuracy of aliasing samples in multi class support vector data description (SVDD) method, a multi classification SVDD algorithm with heterogeneous samples is proposed. The method is based on the common SVDD hyper sphere model, in the presence of aliasing in the regional category, all the samples for the target class, other classes with aliasing samples were heterogeneous, using SVDD algorithm with heterogeneous sample re training, until all the hyper-sphere after optimization. Simulation results show that the algorithm can eliminate aliasing and improve the accuracy of the algorithm, and the algorithm is applied to analog circuit fault diagnosis. Compared with SVDD multi classification algorithm, one to one and one to many SVM algorithms, this method has higher diagnostic accuracy in analog circuit fault diagnosis.

Keywords: SVDD · Aliasing region · Heterogeneous samples · Fault diagnosis

1 Introduction

With the rapid development of electronic technology, the integration degree of the circuit board in the missile electronic protective equipment is more and more high, the structure and function are becoming more and more complex. The test of analog circuit because of the condition and level of staff constraints, are often unable to obtain complete information, coupled with the close current in the circuit, the single fault may cause many faults, resulting in fault diagnosis more difficult, and the traditional fault diagnosis methods have been unable to meet the requirements of existing equipment maintenance. Therefore, it is of great significance to study the suitable fault diagnosis method for analog circuits in order to ensure the effective performance of the equipment and the victory of the war.

Support vector data domain description (Support vector data description, SVDD) is a commonly used single classifier. Literatures [2–4] extended that to multi classification applications, namely multi classification SVDD (Multi-SVDD, MSVDD). MSVDD is an ideal model for each hyper-sphere are independent of each other, but in the actual application is likely to occur in two or more hyper-spheres aliasing situation, how to

© Springer International Publishing AG 2018
F. Qiao et al. (eds.), *Recent Developments in Mechatronics and Intelligent Robotics*,
Advances in Intelligent Systems and Computing 691, DOI 10.1007/978-3-319-70990-1_17

aliasing region samples were correctly classified, and the improved algorithm is applied to practical problems, MSVDD is one of the hot and difficult research method the. At present, many researchers put forward their own solutions and applied to the fault diagnosis, for example, the literature [5] from the SVDD algorithm itself to achieve the purpose of eliminating aliasing. The literature [6] put forward a new method of analog circuit fault diagnosis of space mapping based on SVDD, to improve the standard slack description boundary of SVDD, reduce the misdiagnosis caused by cross regional rate. The literatures [7, 8] uses the relative distance judgment method, puts forward a membership function based on a decision rule, by considering two aliasing hyper-sphere size, to improve the aliasing problem. In literature [9], the absolute distance measure and the relative distance measure are used to realize the recognition of state monitoring and fault classification. Literature [10] based on the SVDD algorithm, the introduction of hyper-sphere nuclear distance metric, the multi parameters into a single parameter, solves the contradiction between many parameters.

By analyzing the reason of mixing the stack, combined with the previous experimental results, confirmed that if a small amount of heterogeneous samples in SVDD training, you can appropriately enhance ability of data description, the hyper sphere is more compact, so as to improve the classification accuracy rate model. Therefore, in this paper, a heterogeneous sample MSVDD algorithm is introduced to solve the problem of multi class classification in the process of MSVDD model training. The simulation data set is used to verify the ability of the algorithm to eliminate aliasing and improve the accuracy, and then the algorithm is applied to the fault diagnosis of analog circuits.

2   MSVDD Algorithm and Its Shortcomings

Multi classification SVDD algorithm has the advantages of simple model, strong scalability and less computation. The basic idea is: for each kind of samples, respectively, SVDD training, get their own hyper-sphere model, and then use the model to classify the test samples.

Suppose there are k samples $X_k = \{x_1^k, x_2^k, \cdots, x_{N^k}^k\}$, where $x_i^k \in R^d$, N^k indicates the number of samples k, the goal of the algorithm is to find k as much as possible each sample contains minimal hyper-sphere S^k, radius R^k and center α_k of S^k can be obtained by formula (1) and (2).

$$R_k^2 = 1 - 2 \sum_{i=1}^{N_k} \alpha_i^k K(x_i^k, x_{sv}^k) + \sum_{i=1}^{N_k} \sum_{j=1}^{N^k} \alpha_i^k \alpha_j^k K(x_i^k, x_j^k) \qquad (1)$$

$$\alpha_k = \sum_{i=1}^{N_k} \alpha_i^k \varphi(x_i^k) \qquad (2)$$

The parameters, $\alpha_i^k \geq 0$ is the Lagrange factor, α_{sv}^k is a support vector sample, φ is a mapping function, $K(x_i^k, x_j^k)$ is a kernel function.

$$K\left(x_i^k, x_j^k\right) = \exp\left(-\frac{\left\|x_i^k - x_j^{k2}\right\|}{2\sigma^2}\right) \tag{3}$$

The parameter in the formula (3) α is a Gauss kernel function.

The distance, between test sample x and the ball center of hyper-sphere S_k, can be expressed as the followings.

$$R_{xk}^2 = \left\|\emptyset(x) - \alpha_k\right\|^2 = 1 - 2\sum\nolimits_{i=1}^{N_k} \alpha_i^k K\left(x_i^k, x\right) + \sum\nolimits_{i=1}^{N_k} \sum\nolimits_{j=1}^{N^k} \alpha_i^k \alpha_j^k K\left(x_i^k, x_j^k\right) \tag{4}$$

By comparing the size of the R_{xk}^2 and R_k^2 can determine which type of x into the hyper-sphere to determine the classification of x. However, the traditional MSVDD algorithm usually has the problem of aliasing, which will greatly affect the classification accuracy of the algorithm. There are two main reasons for the aliasing area. One is that the sample itself, such as the uneven number of samples, the ambiguity between samples, etc. Another case is that the ball is not compact enough. By increasing or decreasing the rejection rate *fracrej* Gauss parameters can improve the compactness of the ball, but also makes the model generalization ability decreased. Before the model training, we need to optimize the pair parameters of (*fracrej*, σ).

3   MSVDD Algorithm with Heterogeneous Samples

Through the analysis of multi classification SVDD reason, based on the premise of the optimal parameters, by increasing the compact hyper-sphere multi classification model to improve the quality of the sample aliasing problem. In order to improve the compactness of hyper-sphere, a SVDD algorithm with heterogeneous samples (SVDD with Negative Samples, NSVDD) is introduced. It is found that adding a small amount of heterogeneous samples in SVDD training can improve the data description ability and the classification accuracy. When the multi classification is in the aliasing region, the NSVDD algorithm can be used to re train the hyper sphere. On this basis, the training model of NMSVDD algorithm is constructed, and the multi classification SVDD (NMSVDD) algorithm with heterogeneous samples is proposed.

3.1   NSVDD

The target of NSVDD is to establish a hyper-sphere that contains all normal samples, and the minimal hyper-sphere fault samples are excluded. Using i, j to represent normal samples, and p, q to represent fault samples, mark normal sample as $y_i = 1$ and fault samples as $y_p = -1$, the number of two types of samples were m and n. Hyper-sphere radius R, center for a, then the optimization problem can be expressed as the followings.

$$\min F\left(R, a, \xi_i, \xi_p\right) = R^2 + \sum\nolimits_i C_i \xi_i + \sum\nolimits_p C_p \xi_p \tag{5}$$

$$\text{s.t.} \left\| \varphi(x_i) - a \right\|^2 \leq R^2 + \xi_i,$$
$$\left\| \varphi(x_i) - a \right\|^2 \geq R^2 + \xi_p, \xi_i \geq 0, \xi_p \geq 0, \forall i, p \tag{6}$$

The parameters, ξ_i, ξ_p are slack variables, C_i, C_p are penalty factors.
The Lagrange function in the formula (5) is as follows.

$$\begin{aligned} L(R, a, \xi_i, \xi_p, \alpha, \beta) = & R^2 + \sum_i C_i \xi_i + \sum_p C_p \xi_p - \sum_i \beta_i \xi_p - \sum_p \beta_p \xi_p \\ & - \sum_i \alpha_i \{ -[\langle \varphi(x_i), \varphi(x_i) \rangle - 2a\varphi(x_i) + a.a] \} + R^2 + \xi_i \\ & - \sum_p \alpha_p \{ -[\langle \varphi(x_p), \varphi(x_p) \rangle - 2a\varphi(x_p) + a.a] \} + R^2 + \xi_p \end{aligned} \tag{7}$$

The parameters, $\alpha_i \geq 0, \alpha_p \geq 0, \beta_i \geq 0, \beta_p \geq 0$ in the formula (7) are Lagrange factors.

According to the extreme condition,

$$\sum_i \alpha_i - \sum_p \alpha_p = 1, \sum_i \alpha_i \varphi(x_i) - \sum_p \alpha_p \varphi(x_p) = a$$
$$\text{s.t. } 0 \leq \alpha_i \leq C_i, 0 \leq \alpha_p \leq C_p, \forall i, p \tag{8}$$

Define new variables $\alpha_i^* = y_i \alpha_i$, i not only represents the normal samples, but also indicates the different samples, and only a small number of support vector samples α_i^* is not 0. With $K(x_i, x_j)$ kernel function instead of inner product operation according to the dual rules, solving form can be obtained with the standard NSVDD algorithm SVDD algorithm is the same, only the α_i^* can replace α_i. Therefore, in the training of NSVDD, just give the Lagrange multiplier a category label, you can get the same SVDD and the general form of solution and results.

3.2   Process of NMSVDD Algorithm

The basic idea of NMSVDD is: if a hyper-sphere and other hyper-sphere are aliased with the area, all the samples for the target class, other classes and mixing samples for heterogeneous re-training sphere, to enhance the compactness of hyper-sphere, reduce or eliminate the aliasing region, and improve the model precision. The detailed steps of the algorithm are as follows.

Step1: The measurement circuit signal is extracted and pretreated to form the fault samples needed by the analog circuit, and the sample set is set up and divided into training samples and test samples.

Step2: Get L initial hyper-sphere $S_1, S_2, \cdots, S_L$ trained by SVDD L samples.

Step3: With the formula (4) to determine the existence of mixed samples between various hyper-spheres and other hyper-sphere, if not, then the class sample initial model optimization is completed, otherwise go to the next step.

Step4: In the case of the existence of the aliasing region, all the samples of this class are the target class, and the samples of the other classes are re trained by the NSVDD algorithm.

Step5: Whether there are aliasing between samples and the other judge all kinds of hyper-sphere, until all the optimization is completed, otherwise it returns the Step4.

Through the simulation experiment, after 2–3 cycles, it will completely eliminate aliasing, so the algorithm will be set to the maximum number of cycles 5 times.

4   Case Analyses

In order to verify the effect of the application of the multi classification SVDD algorithm with different samples to the analog circuit fault diagnosis, the Sallen-Key band-pass filter circuit is taken as an object, as shown in Fig. 1.

(1)   Selection of excitation signal and test point

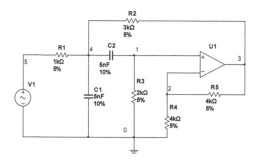

Fig. 1.   Sallen - key band-pass filter circuit

The excitation signal of the input node 5 is the voltage signal with the amplitude of 5 V and the fundamental frequency of 10 kHz, and node 3 is the output test point.

(2)   Fault element and fault mode selection

Multisim software is used to simulate the soft fault which is difficult to judge. The nominal values of each element are shown in the figure, where the tolerance range of the resistance is 5%, and the tolerance range of the capacitance is 10%. Through the sensitivity analysis of the circuit, found that R1, R2, R3, C1 and C2 has the high sensitivity and impact on the output of the circuit is large, so the circuit fault mode settings are shown in Table 1, including five kinds of fault mode and normal mode circuit.

(3)   Acquire fault samples

Collect the signal of output node 3 and use Multisim for each fault mode of the Monte Carlo analysis, a total of 180 fault samples set, 90 samples used to train, the other 90 are used to test.

(4)   Fault diagnosis and result analysis

Using MSVDD algorithm for training, the parameters is $(fracrej, \sigma) = (0.03, 5)$ X. The analysis results show that there are four kinds of aliasing fault, fault aliasing using

NMSVDD algorithm to train the circuit there, heterogeneous samples refused to factor 0.99, and under the same conditions of NMSVDD algorithm based on MSVDD algorithm, one to one, and one to more SVM diagnostic methods of test results, the diagnosis results are shown in Table 2 shown.

Table 1. Circuits fault mode table

Circuit state	Fault kind	Fault mode	Fault type
Normal	——	Normal	F1
R1 fault	↑	Tolerance 50%	F2
R2 fault	↑	Tolerance 50%	F3
R3 fault	↑	Tolerance 50%	F4
C1 fault	↑	Tolerance 50%	F5
C2 fault	↑	Tolerance 50%	F6

Table 2. Fault diagnosis results of algorithm classifier

Aliasing fault	NMSVDD	MSVDD	1 to 1	1 to more
F2F3	95.6%	93.3%	87.8%	84.4%
F2F6	91.1%	87.8%	77.8%	78.9%
F3F5	93.3%	81.1%	76.7%	75.6%
F3F6	96.7%	90%	80%	74.4%

As can be seen from Table 2, the NMSVDD algorithm is applied to the fault diagnosis of analog circuits, and its accuracy is much higher than that of MSVDD algorithm, one to one and one to more SVM multi-classification algorithms. The results show that when the circuit is aliasing region, it will fall into the intersection region of the standard SVDD and the intersection region of the SVM classification. The method to solve this problem, the aliasing of heterogeneous samples as the re training sphere, to eliminate aliasing, experimental results show that this method can achieve rapid aliasing fault diagnosis, fault detection and location of analog circuit fault diagnosis, compared with the traditional method has better diagnosis effect.

5   Conclusions

In order to overcome the shortcomings of the classical diagnosis method in analog circuit fault diagnosis, a multi classification SVDD algorithm with heterogeneous samples is proposed. The simulation results show that the NMSVDD algorithm can eliminate aliasing, enhance the ability to describe the data, make the hyper-sphere more compact, and improve the accuracy of the model classification. In addition, compared the algorithm for fault diagnosis of analog circuit fault diagnosis example, this algorithm with the classic, can be seen from the experimental results, NMSVDD algorithm achieved good results in analog circuit fault diagnosis, has great potential.

References

1. David, M.J.T., Robert, P.W.D.: Support vector data description. Mach. Learn. **54**, 45–66 (2004)
2. Kang, W.S., Im, K.H., Choi, J.Y.: SVDD-based method for fast training of multi-class support vector classifier. In: Third International Symposium on Neural Networks, pp. 991–996. Springer, Chengdu (2006)
3. Lee, D., Lee, J.: Domain described support vector classifier for multi-classification problems. Pattern Recogn. **40**(1), 41–51 (2007)
4. Luo, H., Wang, Y.R., Cui, J.: A SVDD approach of fuzzy classification for analog circuit fault diagnosis with FWT as preprocessor. Expert Syst. Appl. **38**(8), 10554–10561 (2011)
5. Mu, T.T., Nandi, A.K.: Multi class classification based on extended support vector data description. IEEE Trans. Syst. Man Cybern. Part B **39**(5), 1206–1216 (2009)
6. Luo, H., Wang, Y.: A GSM-SVDD method for analog circuits fault diagnosis. Electr. Mach. Control **17**(1), 108–113 (2013)
7. Chiang, J.H., Hao, P.Y.: A new kernel-based fuzzy clus-tering approach: support vector clustering with cell growing. IEEE Trans. Fuzzy Syst. **11**(4), 518–527 (2003)
8. Liu, S., Liu, Y.K., Wang, B. et al.: An improved hyper-sphere support vector machine. In: Third International Conference on Natural Computation, Haikou, China, pp. 497–500. IEEE (2007)
9. Wang, T., Li, A., Wang, X., Cai, Y.: Fault diagnosis method for a gear pump based on SVDD and distance measure. J. Vib. Shock **32**(11), 62–65 (2013)
10. Li, D., Li, B., Wang, Y., Zhao, K.: Aeroengine performance deterioration evaluation using clustering and multi-scaling optimal hyper sphere kernel distance assessment. J. Propuls. Technol. **34**(7), 977–983 (2013)
11. Debnath, R., Takahide, N., Takahashi, H.: A decision based on one-against-one method for multi-class support vector machine. Pattern Anal. Appl. **27**(5), 164–175 (2004)
12. Rifkin, R., Klautau, A.: In defense of one-vs-all classification. J. Mach. Learn. Res. **5**, 101–141 (2004)
13. Li, C., Wang, Y., Luo, H., Cui, J.: Analog circuit fault classification based on all samples support vector data description. Trans. China Electro Tech. Soc. **27**(8), 215–221 (2012)

A Kind of New Method to Measure Tank Capacity

Cheng-Hui Yang[1,2]([⊠])

[1] College of Electrical Engineering,
Northwest University for Nationalities, Lanzhou, China
yangchenghui36@163.com
[2] School of Automation and Electrical Engineering,
Lanzhou Jiaotong University, Lanzhou, China

Abstract. This paper presents the median surface model and the integral model, and we use them to analyses the practical application of a kind of typical oil tank displacement recognition and tank volume calibration problem. Finally, we obtain the more accurate displacement parameters and accomplish the solution of the two questions. Besides, according to the measured data of annex 2, we can find two functional relations that named equation 14 and equation 15. And by these two equations, we can get a set of approximate solution. At the same time, we can get that the calculated model error is less than or equal to 1.61% by MATLAB.

Keywords: Median surface model · Integral · Model · Displacement identification · Identification · Capacity

1 Introduction

Gas stations usually have several underground storage tanks to store fuel oil and generally match the system named "oil level measurement and management". It use flow-meter and oil level meter to measure the data such as the into and out of the oil tank, the oil level height and so on. At the same time, by processing real-time calculation for the advanced calibration of the tank capacity table to get the changes of the tank oil level height and oil capacity. After using a period of time, due to the ground deformation and other reasons, the location of the tanks can happen longitudinal slope and transverse deflection (the following referred to as the displacement) and then leading to tank capacity table change. In accordance with the relevant regulations, gas stations need regular calibrate for the tank capacity table. Figure 1 is a typical oil tanks longitudinal slope with displacement, its main body is a cylinder and both ends are spherical cap.

2 Questions to Solve

2.1 System Design

After using a period of time, due to the ground deformation and other reasons, the location of the tanks can happen longitudinal slope and transverse deflection and then

F. Qiao et al. (eds.), *Recent Developments in Mechatronics and Intelligent Robotics*,
Advances in Intelligent Systems and Computing 691, DOI 10.1007/978-3-319-70990-1_18

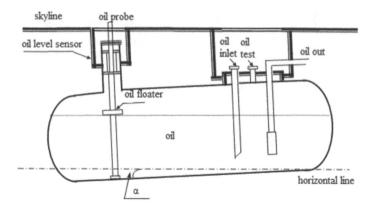

Fig. 1. Storage tanks with a longitudinal slop after displacement

leading to tank capacity table change. In accordance with the relevant regulations, gas stations need regular calibrate for the tank capacity table. So how to identify the deflection and accurately calibrate the storage tank after displacement become a problem to be solved. This paper mainly studies the following two questions:

(1) In order to grasp the impact on the tank capacity table after displacement of tank, we use elliptical storage tanks (flat head on both ends of the elliptical cylinder) to do experiments respectively for tanks without displacement and longitudinal displacement of a slope angle for a = 4.10.

We establish mathematical model to research the tank's influence that after displacement on the tank capacity table. Besides, we give the tank capacity table calibration value after displacement with the oil height interval 1 cm.

(2) For the actual storage tanks, establishing mathematical model of the calibration tank capacity table after displacement of tank, in other words, that is the general relationship between the storage quantity and the height of oil level and displacement parameters (longitudinal horizontal deflection Angle a and b).

Using the actual testing data of tanks after displacement in the process of oil into and out and according to the established mathematical model to determine the displacement parameters, we give the tank capacity table calibration value after displacement with the oil height interval 10 cm.

To analyze and verify the correctness of the test model and the reliability of the method, we use the actual test data to analyze further.

2.2 National Standard and Technical Index of Storage Tank Volume

(1) Storage tank volume calibration system in accordance with national standards
Horizontal gas station oil jars of the calibration system is completely according to national metro-logical verification regulations such as JJG226-1996 and JJG133-2005 for production. The products passed the strict examination and

approval of national metro-logy department and after calibration of volume table completely meeting the standards of measurement, the whole system is according to the petroleum and petrochemical safety operation specification for design, production and inspection.

At the same time, it is in accordance with national standards as follows:

(1) "Laws of the People's Republic of China on explosive dangerous location of electrical safety regulations"
(2) GB3836 "Explosive environment with explosion-proof electrical equipment" standard series
(3) GB50058 "Explosion and fire dangerous environment electrical device design specification"
(4) GB50156 "Specification for design of small oil tanks and filling station"

(2) Storage tank volume calibration system's main technical indicators:

Measured height: up to 3800 mm
Liquid level resolution: 0.01 mm
Liquid level accuracy: ±0.1 mm
Accuracy of measurement: <plus or minus 2‰
Explosion-proof marks: Exd II AT3
Flow range: gasoline: 150–360 l/min; Diesel: 110–400 l/min
Power: AC220V50 Hz or AC380 V 50 Hz
Environment temperature: −25 °C–+55 °C
Relative humidity: 20%–95% RH
Working pressure: 0.5 Mpa or less
Applicator liquid type: diesel, gasoline, kerosene, water (some material for part of the accessories should be replaced by stainless steel).

3 Proposed Methods

3.1 Model Hypothesis

(1) Supposing tank's shape cannot change because of external force;
(2) Supposing tank's wall thickness is equal;
(3) Supposing liquid in the tank leaned left;
(4) Supposing the condition that ground is too soft to make slant angle and is too big so that cannot happen;
(5) Supposing tank cannot excessively vertically descend.

3.2 System Analysis

Now analyzing the problem by the above-mentioned model. The tank whose ends is spherical-capped and its deviation may be both crosswise and vertically leaned. It can be analyzed by dividing into two directions. Firstly processing crosswise analysis and then analyzing vertically leaned. From physical analysis, it is found that crosswise deflection can only change the floater's height but cannot influence the oil's volume. And it can be found that vertically leaned deflection can not only make oil-body in the tank change of its location but also influence tank capacity. According to the above analysis, the mapping between display and equivalent height can be completed. At the same time, combining tank capacity calculation method to analyses the conditions of no deflection and after deflection. We can get the relation among oil reserve, oil height and deflection parameter α、 β is $V_{变位}(h_3,\alpha,\beta)$ and then using the practically measured data to ensure α、 β. According this to make tank capacity after deflection and thus solving practical problems and improving oil volume's measurement accuracy.

4 Analysis of Model's Reliability

Using "initial input", "initial output" and "display oil height" these practical data to estimate original oil reserve before collecting and satisfy $V_{初} = V_{变位}(h_3,\alpha,\beta) - V_\lambda + V_{出}$.

The fitted curve is shown in Fig. 2:

Firstly, fitting result $V_{初} = 57646L$ is a constant and the similarity is high so we can see it fit reality absolutely.

Secondly, it appears singularity in the figure at the time of 6.3×10^2 s. We can find that the place is one-off supplementary feed time inflow and it fit reality too. This can explain model's reliability.

Fitting original oil reserve in oil tank calculated by the model, we can get the linear relationship between original oil's volume and its height and can get relative error that each oil's height corresponded also. Removing singular value engendered by one-off added oil, the relative error range controlled within 1.61%. α and β is estimated by MATLAB and statistics method. Because its error, thus it influence the final result.

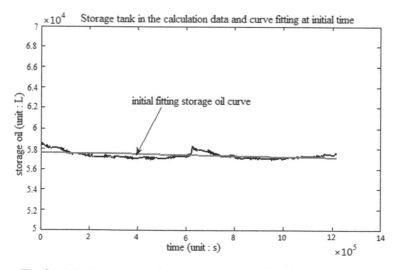

Fig. 2. Calculation data and fitting curve in the oil tank at first occasion

(1) The model established has mature theoretical basis and high accuracy;
(2) The model established has a dedicated software support and by analyzing large amount of data to verify the correct;
(3) Using different methods to construct and improve mathematical model so that to make the accuracy is higher and higher;
(4) The method and thought of modeling is also suitable for other types, so it's easy to spread to other areas.

5 Conclusion

In this paper, median surface model and integral model are used to analyze the practical application of a kind of typical oil tank displacement recognition and tank volume calibration problem. By building and analyzing, we obtain the more accurate displacement parameters and accomplish the solution of the two questions. These two equations, we can get a set of approximate solution. At the same time, we can get that the calculated model error is less than or equal to 1.61% by MATLAB. Finally, we can get that the statistical analysis and data fitting method for mass of data in the model can be applied to some data processing problems such as market forecast, raw material storage and economic growth analysis, etc.

Acknowledgment. This work is supported by Fundamental Research Funds for the Central Universities of Northwest University for Nationalities (Grant 31920160003, Grant 31920170079). And the project also supported by Experiment Funds of Northwest University for Nationalities, Northwest University for Nationalities, China. (Grant NO. SYSKF2017035, Grant NO. SYSKF2017036, Grant NO. SYSKF2017037, Grant NO. SYSKF2017043, Grant NO. SYSKF2017044).

References

Sun, F.J.: A discussion on some difficulties in calibran calculation of horizontal oil tank volume. J. Pet. Prod. Appl. Res. **18**(5), 20–24 (2000)
Sun, H.-D., Guan, J.-B.: Using approximation to calculate tank's volume that cross section is spherical-capped. Fitt. Equip. **3**, 29–30 (2001)
Wu, J.-G., Wang, M.-J., et al.: Mathematical Modeling Choreography Case, vol. 3. China Water Power Press, Beijing (2005)
Li, Y., Ren, J.P., Ning, C.L.: The interpolation's application in oil. Reserves. TecInformation. Development & Economy 16(12) (2006)
Chen, J: MATLAB Valuable book , Beijing, PHEI.2007.1
Wang, J.-W., Qu, Z.-S., Ling, B.: MATLAB 7.X Program Design. China Water Power Press, Beijing (2007)
Thirugnanasambandam, M., Iniyan, S., Goic, R.: A review of solar thermal technologies. Renew. Sustain. Energy Rev. **14**(1), 312–322 (2010)
Liu, L.-F., He, Q., et al.: Using Maple and Matlab to Solve Scientific Calculation Problems. CHEP, Beijing (2017)

Haillot, D., Nepveu, F., Goetz, V., et al.: High performance storage composite for the enhancement of solar domestic hot water system: Part 1: storage material investigation. Sol. Energy **86**(1), 64–77 (2012)

Specification for field welded tanks for storage of production liquids. SY/T 0606-2006

Hu, S.-X., Li, B.-N.: Math Experiment on the Basis of Matlab. Science Press, Beijing (2012)

Code for design of vertical cylindrical welded steel oil tanks. GB, 12(1), pp. 50341–2003 (2003)

Design of a Compact Electrically Tunable Band-Pass Filter with Wide Tuning Range

Chun Li$^{(\boxtimes)}$, Xiaofa Zhang, and Naichang Yuan

College of Electronic Science and Engineering,
National University of Defense Technology, Changsha, China
lichun15@nudt.edu.cn

Abstract. This paper introduces the design of electrically tunable band-pass filter with variable capacitance and microstrip line, which is based on the Chebyshev filter circuit model and improves the coupling circuit. we use the π-type transform coupling circuit with inductance tap transform to reduce the return loss on the basis of wide tuning, and use the design method of zero topology to improve the out-band suppression of the filter, then design the structure of the filter and convert it into microstrip circuit. The varactor diode is used as a part of the tuning circuit to change pass-band movement of the filter by controlling the load voltage. The electrically tunable filter circuit is simulated by using ADS and HFSS software. Through the actual measurement, the filter can be adjusted in the range of 700–1.400 MHz. When the center frequency is 1.34 GHz, $S_{11} = -24.6$ dB, $S_{21} = -5.1$ dB, the out-band suppression is relatively steep, good selectivity, and the size of the filter is small.

Keywords: Electrically tunable band-pass filter · Chebyshev · Wide tuning · Varactor diode · Tuning circuit · Zero · Out-band suppression

1 Introduction

With the rapid development of modern information technology, spectrum resources become more and more tense, and it is reasonable to use the frequency spectrum, which can solve the serious interference problem between different communication users, thus it can protect effectively wireless communications. And high-performance filter ensure the normal operation of electronic equipment in a complex electromagnetic environment. The filter is used as the primary component of the antenna, which is used to restrain the harmful interference signal frequency and make the equipment work normally. However, the space electromagnetic environment is more and more complex. In the environment of electronic warfare and anti-jamming, frequency hopping technology has the advantages of security and anti-jamming [1].

Electrically tunable filter is a band-pass filter which changes quickly the passband in frequency hopping communication system, which allows the signal near the tuning frequency to pass smoothly, while the signal far from the tuning frequency is greatly attenuated [2]. Compared with the filter group, the filter can change quickly the center frequency of bandpass; relative to other band-pass filter, tunable filter has narrower

© Springer International Publishing AG 2018
F. Qiao et al. (eds.), *Recent Developments in Mechatronics and Intelligent Robotics*,
Advances in Intelligent Systems and Computing 691, DOI 10.1007/978-3-319-70990-1_19

frequency band, variable center frequency and flexible, small size, good selectivity and other advantages [3, 4].

2 Principle of Electrically Tunable Filter

In this paper, the Chebyshev filter is used as the base circuit model, and the theoretical model is deduced and analyzed. The mutual coupling mode between the resonators is determined, and the π-transform coupling circuit with inductance tap transform is used to realize requirements of return loss in wide tuning. Using coupling circuit and design method of adding zero topology, then multiple zeros are introduced in the Chebyshev circuit, which can greatly improve the filter suppression [5–7].

2.1 Analysis of Filter Model

According to the design theory of the filter at present, the Chebyshev type filter has a steeper attenuation characteristic [8], the function of the Chebyshev amplitude squared is:

$$A(\Omega^2) = |H_a(j\Omega)|^2 = \frac{1}{1 + \varepsilon^2 V_N^2 \left(\frac{\Omega}{\Omega_C}\right)} \tag{1}$$

In the formula, Ω_C is effective passband cut-off frequency; ε is parameters associated with passband ripples, ε is larger, the larger the ripple, and $0 < \varepsilon < 1$; $V_N(\chi)$ is the N order Chebyshev polynomial.

$$V_N(\chi) = \begin{cases} \cos(N\arccos\chi) & |\chi| \leq 1 \\ \cosh(N\arccosh\chi) & |\chi| \geq 1 \end{cases} \tag{2}$$

2.2 Improvement of Filter Model

Compared to capacitive coupling band-pass filters, the tuning performance of the inductive coupling is better at the low frequency [8, 9]. Adjusting the coupling inductance ratio, which can broaden the tuning range of the filter [10].

Then increasing the transmission zero to improve the filter selectivity, and increasing the tap inductance conversion circuit to improve the resonant impedance, as shown in Fig. 1.

In Fig. 1, the inductance L_2 and L_3 constitute the impedance conversion circuit, and the inductance L_1 achieve impedance matching function. In order to increase the out-band suppression, inductor L_1 in parallel with a capacitor introduces two zeros in the transmission passband.

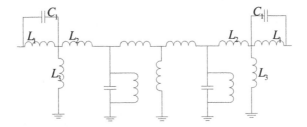

Fig. 1. Structure of electrically tunable filter

3   Modeling of Varactor Diodes

The varactor diode is the most critical tuning element in the electrically tunable filter. The linearity of the filter is affected seriously by the linearity of the varactor diode [11].

3.1   Varactor Diode Model

In the tuning circuit, the junction capacitance can be controlled by applying voltage to the varactor diode. Varactor diode consists of PN junction, epitaxial layer and a resistance substrate, different doping concentration can achieve different voltage-controlled capacitor change. The relationship among the area of the varactor, the voltage and doping concentration of the epitaxial layer is:

$$\frac{c(v)}{A} = K\left(\frac{N}{V + \varphi}\right)^{n} \tag{3}$$

In the formula, $c(v)$ is capacitance value in the diode voltage, A is the area of the varactor diode, N is doping concentration of epitaxial layer, V is diode voltage, φ is a built-in potential, n is slope of C–V curve, K is constant.

Assuming that the intermodulation interference input signal $\sin(w_1 t)$ and $\sin(w_2 t)$, and non-linear products can be expressed as $\sin(2w_1 t - w_2 t)$ $\sin(w_1 t - 2w_2 t)$. Combined with the above formula (1), the intermodulation product is:

$$IM = \frac{n(n+1)v^2}{8(V_0 + \varphi)^2} \tag{4}$$

From the formula (4), the intermodulation product is affected by voltage v, there are changes in the slope of the index, and there must be nonlinear product.

The circuit structure with the nonlinearity of the diode is improved, and the structure of series-parallel connection is used to cancel intermodulation product. This structure can cancel both the two order harmonic and the three order intermodulation product, as shown in Fig. 2.

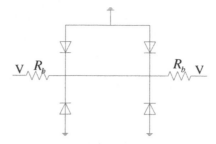

Fig. 2. Series-parallel cancellation structure of diode

3.2 Simulation of Diode Model

In ADS, the C–V characteristic curve of the varactor diode can be simulated in 800 MHz, as the voltage across the varactor diode is increased, the capacitance value decreases. When the tuning range of loaded voltage is 0–6 V, the range of the capacitance value is from 32 to 5 pF, as shown in Fig. 3.

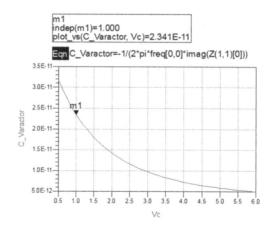

Fig. 3. The C–V characteristic curve of the varactor diode

4   The Results and Analysis of Filter Structure Simulation

After the above analysis, we establish the S2P model of varactor diode BBY-58V in the ADS, which facilitate the simulation. At the same time, the structure of the filter is further improved. It is more accurate to convert the concentrated parameter of the inductive and capacitance elements into the distributed parameter element, because the filter works in the high frequency band. According to the corresponding microstrip theory [12].

Adjust the corresponding variables, then we can get the S_{11} and S_{21} curve characteristics of the filter by simulation, as shown in Fig. 4.

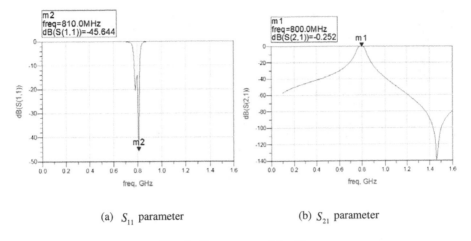

(a) S_{11} parameter   (b) S_{21} parameter

Fig. 4. S parameter of the filter

Through the electromagnetic simulation, we can get corresponding characteristics of the filter, and the change of the voltage makes the varactor diode performance different capacitance values, which makes the tuning passband move (Fig. 5).

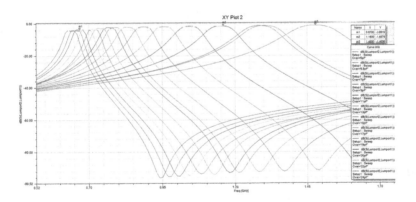

Fig. 5. Model of electrically S_{21} parameter of electrically tunable filter

5 Conclusion Hardware Circuit and Test Results

During the test, the tuning voltage is reverse loaded into the varactor diode in the bias circuit. Through changing the voltage value, the S_{11} and S_{21} characteristic curves of the electrically tunable filter are measured by Agilent's vector network analyzer N5230C.

When the voltage across the varactor diode is 6 V, the center frequency measured is located at 1.34 GHz. Figure 6 shows $S_{11} = -24.6$ dB, $S_{21} = -5.1$ dB, and the out-band suppression is steep which can suppress the out-of-band interference signal. The 3 dB bandwidth of the passband is 92 MHz, and the reflection coefficient at the center

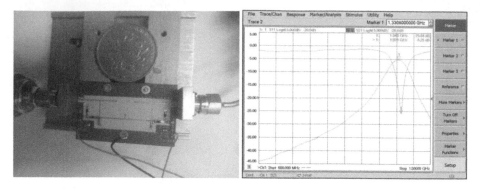

Fig. 6. Actual hardware circuit and test results

frequency is small, indicating that the reflected signal is small, but insertion loss is large. Capacitor diode Q value is low, parasitic resistance value is large and the loss of microstrip line, that resulting in large insertion loss by analysis. In applications, we should be appropriate to increase input signal power.

6 Conclusion

In this paper, the design of the electric filter has the characteristics of small size, low power consumption, good selectivity and so on. It is very convenient to tune filter passband and center frequency, out-band suppression performance, and filter interference signal by adjusting the value of the load voltage. With the development of electronic countermeasures and anti-jamming technology, frequency hopping technology are more and more attention in civilian and military field. Electrically tuning filter commonly is used in frequency hopping technology and other technical fields, so there are broad application requirements.

References

1. Jiang, X.Q., Shi, Y., Zhao, B.L.: Design of hybrid coupled electrically tunable filter. Technol. Appl. **45**(5), 35–38 (2014)
2. Wang, Y.J.: A new design of electrically tunable filter based on ADS. Telecommun. Technol. **52**(3), 367–370 (2012)
3. Lin, Y.Q.: Design of ultra wide band frequency tunable filter for wideband receiver applications. Microw. J. **31**(2), 59–62 (2015)
4. Zhao, Z.B., Mao, G.J., Liu, C.M.: Design of a L band tunable microstrip filter. Electromagn. Field Microw. **42**(7), 44–45 (2012)
5. Mao, J.R., Choi, W.W., Tam, K.W., Che, W.Q., Quan, X.: Tunable bandpass filter design based on external quality factor tuning and multiple mode resonators for wideband applications. IEEE Trans. Microw. Theory Techn. **61**(7), 2574–2583 (2013)
6. Kim, C.H., Chang, K.: Ring resonator bandpass filter with switch-able bandwith using stepped-impedance stubs. IEEE Trans. Microw. Theory Techn. **58**(12), 3936–3944 (2010)

7. Brown, R., Rebeiz, G.M.: A varactor-tuned RF filter. IEEE Trans. Microw. Theory Techn. **48**(7), 1157–1160 (2000)
8. Buisman, K., Vreede, L.C., Larson, L.E.: Distortion-free varactor diode topologies for RF adaptivity. In: IEEE MTT-S Internation Microwave Symposium Digest, pp. 4–8 (2005)
9. Zhang, N., Deng, Z., Hu, C.: Design and analysis of a tunable bandpass filter empolying RF MEMS capacitors. IEEE Electron. Device Lett. IEEE Trans. Microw. Theory Techn. **32** (10), 1460–1462 (2011)
10. Williams, A.B.: Analog filter and circuit design manual. Electronic Industry Press, Beijing (2016)
11. Yan, W.D., Mansour, R.R.: Compact tunable bandstop filter integrated with large deflected actuators. In: IEEE/MTT-S International Microwave Symposium, vol. 32(10), pp. 1460–1462 (2011)
12. Lei, Z.Y.: Microwave Electronic Circuit. Xidian University Press, Xi'an (2009)

Research on Growth Evaluation of New OTC Market of Science and Technology Enterprises

Li Qi$^{(\boxtimes)}$ and Tong Na

School of Management, Wuhan University of Science and Technology, Wuhan, People's Republic of China
liqi@wust.edu.cn

Abstract. The New OTC Market provides an important platform to promote the healthy development of science and technology enterprises. The growth of new technology companies have become the focus of attention of entrepreneurs. This paper analyzes the growth of 146 new technology enterprises in the New OTC Market by using the factor analysis method to know new technology companies the latest comprehensive growth ranking situation. It is of great significance for the managers to formulate an efficient and feasible policy for the development of new technology enterprises by analyzing the latest ranking results.

Keywords: The New OTC Market · The growth of new science and technology enterprises · Factor analysis method

1 Introduction

The New OTC Market is the third national stock exchange, and it was following the Shanghai Stock and Shenzhen Stock Exchange occurred. The New OTC Market has only 9 years old, the domestic research on the New OTC is not much of the growth research of the New OTC, mostly concentrated in market positioning and development scale. Science and technology enterprises symbol the development of the times and the growth trend. So the research of new technology enterprise growth is a relatively new subject.

There are many factors affecting the growth of enterprises in the process of enterprise growth. For example, Shen and Wu studied the growth of small and medium-sized enterprises in the use of panel data, found that the profitability and capital expansion capacity play a role in promoting the growth of enterprises [1]. Lu used empirical analysis to analyze the factors that influence the growth of SMEs in Qinghai Province, and put forward effective suggestions [2]. Qin combined the external economic factors, business factors with comprehensive corporate governance factors researched on the influencing factors on the growth of the GEM listed companies by using the panel data model [3]. Liu and Liu combined profitability, operating capacity, the size of the company, solvency with other financial factors and non-financial factors to study the growth of the 300 GEM listed companies [4]. Dai revealed the impact of the New OTC listed company financial growth of the main factors from the macro and micro two aspects [5]. This article regards 146 new science and technology enterprises

F. Qiao et al. (eds.), *Recent Developments in Mechatronics and Intelligent Robotics*,
Advances in Intelligent Systems and Computing 691, DOI 10.1007/978-3-319-70990-1_20

in 2015 of the New OTC as the research object, uses the factor analysis method to evaluate the growth of the new science and technology enterprises.

2 Construction of Growth Evaluation Index

Taking into account the most enterprise information can be reflected in the financial indicators, financial indicators are easy to be obtained. This paper uses the SPSS statistical analysis and combines with the New OTC listed enterprise development status and analysis the internal structure of the New OTC of the new science and technology enterprises with the four aspects of profitability, growth capacity, solvency, operation ability. It constructs 12 evaluation indicators system as shown in Table 1.

Table 1. The New OTC of the new technology enterprises growth evaluation system

Target	First level index	Two level index
The New OTC of the new technology enterprises growth evaluation system	Profitability	The net interest rate of total assets A_1 Rate of return on net assets A_2 Earnings per share A_3 Net profit margin on sales A_4
	Growth capacity	The interests of the shareholders growth rate A_5 Equity growth rate attributable to shareholders of the parent company A_6 Total assets growth the rate of growth rate A_7 Total revenue growth rate A_8
	Solvency	The current ratio A_9 Quick ratio A_{10}
	Operational ability	Turnover rate of current assets A_{11} Turnover ratio of total assets A_{12}

3 Enterprise Growth Evaluation Model Based on Factor Analysis Method

The traditional methods of evaluating the growth of enterprises are factor analysis, analytic hierarchy process, two dimensional judgment method and fuzzy comprehensive evaluation method. Compared with other analysis method, factor analysis method has the following advantages: (1) factor analysis process is relatively simple and straightforward. It is through the orthogonal transformation will change a number of indicators for a few factors, eliminate the index information overlap between the original index and contain most of the information; (2) it will start by the linear combination of multiple variables into a few factors, which makes the analysis of variables to minimize, then factors also can replace the original multiple variables involved in data modeling in the analysis process, greatly reduce the amount of calculation.

3.1 Mathematical Model of Factor Analysis

The basic idea is that the observed variables are related to the classification of high correlation, more closely linked into the same class, the other is a low correlation. Each class represents a basic structure, namely the public factor. The mathematical model of factor analysis is introduced as follows:

$$
\begin{cases}
x_1 = a_{11}f_1 + a_{12}f_2 + a_{13}f_3 + \cdots + a_{1n}f_n + \varepsilon_1 \\
x_2 = a_{21}f_1 + a_{22}f_2 + a_{23}f_3 + \cdots + a_{2n}f_n + \varepsilon_2 \\
x_3 = a_{31}f_1 + a_{32}f_2 + a_{33}f_3 + \cdots + a_{3n}f_n + \varepsilon_3 \quad n \leq m \\
\qquad\qquad \cdots \\
x_m = a_{m1}f_1 + a_{m2}f_2 + a_{m3}f_3 + \cdots + a_{mn}f_n + \varepsilon_m
\end{cases}
$$

The above model is represented by a matrix is $X = AF + \sigma$, $F = (f_1, f_2, \ldots, f_n)$ is a common factor, $X = (x_1, x_2, \ldots, x_m)$ is variable, and a_{ij} is factor loading (among $i = 1, 2, \ldots, m; j = 1, 2, \ldots, n$). Represents the loading on the first j factor of the first i, namely the variable x_i and factor f_j are correlation coefficient. σ indicates that the original variable cannot be explained by the factor.

3.2 Steps of Factor Analysis Method

(1) To test whether the original variable is suitable for factor analysis.
   Based on the sample data by KMO test and Bartlett test of sphericity, if the value of KMO > 0.5 and Bartlett level test of sphericity of sig < 0.05, indicating the original variables suitable for factor analysis. If KMO < 0.5, namely the original variables do not fit factor analysis.
(2) Construct factor variable.
   The paper used principal component analysis method to get the factor loading matrix. Each column element of factor loading matrix which is the sum of the square to obtain the variance contribution of factor to the original variable, which is recorded as g_j^2, The greater the g_j^2 value, the greater the contribution of the factor to the original variable.
(3) Factor rotation.
   If the absolute value of a_{ij} is larger than the columns in the i line, the original variable x_i has a large correlation with multiple factors; If the absolute value of a_{ij} is larger than the number of rows in column j, the factor f_j is able to interpret the information of multiple variables at the same time; These two kinds of situations can lead to the meaning of factor f_j fuzzy. In order to solve this problem, it is needed to carry out the orthogonal rotation.
(4) Factor variable score.
   Factor score is used to describe the factor with the original variable.

$$F_{ji} = w_{j1}x_{1i} + w_{j2}x_{2i} + \cdots + w_{jm}x_{mi}, (j = 1, 2, 3 \cdots, n)$$

w_{j1}, w_{j2}, $w_{j3}, \ldots, w_{jm}$ is the factor values of the first j factor.

4   Empirical Research

This paper data is from wind information, and the New OTC Market of the new science and technology 146 companies in 2015 are the study sample.

4.1   Suitability Testing of Samples

By Bartlett sphericity test and KMO test, the analysis results are shown in Table 2, KMO = 0.653, Bartlett's test showed significant level of 0, suitable for factor analysis of the original data.

Table 2.   KMO and Bartlett test

Kaiser–Meyer–Olkin metric for sampling sufficient degree		.653
Inspection of Bartlett's degree of roundness	Approximate Chi square	2169.866
	df	66
	Sig.	.000

4.2   Extracting Main Factor

In the original variable correlation coefficient matrix, extracting eigenvalue greater than 1 and the cumulative contribution rate of more than 80% factors by principal component analysis, they are shown in Table 3. The cumulative contribution rate is 81.794%, the effect of factor analysis is better.

Table 3.   The explanation of the total variance

	Initial eigenvalue			Load extraction sum of squares			Load rotation sum of squares		
	T	V	C	T	V	C	T	V	C
A_1	3.814	31.779	31.779	3.814	31.779	31.779	3.104	25.863	25.863
A_2	3.024	25.199	56.979	3.024	25.199	56.979	2.884	24.033	49.896
A_3	1.728	14.404	71.382	1.728	14.404	71.382	2.071	17.259	67.154
A_4	1.249	10.412	81.794	1.249	10.412	81.794	1.757	14.640	81.794

4.3   Name of Factor

The original variables are combined into a few factors, and meaning of factors should be defined, which will be helpful to the analysis of the next step. The results of rotating as shown in Table 4.

Table 4. Rotational component matrix

	Component			
A_1	**.940**	.126	.038	.133
A_2	**.896**	.094	−.013	.068
A_3	**.817**	.150	−.125	.102
A_4	**.783**	.008	.074	.089
A_5	.002	**.940**	.127	−.029
A_6	−.003	**.938**	.130	−.022
A_7	.187	**.842**	.278	−.005
A_8	.189	**.561**	−.095	.124
A_9	−.009	.130	**.976**	−.118
A_{10}	−.007	.160	**.976**	−.107
A_{11}	0.63	−.011	−.112	**.936**
A_{12}	.249	.007	−.101	**.893**

4.4   Factor Score

The four main factors using regression evaluation to get factor scores (S), the calculation formula is as follows:

$$W = 25.86\%F_1 + 24.03\%F_2 + 17.26\%F_3 + 14.64\%F_4$$

It can obtain the comprehensive score, and according to the scores in descending order, namely the New OTC of new technology enterprise growth from high to low ranking (R). Because of the limited space, the comprehensive score in the top 10 ranking table of New OTC of new technology enterprises are shown in Table 5.

Table 5. New OTC of science and technology enterprises in the top 10 ranking table

Company name	F_1		F_2		F_3		F_4		W	
	S	R	S	R	S	R	S	R	S	R
Data hall	−0.74	127	6.37	1	3.29	7	−1.15	139	1.74	1
S. technology	1.28	8	4.43	2	−1.35	146	1.69	9	1.41	2
Business online	0.02	79	1.54	9	3.86	3	0.66	31	1.13	3
E-xunton	1.91	2	2.03	7	−0.44	101	1.00	20	1.05	4
Public shares	1.04	15	1.23	13	0.68	19	1.99	4	0.97	5
K. technology	1.24	11	2.72	5	−0.64	135	−0.11	77	0.85	6
Aerial view	0.49	36	0.85	17	3.43	5	−0.77	114	0.81	7
AI casa-chinco	1.53	5	−0.24	71	0.81	17	1.75	7	0.73	8
B. entertainment	1.34	7	−0.10	57	1.00	13	1.38	11	0.70	9
Huazun science and technology	1.23	12	0.80	18	0.54	21	0.24	51	0.64	10

5   Suggestions and Conclusions

5.1   Suggestions

(1) For the growth of the first comprehensive ranking of the Data Hall company, its growth ability and the repayment ability is very strong, but its profitability and operating capacity is not very ideal. Managers should make strategic planning of the company operating income from the aspects, such as the improvement of products, develop marketable products, promotions, or to low cost in the market dominant position.

(2) Growth ability is poor, but its operational capacity ranked second in the source of Homo sapiens. So business managers should focus on the development of promising products. And at the moment, companies can borrow relatively low borrowing threshold of financial institutions to finance, such as bank or non-bank financial institutions.

(3) Repayment ability is the worst, but the profitability, growth ability and capital operation ability of Saul technology is better, managers need to understand what strategy should be used to solve the short-term shortage of funds. Such as the issuance of bonds and stocks and other methods to improve short-term solvency, or part of the long-term assets as soon as possible to sell cash, the use of its own funds and debt reduction.

5.2   Conclusions

New science and technology enterprises in the economic market position more and more prominent. The New OTC Market is an important development platform for new technology enterprises. This paper analyzes the growth of the New OTC Market of new technology enterprises from the four aspects of profitability: profitability, growth ability, repayment ability and operation ability. Based on the performance characteristics of these four main aspects, the manager has to formulate the highly feasible and feasible policy investors to provide investment reference.

References

1. Shen, H., Wu, Q.: An analysis of the factors affecting the growth of small and medium sized enterprises: an empirical study based on panel data of small and medium sized listed companies. Financ. Dev. Res. 2, 66–70 (2010)
2. Lu, X.: Enterprise growth evaluation method research: an empirical test of the listed companies in Jilin Province. Friends Account. 19, 4–7 (2011)
3. Qin, T.: Study on the growth and influence factors of GEM listed companies. Southwestern University of Finance and Economics, Chengdu (2013)
4. Liu, J., Liu, M.: Analysis of the factors affecting the growth of China's listed companies. Bus. Times. 15, 78–80 (2014)
5. Dai, X.: Research on the growth of listed companies based on entropy method. Econ. Forum 7, 51–54 (2016)

A WSN Semantic Web Service Discovery Method Based on User Context

Daoqu Geng, Qingming Zhang$^{(\boxtimes)}$, Yanping Yu, Chuntang Chen, and Guan Gong

Key Laboratory of Industrial Internet of Things and Networked Control, College of Automation, Chongqing University of Posts and Telecommunications, Ministry of Education, Chongqing, China
zqmmail@163.com

Abstract. As a new application based on network environment, Web services are characterized by loosely coupled, modular and self-describing, and are extremely important in many aspects such as process management and application integration development. In the field of wireless sensor networks (WSNs), the demands for Web services between different users are different because the Web services needed by users are closely related to their situations. In this paper, semantic Web service technologies were applied to wireless sensor network. We proposed a Web service discovery algorithm, which took user context into account and provided Web services they needed.

Keywords: WSN · Semantic Web service · User context

1 Introduction

Web services are software components that are used to support interoperability between machines, with the characteristics of platform-independent, modular, programmable, loosely coupled, self-describing, etc. [1]. With the rapid development of service-oriented computing, cloud computing, big data and other technologies, Web services are widely developed and used as applications to build service objects on the Internet. Semantic Web is rich in semantic description and logical reasoning [2]. Semantic Web service technology combines Web services technology and semantic Web technology, so that the computer can intelligently understand Web services and user requests, to achieve Service discovering, combining, monitoring, and calling automatically [3]. As the increasing number of Web services on the network, Quality of Service (QoS) becomes the main concern of users [4], how to find a high-quality semantic Web service that meets the user's context in a large number of similar Web services has become one of the hot issues of current research.

In [5], a semantic Web service discovery method is proposed to measure semantic similarity. This method integrated multiple conceptual relationships, and achieved more accurate semantic Web service matching by dealing with the relationship between different concepts in ontologies. [6] proposed an improved query process based on WMS (Web Map Service) graphical content, combined with SVW(support vector machine) and user relevance feedback, to solve the difference of how end-users and

F. Qiao et al. (eds.), *Recent Developments in Mechatronics and Intelligent Robotics*,
Advances in Intelligent Systems and Computing 691, DOI 10.1007/978-3-319-70990-1_21

computers understand WMS, and the mismatching between WMS graphical content and semantic description in metadata. Based on the dependency of service discovery on service composition, [7] proposed a novel service composition framework, which integrated fine-grained I/O service discovery, and also had better combination of services and flexibility of system extending. [8] proposed a systematic approach to solve the service discovery mechanism that had little focus on a variety of business quality of service (QoS) attributes, the main features of the method are: Registrating mechanism of business rules, dealing kernel attributes through an application filter, calculating the rank of a service or set according to the QoS constraints and QoS level of importance, using the proposed query mechanism to retrieve the appropriate single and service set capacity. At present, semantic Web services are mostly concerned with the accuracy and service quality of service discovery, and rarely care about the user's context. In this paper, considering the users in different contexts where require different semantic web services, we proposed a semantic Web service discovery method that can meet the specific needs of users based on the user's context.

2 Semantic Web Service Architecture

2.1 Web Services Architecture

As shown in Fig. 1, the Web services architecture includes three elements: roles, operations, and artifacts [9].

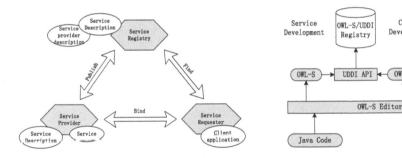

Fig. 1. Web services architecture

Fig. 2. Semantic Web service discovery process.

The Web service architecture is based on the interaction of three roles (service providers, service registries, and service requesters) that involve "publish", "find", and "bind" operations. These roles and operations work together on Web services artifacts: Web services software modules, descriptions of roles (including service provider description) modules, and customer application modules.

2.2 Web Service Discovery Framework

Figure 2 shows the server discovery process in the server and client programs. In the server development process, a code-based or model-based approach is used to describe semantic Web services with OWL-S [10]. When a semantic Web service has a complete OWL-S description, it will be registered to the OWL-S-enabled registry so that the semantic Web service can be discovered and invoked by other applications. In the client development process, OWL-S configuration file developed by OWL-S editor describes the functions of the service, and is used to invoke the OWL-S registry. In the registry, service requests will match with semantic Web services that have been registered and get a return that satisfies the client's needs. Finally, the client can select a Web service that best meets the user's needs from the UDDI's returned results.

3 Semantic Web Service Discovery Algorithm Design

3.1 Cosine Similarity Algorithm

In the process of semantic Web service discovery, it is very important to find out the semantic Web service which can meet the user's needs in specific contexts through the specific keyword query based on the keyword of the text. The cosine similarity algorithm [11] uses the cosine of angle between two different vectors in the vector space to determine the difference on content between two vectors, it mainly base on user's own preference and the level of difference between provided Web services and determine a semantic Web service that ultimately conforms to user's context then feed the semantic Web service back to the user to meet the different needs of the user in different contexts.

In two dimensions, the angle cosine of the vector a and the vector b is calculated as follows:

$$\cos(\alpha) = \frac{a \cdot b}{\|a\| \times \|b\|}$$
$$= \frac{(x_1, y_1) \cdot (x_2, y_2)}{\sqrt{x_1^2 + y_1^2} \times \sqrt{x_2^2 + y_2^2}}$$
$$= \frac{x_1 x_2 + y_1 y_2}{\sqrt{x_1^2 + y_1^2} \times \sqrt{x_2^2 + y_2^2}}$$

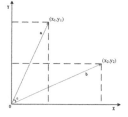

Figure 3 Semantic Web service discovery process

In the case of multidimensional, the angle cosine of the vector a and the vector b is calculated as follows:

$$\cos(\alpha) = \frac{\sum\limits_{i=1}^{n} (x_i \times y_i)}{\sqrt{\sum\limits_{i=1}^{n} (x_i)^2} \times \sqrt{\sum\limits_{i=1}^{n} (y_i)^2}}$$

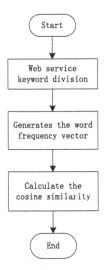

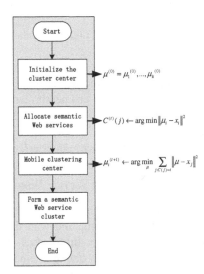

Fig. 3. Cosine similarity algorithm flow chart

Fig. 4. K-means algorithm flow chart

Since the cosine similarity algorithm focuses on the difference of vectors' directions, it is not sensitive to their size, so it is mainly used by users to determine whether the content of the Web service is interested.

3.2 K-Means Algorithm

Different data are classified into different clusters, so that objects in the same cluster have a high similarity, and the objects between different clusters are very different. This method is called clustering. K-means algorithm [12] is used to cluster the existing semantic Web services. Firstly, the first cluster center is selected randomly. After the first cluster center is selected, the selection principle of other cluster centers must guarantee a range from the selected cluster center as far as possible to avoid the existence of relatively close distance cluster center and ensure that the clustering results are defined clearly.

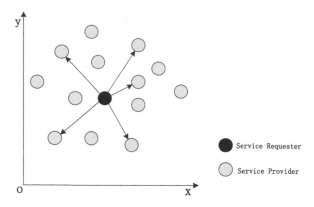

Fig. 5. Get the semantic Web service process

Figure 4 shows the K-means algorithm flow.

In this paper, the cosine similarity is used as the clustering basis, so that all the services in the semantic Web service cluster can meet the demand of service requesters in a certain situation. On this basis, the feeded semantic Web Services can meet the special needs of service requesters.

3.3 Semantic Web Service Discovery

Considering the fact that the number of semantic Web services for the sensor network is relatively small compared with the Internet semantic Web service, while the time requirement of service discovery is low [13]. In this paper, we first use the cosine similarity algorithm to obtain a semantic Web service satisfying the specific context of the service requester's needs (shown in Fig. 5), and then performs the K-means clustering operation to cluster the semantic Web services satisfying some characteristic by the cosine similarity algorithm and proceed into the database after recorded.

After obtaining description document corresponding with the input and output, we can get the corresponding parent classes, subclasses and other information through Jena and Pellet logical reasoning machine. When the service requester sends a request, it first searches whether there is a corresponding semantic Web service request record in the database. If exists, it returns the query result directly to the service requester. If it does not exist, the requested semantic Web service is parsed firstly through description document, and the semantic Web service analyzed by the document will be matched with a cluster center of minimal cosine similarity from cluster centers. Then the semantic distance between semantic Web services from cluster center and the requested services are calculated in turn. Once a semantic distance is less than a certain threshold, the semantic Web service is fed back to the service requester and the matching result is stored in the database.

4   Semantic Web Service Discovery Algorithm Design

In order to evaluate the semantic Web service matching algorithm, the OWL-TC test set contains a total of 1083 semantic Webs from nine different areas (education, health care, food, tourism, communication, economics, weapons, geography and simulation). Those services are taken from the public IBM UDDI registry and are enhanced and expanded by approximately 30 people from the German Center for Artificial Intelligence Research (DFKI), Jena University and University of Thessaloniki to achieve semi-automatic conversion from WSDL to OWL-S Features. In this paper, OWL-TC4 is used as the test version, and 1083 services provided by OWL-TC4 are tested and the results are recorded as shown in the following table.

Areas	Number of services
Education	286
Health care	73
Food	34
Tourism	197
Communication	59
Economic	395
Weapons	40
Geography	60
Simulation	16
Total	1160

All services provided by OWL-TC4 were validated respectively, resulting in 1160 results in different areas, since some of these services belonged not only to one area but to two or more domains, which conform to the domain structure of the service published by OWL-TC4. It indicates that the algorithm proposed in this paper can accurately calculate the domain of a semantic Web service, so the algorithm can feedback to service requester the Web service meets the context within the same field.

5   Conclusion

This paper considers that the semantic Web service discovery process does not take into account the context of user and cannot feed the semantic Web service that satisfies its context to the semantic Web service requester, and then presents a novel semantic Web service discovery method based on user context. This method firstly clusters the Web services provided by semantic Web service provider with the cosine similarity. Secondly, the cluster center of each service cluster is matched with the requested semantic Web service to get the best matching service cluster, and then get the semantic Web service meets users' needs from this cluster, and it is responded to the user. This matching method takes into account users in different contexts, so it is able to provide users with the service they really need.

Acknowledgements. This work is partially supported by the Chongqing Science & Technology Commission, China, with Project No.: cstc2016jcyjA0586.

References

1. Zheng, Z., Ma, H., Lyu, M.R., et al.: Collaborative web service qos prediction via neighborhood integrated matrix factorization. IEEE Trans. Serv. Comput. **6**(3), 289–299 (2013)
2. Jung, H., Yoo, S., Kim, D., et al.: A grammar based approach to introduce the semantic web to novice users. Multimed. Tools Appl. 1–14 (2015)
3. Paulraj, D., Swamynathan, S., Chandran, D., et al.: Service composition and execution plan generation of composite semantic WEB services using abductive event calculus. Comput. Intell. **32**(4), 711–737 (2016)
4. Zheng, Z., Ma, H., Lyu, M.R., et al.: Collaborative web service qos prediction via neighborhood integrated matrix factorization. IEEE Trans. Serv. Comput. **6**(3), 289–299 (2013)
5. Chen, F., Lu, C., Wu, H., et al.: A semantic similarity measure integrating multiple conceptual relationships for web service discovery. Expert Syst. Appl. **67**, 19–31 (2017)
6. Hu, K., Gui, Z., Cheng, X., et al.: Content-based discovery for web map service using support vector machine and user relevance feedback. PLoS ONE **11**(11), e0166098 (2016)
7. Rodriguez Mier, P., Pedrinaci, C., Lama, M., et al.: An integrated semantic web service discovery and composition framework. IEEE Trans. Serv. Comput. 1 (2015)
8. Ma, S.P., Chang, K.Y., Lin, J.H., et al.: QoS-aware query relaxation for service discovery with business rules. Future Gener. Comput. Syst. **60**, 1–12 (2016)
9. Angulo, P., Guzmán, C.C., Jiménez, G., et al.: A service-oriented architecture and its ICT-infrastructure.to.support.eco-efficiency.performance.monitoring in manufacturing enterprises. Int. J. Comput. Integr. Manuf. **30**(1), 202–214 (2016)
10. Mrissa, M., Sellami, M., Vettor, P.D., et al.: A decentralized mediation-as-a-service architecture for service composition. In: IEEE, International Workshop on Enabling Technologies: Infrastructure for Collaborative Enterprises, pp. 80–85. IEEE, New York (2013)
11. Nguyen, H.V., Bai, L.: Cosine similarity metric learning for face verification. In: Asian Conference on Computer Vision, pp. 709–720. Springer-Verlag, Berlin (2010)
12. Jain, A.K.: Data clustering: 50 years beyond K-means. Pattern Recognit. Lett. **31**(8), 651–666 (2010)
13. Mel, G.D., Bergamaschi, F., Pham, T.: Service-oriented reasoning architecture for resource-task assignment in sensor networks. In: Proceedings of SPIE - The International Society for Optical Engineering vol. 8047, no. 23 (2011). 80470X-80470X-12

Network Interconnection of PROFIBUS and MODBUS for Networked Control System

Qiang Wang[1(✉)] and Fang He[2]

[1] School of Mechanical Engineering, University of Jinan, Jinan, Shandong, China
me_wangq@ujn.edu.cn
[2] School of Electrical Engineering, University of Jinan, Jinan, Shandong, China

Abstract. Various of equipments and instruments with different network protocols are used in the same networked control system for integrated manufacturing automation. The communication technology of hybrid network is widely concerned. The interconnection technologies of different network protocols are discussed in this paper. In particular, the interconnection design of hybrid network between PROFIBUS and MODBUS is discussed in depth. A concrete communication design of PROFIBU-DP and MODBUS hybrid network interconnection using NT50-DP-RS intelligent gateway is given. The design has been verified by our communication experiment, and its design method has a good guiding effect on the related engineering design application.

Keywords: Network interconnection · PROFIBU-DP · MODBUS · NCS

1 Introduction

In the networked control system (NCS) of integrated manufacturing automation, it is very common that devices with different network protocols are used in the same system [1, 2]. By network interconnecting, data exchange between devices in different segment can be realized, though they have different transmission rate, different communication protocols and different network transmission medium [3].

The technology of network interconnection can be used for not only networks with same communication protocols, but also heterogeneous network composed of different network protocols. For networks with same bus type, network interconnection generally adopts bridge or repeater. For example, CAN bus and PROFIBUS bus have fast and low speed network segment. The network interconnection between their low speed segment and high speed network segment is realized using bridge. For interconnection of heterogeneous network with different protocols, intelligent network gateway often is used [4].

Except introduction and reference, the principle of network interconnection is analyzed in Sect. 2. Section 3 gives the structure of PROFIBUS and MODBUS hybrid NCS, including PROFIBUS NCS with MODBUS devices and MODBUS NCS with PROFIBUS devices. In Sect. 4, the NT50-DP-RS intelligent gateway is used for interconnection design of PROFIBU-DP NCS with MODBUS device, in which a kind of temperature detection device with MODBUS communication protocol is taken as the experimental device. Finally, some conclusions are stated in Sect. 5.

© Springer International Publishing AG 2018
F. Qiao et al. (eds.), *Recent Developments in Mechatronics and Intelligent Robotics*,
Advances in Intelligent Systems and Computing 691, DOI 10.1007/978-3-319-70990-1_22

2   The Principle of Interconnection of Different Network Protocols

There are different devices with different network protocols in industrial NCS. They are not compatible with each other and can not directly be connected to realize data communications. This problem can be solved by protocol conversion to realize data transmission between different network protocols devices.

Figure 1 shows the interconnection conversion principle of different protocol network. For different protocol networks, their parameters are different such as data transmission rate, size of the data stream, type of transmission medium, connection mode with communication line, transmission distance and number of nodes, etc. According to the reference model of fieldbus, different protocol network is defined itself data link layer. The principle of protocol conversion is to match two different network protocols, so that the data transmitted by the industrial network can be recognized in different protocols network. Using the analytical mechanism of each protocol, we can restore the data from different protocol device, receive information from one protocol, and then convert the information to another network protocol.

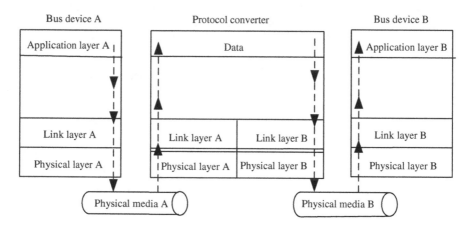

Fig. 1. Interconnection conversion principle of different protocol network

3   Hybrid NCS of PROFIBUS and MODBUS

3.1   PROFIBUS NCS with MODBUS Devices

Many traditional instruments and equipments in plant are based on MODBUS communication protocol without PROFIBUS communication ability. In this case, the intelligent gateway is needed to realize the interaction and data exchange access for PROFIBUS NCS. Figure 2 shows the structure of PROFIBUS NCS with MODBUS devices including: temperature transmitter, liquid level sensor and so on.

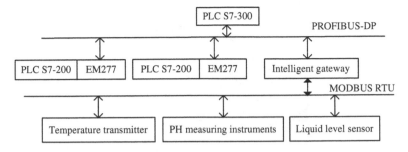

Fig. 2. Structure of PROFIBUS NCS with MODBUS devices

3.2 MODBUS NCS with PROFIBUS Devices

Sometime, it is also necessary to access the PROFIBUS bus devices into MODBUS NCS in the industrial field. For example, a certain temperature monitoring network control system is that their several slave stations are VLT6000 series inverters with PROFIBUS communication interface and their master station is EC20 series PLC based on the MODBUS communication protocol. The structure of this NCS is shown in Fig. 3.

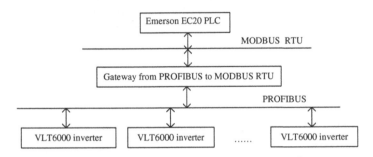

Fig. 3. Structure of MODBUS NCS with PROFIBUS devices

4 Design of Hybrid NCS of PROFIBUS-DP with MODBUS Device Based NT50-DP-RS Gateway

4.1 Establishment of Communication Experiment System

Communication experiment of our hybrid network interconnection uses Hilscher NT50-DP-RS intelligent gateway, which is an intelligent gateway realizing exchange protocols and data transmission between PROFIBUS and MODBUS each other. Its structure is shown in Fig. 4.

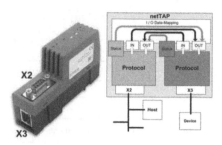

Fig. 4. Structure of NT50-DP-RS

Figure 5 shows the hybrid network structure of communication experiment system with PROFIBUS-DP and MODBUS RTU. NT50-DP-RS Gateway is used. A kind of temperature transmitter with MODBUS RTU protocol is used as MODBUS device.

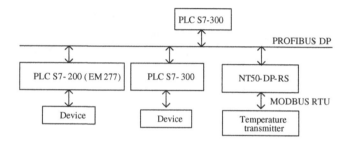

Fig. 5. The hybrid network structure of PROFIBUS-DP and MODBUS RTU

Table 1 shows some parameters of this temperature transmitter which should be particularly concerned for parameter setting of NT50-DP-RS intelligent gateway.

Table 1. Parameters of temperature transmitter

Code	Interface type	Parity	Stop bits	Baud rate	Error checking
Binary, 8 bits	Rs. 485	None	1	4800 bit/s	CRC

4.2 Configuration of Network and Setting of Communication Parameters

(1) Configuration of PROFIBUS-DP NCS

In this experiment system, PLC S7-300 is selected as the master station, which CPU module is CPU314C-2DP. Using STEP7 software, a new project file can be set up. Selecting the "Properties" option in the "General" tab, the address of master station and its main parameters are set.

The GSD file of NT50-DP-RS intelligent gateway must be installed in STEP7 software firstly. After successfully importing the GSD file, the NT50-DP-RS intelligent gateway is configured as a slave station of the PROFIBUS-DP network.

(2) Setting of NT50-DP-RS

Before communicate, IP address of the NT50-DP-RS gateway must manually be set the through the Ethernet device setup software. It gets ready to download the configuration file through the SYCON.net software. Only this IP address is set on the same network segment as the gateway computer's IP address, SYCON.net software can detect the intelligent gateway.

After completing the IP address setting of intelligent gateway, using the SYCON.net software, the firmware of protocol convert from PROFIBUS-DP to MODBUS is downloaded into the gateway.

Using the SYCON.net software, the MODBUS RTU device (temperature transmitter) is set the gateway as the master station.

According to Table 1, serial port parameters of the gateway are set such as: type of serial port, baud rate, parity, etc. It is very important to set these parameters consistent with parameters of temperature transmitter. Figure 6 shows parameters setting of serial port of NT50-DP-RS intelligent gateway.

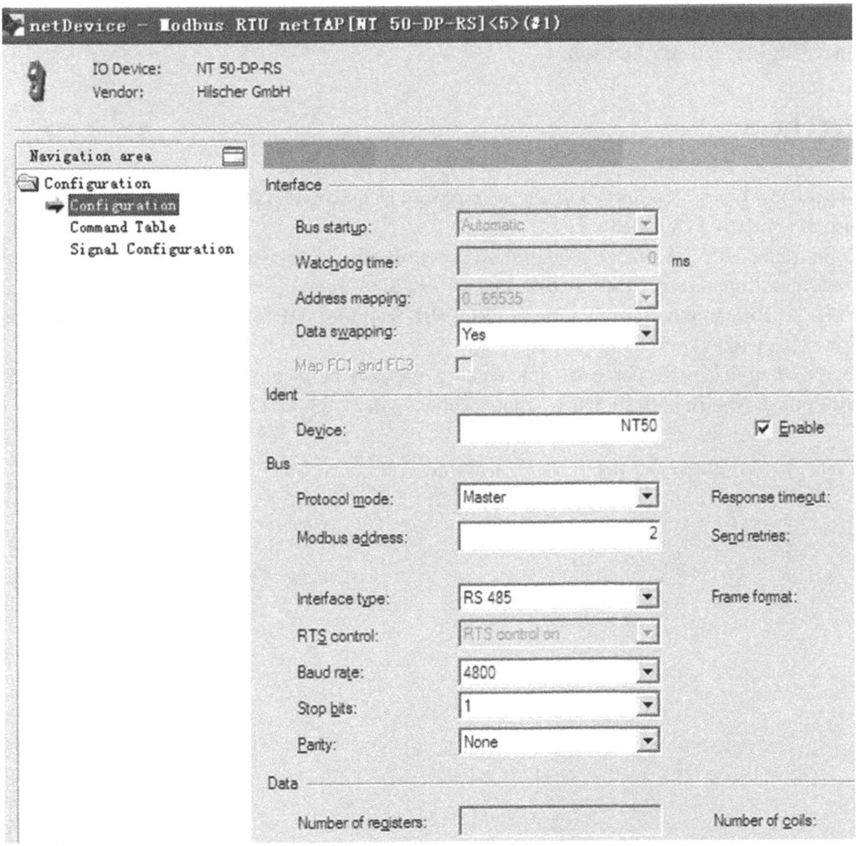

Fig. 6. Parameters setting of serial port of NT50-DP-RS intelligent gateway

The data exchange between PROFIBUS-DP with MODBUS for their interconnection communication is accomplished by a "mapping" relationship. Using SYCON.net software, the data mapping of the gateway and temperature transmitter is completed. During communication, the read command of MODBUS writes the data into the input buffer of the PROFIBUS-DP network, and then PROFIBUS-DP network can read it.

5 Conclusion

The technology of network interconnection design of hybrid network between PROFIBUS and MODBUS is discussed in this paper. Configuration of hybrid NCS of PROFIBUS-DP with MODBUS device based NT50-DP-RS intelligent gateway is presented in detail. After finishing configuration of the SIEMENS PLC S7-300 master station and intelligent gateway slave station, PLC data transmission program is designed and edited using STEP7 software, communication state monitoring program can be designed using WinCC configuration software.

The network interconnection design above has been verified by our communication experiment. Its communication design method has a good guiding effect on the related engineering design application.

References

1. Keyou, Y., Lihua, X.: Survey of recent progress in networked control systems. Acta Autom. Sin. **39**(2), 101–118 (2013)
2. Camarnha-Matos, L.M.: Collaborative networked organizations: status and trends in manufacturing. Annu. Rev. Control **33**(2), 199–208 (2009)
3. Pang, Z.: Research and application of PROFIBUS and MODBUS protocol conversion. Chem. Eng. Mach. **42**(1), 151–153 (2015)
4. Li, N., Fang, Y.: Research on interlinking method between PROFIBUS and MODBUS protocol. J. Nanjing Norm. Univ. (Eng. Technol.) **4**(3), 23–26 (2004)

Research on 3D Technology Based on Power Continuation

Guo Zhiqiang[1,3(✉)], Zhou Yiren[1,3], Zhou Chao[2,3], Li Kai[1,3], Wu Yunqiang[1,3], and Wu Ting[1,3]

[1] Nanchang Institute of Technical, Nanchang 330044, China
1053537388@qq.com
[2] Jiangxi Province Energy Conservation and Environmental Protection Investment Co., LTD, Nanchang, China
[3] Nanchang Aeronautical University, Nanchang 330063, China

Abstract. In this paper, we study a 3D technology with power-off function, which includes energy storage power module, power-off detection module of power supply, power supply and power-off switching control module. When the power is turned off, the X, Y, Z coordinates are recorded by the displacement measurement module, and when the power supply is restored, the new g code is generated into the 3D printer, and the last work is continued, Thus avoiding the re-printing, effectively save supplies, time, which greatly improve the printing success rate, and then with the displacement detection module to achieve closed-loop control, and further improve the power to continue to print the printing accuracy.

Keywords: Rapid prototyping · Energy storage · Displacement detection · Data record

1 Background Introduction

FDM Rapid Prototyping Technology (Fused Deposition Modeling), also known as melt deposition type, its working principle is the first through the 3D modeling software for three-dimensional model design such as: 3Dmax, UG, pro will be designed a good three-dimensional model through the slice software: Such as cure, kisslicer, slic3r and so on, the resulting G code data into the 3D printer, and then print the wire (PLA, ABS, etc.) heated to the molten state, by feeding the material from the nozzle at the molten state of the wire into a filamentous, and finally in accordance with the section of the cross-sectional trajectory layer by layer to print until the completion of a three-dimensional objects.

Compared with the traditional process technology, the use of FDM rapid prototyping technology of the 3D printer in the processing costs, the complexity of forming parts and product testing and other advantages obvious, in the mechanical, civil engineering, military, medical and other aspects have also been widely used, has a good The development prospects and great business value. As FDM rapid prototyping technology is a multidisciplinary integration of a comprehensive technology, covering areas such as

F. Qiao et al. (eds.), *Recent Developments in Mechatronics and Intelligent Robotics*,
Advances in Intelligent Systems and Computing 691, DOI 10.1007/978-3-319-70990-1_23

machinery, electronics, materials, software, so its printing process in the problems encountered is also complex.

In the 3D printing rapid prototyping process, will inevitably encounter power outages, when printing large models, if you encounter a power outage, then the front of the semi-finished products, all void, when the call, only to re-print! While the conventional 3D printer are not cut off the function of repeated, re-print not only a waste of 3D printing materials, and very waste of time.

The main content of this paper is for the 3D printer in the work of a power outage protection technology, its function is to save power when the 3D print storage power before the work data, to avoid loss of data, when the resumption of power supply, Can continue to power off the print before the task, to avoid the waste of all aspects of resources.

2   Design Program

(1) Install the power storage buffer module in the control circuit board, as shown in Fig. 1.

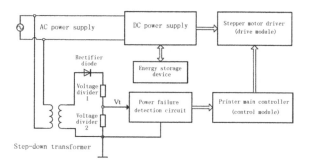

Fig. 1. Overall block diagram

① power off the moment before the code to save the code at the fault, to continue to continue to print the code editor.

② can be achieved after the power of the machine reset to prevent the nozzle after a long time to stay in the printing material, the waste heat on the nozzle will be attached to the adhesion of the print together. When the nozzle temperature is cooled, it will be tightly integrated with the print to avoid mechanical damage to the print and the instrument in the recovery circuit, as shown in Fig. 1.

The power supply module includes a DC power supply and an energy storage device. Wherein the DC power supply supplies the alternating current of the external power supply into the DC power supply drive module; the energy storage device is used for the switching between the power-off state and the power supply state under the action of the control module, and the power supply is stored when the circuit is in the normal state; When the power outage state, to the drive module, the control module power supply for memory preservation and machine reset. The energy storage device is a battery or a capacitor (Fig. 2).

Fig. 2. Damage of the print head to the print

"Stop and Save" the principle of the design: When you pause the print job, the design will be at this time the printhead coordinates (X, Y, Z) stored in the SD card. When the user wants to resume the print job, he can start printing at the location where it leaves. This data file, which is automatically generated by the design, is named "resurrection.g".

A buffer battery designed to prevent accidental power failure, which uses a resistor divider to read the current voltage and program, which can command the 3D control circuit (Arduino) to immediately perform the "Stop and Save" function. The design also includes a diode to ensure that energy can only be used in Arduino. "If the voltage suddenly becomes zero, the engine can not use the energy stored in the diode-protected capacitor, which acts as a valve to prevent the current from flowing in the opposite direction, so the energy stored in the small capacitor is sufficient When necessary, support Arduino."

(2) Code editing

The establishment of the 3D model generated by the slice software g code, the code into the printer control software in the machine will be implemented in layers according to the code, power recovery is normal when the preservation of the code through the code editor, and then start the code at the beginning of the code carried out.

First Z-axis nozzle to raise, if you want to drop the table, then the relative increase in the nozzle so the input value is positive value, if the input negative, the nozzle will be directly collided with the model, there will be collision machine!

After the nozzle leaves the model, the X-axis and Y-axis can be reset to return to the origin, because in the re-execution of the program in the homing point can call less inertia caused by the error, after the reset with the utility knife to 3D Print the model because of the power outage and scars to be removed, be light, be sure to make sure that the model shift does not move any, after clearing the scars, we open the SD card (this model is printing), find the model and then Use code editing software to open the code file, here with you recommend a code text editing tools: Ultra Edit, you can also use free open source editing software: Notepad++ and so on, of course, you can use the computer itself with the text editing software.

In the process of printing the machine will be carried out in accordance with the code layer, when the stop appears when the code will stop the progress of the current layer, the command will contain X, Y, Z coordinates. The printer stops the layer code as shown in Fig. 3.

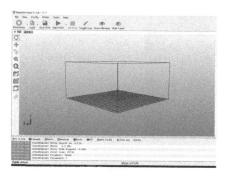

Fig. 3. Stop Code Layer Display

Copy the last line of the traffic in the order of the order G1 X98.059 Y99.196 E14.41442, open the code editing software to edit, find and then this line of code above the second G92 E0 code all deleted, because the front The code has been executed (Fig. 4).

Fig. 4. G code to find and edit

Modify the first layer of code to re-implementation, will not delete the G92 E0 to G92 EXXX, XXX from the next line to copy, add a line "G1 ZXX" XX for the height of Z when the stop, then use the vernier caliper measurement actual print the height of the assignment.

We can use the vernier caliper, depth ruler, height gauge and other measurement tools to measure the value (measure the time to try to clamp, but do not over, moderate efforts. Because the measurement height is too large, in the latter part of the time will be printed when the compartment, and can not be bonded when the second print). Next to modify it, modify the X value, Y value and Z value, X value and Y value set Into the value of 0, of course, this value can not be set, the most important is the Z value, be sure to set, set to be larger than we measured the number of fans and open the code.

After completing the above settings, the edited program slice file is downloaded to the 3D printer to start printing, one hand on the model, one hand touch on the 3D printer power switch, to prevent the data error so that the nozzle and model collision! If the nozzle directly hit the model, and directly turn off the 3D printer power! Or empty in the top of the model, indicating that your program is deleted by you or less deleted, we must return to the computer to re-modify the correct value, the code to run (Fig. 5).

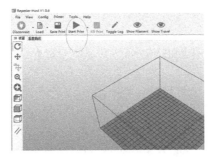

Fig. 5. Task re-run

During the operation of a careful observation, when the implementation of the three-tier order after the problem did not occur, this time has completed the entire power to continue playing technology. Figure 6 shows the modified print.

Fig. 6. Print the work

3 Conclusion

It is possible to record the state of the printer by using the power stored in the power module when the power is off, and the printer can use the power storage module to detect whether the external power supply is normally powered by the power failure detection module. Restore the power supply, combined with the code editing process and then start printing at the fault, as shown in Fig. 7, so as to avoid re-printing, saving printing supplies, a substantial increase in print success rate.

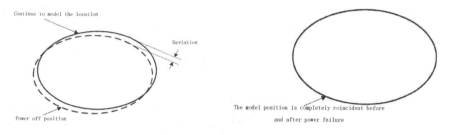

Fig. 7. Breakdown of power off

Based on the power of the continued 3D technology to solve the normal operation of the 3D printer, even if the external power supply accidentally interrupted, the 3D printer can also restore power after the power supply to continue after the power of the print job tasks, the perfect model print, Thus saving time for printing models, saving materials, improving efficiency, and extending machine life. Power-off technology has a simple structure, simple control, easy to promote the characteristics.

Acknowledgements. The work was supported by Jiangxi Provincial Education Foundation (GJJ16115).

References

1. Qi, X., Li, H., Qi, Z.: Summary of voltage equalization technology of electric double layer capacitor. High Volt. Technol. 02 (2008)
2. Li, H., Qi, Z., Feng, Z.: Voltage balance strategy for supercapacitor power storage system. Power Grid Technol. 03 (2007)
3. Niu, X.: Research on desktop 3D printer structure analysis and body design. Design 04 (2015)
4. Chen, L.: 3D printing technology: to add vitality to the classroom innovation. Manag. Prim. Second. Sch. 04 (2015)
5. Li, X., Ma, J., Li, P., Chen, Q., Zhou, W.: 3D printing technology and application trends. Autom. Instrum. 01 (2014)
6. Wu, Z.: Study on the influence of FDM process parameters on the dimensional precision of molding products. J. Chifeng Univ. (Nat. Sci. Edn.). 01 (2015)
7. Li, X., Li, Z., Li, L.: Accuracy analysis and study of melt deposition molding process. Manuf. Technol. Mach. Tools. 09 (2014)
8. Wang, H.: Research status and development trend of material production. J. Beijing Inf. Technol. Univ. (Nat. Sci. Edn.). 03 (2014)
9. Chen, X., McMains, S.: Polygon offsetting by computing winding numbers. In: ASME 2005 International Design Engineering Technical Conferences & Computers and Information in Engineering Conference (2005)
10. EVANS B. Practical 3D printers: the science and art of 3D printing (2012)

Research on Electric Power Steering Test System of Vehicle Steering Device Based on Industrial Personal Computer

Ying Wang[1(✉)], Xiuqing Mu[2], Shaochun Zhai[3], Fengyang Gao[1], Xiaoqiang Chen[1], Qiyao Li[1], Jiarong Wang[1], Shiwen Liang[1], and Wei Yan[1]

[1] Key Laboratory of Opto-Technology and Intelligent Control, School of Automation and Electrical Engineering, Lanzhou Jiaotong University, Lanzhou, China
wangying01@mail.lzjtu.cn
[2] School of Electrical Engineering, Southwest Jiaotong University, Chengdu, China
[3] Guangzhou Mechanical Engineering Research Institute, Guangzhou, China

Abstract. Electric power steering (EPS) is a new technical application on steering system. And EPS is one superior technique which supply power to steering system of automobile with motor directly. Based on industrial personal computer (IPC), this paper introduces the hardware structure and operating principle of EPS test system. The control software for the system is developed using visual basic language. EPS test system adapts IPC as the main control system, combined with data acquisition card, motion control card and timer/counter card for automatic control and data acquisition. Also, the EPS system tests are carried out and the corresponding performance parameters are determined in this paper. The result reveals that EPS test system is capable of carrying out accurate EPS performance testing.

Keywords: Vehicle steering device · Electric power steering (EPS) · Test system · Industrial personal computer (IPC)

1 Introduction

Turning performance is very important for all vehicles. In recent years, more and more vehicles have been equipped with electric power steering system (EPS) [1] for its energy saving, environmental protection, active safety and good manipulation stability. EPS industry is a more popular and important research value and great potential application prospects.

The EPS system was first applied in Japan [2]. Since then, the electric power steering technology has been rapid development [3, 4]. So far, the EPS possesses various distinct advantages over traditional steering devices, which is popular amongst steering industry since its existence. Some domestic scholars and research institutions also got some research results of EPS [5–7]. In addition, in view of the EPS test ways, in [8] a PID closed-loop feedback control was designed for monitoring the boost motor output current. The servo motor and the DMA data acquisition methods were applied for the EPS test [9]. Also, in [10] a combined inertial compensation current and damping

© Springer International Publishing AG 2018
F. Qiao et al. (eds.), *Recent Developments in Mechatronics and Intelligent Robotics*,
Advances in Intelligent Systems and Computing 691, DOI 10.1007/978-3-319-70990-1_24

compensation current compound control strategy were developed for electric power steering system.

Generally, the performance of vehicle steering device is closely related to the safety of steering operation. EPS test system is a tool for testing the performance of EPS. Under this background, the EPS performance and core components of controller need to be tested, so it is very necessary to establish the performance test bed of electric power steering.

The outline of this paper is as follows. The structure and operation principle of EPS testing system are theoretically described in Sect. 2. The peripheral circuit of digital and pulse output signal is selected and designed in Sect. 3. Section 4 introduced the test program using Visual Basic language with modularized structure. The EPS system tests are carried out and the corresponding performance parameters are determined in Sect. 5. Section 6 draws a conclusion of this paper and suggests some further improvements.

2   Structure and Operating Principle of EPS Test System

IPC is designated for in-site operation in industry, which is characterized in resisting magnetism, dust, impact and interference. Industrial personal computer (IPC) was adapted as the main control system for developing EPS test system. The operating principle of the system is: test data is acquired using PCL-818HD multifunctional data acquisition card; steering resistance subjected to EPS is simulated by magnetic brake and controlled by a D/A channel on the board; digital value signal is controlled by the integrated 32 digits I/O channel on the board; the manipulation of driver on steering is simulating using servomotor, whose motion is controlled by motion control card.

Other test conditions required during EPS testing are vehicle speed signal and engine rotational speed signal. Vehicle speed signal is obtained from speed transducer on headstock or electronic panel, while engine rotational speed signal is obtained from ignition coil. Both signals are pulse signals with voltages above 10 V. If a device similar to headstock or ignition coil were used to generate these two pulse signals, the structure of the test system would become much more complicated. For the sake of convenience and accuracy, PCL-836 was used to generate the required signals, for its ability to generate desired frequency and pulse signals via programming.

When EPS reaches operating conditions, start collecting test data through data acquisition system consisting of IPC, data acquisition card and transducers, which is analyzed to obtain test results. The structure of EPS test system is shown in Fig. 1.

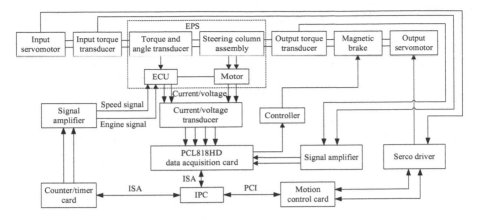

Fig. 1. The structure diagram of EPS testing system

3 Peripheral Circuit Design for Digital and Pulse Output Signal

The high voltage of digital quantity output channel on the data acquisition card is 5 V, and the channels are used to control the 12 V EPS ignition switch signal, meanwhile some of the channels are used to control magnetic brake via the 220 V AC power supply and a motor. Due to the limited electric power of the computer and the 5 V output voltage of the data acquisition, peripheral circuit is required to satisfy the demand on control. BTS621L1 chip was selected to amplify the control signal from the digital quantity output of data acquisition card. The first digital quantity output port selected consists of 8 digits. The peripheral circuit is shown in Fig. 2.

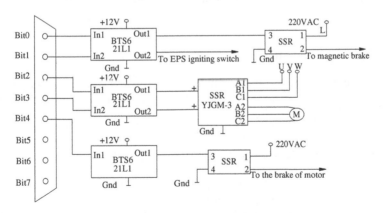

Fig. 2. The diagram of peripheral circuit

The pulse signals of speed and engine rotational speed that EPS could receive is approximately 12 V, obviously the 5 V output from PCL-836 is out of useful range. Similarly, BTS621L1 was used to amplify the 5 V signal into 12 V. The two pulse output ports of PCL-836 were connected to IN1 and IN2 of BTS621L1 correspondingly, while

the +Vbb port of BTS621L1 connected to the 12 V power supply, and the OUT1 and
OUT2 ports connected to EPS input ports for speed and engine signals correspondingly.
EPS is then able to work in various speeds and engine rotational speeds.

4   Programmed Software Design

The testing system was programmed using Visual Basic language with modularized
structure. The program is mainly composed of initializing module, motion control
module, data acquisition and processing module, data and curve display module and
database. The modularized structure is as shown in Fig. 3.

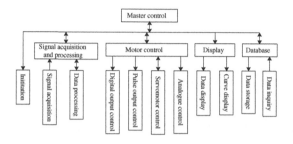

Fig. 3. The programmed structure diagram of EPS test system

4.1 Initializing Module

On entering EPS test system, the software will start initialization. The process mainly
involves initializing the data acquisition card, the motion control card and counter/timer
card installed in IPC, ensure the normal operation of the hardware.

4.2 Motion Control Module

The main function of motion control module is to control the motion control card in IPC,
the digital quantity channel in data acquisition card and the pulse output in timer/counter
card.

4.3 Signal Acquisition and Processing Module

The parameters required are: input torque, output torque, EPS current and rotating angle.
Among these, torque and current are measured via transducers and exported as analogue,
which is read by computer via data acquisition card. Since the signal from transducer is
analogue, so the test data is processed through filtering to obtain true values.

4.4   The Display Module for Test Data and Curve

After analyzing and processing, test data is presented in the forms of data and curves, and displayed on the screen in real time. This is achieved through the display module.

4.5   Database Module

Database module is used for the storage and inquiry of test data, through which test data collected from each measuring point can be stored, and historical data can be quoted.

5   EPS Test Results

In order to validate whether or not the EPS test system developed meets design requirements, an EPS testing was carried out using the system and each performance parameters was determined. Through comparing the standard parameters provided by the manufacturer, it is concluded that the EPS test system developed is capable of determining the performance of EPS with high test precision.

Figure 4 is the curve gained for EPS input–output performance testing. The maximum torque to the right was measured to be 24.652 N.m, the maximum torque to the left was 24.637 N.m, and the symmetry of the curve was 96.7%.

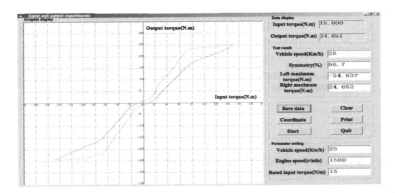

Fig. 4. The input–output characteristic test

Also, Fig. 5 shows the EPS current characteristic testing result. The maximum current to the right was measured to be 18.21A, the maximum current to the left was 17.91A, and the symmetry of the curve was 97.7%.

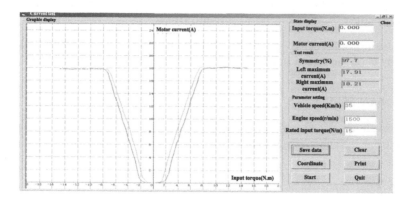

Fig. 5. The current characteristic test

Through carrying out EPS testing, the test system developed is able to determine the parameters of an EPS, with real time display function for test data and curve.

6 Conclusion

In this paper, the structure and operation principle of EPS testing system are theoretically analyzed. EPS test system adapts IPC as the main control system, equipped with multifunctional data acquisition card, motion control card and timer/counter card for collecting test data and controlling test status. The software used for test system adapts modularized design. In this paper, the TPS test platform is user-friend, easy to manipulate, and the system function is easily scalable, highly precise and more stable. Also, the EPS test system developed is able to satisfy demand on EPS testing.

References

1. Han, J.G., Shen, R.W., Tai, X.H., He, Z.G.: Research on assist characteristic of electric power steering system based on adaptive neuro-fuzzy control. Appl. Mech. Mater. **347**(1), 357–361 (2013)
2. Nakayamat, S.: The present and future of electric power steering. Int. J. Veh. Des. **15**(3), 243–254 (1994)
3. Osuka, A., Matsuoka, Y., Tsutsu, T., Obata, Y., Droulers, A.: Development of pinion-assist type electric power steering system. KOYO Eng. J. **1**(161), 46–51 (2002)
4. Takahashi, T., Suzuki, H., Nakayama, T.: Application technology regarding high-powered electric power steering system. JTKET Eng. J. **1**(1006), 47–52 (2009)
5. Liu, J., Chen, W.W.: A study on electric power steering control based on rapid control prototype. Automot. Eng. **31**(7), 634–639 (2009)
6. Wang, W., Yun, H.: Parametric study and optimal design on electric power steering systems. Chin. J. Constr. Mach. **12**(1), 28–33 (2014)
7. Zhang, H., Zhang, J.W., Guo, K.H., Li, Y.: Robust prediction current control of permanent magnet synchronous motor for EPS based disturbance observer. J. Jilin Univ. **45**(3), 711–718 (2015)

8. Sun, T., Du, P.P., Zheng, S.L., Zhang, Z.D.: Multi-disciplinary optimization for an electric power steering system and evaluation of vehicle handling stability. J. Highway Transp. Res. Dev. **33**(7), 142–152 (2016)
9. Zhai, S.C., Min, X.H., Huang, Z.J., Ta, Z.Y.: Research and development on testing and control system for performance test-bed of electric power steering. Mach. Tool Hydraul. **36**(12), 141–143 (2008)
10. Xiang, D., Chi, Y.B., Li, W.B., Yang, Y.: Study on control strategy and simulation for electric power steering system. Control Eng. China. **12**(2), 254–258 (2013)

Research on Fault Diagnosis of Power Transformer Equipment Based on KNN Algorithm

Huadong Yu[1(✉)], Qing Wu[1], Yongling Lu[2],
Chengbo Hu[2], Yubo Wang[3,4], and Guohua Liu[3,4]

[1] State Grid Information and Communication Industry Group, Beijing 100031, China
453682740@qq.com
[2] State Grid Jiangsu Electric Power Company Research Institute, Nanjing 211103, China
[3] State Grid Key Laboratory of Power Industrial Chip Design and Analysis Technology,
Beijing Smart-Chip Microelectronics Technology Co., Ltd., Beijing 100192, China
[4] Beijing Engineering Research Center of High-Reliability IC with Power Industrial Grade,
Beijing Smart-Chip Microelectronics Technology Co., Ltd., Beijing 100192, China

Abstract. Power equipment is of great importance to the power of things, the operation of equipment directly impact on the safety and stability of the power of things. Therefore, it is important to ensure secure and stable operation of the power grid by prejudging monitoring, observing and fault diagnosis of power equipment. A fault diagnosis method of power equipment based on the technology of power internet of things and KNN algorithm is proposed in this paper. Through the real-time sensing and data acquisition of the basic equipment status in power network operation, the equipment status is analyzed and prejudged, which can provide alarm signals of the equipment to avoid greater security risk.

Keywords: Internet of things · KNN · Security · Fault diagnosis · Transformer

1 Introduction

Electricity is the lifeline of national economic and supporting the national economy. It is not only one of the most important basic industries but also the basis of national energy security. In general, the power system is generated by the power generation, transmission, substation and sale. Among them, the transmission and substation in power system plays an extremely important role. Power transmission equipment, especially large-scale high-voltage equipment will cause huge economic losses and bad social impact when they generate faults. Therefore, the reliability of power transmission equipment is a basic guarantee of the security and stability of a power system.

With the comprehensive construction of smart grid, Internet of Things technology are widely used in the production segments. The application of Internet of Things to smart grid is inevitable when the information and communication technology develop to a certain degree. It will effectively integrate the communication infrastructure resources and power system infrastructure resources. It can improve the level of power system information and the existing infrastructure utilization and efficiency of power systems. It also can provide important technical support for the power grid [1]. Through

© Springer International Publishing AG 2018
F. Qiao et al. (eds.), *Recent Developments in Mechatronics and Intelligent Robotics*,
Advances in Intelligent Systems and Computing 691, DOI 10.1007/978-3-319-70990-1_25

the use of the power system and its protective equipment, we can diagnosis the fault of power transmission equipment using extensive knowledge, or using relay protection means and other information to identify the failure of the component location and type. The fault diagnosis of power transmission equipment has important practical significance. With the continuous expansion of the scale and the increasingly complex structure of the power system, a large number of incorrect information will reach the dispatch center in a short period of time. They are beyond the operational capacity of the operator and cause dispatcher misjudgment. In order to adapt to various accident case of fault quickly and accurately identify, the power fault diagnosis system for decision-making is needed [2]. At the same time, due to the continuous improvement of the power system dispatching automation level, more and more alarm information is transmitted to the power grid dispatching center through the remote terminal device (RTU) of each substation, so it is possible to solve these problems. At present, there are many techniques and methods for fault diagnosis of power transmission equipment based on the expert system of fault diagnosis, which uses expert reasoning method to solve these problems [3].

However, most of the current fault diagnosis methods based on the information obtained must be completely correct. It is needed to give a fault diagnosing method of power equipment with incomplete information.

2 An Improved KNN Algorithm Based on Nearest Neighbor Metric Function

K-Nearest Neighbor (KNN) is a basic classification and regression method proposed by Cove and Hart. The input of the k nearest neighbor method is the eigenvector of the instance, and the output is the result of instance corresponding to the point of the feature space.

2.1 Nearest Neighbor Metric Function

Information Offset: for data set T, the collection of the system is $\psi = \{c_1, c_2, \cdots, c_r\}$. There are r classes in this collection. The eigenvector X has m attribute $\{v_1, v_2, \cdots, v_m\}$. The occurrences number of the attribute v_i in T is denoted by $|v_i|$ and the number of instances belonging to the jth class is denoted by $|v_{ij}|$, we note that the information deviation of the attribute $|v_i|$ is

$$T(v_i) = -\sum_{j=1}^{n} P_{ij}\ln(P_{ij}) \tag{1}$$

Where $P_{ij} = \dfrac{|v_{ij}|}{|v_i|}$ is the probability that the sample with the attribute value v_i belongs to class c_i on V. $T(v_i) = 0$ when $|v_i| = |v_{ij}|$.

Let X and Y have two samples, both of which have the same attributes $v_1, v_2, \cdots, v_n$. The distance metric for X and Y is defined as follows:

$$D(Y, X) = \frac{1}{n} \sum_{i=1}^{n} T(v_i) \tag{2}$$

This kind of distance measurement is considered in this classifying method. The average of distance deviation from the traditional KNN method to the degree of information deviation of multiple attribute characteristics is defined.

3   Fault Detection Method of Power Transmission and Transformation Equipment

The fault occurrence of power transmission equipment has complicated causal relationship with various state information perceived by electricity object network, which constitutes the information source of fault diagnosis.

Application of KNN algorithm for fault detection requires the following steps:

a. **The preprocessing of monitoring information:** The information is include of the values related to color, sound, temperature, odor, vibration and so on. Before using these information, these values must be quantized. For example, the values from 1 to 5 express the changing degrees of state information from the equipment.

b. **The construction of characteristic matrix:** According to the results of various status, we can determine the fault type T according to the specific fault characteristics. Therefore, you can get the whole fault vector X from all the fault feature vectors.

c. **The analyzing of classification:** The training is done with using the much data in KNN algorithm. So the classifying result of data can be gotten.

d. **The fault precaution of equipment:** According to the classifying result, the characteristics of data are analyzed and judged. So the fault precaution information to the equipment can be given.

4   An Example for Fault Diagnosis of Power Transformer

The role of transformer infrared on-line monitoring system in the power system is to promote power production automation (Fig. 1 and Table 1).

Table 1. Gas content in normal transformers

Gas composition	H_2	CO	CO_2	CH_4	C_2H_2	C_2H_4	C_2H_6
Normal limit value μL/L	100	45	35	55	5	10	15

We use gas sensor to collect the gas in the transformer, and get the data shown in Fig. 2. The gas content for each fault type is shown in the Fig. 2.

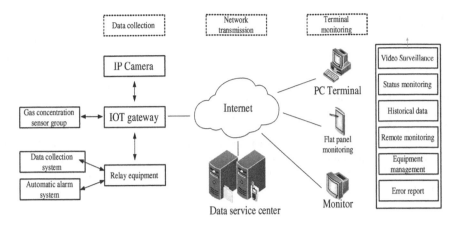

Fig. 1. Power network transformer diagnostic system topology

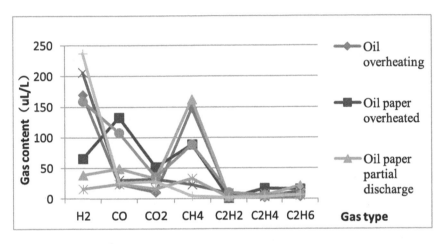

Fig. 2. Comparison of the type of failure and gas content

5   Conclusion

The essence of fault diagnosis in power transmission and transformation equipment is the comprehensive use of effective methods and comprehensive information on the equipment to conduct a comprehensive status conclusion. A fault diagnosis method of power equipment based on the technology of power internet of things and KNN algorithm is proposed in this paper. The feasibility and correctness of this method is shown by experiment.

Acknowledgements. This work is supported by the Science and Technology Research Project of State Grid Corporation of China (526816160024).

References

1. Mengchun, J., Jiye, W., Zhihua, C., et al.: Research and application of information model of power object network. Power Syst. Technol. **38**(2), 532–537 (2014)
2. Ye, H., Tao, Z., Yufei, W., et al.: Study on location privacy protection of wireless sensor network under power network. Comput. Digit. Eng. **40**(10), 91–94 (2012)
3. Ou, H., Zeng, L., Li, X., et al.: Electric power network overview and development status. Digit. Commun. **39**(5), 62–64 (2012)
4. Zhang, H.: Study on the application of diagnostic technology based on expert system in power system condition monitoring. Huazhong University of Science and Technology (2003)
5. Lingyun, W.: Fault diagnosis expert system based on neural network. Mod. Electron. Technol. **1**, 41–43 (2003)

Research on Identification Method of Transformer Based on Improved Hu Invariant Moments

Hua-dong Yu[1](✉), Qing-hai Ou[1], Qing Wu[1], Zhe Zhang[1], Wen-jing Li[1], Yubo Wang[2,3], Guohua Liu[2,3], Yongling Lu[4], Chengbo Hu[4], Na Liu[5], and Rong Wang[5]

[1] State Grid Information and Telecommunication Group Co., Ltd., Beijing, China
1091006271@qq.com
[2] State Grid Key Laboratory of Power Industrial Chip Design and Analysis Technology, Beijing Smart-Chip Microelectronics Technology Co., Ltd., Beijing 100192, China
[3] Beijing Engineering Research Center of High-Reliability IC with Power Industrial Grade, Beijing Smart-Chip Microelectronics Technology Co., Ltd., Beijing 100192, China
[4] State Grid Jiangsu Electric Power Company Research Institute, Nanjing 21110, China
[5] Control and Computer Engineering College, North China Electric Power University, NCEPU, Beijing, China

Abstract. With the rapid development of electric power industry and the expansion of power grid scale, more and more attentions have been paid to the safe operation of electrical equipment. The target recognition method we used now couldn't recognize power transformer well, which is based on gray information. In this paper, a transformer identification algorithm based on improved Hu moment invariants is proposed. The transformer identification algorithm proposed in this paper has the advantages of good recognition effect and high recognition accuracy. The correctness and feasibility of the proposed algorithm have been verified by experiments.

Keywords: Power transformer · Hu invariant moments template matching · Image recognition

1 Introduction

With the continuous progress of human society and the increasing development of economy, there is a greater demand for industrial and agricultural use of electricity, thus contributing to the rapid development of power industry and power-supply network. So the security and reliability of power supply are required more heavily than before, while the safe operation of power equipment will affect the security and stability of power system directly [1].

The pattern recognition system can be divided into two directions: pattern classification and pattern matching. Correspondingly, the target recognition system of images will be composed of two processes [2]. With the similarity of pattern recognition, the methods for image recognition is various too [3], image recognition based on template matching, and identification method using artificial neural network [4].

© Springer International Publishing AG 2018
F. Qiao et al. (eds.), *Recent Developments in Mechatronics and Intelligent Robotics*,
Advances in Intelligent Systems and Computing 691, DOI 10.1007/978-3-319-70990-1_26

2   Transformer Equipment Identification Base Gray Information

2.1   Fundamentals

Image matching based on gray information is the most popular method of template matching. It can generally use the following two methods to measure the similarity degree.

$$D(i,j) = \sum_{m=1}^{M_1} \sum_{n=1}^{N_1} \left[S^{i,j}(m,n) - T(m,n) \right]^2 \tag{1}$$

Then their expansion can be expressed by the following formula.

$$D(i,j) = \sum_{m=1}^{M_1} \sum_{n=1}^{N_1} \left[S^{i,j}(m,n) \right]^2 - \\ 2 \sum_{m=1}^{M_1} \sum_{n=1}^{N_1} \left[S^{i,j}(m,n) T(m,n) \right] + \sum_{m=1}^{M_1} \sum_{n=1}^{N_1} \left[T(m,n) \right]^2 \tag{2}$$

In Eq. (3), the first term on the right side of the equal sign represents the energy of the sub-region in the image to be matched, whose value changes continuously with the movement of the template. And the third term represents the energy of the template, its value is a constant that does not change [5]. The second term on the right is the cross-correlation between the template and the sub-region.

2.2   Experimental Tests

Since the template is cut out from the original image, the recognition effect is good. However, we set the threshold to 0.5 by experimentation, the image recognition effect of the transformer as shown in Fig. 1, only two high-voltage casing are identified.

Fig. 1. Match results when the threshold is 0.5

3   Transformer Identification Based on Hu Moment Invariants

3.1   Improvement of Hu Moment Invariants

In 1962, Hu M. K made use of two order and three order normalized central moments to construct the Hu invariant moments, and proved that they have the properties of rotation, scale and translation invariant, which is based on the target shape representation of the region. The seven Hu invariant moments are defined as follows.

Since the great difference between the orders of magnitude of eigenvalues calculated directly, the invariant moments above which are actually based on the region and defined in the condition of continuous function. The expression are shown below. In addition, since the range of the seven invariant moments after correction is still large, in order to facilitate the calculation and comparison, the logarithmic method will be used to compress the data.

3.2   Similarity Measure

In order to reflect the degree of similarity between the two images accurately, an objective measure of the similarity between the images is required. In the process of matching, the modified invariant moments of the sub-region and the template are obtained respectively. We use the Euclidean distance to measure the similarity between the two sets of characteristics.

$$d = \sqrt{\sum_{i=1}^{7} \left(M_i - M_i' \right)^2} \tag{3}$$

Based on the improved Hu moment invariant, the transformer is identified by the template matching method. Firstly, the image to be matched need to be preprocessed, and the selected template is matched with the image to be recognized. Then the sub-region of the minimum Euclidean distance is found by the similarity measure, which is the most similar part of the template. Finally, the matching is successful as long as the Euclidean distance between the most similar sub-region and the template is less than the threshold according to the experiment. The template matching method based on Hu invariant moments can match the devices with obvious shape features more effectively, which make up for the lack of template matching based on gray information in some ways.

4   Experiment and Analysis

The transformer has remarkable characteristic, there are three columns of high pressure casings. We select the high-pressure casing as a template (shown in Fig. 1) and Table 1 shows the seven eigenvalues of the improved Hu invariant moments of the template image.

Table 1. Invariant moment characteristics of high pressure casing templates

$|\log M_1|$	$|\log M_2|$	$|\log M_3|$	$|\log M_4|$	$|\log M_5|$	$|\log M_6|$
6.5456	3.2671	8.7121	7.0378	15.1773	9.2292

For those different angles of the transformer picture, though there exists a slight deviation in the location of the template registration, the location of the template area can be generally matched through the experiment. And for the image that the transformer device doesn't exist, template will not be matched, which shows that the proposed method is feasible and effective for the identification of transformer equipment. The background of the transformer device picture mentioned above is relatively simple. When we use the image of the complex background in the scene as the image to be matched, the experimental results are shown in Fig. 2.

Fig. 2. Results of field transformer identification

In addition, the template matching depends on the angle of the shooting direction of the image to be matched. If an image contains a transformer device, we can see that the transformer's oil pillow is also a more prominent feature, and the invariant moments of the template are shown in Table 2.

Table 2. Invariant moment characteristic of the oil pillow template

$|\log M_1|$	$|\log M_2|$	$|\log M_3|$	$|\log M_4|$	$|\log M_5|$	$|\log M_6|$
7.0346	20.2842	32.5687	34.7503	68.4174	70.5023

5   Conclusion

The method of recognition based on template matching is simple in principle and intuitive in process of image recognition, and it is a commonly used method in pattern recognition. In this paper, the template matching method based on Hu's invariant moments shows better effect and higher accuracy in matching and identifying different devices by using different templates.

Acknowledgements. This work is supported by the Science and Technology Research Project of State Grid Corporation of China (526816160024).

References

1. Hu, W.: Research on New Fault Diagnosis Technology of Power Equipment Based on Intelligent Information Fusion. Huazhong University of Science and Technology, Wuhan (2005)
2. Bian, Z.: Pattern Recognition. Tsinghua University Press, Beijing (2000)
3. Liu, X., Luo, B., Chen, Z.: Study on optimal model selection of support vector machines. J. Comput. Res. Dev. **42**(4), 576–581 (2005)
4. Qiang, W., Xiaoxi, Z., Yihong, H.: Image recognition based on neural network. Electron. Design Eng. **20**(9), 187–189 (2012)
5. Hu, M.: Visual pattern recognition by moment invariants. IRE Trans. Info. Theor. **8**(2), 179–187 (1962)

Simulation Analysis of Error Separation Precision of Three-Probe Method

Lei Zhang[1(✉)], Rui Ma[1], Dianqing Yu[2], and Jun Cheng[1]

[1] School of Mechanical Engineering and Automation,
Northeastern University, Shenyang, China
leizhang@mail.neu.edu.cn
[2] Institute of Length, Liaoning Institute of Metrology,
Shenyang, China

Abstract. The error separation with three-probe method mathematical model of the rotary parts in the frequency domain is established, and the basic equation of error separation is solved by discrete Fourier transform, the computer simulation is used to analyze and compare the effect and precision of the workpiece body shape error, rotary motion error and straight motion error in frequency domain adapted by three-probe error separation method, the simulation results show the correctness of the three-probe error separation theory and of the mathematical model, and draw the conclusion: Frequency domain three-probe error separation method can basically achieve the complete separation of each error, it is suitable for in-situ measurement and can get a high measurement accuracy.

Keywords: Error Separation technique · Three-probe method in frequency domain · Precision of error separation · Simulation analysis

1 Introduction

Among the methods of error separation, the three-probe method is obviously one of the most recognized and studied methods. At present, the three-probe error separation technology has been mature to apply to the high precision measurement of roundness, but it cannot be considered that the cylindricity error separation is a simple extension of the roundness error separation. There are two main reasons: The composition of the cylindricity error is more complex than the roundness error. It is composed of different error factors such as the roundness error of the section, the straightness error of the axis line and the parallelism error of the generatrix. Since the cylindrical surface of the measured shaft part is a three-dimensional continuous surface, it is very difficult to actually measure the true appearance of the whole surface in practice. In the sampling process, the measuring frame equipped with the sensor not only has the relative circular motion on each section of the measured part, but also has the relative axial linear motion to the measured part. Therefore, the process of cylindricity error separation needs to take full account of two motion error, that is the linear motion error of the measuring frame along the guide rail and the space rotation error of the measured part.

© Springer International Publishing AG 2018
F. Qiao et al. (eds.), *Recent Developments in Mechatronics and Intelligent Robotics*,
Advances in Intelligent Systems and Computing 691, DOI 10.1007/978-3-319-70990-1_27

2 Mathematical Model of Three-Probe Error Separation in Frequency Domain

2.1 The Establishment of Mathematical Models

As shown in Fig. 1, Let three axis of sensors V_1, V_2, V_3 are in the same plane and intersect at one point o, the plane is perpendicular to the axis of the measured rotating body. Establish a static coordinate system $oxyz$ fixed with earth, the axis oz is perpendicular to the plane where the axis of three sensors V_1, V_2, V_3 is located. At the initial position, the angles between three sensors V_1, V_2, V_3 and axis ox are ϕ_1, ϕ_2, ϕ_3 respectively. Depending on the measured workpiece as rigid body, take the axial coordinate $z = 0$ of the section as the reference basic section. The polar o' is taken as the basic point in the basic section. If the axial motion of the workpiece is not considered, the motion of any points on workpiece can be decomposed into three kinds of absolute motion in five directions: (1) Workpiece counterclockwise rotation around the axis itself, rotation angle is $\theta = \int \varpi(t)dt \approx \varpi t$; (2) Translational motion $\vec{x}(\theta) + \vec{y}(\theta)$ followed with basic point o'; (3) The angular motion $\vec{\varphi}_y(\theta) + \vec{\varphi}_x(\theta)$ of the workpiece axis of rotation (moving axis) relative to its average position (static axis), and the stationary random motion superimposed on three kinds of absolute motion. There is a minimum common multiple $2l\pi$ for the period of these five motions, that is a public period of $2l\pi$.

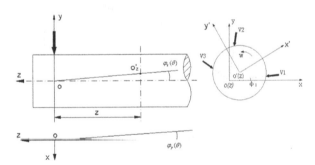

Fig. 1. The relationship of rotary errors between different sections of the workpiece

During the sampling process, keep the initial sampling points on different sampling sections located on the same prime line, re-adjustment the initial values of the sensor probes is forbidden to ensure that zero harmonic component in the sampled signal is not lost. Discrete the sampled signal, apply to uniform angle sampling, the sampling uniform angle $\Delta\theta$ is chosen to satisfy the sampling theorem, sampling points N meet 2FFT requirements, that is $N = 2^s$ (s is an integer), $\Delta\theta = 2\pi/N$, the angle of the kth sampling point is $\theta_k = k \cdot \Delta\theta = 2k\pi/N$ ($k = 0, 1, \cdots, N-1$). Set the angles between three sensors V_1, V_2, V_3 and axis ox are $\phi_1 = m_1 \cdot \Delta\theta$, $\phi_2 = m_2 \cdot \Delta\theta$, $\phi_3 = m_3 \cdot \Delta\theta$, and let m_1, m_2, m_3 be integer.

Taking the polar o' in the basic section as the origin, the dynamic coordinate system $o'x'y'z$ fixed with the workpiece is established, the absolute motion of the least squares center $o'(z)(x'(z), y'(z), z)$ of the sectional profile in the basic cross section z can be expressed as [1, 2] (see Fig. 1) [1]

$$\left. \begin{array}{l} x(k, z) = x(k) - z \cdot \varphi_y(k) + x'(z) \cdot \cos\theta - y'(z) \cdot \sin\theta \\ y(k, z) = y(k) + z \cdot \varphi_x(k) + x'(z) \cdot \sin\theta + y'(z) \cdot \cos\theta \end{array} \right\} \tag{1}$$

Let the measuring frame drive the probe along the axial motion, and get the roundness error measured by the three-probe method in different sections of the workpiece. Set the measurement plane along the axial motion is always approximately perpendicular to the axis z (the guide rail angle $\gamma(z)$ can be ignored). The components x, y of straight motion error in the static system are $u(z), v(z)$, the coordinate z of each sampling section is $z(j_0)(j_0$ is the number of each sampling section, $j_0 = 1, 2, \cdots, M)$, then the original equation of signal in the three-probe method is [1, 2]:

$$\left. \begin{array}{l} V_1(k, j_0) = S(k + m_1, j_0) + [x(k) - z(j_0) \cdot \varphi_y(k) + x'(j_0) \cdot \cos\theta_k \\ \quad - y'(j_0) \cdot \sin\theta_k - u(j_0)] \cdot \cos\phi_1 + [y(k) + z(j_0) \cdot \varphi_x(k) \\ \quad + x'(j_0) \cdot \sin\theta_k + y'(j_0) \cdot \cos\theta_k - v(j_0)] \cdot \sin\phi_1 \\ V_2(k, j_0) = S(k + m_2, j_0) + [x(k) - z(j_0) \cdot \varphi_y(k) + x'(j_0) \cdot \cos\theta_k \\ \quad - y'(j_0) \cdot \sin\theta_k - u(j_0)] \cdot \cos\phi_2 + [y(k) + z(j_0) \cdot \varphi_x(k) \\ \quad + x'(j_0) \cdot \sin\theta_k + y'(j_0) \cdot \cos\theta_k - v(j_0)] \cdot \sin\phi_2 \\ V_3(k, j_0) = S(k + m_3, j_0) + [x(k) - z(j_0) \cdot \varphi_y(k) + x'(j_0) \cdot \cos\theta_k \\ \quad - y'(j_0) \cdot \sin\theta_k - u(j_0)] \cdot \cos\phi_3 + [y(k) + z(j_0) \cdot \varphi_x(k) \\ \quad + x'(j_0) \cdot \sin\theta_k + y'(j_0) \cdot \cos\theta_k - v(j_0)] \cdot \sin\phi_3 \end{array} \right\} \begin{array}{l} k = 0, 1, \cdots, N - 1, \\ j_0 = 1, 2, \cdots, M \end{array}$$

$$\tag{2}$$

The measurement signals picked up by the three sensor probes with weighting are added to obtain three-probe combined signal

$$V(k, j_0) = c_1 \cdot V_1(k, j_0) + c_2 \cdot V_2(k, j_0) + c \cdot V_3(k, j_0)$$
$$k = 0, 1, \cdots, N - 1; \quad j_0 = 1, 2, \cdots, M \tag{3}$$

$c_i(i = 1, 2, \cdots, n)$ is the sensitivity coefficient of each sensor probe. In order to eliminate the rotational motion error and the straight motion error information included in the combined signal of the Eq. (3), set c_1 is equal to 1, and

$$\left. \begin{array}{l} 1 + c_2 \cos\phi_2 + c_3 \cos\phi_3 = 0 \\ c_2 \sin\phi_2 + c_3 \sin\phi_3 = 0 \end{array} \right\} \tag{4}$$

So the Eq. (3) can be expressed as:

$$V(k,j_0) = S(k+m_1,j_0) + c_2 \cdot S(k+m_2,j_0) + c_3 \cdot S(k+m_3,j_0)$$
$$k = 0,1,\cdots,N-1; \quad j_0 = 1,2,\cdots,M \tag{5}$$

Equation (5) is the basic equation for the error separation of the three-probe method.

The basic equation of error separation is solved by discrete Fourier transform. $V(k,j_0)$ discrete Fourier transform (DFT) is recorded as

$$F_V(r,j_0) = DFT[V(k,j_0)] = \sum_{k=0}^{N-1} V(k,j_0) \cdot W_N^{k-r}$$
$$r = 0,1,2,\cdots,N-1; \quad j_0 = 1,2,\cdots,M \tag{6}$$

The discrete values of the section shape error functions in the time domain can be obtained [2]:

$$S(k,j_0) = IDFT[\frac{F_V(r,j_0)}{W(r)}] \qquad r = 0,1,2,\cdots,N-1; \quad j = 1,2,\cdots,M \tag{7}$$

$W(r)$ is called a harmonic weighting function. While $W(r)$ is equal to zero means the loss of the shape error function of the harmonic components will make the shape error function distortion.

2.2 The Separation of the Motion Error of the Measuring Frame (or Workpiece) and the Description of the Rotation Error of Each Sampling Section

In the period of $2l\pi$, calculate the average value of each variable respectively, and recorded as $\bar{V}_1(j_0)$, $\bar{V}_2(j_0)$, $\bar{V}_3(j_0)$, $\bar{S}(j_0)$, $\bar{x}$, $\bar{y}$, $\bar{\varphi}_y$, $\bar{\varphi}_x$. Set the first probe mounting angle $\phi_1 = 0$, $m_1 = 0$, according to the original Eq. (2), the error motion $u(j_0)$, $v(j_0)$ of the measuring frame (or measured workpiece) along the guide rail is obtained:

$$\left.\begin{array}{l} u(j_0) = \bar{S}(j_0) - \bar{V}_1(j_0) + \bar{x} - z(j_0) \cdot \bar{\varphi}_y \\[2mm] v(j_0) = [\bar{S}(j_0) - \bar{V}_2(j_0)] \cdot \dfrac{1}{sin\phi_2} \\[2mm] \qquad + [\bar{x} - z(j_0) \cdot \bar{\varphi}_y - u(j_0)] \cdot ctg\phi_2 + \bar{y} + z(j_0) \cdot \bar{\varphi}_x \end{array}\right\} \quad j_0 = 1,2,\cdots,M \tag{8}$$

According to the least squares principle, the radial error motion of the polar o'_z on each sampling section of the cylinder can be calculated [1].

2.3   Separation of the Least Square Centers of Each Sampling Section

The coordinates C_z of each section moving center in moving coordinate relative to the least squares center are $(a(j_0), b(j_0))$, the discretization formula is:

$$
\left.
\begin{aligned}
a(j_0) &= -\frac{1}{IN} \sum_{k=0}^{IN-1} [(x(k) - z(j_0) \cdot \varphi_y(k)) \cdot \cos\theta_k + (y(k) + z(j_0) \cdot \varphi_x(k)) \cdot \sin\theta_k] - x'(j_0) \\
b(j_0) &= -\frac{1}{IN} \sum_{k=0}^{IN-1} [-(x(k) - z(j_0) \cdot \varphi_y(k)) \cdot \sin\theta_k + (y(k) + z(j_0) \cdot \varphi_x(k)) \cdot \cos\theta_k] - y'(j_0)
\end{aligned}
\right\}
\quad (9)
$$

Set the basic section $x'(0) = 0, y'(0) = 0, z(0) = 0$ and according to the least square, we can get the coordinates $x'(j_0), y'(j_0)$ of the least squares center $o'(j_0)$ of the j_0th section in moving coordinate [1, 2].

3   Simulation Analysis of Error Separation Precision of Three-Probe Method

3.1   Design of Simulation Scheme

According to the processing of the rotary workpiece, the cylindrical circumference of each section is sampled at equal intervals. Using the original Eq. (2) of the three-probe method, and the sampling value of the probe is fitted as $V_i(k, j_0)$ ($i = 1, 2, 3$; $k = 0, 1, \cdots, N-1$; $j_0 = 1, 2, \cdots, M$). The number of sampling points are $N = 128$, the number of sections are $M = 10$, each mounting angle of the probe is $0°$, $90°$, $160.3125°$. The fitted sampling values shall include the roundness error function $S(k, j_0)$ on the section of the measured workpiece, mounting eccentricity of workpiece $x'(j_0), y'(j_0)$ the guide rail moving error $u(j_0), v(j_0)$ in x, y direction of coordinate, angular motion error $\varphi_x(k)$ $\varphi_y(k)$ of rotation axis of workpiece, random radial vibration and runout error $x(k), y(k)$. The sample fitting value containing a large number of error signals is made into a data file, and then the combined signal $V(k, j_0)$ is synthesized by the computer, and the error separation calculation is performed with computer. By changing the various factors to find the impact in the actual situation are more objective.

In order to facilitate the analysis and comparison, the dS_{max}, dS_{min} and dS are respectively taken as the maximum value, the minimum value and the average value of the difference between separating value and the setting value.

3.2   Comparison of Simulation Results

See Fig. 2, Tables 1 and 2

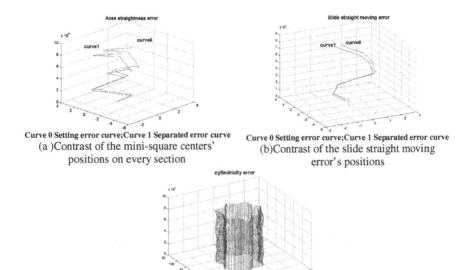

Curve 0 Setting error curve;Curve 1 Separated error curve
(a)Contrast of the mini-square centers'
positions on every section

Curve 0 Setting error curve;Curve 1 Separated error curve
(b)Contrast of the slide straight moving
error's positions

(c) Cylindricity error separation result

Fig. 2. The error curves of simulation.

Table 1. The first table of comparison of simulation results (unit: μm)

Section	Setting roundness error value f_R	Separated roundness error value f_R'	$dS(10^{-6})$	$dS_{max}(10^{-3})$	$dS_{min}(10^{-6})$
1	20.0002	20.0004	3.549	0.908	5.3011
2	19.922	19.9218	2.8571	1.6474	3.3642
3	19.9041	19.905	8.99	1.7599	6.304
4	19.4039	19.4038	2.8344	1.4016	2.7148
5	20	19.9996	0.5242	1.8433	2.467
6	18.155	18.1553	1.7723	1.6083	3.1008
7	19.9041	19.9048	3.3977	1.3289	5.0484
8	18.536	18.5377	7.26	1.5126	10.6779
9	19.689	19.6915	11.4648	1.5248	2.5248

Table 2. The second table of Comparison of simulation results (unit: μm)

		Setting value	Separated value
Cylindricity error value		$f_C = 20.0048\,\mu m$	$f'_C = 20.0123\,\mu m$
Axis straightness error value		$f_A = 1.3956 \times 10 - 4\,\mu m$	$f'_A = 2.0211 \times 10 - 4\,\mu m$
Linear straightness error value	Position1 (k = 0)	$f = 9.88303\,\mu m$	$f' = 9.88303\,\mu m$
	Position 2 (k = 16)	$f = 8.30366\,\mu m$	$f' = 8.30366\,\mu m$
	Position 3 (k = 32)	$f = 14.2281\,\mu m$	$f' = 14.2281\,\mu m$
	Position 4 (k = 64)	$f = 18.8421\,\mu m$	$f' = 18.8421\,\mu m$
Guide straight motion error	Direction: x	$f_x = 1.72333\,\mu m$	$f'_x = 1.72333\,\mu m$
	Direction: y	$f_y = 3.77\,\mu m$	$f'_y = 3.77002\,\mu m$

4   Precision Analysis of EST by Three-Probe Method and Conclusion

In contrast to the above error results, it can be considered that the EST by three-probe method can basically achieve the separation of following errors completely: (1) Separation of the circular shape error function of the section; (2) Separation of linearity error; (3) The straight motion error of the guide rail in the x, y direction.

The deviation analysis of the axis straightness. The error in the separation of the least squares centers is mainly due to the error of the rotation angle of the spindle.

Analysis of the straight motion error of guide rail. The straight motion error of the guide after separation is still accurate, which is still consistent with the setting value, and the shape of the error curve is consistent with the setting error curve shape in the overall trend, only the slope of the least square line has changed.

Acknowledgement. This project was supported by National Natural Science Foundation of China in 2015 (No. 51575096).

References

1. Guoshong, Z.X.G.L., Jiyao, B.: A study of the theory for cylindricity and accuracy of rotational motion and straight motion error. J. Shanghai Jiaotong University **26**(4), 8–13 (1992)
2. Zhang, L., Zhao, Y., Zhang, Y.: Theory and experiment of three-probe error separation technique. J. Mech. Eng. **45**(6), 256–261 (2009)

A Modeling Method Supporting Turbine Blade Structure Design

Tian Wang, Bifu Hu[✉], Ping Xi, and Jixing Li

School of Mechanical Engineering and Automation, Beihang University, Beijing 100191, China
hoobye@buaa.edu.cn

Abstract. Traditional feature-based turbine blade models can match the needs of geometric modeling but could hardly meet the requirements of structural design model for simulation analysis and manufacturing. These requirements are taken into consideration as well as geometric representation in parametric design process. An improved turbine blade parametric modeling method is proposed and a system structure of blade modeling process supporting turbine blade structure design is devised. Eventually, a turbine blade parametric modeling system is constructed to verify the feasibility of the proposed modeling method and system structure. Experiments show that the parametric modeling method proposed can make geometric models better support structure design and has certain reference value to the creation of high quality digital models.

Keywords: Parametric modeling · Turbine blade · Structure design

1 Introduction

Product design is first step of product development. Its quality will directly affect product quality, performance and economic benefits [1]. Product design can generally be divided into product modeling and product evaluation. Product modeling and product evaluation constitute the product design process. The former solves problem of product "whether or not", and the latter solves problem of product design results "good or bad" [2]. As the final product of modern digital design process, product model is the specific carrier of design requirements and engineering constraints, and it has become the data source of the following links in engineering analysis, optimization design, process planning, NC (Numerical Control) programming and virtual assembly [3]. Product modeling, as the basis of product design, its quality determines product design quality and the follow-up analysis and manufacturing.

Turbine blade is one of the most important parts of aero engine, and it is also a typical product with complicated function and structure. Figure 1 shows main structure of turbine blade. Its structure design process [4] include: design department firstly provides performance index that turbine blade needs achieve, then model is designed for aero-dynamic analysis, and then strength calculation department evaluates strength, vibration and life check, if qualified, structure design is finished, if not, designer needs redesign. This process involves multiple disciplines and a number of departments to work together. Downstream departments may modify the model provided by structural design

© Springer International Publishing AG 2018
F. Qiao et al. (eds.), *Recent Developments in Mechatronics and Intelligent Robotics*,
Advances in Intelligent Systems and Computing 691, DOI 10.1007/978-3-319-70990-1_28

department according to the analysis of simulation and manufacturing needs. To meet these needs, the structural design department put forward higher requirements for the structural modeling of turbine blades. A modeling method, therefore, is needed for structure department that can provide a high-fidelity blade model directly for downstream department.

Fig. 1. Main structure of turbine blade

2   Literature Survey

With the extensive application of digital technology in the field of turbine blade, the blade parametric design has been widely developed. Yu [5] realized design of blade body, flange plate, Tenon. Zhao [6] analyzed blade features and proposed a blade design structure system. According to Tenon design requirements, Yao [7] realized Tenon parametric design. Tian [8] developed the leaf body modeling system focused on the profile curve of the section line. Research group [9–11] analyzed turbine blade features and adopted feature-based parametric modeling method to achieve turbine blade structure design. Taking analysis and manufacturing into account, there are still problems for turbine blade structure design as follow:

Lack of correction for surface modeling initial data to ensure the continuity and smoothing of blade model. In the Fig. 2(a) on the left, if directly modeling the turbine blade body by the blade profile data, it may cause that the blade section line shape can't reach G1 continuity. As the initial section line is not smooth, blade surface folds in "wrinkles" shown in Fig. 2(a) on the right.

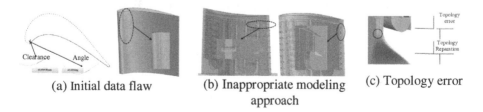

(a) Initial data flaw   (b) Inappropriate modeling   (c) Topology error
                                      approach

Fig. 2. Problems to be solved for turbine blade structure design

Intermediate data in the geometric modeling process is not saved. For example,computational domain model used for numerical simulation need the auxiliary face sheets related to middle cambered surface so that it can be generated successfully.

Inappropriate modeling approach: feature interference, filleting failure, feature dependency problem. Impingement hole and film hole interfere each other shown Fig. 2(b) on the right, which makes blade model useless. Shown in Fig. 2(c) on the right trailing edge slots can't be filleted for missing edges.

Data interaction problem: dependencies between features affects the model updating; broken edges and faces cause the model useless. Shown in Fig. 2(c), blade can't be used for analysis for its topology error and manufacturing because of broken edges and faces causing by trimming or other Inappropriate modeling approach.

3   Parametric Modeling Technique for Turbine Blade Structure Design

3.1   Geometric Modeling for Turbine Blade Structure Design Requirement Analysis

Requirement of the geometric modeling process is mainly reflected in three aspects: structure integrity, modeling rationality, data exchange (Fig. 3).

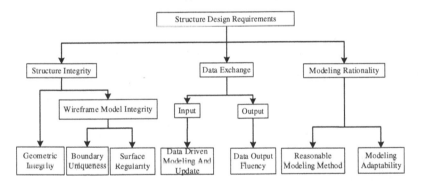

Fig. 3. Requirement of the geometric modeling

Structure integrity includes wireframe model integrity and geometric integrity. Wireframe model integrity includes boundary uniqueness and surface regularity. First, a continuous line of wireframe model must only correspond to a line. Secondly, the geometric model of surface cannot contain infinite curvature point. Table 1 illustrates wireframe model integrity. Geometric integrity refers to models avoiding broken edges and surfaces.

Table 1. Wireframe model integrity explanation

Geometric model	Correct wireframe	Wrong wireframe 1	Wrong wireframe 2
		Two Edges	

Data exchange includes data input and output. Data input should guarantee blade can be rapidly modeling, editing and updating when data changed. Data output mainly include a high-fidelity blade model is provided.

Modeling rationality includes adopting reasonable modeling method and modeling adaption. Reasonable modeling method should be taken into consideration to solve problems of feature interference, filleting failure and dependency problem. The program modeling module should have strong adaptability.

3.2   Parametric Modeling Method for Turbine Blade Structure Design

According to the requirements in 3.1, we proposed an improved parametric modeling method for turbine blade structure design shown in Fig. 4. The part outside the dotted line is shown to be a supplement and improvement to the traditional method.

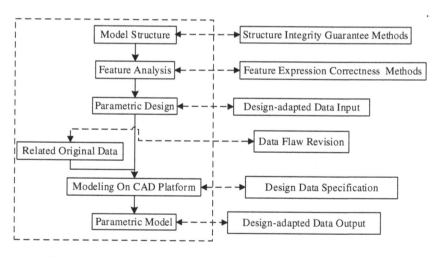

Fig. 4. Parametric modeling method for turbine blade structure design

Structure integrity guarantee methods: choosing appropriate modeling method for the features extracted from structure model (adopting Boolean rather than cutting and trimming operations) and repairing broken edges and surface.

Proper expression of features: in the stage of feature analysis defining the parent-child dependency guarantees blade model be updated automatically when the leaf data changes. Featured-based modeling should guarantee different features adapt to each other.

Data flaw revision: curve smoothing algorithm should be made to adjust the underlying data to ensure input section line data meet the geometric continuity.

Design data specification: mid-data should also be preserved. Mid-data includes auxiliary geometric elements for meshing and auxiliary recognition element.

Data interaction fluency: design-adapted data input and output method should be made meet the frequent modification of leaf data and follow-up work.

In summary, Fig. 5 shows functional structure of the blade modeling system. The frame structure of structure design supported parametric modeling system is proposed (Fig. 6). The dashed line represents the direction of the information flow. The filter and driver are the core of the system. The constraint library, method database and rule sets are the external support for the design of supporting structure.

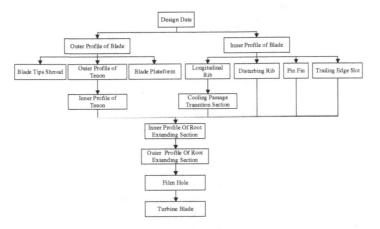

Fig. 5. Functional structure of the blade modeling system

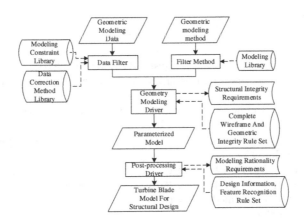

Fig. 6. Frame of structure design supported parametric modeling system

4 Experimental Results

A parametric modeling system of some turbine blade is developed based on UG to verify proposed modeling technology and system structure feasibility.

Figure 7(a) shows model of turbine blade via the modeling system and its complete wireframe model adapted to analysis and manufacturing; Fig. 7(b) on the left shows the auxiliary face sheets related to middle cambered surface of turbine blade; Fig. 7(b) on the right demonstrates the model for follow-up analysis.

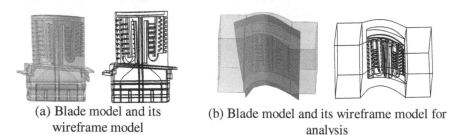

(a) Blade model and its   (b) Blade model and its wireframe model for
    wireframe model   analysis

Fig. 7. Turbine blade model supporting structure design

It can be seen from figures above that the geometric model of turbine blade generated by the system not only represents geometric shapes of product precisely, but also well supported blade structure design.

Acknowledgements. The authors wish to express their thanks to Aviation Industry Corporation of China, Shenyang Aero Engine Design and Research Institute for funding this project and for technological guidance about turbine blade design. The authors would also like to express their gratitude to Academician Yin Zeyong and Senior Scientist Wang Ming, for their expertise in Aero Engine.

References

1. Wang, Y.: Digital design. Mechanical Industry Press, Beijing (2003)
2. Wang, T.: Research on generalized product modeling method. Tsinghua University, Beijing (2004)
3. Xi, P., Zhang, B.: Development of product structure modeling technology. Aeronaut. Manufact. Technol. **493**(23–24), 36–41 (2015)
4. Wu, L., Yin, Z., Cai, X.: Multidisciplinary design optimization of aero engine turbine blade. J. Aerosp. Power **20**(5), 795–801 (2005)
5. Yu, K.H., Li, L.Z., Yue, Z.F.: Parametric design of turbine cooling blade based on analysis and feature modeling. Propul. Technol. **28**(6), 637–640 (2007)
6. Zhao, W., Ma, M.: Feature modeling of aero engine blade based on STEP. J. Beihang Univ. **25**(5), 535–538 (1999)
7. Yao, L.B., Mo, R., Liu, H., et al.: Tenon parametric design based on constraints of strength. Aeronaut. Manufact. Technol. **2007**(8), 93–95

8. Tian, Q., Mo, R., Xia, Y., et al.: CAD modeling method of aero engine blade. Aeronaut. Manufact. Technol. **2007**(2), 78–81 (2007)
9. Song, Y., Xi, P.: Parametric design of turbine blade based on feature modeling. J. Beihang Univ. **30**(4), 321–324 (2004)
10. Zhang, B., Xi, P., Yang, J.: A rapid modeling of composite curved turbine blade trailing edge. J. Beihang Univ. **1**, 009 (2014)
11. Li, J., Xi, P., Guo, Y., et al.: A turbine blade parametric modeling method considering 1-D heat transfer analysis. Comput. Aided Draw. Des. Manuf. **25**(3), 1–9 (2015)

Simulation Analysis of Influencing Factors for Resonant Frequency of PCB Coil Based on HFSS

Haokun Chi[✉], Zhiqiang Wei, Yanping Cong, Bo Yin, and Feixiang Gong

College of Information Science, The Ocean University of China,
No. 238, Songling Road, Laoshan District, Qingdao, Shandong, People's Republic of China
chihaokun@126.com

Abstract. In recent years, the application of implantable medical devices has become increasingly popular, and its power supply problem has become a hot research, and the problem of its energy supply is also a research hot topic. The PCB (printed circuit board) coils can be used to achieve cascading and stacking between adjacent layers in very small volumes, so they are very suitable for implantable applications.

In this paper, the resonant frequency and inductance of the PCB coil with specific shape are theoretical analyzed. The influence of line width, pitch, and turns on the resonant frequency of the coil is studied through the HFSS simulation results. Finally, we use the resonant frequency as the dependent variable, the linewidth and spacing as the independent variables for multiple nonlinear fitting on MATLAB, and the results show that the fitting effect is ideal.

Keywords: Resonant frequency · PCB coil · HFSS

1 Introduction

Nowadays implantable medical has developed rapidly, however, for some specific implantable medical devices, such as pacemakers, energy supply has become an important issue. It must replace the battery every few years traditionally, which adds an economic burden, surgical risk and psychological stress and pain to the patient. The WPT (wireless power transmission) technology can just solve this problem, it can charge the battery of heart pacemaker through the non-contact way and make the pacemaker work permanently. In 2007, the team of MIT realized magnetic coupling wireless power transmission, they light a 60-watt bulb at a distance of 2 m and the efficiency of the entire system is 40% [1], which provides the possibility for medium-range wireless power transmission.

Compared to traditional coils, the PCB coils (Fig. 1) are small in total volume and thin in thickness, making it ideal as a power unit for implantable medical equipment. Amato M conducted a theoretical analysis and experimental comparison for the resistance and inductance of a circular micro PCB coil, and finally proposed a system which was suitable for wireless recharging of biomedical implants [2]. Ellstein D researched a single layer and a double layer PCB spiral coils using circuit models, then compared the theoretical analysis results with experimental measurements and numerical

© Springer International Publishing AG 2018
F. Qiao et al. (eds.), *Recent Developments in Mechatronics and Intelligent Robotics*,
Advances in Intelligent Systems and Computing 691, DOI 10.1007/978-3-319-70990-1_29

simulations [3]. Sonntag C proposed a method for estimating the mutual inductance between two planar PCB coils which can only be used to solve for relatively simple geometries, and the measured and calculating results show excellent agreement [4]. Meyer P estimated values of resistance and internal self inductance of single layer PCB coils basing energy considerations [5]. Liu JQ researched the coupling coefficient between two PCB coils, designed a complete wireless power transmission system which achieved a power transmission efficiency of approximate 50% at the frequency of 500 kHz [6]. Gao J proposed a three-dimensional receiving coil to improve space utilization, which could successfully be integrated to a capsule robot [7]. Mutashar S analysed transmission efficiency of a wireless power transfer system which using PCB coils based on coupling. Then he simulated efficiency and SAR in the dry and wet-skin situation, respectively [8]. Kong S presented a prediction model of the near-field intensity of a WPT system and then validated the model using PCB coils [9]. There were many other people who have studied inductance, mutual inductance, resistance or transmission efficiency of the WPT system using different PCB coils [10–17], but none of them proposed a fitting formula for the resonant frequency of the PCB coil.

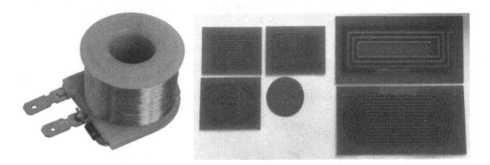

Fig. 1. Traditional coil (left) and PCB coils (right)

In the second section of this paper, the inductance of rectangular PCB coil is solved by using the formula of paper [18]. The influence of the three parameters, linewidth, track separation and number of turns, on the resonant frequency of PCB coil is simulated by HFSS software in the third section. Then, the nonlinear equation of resonant frequency about linewidth and track separation is fitted according to the simulation results and the results show that the fitting of the equation is effective.

2 Theoretical Analysis of Resonant Frequency of PCB Coil

The structure of PCB coils discussed in this paperi is as shown in the Fig. 2, copper traces are printed on each FR_4 layer, which can increase the inductance through cascading adjacent copper wires. In general, the total thickness of the coil is about 1 mm, and the plane size is about a few centimeters.As according to the Kirchhoff law, when a circuit is in a resonant state, there is:

$$f_0 = \frac{1}{2\pi\sqrt{LC}} \tag{1}$$

where f_0, L and C represent resonant frequency, inductance and capacitance of the circuit, respectively. For the inductance, there is:

$$L_{total} = L_0 + L_+ - L_- \tag{2}$$

In Eq. (2), L_{total} is the total inductance of the coil, L_0 represents the sum of the inductance of all straight conductors in the coil, L_+ represents the sum of the mutual inductance which is positive between the two straight conductors, L_- represents the sum of the mutual inductance which is negative between the two straight conductors. For example, a two-layer two-turn PCB coil can be divided into 16 straight wires (Fig. 3 left).

$$L_0 = \sum_{i=1}^{16} L_i \tag{3}$$

$$L_i = 0.002l_i \left[\ln \frac{2l_i}{w + h\_c} + 0.50049 + \frac{w + h\_c}{3l_i} \right] \tag{4}$$

Here, L_i indicates the inductance of section i, l_i indicates the length of section i, w represents the width of copper wiring, h_c represents the thickness of wiring (it is 0.035 mm in this paper). Beside, in this paper, d is the track separation on one layer, nn represents the number of turns for each lay of the coil (Fig. 3 right).

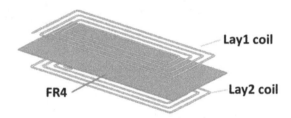

Fig. 2. Structure of a two-layer PCB coil

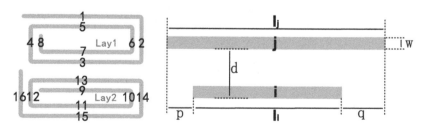

Fig. 3. Number of two-layer two-turn PCB coil (left) and coil parameters (right)

Since the coils of the first and second layers are connected in series, the current directions in the straight conductors numbered 1 and 5, 9, 13 are the same, and the mutual inductance between them is positive; while the current directions in the straight conductors numbered 1, 5 and 3, 7, 11, 15 are the same, and the mutual inductance between them is opposite, the mutual inductance between them is negative. So there is:

$$L_{+} = \sum_{i,j=1,5,9,13}^{i \neq j} M_{i,j} + \sum_{i,j=2,6,10,14}^{i \neq j} M_{i,j} + \sum_{i,j=3,7,11,15}^{i \neq j} M_{i,j} + \sum_{i,j=4,8,12,16}^{i \neq j} M_{i,j} \qquad (5)$$

$$L_{-} = \sum_{i=1,5}^{j=3,7,11,15} M_{i,j} + \sum_{i=2,6}^{j=4,8,12,16} M_{i,j} + \sum_{i=3,7}^{j=9,13} M_{i,j} + \sum_{i,j \in 4,8}^{j=10,14} M_{i,j} \qquad (6)$$

For the mutual inductance $M_{i,j}$ between two conductors (Fig. 3 right), there is

$$M_{i,j} = \frac{1}{2}((\bar{M}_{i+p} + \bar{M}_{i+q}) - (\bar{M}_{p} + \bar{M}_{q})) \qquad (7)$$

$$\bar{M}_{k} = 2l_{k} \left\{ \ln \left[\frac{l_{k}}{d_{k,j}} + \sqrt{1 + \left(\frac{l_{k}}{d_{k,j}} \right)^{2}} \right] - \sqrt{1 + \left(\frac{d_{k,j}}{l_{k}} \right)^{2}} + \frac{d_{k,j}}{l_{k}} \right\} \qquad (8)$$

3   Simulation Analysis of Resonant Frequency of PCB Coil

3.1   HFSS Simulation Calculation

According to the Kirchhoff law, when a circuit is in a resonant state, the imaginary part of the impedance in the circuit is equal to 0, the real part is the largest. So we use HFSS to simulate the impedance of the coil about the frequency, then we find the frequency where the impedance is the largest, that is the resonant frequency of the coil.

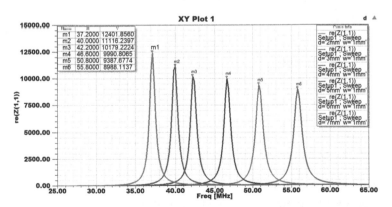

Fig. 4. The coil resonant frequency simulated by HFSS

We fixed the outer edge of each coil is 40 mm, the coil is wound from the outside, the coil resonant frequency (Fig. 4) is simulated when the linewidth - w, track separation - d, turns - nn are changed. The parameters of all the coils simulated are as follows in Table 1: For example, the parameters of NO. 9 coil are nn = 2, d = 4 mm, w = 2 mm.

Table 1. The parameters of all the coils

Number	nn	d (mm)	w (mm)
1, 2	2	2	1, 1.5
3, 4, 5, 6	2	3	1, 1.5, 2, 2.5
7, 8, 9, 10, 11, 12	2	4	1, 1.5, 2, 2.5, 3, 3.5
13, 14, 15, 16, 17, 18, 19, 20	2	5	1, 1.5, 2, 2.5, 3, 3.5, 4, 4.5
21, 22, 23, 24, 25, 26, 27, 28, 29, 30	2	6	1, 1.5, 2, 2.5, 3, 3.5, 4, 4.5, 5, 5.5
31, 32, 33, 34, 35, 36, 37, 38, 39, 40, 41, 42	2	7	1, 1.5, 2, 2.5, 3, 3.5, 4, 4.5, 5, 5.5, 6, 6.5
1, 43, 44, 45, 46, 47	2, 3, 4, 5, 6, 7	2	1

By comparing the coil NO. 1–42, we can see the effect of linewidth on the resonant frequency of the PCB coil (Fig. 5) when track separation is different. By comparing the coils NO. 1, 3, 7, 13, 21, 31, it can be seen that the effect of track separation on the resonant frequency (Fig. 5 right) when linewidth is the same. By comparing the coils NO. 1.43–47, we can see the effect of the turns of the coil on the resonant frequency (Fig. 6 left).

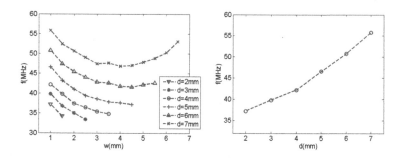

Fig. 5. Effect of w, d to resonant frequency

Through the above simulation results, it can be seen that the resonant frequency of coil will increase when the track separation increases, that is because the increase in track separation will reduce the inductance of coil, and the resonant frequency will be reduced according to formula (1); when the number of turns of the coil increases, the inductance of the coil will increase significantly, so the resonant frequency will be reduced; When we keep the track separation constant and increase the linewidth, the resonant frequency will first decrease, but there may be an increasing trend when reduced to a certain extent, the greater the track separation the more obvious the trend. The possible reason for this phenomenon is that the inductance of the coil is gradually

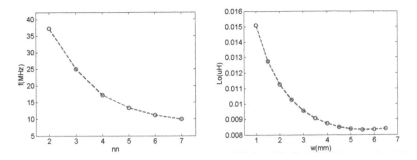

Fig. 6. Effect of nn to resonant frequency (left) and w to inductance (right)

reduced (Fig. 6 right) with the increase in linewidth, and the capacitance of coil will first increase and then reduce, resulting in the change in the resonant frequency.

3.2 Fitting Analysis Based on MATLAB

We use MATLAB for multi-nonlinear fitting on the resonant frequency of coils NO. 1–42 with linewidth and track separation as independent variables, the resonant frequency as the dependent variable. No. 2, 9, 16, 23, 30, 37 coils are used as test groups and others as fitting groups. The fitting results are as follows:

$$f_0 = 41.7994 + 0.3687 * d^2 + 0.9519 * w^2 + 0.3397 * d - 7.3677 * w - 0.0445 * d * w$$

The fitting parameter $P = 7.4959*10^{-37} < 0.01$, which shows that the fitting is reasonable. The fitting effect is shown in Fig. 7:

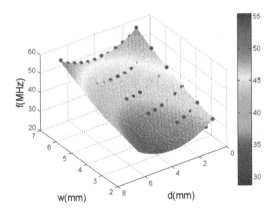

Fig. 7. The fitting effect

We obtain the fitting resonant frequency by the test group data and the fitting formula, then results are compared with the simulation results (shown in the Table 2), the comparison results shows that the fitting is very ideal.

Table 2. Comparison results

Number	nn	d (mm)	w (mm)	Simulation resonant frequency (MHz)	Fitting resonant frequency (MHz)	Error (%)
2	2	2	1.5	34.2	34.9	−2.05
9	2	4	2	37.4	37.8	−1.07
16	2	5	2.5	39.4	39.7	−0.76
23	2	6	2	45.4	45.6	−0.44
30	2	6	5.5	42.6	43.9	−3.05
37	2	7	4	46.8	46.8	0.00

4   Conclusion

In this paper, the resonant frequency and inductance of the PCB coil with specific shape are theoretical analyzed. The influence of line width, pitch, and turns on the resonant frequency of the coil is studied through the HFSS simulation results. The conclusion is as follows: increasing the number of turns and reducing the track separation will reduce the resonant frequency of the PCB coil, and with the increase of the linewidth, the resonant frequency will decrease first and then increase. Finally, we use the resonant frequency as the dependent variable for multiple nonlinear fitting on MATLAB, which provides the possibility to estimate the resonant frequency of the specific shape of the PCB coil.

There are still many deficiencies in this paper. We fail to give an arithmetic expression of capacitance for PCB coils, and the amount of data that is used to fit and test is too little. This provides guidance for our next work.

Acknowledgment. This paper received funding from Qingdao innovation and entrepreneurship leading talent project (13-cx-2), Qingdao national laboratory for marine science and technology Aoshan science and technology innovation project (2016ASKJ07) and China International Scientific and Technological Cooperation Special (2013DFA10490).

References

1. Kurs, A., Karalis, A., Moffatt, R., et al.: Wireless power transfer via strongly coupled magnetic resonances. Science **317**(5834), 83–86 (2007)
2. Amato, M., Dalena, F., Coviello, C., et al.: Modeling, fabrication and characterization of micro-coils as magnetic inductors for wireless power transfer. Microelectron. Eng. **111**(1), 143–148 (2013)
3. Ellstein, D., Wang, B., Teo, K.H.: Accurate models for spiral resonators. In: Radar Conference, pp. 461–464 (2012)
4. Sonntag, C., Lomonova, E.A., Duarte, J.L.: Implementation of the Neumann formula for calculating the mutual inductance between planar PCB inductors. In: International Conference on Electrical Machines, pp. 1–6. IEEE (2008)
5. Meyer, P., Perriard, Y.: Skin and proximity effects for coreless transformers, pp. 1–5 (2011)

6. Liu, J.Q., Wang, L., Pu, Y.Q., et al.: A magnetically resonant coupling system for wireless power transmission. In: International Symposium on Antennas, Propagation & Em Theory, pp. 1205–1209 (2012)
7. Gao, J., Yan, G., Wang, Z., et al.: A capsule robot powered by wireless power transmission: design of its receiving coil. Sens. Actuators A **234**, 133–142 (2015)
8. Mutashar, S., Hannan, M.A., Samad, S.A., et al.: Analysis and optimization of spiral circular inductive coupling link for bio-implanted applications on air and within human tissue. Sensors **14**(7), 11522–11541 (2014)
9. Kong, S., Kim, J.J., Park, L. et al.: Near-field intensity prediction model at maximum transferred power frequency in mutual-coupled rectangular coils for WPT system. In: Symposium on Electromagnetic Compatibility, pp. 45–48. IEEE (2012)
10. Peters, C., Manoli, Y.: Inductance calculation of planar multi-layer and multi-wire coils: an analytical approach. Sens. Actuators A **s145–s146**(1), 394–404 (2008)
11. Tang, S.C., Hui, S.Y.R., Chung, H.: Characterization of coreless printed circuit board (PCB) transformers. In: Power Electronics Specialists Conference, Pesc 1999, vol. 2, pp. 746–752. IEEE (1999)
12. Lee, K.H., Jun, B.O., Kim, S., et al.: A study on geometry effect of transmission coil for micro size magnetic induction coil. Solid-State Electron. **119**, 45–49 (2016)
13. Tavakkoli, H., Abbaspour-Sani, E., Khalilzadegan, A., et al.: Analytical study of mutual inductance of hexagonal and octagonal spiral planer coils. Sens. Actuators A **247**, 53–64 (2016)
14. Stęplewski, W., Dziedzic, A., Kłossowicz, A., et al.: Reactance components embedded in printed circuit boards. Circ. World **41**(3), 125–132 (2015)
15. Tao, T., Zhao, Z., Ma, W., et al.: Design of PCB Rogowski coil and analysis of anti-interference property. IEEE Trans. Electromagn. Compat. **58**(2), 344–355 (2016)
16. Ho, G.K.Y., Zhang, C., Pong, B.M.H., et al.: Modeling and analysis of the bendable transformer. IEEE Trans. Power Electron. **31**(9), 6450–6460 (2016)
17. Lope, I., Carretero, C., Acero, J., et al.: Frequency-dependent resistance of planar coils in printed circuit board with Litz structure. IEEE Trans. Magn. **50**(12), 1–9 (2014)
18. Greenhouse, H.M.: Design of planar rectangular microelectronic inductors. IEEE Trans. Parts Hybrid Packag. **PHP-10**(2), 101–109 (1974)

Research on Combinatorial Test Case Generation of Safety-Critical Embedded Software

Ge Zhou$^{(\boxtimes)}$, Xiang Cai, Chi Hu, Jun Li, Wenxiang Han, and Xiaopin Wang

National University of Defense Technology, Changsha, China
zhouge12@nudt.edu.cn

Abstract. With the increasing importance of embedded software in type of weapon equipment system, the complexity increasing, in many aspects, traditional design method of test cases cannot meet such kind of software measurement process of the demand for quality evaluation and testing cycle. To implement safety-critical embedded software automatic generation technology combination test research, extract the key parameters modeling combination test model, using search algorithm to automatically generate multi-dimensional parameter to cover the high quality of test cases, is a mature research direction. The purpose of this paper is to suit our own characteristics of embedded software in strategic weapons model.

Keywords: Combination test · Safety-critical embedded software

1 Introduction

In the field of combination test research, the earliest, Mandl [1] work in the testing of Ada compiler with the strength of two way of combination test. If using the exhaustive testing method, total need 256 test cases. Mandl using the orthogonal matrix based on the mathematical thinking method, finally using a set of test cases for a total of 16 test covers all possible. In this case, you can think combination test can capture 100% defects. In addition, the rapid development in recent years, artificial intelligence technology, high attention by a large number of researchers, using the multiple parameters r searching generation algorithm has a good technical foundation [2], combined with the artificial intelligence technology, can improve the strength of combination test, and enhance the degree of automation.

Combination test cases set can be represented in a matrix, the matrix of each row represents a test case, and each column represents a parameter of the system, each item on behalf of the corresponding parameter values of the test cases. According to Cohen and others are used to describe combination cover an array of test cases set CA (covering array) [3], the definition of general, able to cover all the combination of the parameters of any two values of covering arrays, called two-dimensional combination (2-way) or pairs; Able to cover the t value of a parameter combination is called t d combination (t-way) [4]; Obviously, when t is equal to the number of parameter n, covering an array

F. Qiao et al. (eds.), *Recent Developments in Mechatronics and Intelligent Robotics*,
Advances in Intelligent Systems and Computing 691, DOI 10.1007/978-3-319-70990-1_30

parameter system can cover all the combinations, complete coverage of 100%. There-fore, the combination to solve the mathematical model of the test case set tends to be continuous, multi-objective, nonlinear constraints to solve the problem.

At present, the combination of domestic and foreign research test varieties have dozens of all kinds of method, but no one is able to gm in all types of software.

2   MESCT Combination Test Method

According to NASA and other groups, the study found 20% to about 40% of the failures are caused by the interaction of a two parameter, and account for about 70% of software failure is caused by the action of one or two parameters, more than 90% of the error is caused by the three parameters interact within [5, 6]. Therefore, pairs coverage testing is widely used in the combination test. In this paper, we proposed a combination-test method of safety-critical embedded software.

MESCT (he was Embedded Software test generation method for Combinatorial Testing) is proposed in this paper to Safety-critical software as the combination of the measured target Embedded Software test case generation method. Combining with the characteristics of its high reliability, safety requirements and MESCT method is the main focus is on, take reasonable strategy, solve the problem of test process in case of swelling, at the same time, efforts to ensure that won't miss important values. As to generate test cases set of optimality, the duration of generation and the generation algorithm of generality and so on, is not MESCT method is the most important consideration. As shown in Fig. 1, is the flow chart of combination test method in this paper.

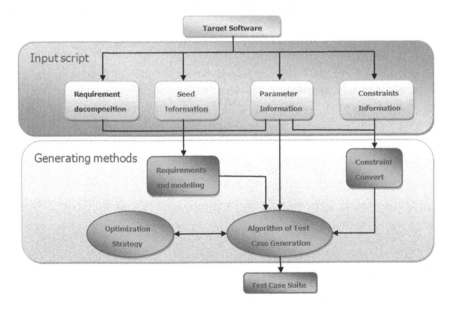

Fig. 1. Combination test method

2.1 Seeds

In the model of high reliable weapon embedded software, there are some parameters combination is particularly important, if these basic functions, performance problems will lead to the survival of huge property losses. Join seed portfolio strategy, that is, in the software combination test modeling stage, people join the combination of software applications is often used in the scene or the combination of very important, as seeds, ensure that in the final generated cases concentrated MESCT algorithm will filter out these important combination.

2.2 Variable Intensity

Is also to ensure the coverage degree of software under test, for safety-critical embedded software, about 70% of the pairs of coverage is obviously stretched. In input, influencing factors and software pattern categories multifarious many parameters, such as system, must adopt the strategy of the strength of the flexible variable, can ensure that important parameter combination of high-dimensional, secondary parameters matching combination. In an onboard software, for example, has been clear about the impact on functional demand software output is the most important parameters including emission model Mod, Tem temperature, wind WindD or WindF, wind conditions. In the parameters of system modeling, should adopt the strategy of the variable intensity of coverage is, the four parameters with 3-way or 4-way combination cover. At the same time, the other little impact on emission features of parameters such as humidity, launch time, altitude, etc., adopt pairs. Input in modeling the script, Algorithm to some scripts as input, the different levels of important parameters in the form of variable intensity combination.

2.3 MESCT Algorithm

As mentioned above, MESCT USES is joined the combination of flexible and practical strategies of the greedy algorithm. Adopts the classic line by line search.

Algorithm framework as follows (Fig. 2):

Algorithm   Kernel algorithm of MESCT
1: init(combination_set);
2: **while** true **do**
3:     gen_opt_problem();
4:     **if** solve() == OK **then**
5:         new_test_case= translate_solver_output();
6:         test_suite.add(new_test_case);
7:         update(combination_set, new_test_case);
8:     **else**
9:         **break;**
10:   **end if**
11: **end while**

Fig. 2. Algorithm of MESCT framework

3   Experimental Results

To experiment shows that this combination test strategy method, and compared with existing main tool to generate effect, a model of embedded data processing software in safety-critical software, for example. The software in the requirements analysis phase, the input situation types classified as shown in the Table 1.

Table 1. Combination test parameter model table

Parameters	Missile_type				
	A_P	A_M	B_P	B_M	No_data
V_1	1_A_P	1_A_M	1_B_P	1_B_M	1_No
V_2	2_A_P	2_A_M	2_B_P	2_B_M	2_No
V_3	3_A_P	3_A_M	3_B_P	3_B_M	3_No

Data processing software, there are three different types of data sources, various data sources are independent of each other, the best software can also accept three sources of data. Each data source data format is A play, play A_M or A_P data, play B_ P, B_ M data is one of the four. Among them, the data of 1 A bomb P data and data sources of 3 B_P combination is an important data source, is also the most often data processing software. And when the data source 3 B_M data arrives, the data source 1 and 2 not to play B_ M data sources; In the same way, the data source 3 B_P data arrives, data source and data source 1,2 will not measured data processing software to send B_P data.

Modeling after the parameters of the model are shown in Table 1 below. In this instance data processing model, there are three input parameters, V1, V2, V3 respectively three data sources of different values, will these situation is all the data processing software to be tested, if guarantee 100% coverage, to all combination of three parameters, need 5 * 5 * 5, a total of 125 kinds of input conditions, assuming that in each input case, and the combination of the four parameters to need to test each parameter domain has three parameter values, then the resulting in the number of cases is 125 * 3 * 3 * 3 * 3, a total of 10125 test cases, as a result of this type weapons being measured data processing software of each data preparation and data processing test takes a long time, one by one to complete these test cases test about consumes a lot of resources, serious is beyond the scope of the test costs to withstand.

MESCT combination test method, the measured parts demand analysis, demand integration and modeling. Can get, seed portfolio: V1 = 1 _a_p, V2 = 2 _no, V3 = 3 _b_p; Parameter constraint is: the V3 = 3 _b_m, V2! V1 = 2 _b_m! = 3 _b_m and V3 = 3 _b_p, V2! = 2 _b_p, V3! = 3 _b_p; In the diagram below form script as a combination of test case generation algorithm input:

Perform the result of the combined generator to generate test cases and PICT the comparison of the results as below in Fig. 3, you can see the method of the generated cases have 29, PICT tool to generate the 31 test cases. And the script syntax more concise, to avoid the PICT tools such as the use of judgment statement and other complex syntax to express the parameter constraint.

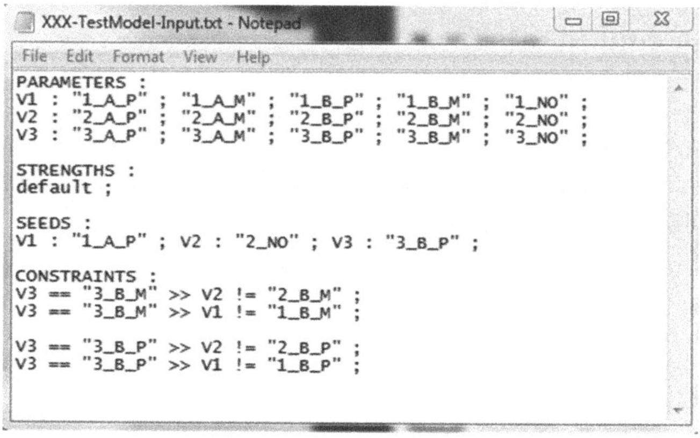

Fig. 3. Combination test case generation algorithm input

4   Conclusion

Method in this paper, according to the characteristics of the safety-critical embedded software test, in the traditional based on parameter values of all matched to cover the high resource cost test mode, in certain conditions or inadequate combination explosion of software, the use of intelligent automatic generation based on line by line search method, a set of test cases generated to cover higher value, to make up for some experience for affecting the quality of test cases to extract test omission. The future can also be on the basis of the existing work, strengthen the further research of test case selection strategy. For example, can under certain conditions, the multiple parameters as a parameter, in the form of this kind of variable granularity screening test cases. In the future, will conduct more research in these areas, further improve the embedded software use cases generated covering the strength and the degree of automation.

References

1. Mandl, R.: Orthogonal Latin squares: an application of experiment design to compiler testing. Commun. ACM **28**(10), 1054–1058 (1985)
2. Qi, Y., Mao, X., Lei, Y.: Efficient automated program repair through fault-recorded testing prioritization. In: 2013 29th IEEE International Conference on Software Maintenance (ICSM), pp. 180–189. IEEE, New Jersey (2013)
3. Cohen, M.B., Gibbons, P.B., Mugridge, W.B., et al.: Constructing test suites for interaction testing. In: International Conference on Software Engineering, Proceedings, pp. 38–48. IEEE, New Jersey (2003)
4. Chen, X., Gu, Q., et al.: Combination test research progress. J. Comput. Sci. **5**(3), 1–5 (2010)
5. Kuhn, D.R., Reilly, M.J.: An study of the applicability of the design of experiments to software testing. In: Caulfield, M. (ed.) Proceedings of the Annual NASA/IEEE software Engineering Workshop (the SEW), pp. 91–95. IEEE, Los Alamitos (2002)
6. Yan, J., Jian, Z.: Combination test: the principle and method. J. Softw. **20**(6), 1393–1405 (2009)

Robust Stability Testing of Time-Delay Bilinear Systems with Nonlinear Norm-Bounded Uncertainties

Chien-Hua Lee[✉]

Cheng-Shiu University, Kaohsiung, Taiwan
k0457@gcloud.csu.edu.tw

Abstract. Here, the stability test criteria for bilinear systems subjected to nonlinear norm-bounded uncertainties and non-commensurate time delays is treated. By using differential inequality techniques, we develop two sufficient robust stability testing conditions for assuring the above systems are robustly stable. Moreover, the decay rate of the aforementioned systems is also measured.

Keywords: Robust stability · Decay rate · Homogeneous bilinear system · Nonlinear norm-bounded uncertainty · Non-commensurate time delays

1 Introduction

It is known that bilinear systems arise as natural models for a variety of practical systems such as nuclear, heat transfer, thermal processes and biology, socio-economics, and ecology processes [3, 18]. Since bilinear systems are subsystems of nonlinear systems. Compare to linear systems, the nonlinearity results in the difficulty of analysis and design of this kind of systems. However, over the past decades, many researches have been proposed to the investigation of bilinear systems. They include the stability and stabilizability analysis and different kinds of controller design [1, 2, 4–8, 14, 15, 17, 18, 20, 22].

In fact, bilinear systems always are used accurately to approximate other nonlinear systems. Furthermore, by redefining state variables appropriately, nonlinear systems may also be transformed to bilinear systems. Therefore, uncertainties must exist in bilinear systems and may become an unstable source of the mentioned systems. Recently, several works have treated with uncertain bilinear systems. Some discussed the robust stability and the other explore the stabilizability analysis and stabilization control design [8, 22]. Besides, time delay which is caused by the information transmission, computation of variables, and so on also exists in real-life systems [10, 13, 16, 23]. Then, both uncertainty and time delay should be added to the mathematic model of bilinear systems. Some investigation results for time-delay bilinear systems have been obtained in the literature [11, 12, 17–19, 21]. However, as mentioned in the above descriptions, both uncertainties and time delays must be added to bilinear systems. By extending the method presented in [17], this work therefore studies the robust stability test problem of bilinear systems with nonlinear norm-bounded uncertainties and non-commensurate time-delays. Delay-independent stability testing conditions will be derived. Besides, we will also measure the decay rate of the closed-loop system. Finally,

© Springer International Publishing AG 2018
F. Qiao et al. (eds.), *Recent Developments in Mechatronics and Intelligent Robotics*,
Advances in Intelligent Systems and Computing 691, DOI 10.1007/978-3-319-70990-1_31

the author will apply the obtained result to develop some robust stability test criteria for uncertain time-delay systems.

2 Results for Uncertain Bilinear Time-Delay Systems

Let a kind of uncertain bilinear time-delay system be shown as follows.

$$\dot{x}(t) = A_0 x(t) + \sum_{d=1}^{k} A_d x(t - \tau_d) + \sum_{i=1}^{m} u_i(t) \sum_{d=0}^{k} B_{id} x(t - \tau_d) + \sum_{d=0}^{k} f_d\big(x(t - \tau_d), t\big) \tag{1}$$

where $x(\cdot) \in \Re^n$ is the state vector, $\tau_0 = 0 < \tau_1 < \tau_2 < \ldots < \tau_k$, are the known finite delay durations, scales $u_i(\cdot) \in \Re$, $i = 1, 2, \ldots, m$, represent the inputs. Furthermore, $A_0 \in \Re^{n \times n}$, $A_d \in \Re^{n \times n}$, and $B_{id} \in \Re^{n \times n}$ denote constant matrices and $f_d\big(x(t - \tau_d), t\big)$ are nonlinear norm-bounded uncertainties. We assure that $f_d\big(x(t - \tau_d), t\big)$ possess the following properties:

$$\left\| f_d\big(x(t - \tau_d), t\big) \right\| \leq \delta_d \big\| x(t - \tau_d) \big\|, \ d = 0, 1, 2, \ldots, k \tag{2}$$

where δ_d are positive constants respectively.

Furthermore, system matrix A_0 is assured to have the property

$$\left\| \exp[A_0 t] \right\| \leq \rho \exp(-\psi t), \tag{3}$$

where constants $\rho \geq 1$ and $\psi > 0$.

Define the following constants

$$U_i \equiv \max_{t \in [-2\tau_k, \infty)} \left| u_i(t) \right|, \ i = 1, 2, \ldots, m \tag{4}$$

$$\beta \equiv \sup_{t \in [-2\tau_k, 0]} \| x(t) \|. \tag{5}$$

By adopting differential inequality technique used in [17], one can develop the following result.

Theorem 1. If inputs $u_i(t)$ are selected such that the condition (6) or (7) is met, then the uncertain bilinear time-delay system (1) is robustly stable.

$$\psi > \rho \left[\sum_{d=1}^{k} \left(\|A_d\| + \sum_{i=0}^{m} U_i \|B_{id}\| + \delta_d \right) + \delta_0 \right] \tag{6}$$

$$\mu(A_0) + \sum_{d=1}^{k} \left(\|A_d\| + \sum_{i=0}^{m} U_i \|B_{id}\| + \delta_d \right) + \delta_0 < 0. \tag{7}$$

Furthermore, one can measure the system's decay rate α as $\alpha = \max(\alpha_1, \alpha_2)$ where α_1 and α_2, respectively satisfies

$$\psi - \rho\left[\sum_{i=1}^{m} U_i\|B_{i0}\| + \delta_0\right] - \alpha_1 = \rho\left[\sum_{d=1}^{q}\left(\|A_d\| + \sum_{i=1}^{m} U_i\|B_{id}\| + \delta_d\right)e^{\alpha_1\tau_d}\right] \tag{8}$$

$$\mu(A_0) + \sum_{i=1}^{m} U_i\|B_{i0}\| + \delta_0 + \alpha_2 = -\sum_{d=1}^{k}\left(\|A_d\| + \sum_{i=1}^{m} U_i\|B_{id}\| + \delta_d\right)e^{\alpha_2\tau_d} \tag{9}$$

Proof: The solution of the system (1) can be expressed as

$$x(t) = e^{A_0 t}x(0) + \int_0^t e^{A_0(t-\tau)}\left[\sum_{d=1}^{k} A_d x(\tau - \tau_d) + \sum_{i=1}^{m} u_i(t)\sum_{d=1}^{k} B_{id}x(\tau - \tau_d)\right]d\tau$$

$$+ \int_0^t e^{A_0(t-\tau)}\left[\sum_{i=1}^{m} u_i(\tau)B_{i0}x(\tau) + \sum_{d=0}^{k} f_d\big(x(\tau - \tau_d), \tau\big)\right]d\tau \tag{10}$$

Taking norm on both sides of (10) and then using the known inequalities $\|AB\| \le \|A\|\|B\|$ and $\|A + B\| \le \|A\| + \|B\|$ yields

$$\|x(t)\| \le \left\|e^{A_0 t}\right\|\|x(0)\| + \int_0^t \left\|e^{A_0(t-\tau)}\right\|\left[\sum_{d=1}^{k}\|A_d\|\|x(\tau - \tau_d)\|\right.$$

$$+ \sum_{i=1}^{m}|u_i(t)|\sum_{d=1}^{k}\|B_{id}\|\|x(\tau - \tau_d)\|\Bigg]d\tau \tag{11}$$

$$+ \int_0^t \left\|e^{A_0(t-\tau)}\right\|\left[\sum_{i=1}^{m}|u_i(\tau)|\|B_{i0}\|\|x(\tau)\| + \|f_0(x(\tau), \tau)\| + \sum_{d=1}^{k}\left\|f_d\big(x(\tau - \tau_d), \tau\big)\right\|\right]d\tau$$

By properties (2), (3), and the definition (5), inequality (11) becomes

$$\|x(t)\| \le \int_0^t \rho e^{-\psi(t-\tau)}\left[\sum_{d=1}^{k}\|A_d\|\|x(\tau - \tau_d)\| + \sum_{i=1}^{m} U_i\sum_{d=1}^{k}\|B_{id}\|\|x(\tau - \tau_d)\|\right]d\tau + \beta\rho e^{-\psi t}$$

$$+ \int_0^t \rho e^{-\psi(t-\tau)}\left[\sum_{i=1}^{m} U_i\|B_{i0}\|\|x(\tau)\| + \delta_0\|x(\tau)\| + \sum_{d=1}^{k}\delta_d\|x(\tau - \tau_d)\|\right]d\tau$$

$$\le \rho\beta e^{-\psi t} + \int_0^t \rho e^{-\psi(t-\tau)}\sum_{d=1}^{k}\left(\|A_d\|\sum_{i=1}^{m} U_i\|B_{id}\| + \varepsilon_d\right)\|x(\tau - \tau_d)\|d\tau \tag{12}$$

$$+ \int_0^t \rho e^{-\psi(t-\tau)}\left(\sum_{i=1}^{m} U_i\|B_{i0}\| + \varepsilon_0\right)\|x(\tau)\|d\tau$$

Now, to prove the condition (6) can work. A scalar function $w(t)$ then is defined as

$$w(t) \equiv \rho \beta e^{-\alpha_1 t}, \quad t \geq -2\tau_k \tag{13}$$

where ρ, β, respectively, is defined by (3) and (5) and constant α_1 is the unique solution of (8). It is shown that the satisfaction of the condition (6) can infer that the solution of the following equation is $w(t)$.

$$w(t) = \int_0^t \rho e^{-\eta(t-\tau)} \sum_{d=1}^k \left(\|A_d\| + \sum_{i=1}^m U_i \|B_{id}\| + \varepsilon_d \right) z(\tau - \tau_d) d\tau + \rho \beta e^{-\eta t}$$
$$+ \int_0^t \rho e^{-\eta(t-\tau)} \left(\sum_{i=1}^m U_i \|B_{i0}\| + \varepsilon_0 \right) z(\tau) d\tau \tag{14}$$

Furthermore, define an error function as

$$e(t) \equiv \|x(t)\| - w(t). \tag{15}$$

From (5), one can obtains

$$e(t) \leq \beta - \beta \rho e^{-\alpha_1 t} \leq 0 \text{ for } t \in [-2\tau_q, 0]. \tag{16}$$

From (12), (14), and (15), we have

$$e(t) \leq M(e(\cdot), t) = \int_0^t \rho e^{-\psi(t-\tau)} \sum_{d=1}^k \left(\|A_d\| + \sum_{i=1}^m U_i \|B_{id}\| + \varepsilon_d \right) e(\tau - \tau_d) d\tau$$
$$+ \int_0^t \rho e^{-\psi(t-\tau)} \left(\sum_{i=1}^m U_i \|B_{i0}\| + \varepsilon_0 \right) e(\tau) d\tau \quad \forall t > 0 \tag{17}$$

Now, it is assumed that $e(t_i) > 0$ for some t_i. Then, one can find a time $t_0 > 0$ is the smallest value so that $e(t_0)$ is positive. This implies that $e(t)$ is less than or equals to zero for $E(t_0, e(\cdot)) \leq 0$. According to (16) and the definition of (17), one can obtain $E(t_0, e(\cdot)) \leq 0$ and $e(t_0) \leq 0$. However, this contradicts $e(t_0) > 0$. Then, the fact $e(t) \leq 0$ for all $t > 0$ can be confirmed. Furthermore, one can also estimate the transient behavior of this system as

$$\|x(t)\| \leq z(t) = \beta \rho e^{-\alpha_1 t}, t \geq 0. \tag{18}$$

Therefore, condition (6) indeed can guarantee the robust stability of (1).

Furthermore, by the well-known fact that $\|e^{At}\| \leq e^{\mu(A)t}$ [9], we then can let $\rho = 1$ and $\psi = -\mu(A_0)$ in (3) and obtain the robust stability testing condition (7) and the transient behavior of this system now becomes

$$\|x(t)\| \leq \beta e^{-\alpha_2 t}, t \geq 0. \tag{19}$$

Now, from the definition of decay and (18) and (19), the decay rate α can be defined as $\alpha = \max(\alpha_1, \alpha_2)$. Thus, the proof of Theorem 1 is accomplished.

Remark 1. If $u_i(t) = 0$, then system (1) become a kind of uncertain time-delay system as follows.

$$\dot{x}(t) = \sum_{d=0}^{k} A_d x(t - \tau_d) + \sum_{d=0}^{k} f_d\big(x(t - \tau_d), t\big), \quad t > 0. \tag{20}$$

Then, the stability conditions (6) and (7) now becomes respectively

$$\eta > \rho \left[\sum_{d=1}^{k} \big(\|A_d\| + \varepsilon_d\big) + \varepsilon_0 \right] \tag{21}$$

$$\mu(A_0) + \sum_{d=1}^{k} \big[\|A_d\| + \varepsilon_d\big] + \varepsilon_0 < 0 \tag{22}$$

For these conditions, also note that all obtained conditions are delay-independent. Furthermore, the conditions (21) and (22) were given in [24].

3   Conclusion

By utilizing differential inequality techniques and linear algebraic inequalities, the robust stability of bilinear systems with nonlinear norm-bounded uncertainties and non-commensurate state delays has been discussed in this work. The author has developed two delay-independent testing conditions that pledge the above systems are robustly stable. Besides, according to the estimation of the transient behavior of system, we also measure the decay rate of the mentioned systems. Finally, the proposed result has been used to derive the robust stability testing conditions for a kind of systems subjected to time delay(s).

References

1. Bacic, M., Cannon, M., Kouvaritakis, B.: Constrained control of SISO bilinear system. IEEE Trans. Autom. Control **48**, 1443–1447 (2003)
2. Berrahmoune, L.: Stabilization and decay estimate for distributed bilinear systems. Syst. Control Lett. **36**, 167–171 (1999)
3. Bruni, C., Pillo, G.D., Koch, G.: Bilinear system: an appealing class of nearly linear systems in theory and applications. IEEE Trans. Autom. Control **19**, 334–348 (1974)
4. Chabour, O., Vivalda, J.C.: Remark on local and global stabilization of homogeneous bilinear systems. Syst. Control Lett. **41**, 141–143 (2000)
5. Chen, Y.P., Chang, J.L., Lai, K.M.: Stability analysis and bang-bang sliding control of a class of single-input bilinear systems. IEEE Trans. Autom. Control **45**, 2150–2154 (2002)

6. Chen, L.K., Mohler, R.R.: Stability analysis of bilinear systems. IEEE Trans. Autom. Control **36**, 1310–1315 (1991)
7. Chen, M.S., Tsao, S.T.: Exponential stabilization of a class of unstable bilinear systems. IEEE Trans. Autom. Control **45**, 989–992 (2000)
8. Chiou, J.S., Kung, F.C., Li, T.H.S.: Robust stabilization of a class of singular perturbed discrete bilinear systems. IEEE Trans. Autom. Control **45**, 1187–1191 (2000)
9. Coppel, W.A.: Stability and asymptotic behavior of differential equations. D. C. Heath, Boston (1965)
10. Goubet-Bartholomeus, A., Dambrine, M., Richard, J.P.: Stability of perturbed systems with time-varying delays. Syst. Control Lett. **31**, 155–163 (1997)
11. Guojun, J.: Stability of bilinear time-delay systems. IMA J. Math. Control Inf. **18**, 53–60 (2001)
12. Ho, D.W.C., Lu, G., Zheng, Y.: Global stabilization for bilinear systems with time delay. IEEE Proc. Control Theory Appl. **149**, 89–94 (2002)
13. Jamshidi, M.: A near-optimum controller for cold-rolling mills. Int. J. Control **16**, 1137–1154 (1972)
14. Jerbi, H.: Global feedback stabilization of new class of bilinear systems. Syst. Control Lett. **42**, 313–320 (2001)
15. Kotsios, S.: A note on BIBO stability of bilinear systems. J. Franklin Inst. **332B**, 755–760 (1995)
16. Lee, C.S., Leitmann, G.: Continuous feedback guaranteeing uniform ultimate boundness for uncertain linear delay systems: an application to river pollution control. Comput. Math. Appl. **16**, 929–938 (1983)
17. Lee, C.H.: On the stability of uncertain homogeneous bilinear systems subjected to time-delay and constrained inputs. J. Chin. Inst. Eng. **31**, 529–534 (2008)
18. Lee, C.H.: New results for robust stability discrete bilinear uncertain time-delay systems. Circ. Syst. Sig. Process. **35**, 79–100 (2016)
19. Lu, G., Ho, D.W.C.: Global stabilization controller design for discrete-time bilinear systems with time-delays. Proceedings of the 4th World Congress on Intelligent Control and Automation, pp. 10–14 (2002)
20. Mohler, R.R.: Bilinear control processes. Academic, New York (1973)
21. Niculescu, S.I., Tarbouriceh, S., Dion, J.M., Dugard, L.: Stability criteria for bilinear systems with delayed state and saturating actuators. Proceedings of the 34th Conference on Decision & Control, pp. 2064–2069 (1995)
22. Tao, C.W., Wang, W.Y., Chan, M.L.: Design of sliding mode controllers for bilinear systems with time varying uncertainties. IEEE Trans. Syst. Man Cybern. Part B **34**, 639–645 (2004)
23. Smith, H.W.: Dynamic control of a two-stand cold mill. Automatica **5**, 183–190 (1969)
24. Wang, S.S., Chen, B.S., Lin, T.P.: Robust stability of uncertain time-delay systems. Int. J. Control **46**, 963–976 (1987)

TMM: Entity Resolution for Deep Web

Chen Lijun[✉]

Zhejiang Yuexiu University of Foreign Languages, Shaoxing, China
chenlj.nbu@foxmail.com

Abstract. Entity resolution is a technique to find different records that belong to the same entity. In this paper, based on clustering idea, we propose a novel framework—twice-merging model (TMM). We introduce self-matching and discuss some novel approaches to automatic blocking, matching evaluation, self-matching detection, similarity calculation, as well as cluster generating and merging. Experimental results show that our method can effectively reduce matching space, improve matching accuracy and system efficiency.

Keywords: TMM · Entity resolution · Deep web · Clustering · Merging

1 Introduction

Entity resolution (also known as object consolidation, duplicate detection, record linkage etc.), which identifies the records that represent the same real-world entity, is particularly important when cleaning or integrating data from multiple sources. Consequently, numerous solutions have been developed to tackle this problem.

Early entity resolution mainly focused on pair-matching [1]. In recent years, it has gradually developed into collective entity matching [2–5], related research and application also extends to many areas such as automatically construct knowledge base etc. Currently, object's attributes are still the main evaluation criteria, and their weights are calculated automatically or semi-automatically [6, 7]. Matching method mainly involves machine learning, graph theory and heuristic methods [8, 9], such as Back Propagation neural network, Markov logic network etc. There yet have blocking and pruning for matching space reduction [10, 11].

In this paper, we divide matching relationship into three types: matching (Y), not matching (N), possible matching (P), and respectively called Y matching, N matching, P matching. Simultaneously we think of entity resolution as clustering, and propose a twice-merging model for deep Web entity resolution, written TMM.

2 Framework Overview

The system framework of TMM is illustrated in Fig. 1, it consists of three parts: Firstly, **Structure Converter** rearranges objects stored in entity object set R(O), which is the input of TMM, and stores them into object attribute set R(A), thus the objects with the same attribute value are gathered into the same object list A_i in R(A). Secondly, **Cluster Generator** scans object list A_i in R(A) and evaluates matching degree between

F. Qiao et al. (eds.), *Recent Developments in Mechatronics and Intelligent Robotics*,
Advances in Intelligent Systems and Computing 691, DOI 10.1007/978-3-319-70990-1_32

objects to prune N matching and merge Y matching, which generates object cluster set R(C). Finally, **Entity Discriminator** continuously evaluates matching degree of P matching in R(A) with the help of internal message passing within R(C) and weighted similarity between objects, then refresh R(C). When all P matching are evaluated, the output of TMM, i.e., entity cluster set R(E) is achieved.

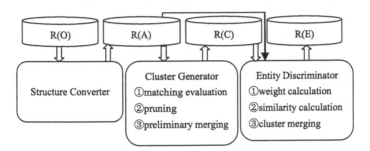

Fig. 1. System framework of TMM

3 Structure Converter

According to the observation [10], the objects representing the same entity prefer to have the same attribute value. Therefore, the objects with the same attribute value are more likely to represent the same entity. The function of structure converter is to alter the organization form from object to attribute, which can aggregate objects with the same attribute value together, so the matching evaluation is performed only among potential objects and the time complexity is reduced effectively.

The alter algorithm for **structure Converter (R(O))** is as follows:

> *(1) objList(s,list)← Ø*
> *(2) for each object O_i in R(O) and each attribute a_{ij} in O_i:*
>   *if a_{ij}∉objList.s then objList.s.add(a_{ij})*
>   *objList.list(a_{ij}).add(O_i)*
> *(3) return objList(s,list)*

As we can see that the structure converter essentially blocking the objects within R(O) into different A_i in R(A), because it just to decide whether the attribute values are the same or not, and do not involve the scheme matching problem, the computational complexity can be greatly reduced.

4 Cluster Generator

Previous studies have shown that if the similarity between objects is calculated and sorted based on the TF-IDF thought, then Y matching and N matching are often distributed at both ends of the sort list [7]. The function of cluster generator is to establish preliminary cluster by pruning N matching and merging Y matching after obtaining matching degree through calculating co-occurrence frequency of the same attribute value between objects. The advantage of cluster generator is that it quickly distinguish matching type in a simple, efficient way to reduce matching space and improve efficiency of the system.

Matching Evaluation: In order to distinguish Y matching and N matching, we define matching degree as the relative co-occurrence frequency of the same attribute values between two objects O_1 and O_2, where $\|$ represent the number of attributes:

$$Mat(O_1, O_2) = (|O_1 \cap O_2|)/(|O_1| + |O_2| - |O_1 \cap O_2|) \qquad (1)$$

If an object O_i in R(O), we cannot find another object in R(O) matched with it, we say O_i is an **self-matching** object. We define self-matching degrees as the number of its own unique attributes against its all attributes when object O_i have own unique attributes:

$$Mat(O_1, O_1) = |self(O_1)|/|O_1| \qquad (2)$$

Or the average of mismatching value with other objects when object O_i have no own unique attributes:

$$Mat(O_1, O_1) = (\sum_{i:1 \sim n} (1 - Mat(O_1, O_i)))/n \qquad (3)$$

Pruning: As mentioned above, when an object has only few co-occurrence attribute value with other objects, they usually belong to N matching, for example, a paper has the same pub date or publisher with other papers, but they may not refer to the same paper, matching degree among these objects is often very low, therefore, we can select a matching threshold θ to prune the list based on experience.

Preliminary Merging: Pruning deal with N matching, preliminary merging processing Y matching. high matching degree between objects often belong to Y matching, when matching degree is lower than the matching value between an object and a self-matching object, matching relationship becomes uncertain. It requires more information for further analysis and judgment.

The algorithm of **cluster Generator (R(A))** is described as follows:

(1) simList, Clusters, selfCluster ← ∅
(2) for each attribute A_k in R(A) and each object O_i in A_k:
      calculate mat(O_i, O_j) and save them to simList
(3) simList.cut(θ), there θ is a threshold for pruning
(4) repeat following steps:
      (O_x, O_y)←simList.getMax()
      if O_x ∈selfCluster ‖ O_y ∈selfCluster then goto step(5)
      else if x==y then selfCluster.add(O_x)
      else if O_x & O_y are both processed then Merge(Clusters(O_x),Clusters(O_y))
      else if O_x & O_y are both untreated then
            Clusters.add(cluster.createNew().add(O_x,O_y))
      *else if O_y **untreated** then Clusters(O_x).add(O_y); vice versa*
(5) return Clusters

Cluster generator, by introducing self-matching, realizes automatically classify P matching and N matching, as well as automatically switch to subsequent cluster merging process. The output R(C) contains only matched objects, other unmatched objects will be transferred to entity discriminator for further processing by data sharing.

5   Entity Discriminator

The message passing among objects can effectively improve the recall of object matching [3]. Entity discriminator is mainly processing P matching to further explore potential matching and merge identical cluster by similarity calculation and message passing within the cluster, finally output the entity cluster set R(E). Its goal is to improve the matching precision.

Weight calculation: Different attributes have different distinguishing ability and should be given different weight. Usually, the more common attributes often have stronger distinguish ability and should be given a higher weight. To this end, we firstly count the occurrence frequency t_i for each attribute, and then select the highest frequency t_{max} for normalization, so the weight of each attribute is:

$$w_i = t_i/t_{max} \tag{4}$$

Similarity calculation: In entity discriminator, we define the similarity between two objects O_1 and O_2 as:

$$Sim(O_1, O_2) = \left(\sum w_i * \max(Sim_i(O_1, O_2))\right)/\min(|O_1|, |O_2|) \tag{5}$$

Where $Sim_i(O_1,O_2)$ is calculated using the method of the instance matcher [see 12]. Max($Sim_i(O_1, O_2)$) represents the maximum similarity value of the ith attribute of two objects, because when the object O_1 has been merged into the cluster C_j, O_1 can take the corresponding attribute value within C_j as its own attribute value.

Cluster merging: Based on preliminary merging, cluster merging select objects to be evaluated according to the matching degree and calculate their similarity using formula (5) to further explore matching type between cluster and cluster, cluster and object, object and object, finally output the entity cluster set R(E).

The process description of **entity Discriminator (R(A), R(C), *limit*)** is as follows:

(1)   when processed objects less then |R(O)|, repeat following steps:
  $(O_x, O_y) \leftarrow simList.getMax()$
  if x==y then selfCluster.add(O_x)
  else if O_x, O_y are both processed & sim(O_x, O_y)≥limit then
    Merge(Clusters(O_x),Clusters(O_y))
  else if O_x, O_y are both untreated then
    if sim(O_x, O_y) ≥limit then
      Clusters.add(cluster.createNew().add(O_x, O_y))
      selfCluster.remove(O_x); selfCluster.remove(O_y)
    else sellfCluster←sellfCluster∪O_x∪O_y
  else if O_x processed then //if O_y processed, then $O_x ≤ O_y$
    if sim(O_x, O_y)≥limit then Clusters(O_x).add(O_y); selfCluster.remove(O_y)
    else sellfCluster←sellfCluster∪O_y
(2)   return Clusters+selfCluster

6   Experimental Results

Experimental data set are selected from the Cora [13], which including three sub sets: utgo-labeled, kibl-labeled and fahl-labeled. Evaluation metrics are precision, recall and F-measure; they are all defined in the interval [0, 1]. According to Fig. 2, we set *limit* to 0.85. From Fig. 3 we can see that preliminary merging has a high precision, but low recall. After cluster merging the recall has been greatly improved, but the precision decreased. Further analysis shows that: ① Data set contains dirty data, some actually same entity marked as different entities; ② Some attribute values appear in different entity, but with extremely high similarity, which interference matching obviously and reduce the precision; ③ Overall performance of *fahl* is lower than the others, because the attribute "date" appear more than once in some objects, which reduces the weight of other attributes and result in false matching.

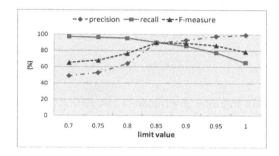

Fig. 2. *Limit* value effect on TMM

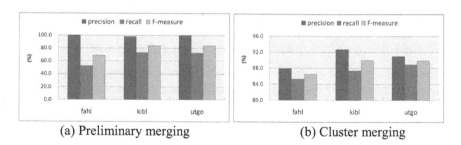

(a) Preliminary merging   (b) Cluster merging

Fig. 3. Performance evaluation for TMM

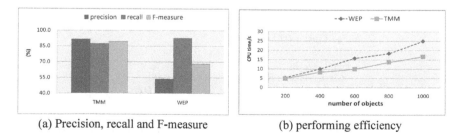

(a) Precision, recall and F-measure   (b) performing efficiency

Fig. 4. Comparison between TMM and WEP

In order to validate the efficiency and effectiveness of TMM, we compared it with WEP [11], in which Jaccard Scheme is adopted to calculate the weight of edges. The results are shown in Fig. 4. The recall of TMM is lower than WEP, but the precision is obviously higher than that of WEP, in terms of F-measure, TMM is still better than WEP. When considering efficiency, TMM and WEP are both growing linearly with the increasing of objects, but TMM raises more slowly, the main reason is that cluster generator adopted an efficient and effective reduction strategy to effectively reduce matching space and greatly reduce the amount of data to be processed in entity discriminator, which required relatively high matching cost.

7 Conclusion

Based on the idea of clustering, a twice-merging model TMM is proposed to refine matching process by handling different matching type with different method in different stage, realize automatic switch merging from one to another and select different matching strategies by introducing self-matching. The experimental results show that TMM can improve performance and efficient by reducing matching space, reducing the complexity of data processing and improving the matching accuracy.

References

1. Hanna, K., Erhard, R.: Frameworks for entity matching: a comparison. Data Knowl. Eng. **69** (2), 197–210 (2010)
2. Han, X.P., Sun, L., Zhao, J.: Collective entity linking in Web text: a graph-based method. In: Proceedings of the 34th Annual ACM SIGIR Conference, pp. 765–774 (2011)
3. Vibhor, R., Nilesh, D., Minos, G.: Large-scale collective entity matching. Proc. VLDB **4**(4), 208–218 (2011)
4. Wang, Z.C, Li, J.Z, Wang, Z.G., et al.: Cross-lingual knowledge linking across Wiki knowledge bases. In: Proceedings of the 21st International Word Wide Web Conference, pp. 459–468 (2012)
5. Fan, J., Lu, M.Y., Ooi, B.C., et al.: A hybrid machine-crowdsourcing system for matching Web tables. In: Proceedings of the 30th International Conference on Data engineering (ICDE), pp. 976–987 (2014)
6. Cui, X.J., Xiao, H.Y., Ding, L.X.: Distance-based adaptive record matching for Web database. J. Wuhan Univ. **58**(1), 89–94 (2012)
7. Liu, W., Meng, X.F.: A holistic solution for duplicate entity identification in deep Web data integration. In: Proceedings of the 6th International Conference on Semantics, Knowledge and Grids (SKG), pp. 267–274 (2010)
8. Xu, H.Y., Dang, X.W., Feng, Y., et al.: Method of Deep Web entities identification based on BP network. J. Comput. Appl. **33**(3), 776–779 (2013)
9. Liu, W., Meng, X.F., Yang, J.W., et al.: Duplicate identification in Deep Web data integration. In: Proceedings of the 11th International Conference on Web-Age Information Management, pp. 5–17 (2010)
10. Li, Y.K., Wang, H.Z., Gao, H., et al.: Efficient entity resolution on XML data based on entity-describe-attribute. Chin. J. Comput. **34**(11), 2131–2141 (2011)

11. Vasilis, E., Geroge, P., Geroge, P., et al.: Parallel meta-blocking: realizing scalable entity resolution over large, heterogeneous data. In: Proceedings of the 4th International Conference on Big Data, pp. 411–420 (2015)
12. Chen, L.J., Lin, H.Z.: Pattern matching method for Deep Web interface integration. Comput. Eng. **38**(12), 42–44 (2012)
13. Mccallum, A.: Cora citation matching (2004-2-9) (2004). http://www.cs.umass.edu/~mccallum/data/cora-refs.tar.gz. Accessed 22 Aug 2015

The Development and Application of Information Security Products Based on Cloud Computing

Haohao Song[✉]

The Third Research Institute of Ministry of Public Security, Shanghai, China
songhh@mctc.org.cn

Abstract. With the development of cloud computing information technology, more and more enterprises deploy their information systems in the cloud computing environment or deploy/migrate to the public cloud environment. In order to protect the security of the cloud computing information system, the information security products need to be deployed in cloud computing environment. Due to the characteristics of the cloud computing environment, the traditional information security products cannot meet the security needs of cloud computing environment. In recent years, information security product manufacturers and cloud computing service providers are committed to research and development the new cloud computing information security products. This paper introduced the types of information security product types, analyzed the main function and application scenarios of these cloud computing information security products, and provide some guidance to the units that deploy cloud computing information systems.

Keywords: Cloud computing · Information security product · Virtualization

1 Cloud Computing

The definition of cloud computing, depending on the perspective. From the user's point of view, cloud computing is to allow users to do not need to understand the specific situations of resources to on-demand be assigned, virtualizing computing resources as a cloud. Overall, the current mainstream cloud computing is closer to cloud services, it can be understood as an extension of the rental services for server data center provided by operators previously. In the past, the physical servers were rented by user, now the virtual machines, software platforms or even applications were rented. Three cloud computing service level is recognized as IaaS (Infrastructure as a Service), PaaS (Platform as a Service) and SaaS (Software as a Service), corresponding to the hardware resources, platform resources and application resources (Table 1).

Typical foreign cloud service providers and cloud services provided include EC2 and SimpleDB of Amazon, AppEngine of Google, Azure of Microsoft and Hadoop of Yahoo. Domestic well-known cloud service providers and cloud services provided include Ali cloud Alibaba, AppEngine of Sina and Tencent cloud of Tencent.

© Springer International Publishing AG 2018
F. Qiao et al. (eds.), *Recent Developments in Mechatronics and Intelligent Robotics*,
Advances in Intelligent Systems and Computing 691, DOI 10.1007/978-3-319-70990-1_33

Table 1. The unit needing to use information security products based cloud computing

Information security products based cloud computing	Application units
Security cloud operating system	The unit that have the needs of building cloud computing data center, and managing the overall resources
Security desktop cloud system	The unit that have the needs of desktop virtualization, centralized operation management
Security cloud storage system	The unit that have the needs of mobile access storage data
Cloud security comprehensive defense product	The unit that have the needs of comprehensive defense for cloud data center
Cloud security defense product for website	The unit that have the needs of security protection for Internet websites

For users:

- When the provider provides you a set of a-core CPU, b G memory host, c M network bandwidth and d G storage space, you need to install the system and fix the application, then this is IaaS, for example, EC2 of Amazon.
- When a complete system including basic database and middleware program is provided, but you need to write your own application interface, there is a PaaS, for example, AppEngine of Google, Azure of Microsoft and SimpleDB, SQS of Amazon.
- The most stupid way is even application written, for example, you only need to tell the service providers what you want is the salary management system for 500 people, the returned service is a HTTPS address, set the account password is set, and service is accessed directly, this is SaaS, such as SalesForce, Hadoop of Yahoo and Collaboration SaaS of Cisco Webex.

With the development of cloud computing and information technology, more and more enterprises are deploying cloud computing information systems. In the face of more and more cloud computing information systems and more and more cloud computing users, its security is a matter of particular concern to system builders and users. How to ensure the security of cloud computing information system has become an urgent problem to be solved.

2 Cloud Computing Information Security Products and Their Applications

In order to guarantee the security of cloud computing information system, it is necessary to deploy information security products for the cloud computing environment. Due to the characteristics of cloud computing environment, the traditional information security products have been unable to meet the security requirements of cloud computing environment. In recent years, information security vendors and cloud computing service providers are committed to the development and introduction of new cloud computing information security products.

At present, mainstream cloud computing information security products in the security market include the following categories: security cloud operating system, cloud security storage system, desktop cloud security system, cloud security comprehensive defense products, website Cloud security defense product etc.

2.1 Security Cloud Operating System

The cloud operating system is the operating system based on cloud computing, cloud storage technology, is the Integrated management system of cloud computing data center, it is the cloud platform integrated management system based on the infrastructure hardware resources such as server, storage, network, the basic software such as middleware and database, managing of mass foundational software and hardware resources. Large cloud service providers are using their own cloud computing operating system for cloud computing data center to manage the overall operation of the background.

The general cloud operating system has the functions of log centralized management, security alarm management, security monitoring management, virtual machine isolation, virtual network security, virtual machine backup and recovery. Security cloud operating system, we mentioned here, is the cloud computing overall management operation system of backstage data center having more security features, it also has exclusive memory isolation, anti-denial of service attacks, user data protection, the remaining information protection function.

At present, there is no national standards or industry standards for security cloud operating system. The MPS Quality Supervision and Testing Center of Security Products for Computer Information System compiled "MSTL_JGF_04-035 Information security technology- Security testing requirements for cloud operating system", there are many products that passed the test by The MPS Quality Supervision and Testing Center of Security Products for Computer Information System.

Security cloud operating system is mainly used for the unit that have the needs of building cloud computing data center, and managing the overall resources.

2.2 Security Desktop Cloud System

Security desktop cloud system virtualizes the desktop, centralizes the virtual desktop on remote server, to achieve the isolation of the desktop, on-demand distribution and rapid delivery. The user uses the remote desktop protocol to access the virtual desktop through different access modes, such as a thin terminal, a fat terminal, a zero terminal or a mobile intelligent terminal.

General desktop cloud system generally provides the functions such as access terminal authentication, virtual desktop access control, user data protection, virtual desktop backup. The security desktop cloud system, we mentioned here, is a desktop virtualization management system has more security features, it has the other functions such as session control, important data integrity testing, virtual desktop restore initial state, anti-malware software loading and patch management.

At present, there is no national standards or industry standards for security desktop cloud system. The MPS Quality Supervision and Testing Center of Security Products for Computer Information System compiled. "JCTJ 001-2015 Information security technology- Security testing requirements for Desktop Cloud access manage System", there are many products that passed the test by The MPS Quality Supervision and Testing Center of Security Products for Computer Information System.

Security desktop cloud system is mainly used for the unit that have the needs of desktop virtualization, centralized operation management.

2.3 Security Cloud Storage System

Security cloud storage system refers to the product that through the cluster, grid technology or distributed file system, different types of storage devices in the network (including host server) is set up by the application software to work together to provide access to data storage and business function products. Data storage types are not limited to block storage, object storage, file storage, etc. Large cloud service providers all store user data with their own security cloud storage systems.

At present, there is no national standards or industry standards for security cloud storage system. Net-drive (or cloud drive) is a typical cloud storage system that provide the cloud storage service. The industry's well-known cloud storage service providers includes Tencent cloud, Ali cloud, and Baidu cloud.

Security cloud storage system is mainly used for the unit that have the needs of mobile access storage data.

2.4 Cloud Security Comprehensive Defense Product

Cloud security comprehensive defense product is the integrated defense product based on the cloud computing, elastic expansion, mainly on the cloud computing platform and applications of cloud computing services and upper business, preventing malicious attacks from external cloud platform, Inside virtual machine or virtual machine. The security functions of product mainly include the linkage and response with the cloud computing platform and between functions, the security function of the product is scalable, including but not limited to anti-attack ability, to malicious virtual machine monitoring and recognition capability, WEB application attack protection ability.

At present, there is no national standards or industry standards for cloud security comprehensive defense product. The MPS Quality Supervision and Testing Center of Security Products for Computer Information System compiled "JCTJ.006-2015 Information security technology- Security testing requirements for cloud security comprehensive defense product", there are many products that passed the test by The MPS Quality Supervision and Testing Center of Security Products for Computer Information System.

Cloud security comprehensive defense product is mainly used for the unit that have the needs of comprehensive defense for cloud data center.

2.5   Cloud Security Defense Product for Website

Cloud security defense product for website is based on the SaaS model, transfers the target site access request to the designated protection node using DNS domain orientation or port forwarding technology, transfers secure access requests to the target site after protection cleaning, and transfers the response from the target site to visitors through designated protection node, so as to achieve security protection for the target site.

At present, there is no national standards or industry standards for Cloud security defense product. The MPS Quality Supervision and Testing Center of Security Products for Computer Information System compiled "JCTJ 002—2016 Information security technology- Security testing requirements for cloud security defense products", there are many products that passed the test by The MPS Quality Supervision and Testing Center of Security Products for Computer Information System.

Cloud security defense product is mainly used for the unit that have the needs of security protection for Internet websites.

3   Development of Cloud Computing Information Security Products

With the development of cloud computing technology and the mature of virtualization technology, more and more enterprises deployed their information system in the cloud computing environment or migrated their information system to the public cloud environment. The Information security products according to the characteristics of cloud computing environment are need continually, R & D and produced. The second chapter introduce the mainstream cloud computing and information security products on the market currently, there are some new cloud computing and information security products, such as cloud fortress machine, cloud security management platform, this paper did not do is introduced in detail.

With the development of cloud computing technology, the boundaries of products and services become blurred. Some cloud computing information security products are provided in the form of SaaS services to users or businesses. In the future, the purchase demand of virtual cloud computing information security products (services) will be more and more.

Virtual cloud computing and information security products (services) provides a large collection of data, analysis of cloud computing and information security products operators, and collaborative response facilities, to ensure the safety of the cloud computing environment.

4   Conclusion

With the development of cloud computing information technology, more and more enterprises deploy their information systems in the cloud computing environment or deploy their information systems to the public cloud environment. In order to guarantee the security of cloud computing information system, it is necessary to deploy information security products for the cloud computing environment. Due to the characteristics of cloud computing environment, the traditional information security products have been unable to meet the security requirements of cloud computing environment. In recent years, information security vendors and cloud computing service providers are committed to the development and introduction of new cloud computing information security products. This paper introduces the mainstream cloud computing information security products, analyzes the main features and application scenarios of these cloud computing information security products, expounds the development trend of cloud computing information security products, as the reference for the units that have deployed and will deploy the cloud computing information system.

References

1. MPS Information Classified Security Protection Evaluation Center. Information security Classified evaluator training course. Electronic Industry Press (2011)
2. Feng, D.G., Zhang, M., Zhang, Y.: Cloud computing security research. J. Softw. **22**(1), 71–83 (2011)
3. Zhang, C., Yu, J., et al.: Cloud computing security architecture. Science Press, Beijing (2014)
4. Jing, F., Hao, W., Songlin, B.: Virtualization security issues in cloud computing. Telecommun. Sci. **4**, 135–140 (2012)
5. Brunette, G., Mogull, R.: Security Guidance for Critical Areas of Focus in Cloud Computing v2.1. Cloud Security Alliance (2009)
6. Liu, F., Tong, J., Mao, J., Bohn, R., Messina, J., Badger, L., Leaf, D.: Cloud computing reference architecture, National Institute of Standards and Technology (2011)

A Discussion on the Application
of the Smoothing Function of the Plus Function

Shu-ting Shao[✉] and Shou-qiang Du

School of Mathematic and Statistics,
Qingdao University, Qingdao, Shandong 266071, China
13070810539@163.com

Abstract. In this paper, we analyze the smooth approximation property of a smoothing function of plus function. And these smooth approximation properties of the smoothing function can be applied to the complementarity problems, penalty functions, optimal control, and support vector machines. Thus, we can use smooth method to solve these problems, and these smooth approximation properties have a good effect in proving the convergence of the smooth method. Subsequently, the application of these good properties in solving practical problems is given.

Keywords: Plus function · Smoothing function · Gradient consistency

1 Introduction

The plus function $(x)_+$ is convex and globally Lipschitz continuous. Moreover, it is widely used in complementarity problems, penalty functions, optimal control, and support vector machines. However, because the plus function is nonsmooth, the smoothing function of the plus function is proposed in [1–6]. Smoothing function is used to smooth the nonsmooth problem, and then the smooth algorithm is used to solve the approximate optimal solution of the original problem.

A class smoothing function of $|x|$ by integration convolution has been proposed in [1]. Let us suppose that $\rho^{[1]} \in [0, \infty)$ is a piecewise continuous density function, and satisfy the following

$$\rho^{[1]}(\tau) = \rho(-\tau), \quad \kappa := \int_{-\infty}^{+\infty} |\tau|\rho(\tau)d\tau. \tag{1}$$

By [1], we know that the smoothing function of $|x|$ is $s_p^{[1]}(x) = \int_{-\infty}^{+\infty} \left|x - \frac{\tau}{p}\right|\rho(\tau)d\tau$, then by $(x)_+ = \frac{x+|x|}{2}$, we can get the smoothing function of $(x)_+$ as follows

$$\tilde{f}_p(x) = \frac{x}{2} + \frac{1}{2}\int_{-\infty}^{+\infty} \left|x - \frac{\tau}{p}\right|\rho(\tau)d\tau \tag{2}$$

© Springer International Publishing AG 2018
F. Qiao et al. (eds.), *Recent Developments in Mechatronics and Intelligent Robotics*,
Advances in Intelligent Systems and Computing 691, DOI 10.1007/978-3-319-70990-1_34

where $p > 0$ is smoothing parameter, and we have that $\tilde{f}_p(x)$ uniformly approximates to $(x)_+$ with $p \to +\infty$.

The structure of this paper is as follows. In Sect. 2, smooth approximation properties of $\tilde{f}_p(x)$ are described. In Sect. 3, the application of the smoothing function (2) are discussed, and an application example of support vector machines is also given.

2 Smooth Approximation Properties of Smoothing Function

Proposition 2.1. Some smooth approximation properties of smoothing function $\tilde{f}_p(x)$ as follows.

(I). $\tilde{f}_p(x)$ is continuously differentiable on R, then we can have that

$$\nabla_x \tilde{f}(x, p) = \frac{1}{2} + \int_0^{px} \rho(\tau) d\tau.$$

(II). $\tilde{f}_p(x)$ converges uniformly to $(x)_+$ on R.

(III). $\tilde{f}_p(x)$ satisfies the gradient consistency as follows

$$G\tilde{f}_p(x) = \partial(x)_+.$$

where $G\tilde{f}_p(x)$ is the subdifferential associated with $\tilde{f}_p(x)$ at a point x.

(IV). $\nabla_x \tilde{f}(x, p)$ is Lipschitz continuous on R for any $p > 0$.

Proof. Similar to [1], we give the following proofs.

(I).

$$\tilde{f}(x, p) = \frac{x}{2} + \frac{1}{2} \int_{-\infty}^{+\infty} \left| x - \frac{\tau}{p} \right| \rho(\tau) d\tau$$

$$= \frac{x}{2} + \frac{1}{2} \left(2x \int_0^{px} \rho(\tau) d\tau + \frac{1}{p} \left(\int_{px}^{+\infty} \tau\rho(\tau) d\tau - \int_{-\infty}^{px} \tau\rho(\tau) d\tau \right) \right),$$

where the last equality can be obtained by formula (1). We need to point out that $\rho(\tau) \geq 0$ and $\int_{-\infty}^{+\infty} |\tau| \rho(\tau) d\tau < \infty$. By the integral mean value theorem, we can get that

$$\nabla_x \tilde{f}(x, p) = \frac{1}{2} \left(\lim_{\Delta x \to 0} \frac{s_p(x + \Delta x) - s_p(x)}{\Delta x} \right) + \frac{1}{2}$$

$$= \frac{1}{2} \left(\lim_{\Delta x \to 0} 2 \int_0^{p(x + \Delta x)} \rho(\tau) d\tau + \lim_{\Delta x \to 0} \frac{2x \int_{px}^{p(x + \Delta x)} \rho(\tau) d\tau - \frac{2}{p} \int_{px}^{p(x + \Delta x)} \tau\rho(\tau) d\tau}{\Delta x} \right) + \frac{1}{2} \text{ where}$$

$$= \frac{1}{2} \left(2 \int_0^{px} \rho(\tau) d\tau + \lim_{\Delta x \to 0} \frac{2px - \xi}{p} \frac{1}{\Delta x} \int_{px}^{p(x + \Delta x)} \rho(\tau) d\tau \right) + \frac{1}{2}$$

$$= \int_0^{px} \rho(\tau) d\tau + \frac{1}{2}.$$

$$\xi \in [px, p(x + \Delta x)].$$

(II). By $(x)_+ = \frac{x+|x|}{2}$, we can get

$$
\begin{aligned}
|\tilde{f}(x,p) - (x)_+| &= \left| \frac{1}{2} \int_{-\infty}^{+\infty} \left| x - \frac{\tau}{p} \right| \rho(\tau) d\tau - \frac{|x|}{2} \right| \\
&= \frac{1}{2} \left| \int_{-\infty}^{+\infty} \left| x - \frac{\tau}{p} \right| \rho(\tau) d\tau - |x| \right| \leq \frac{1}{2} \int_{-\infty}^{+\infty} \left| \frac{\tau}{p} \right| \rho(\tau) d\tau \\
&= \frac{1}{2} \int_{-\infty}^{+\infty} \frac{1}{p} |\tau| \rho(\tau) d\tau = \frac{\kappa}{2p}.
\end{aligned}
$$

where $\kappa := \int_{-\infty}^{+\infty} |\tau| \rho(\tau) d\tau$.

(III). By $\nabla_x \tilde{f}(x,p) = \int_0^{px} \rho(\tau) d\tau + \frac{1}{2}$ and density function ρ defined by formula (1), we know

$$
\begin{aligned}
\lim_{p \to +\infty, x \to x^*} \nabla_x \tilde{f}(x,p) &= \lim_{p \to +\infty, x \to x^*} \left(\frac{1}{2} + \int_0^{px} \rho(\tau) d\tau \right) \\
&= \frac{1}{2} + \lim_{p \to +\infty, x \to x^*} \int_0^{px} \rho(\tau) d\tau.
\end{aligned}
$$

(a). If $|x^*| \neq 0$, we can get

$$
\lim_{p \to +\infty, x \to x^*} \nabla_x \tilde{f}(x,p) = \begin{cases} 1, & x^* > 0 \\ 0, & x^* < 0. \end{cases} \tag{3}
$$

If $|x^*| = 0$, we obtain that

$$
\lim_{p \to +\infty, x \to 0} \nabla_x \tilde{f}_p(x) = \begin{cases} \{1\}, & px \to +\infty \\ \{\alpha\}, & \lim_{p \to +\infty, x \to 0} |px| < |\infty \\ 0, & px \to -\infty. \end{cases} \tag{4}
$$

where $\alpha \in [0, 1]$, by (3)–(4), we have

$$
\lim_{p \to +\infty, x \to 0} \nabla_x \tilde{f}(x,p) \subseteq \partial(x)_+ = \begin{cases} 1, & x > 0 \\ [0, 1], & x = 0 \\ 0, & x < 0. \end{cases} \tag{5}
$$

where $\partial(x)_+$ is the Clarke subdifferential of $(x)_+$.

(b). Define $\alpha(\lambda) = \frac{1}{2} + \int_0^{\lambda} \rho(\tau) d\tau$, then we have $\alpha(\lambda) \in [0, 1]$ is continuous in R by ρ is defined as formula (1). Thus, there exists λ_0 such that $\alpha_0 = \alpha(\lambda_0)$ for any $\alpha_0 \in (0, 1)$. When we choose $p_k x_k = \lambda_0$ and $p_k \to +\infty$, we get $\alpha_0 = \lim_{p_k \to +\infty, x_k \to 0} \nabla_x \tilde{f}(x_k, p_k)$. This means that

$$[0, 1] \subseteq \lim_{p \to +\infty, x \to 0} \nabla_x \tilde{f}(x, p).$$

The subdifferential $G\tilde{f}(x, p)$ is as follows

$$G\tilde{f}(x, p) = con\{\gamma | \nabla_x \tilde{f}(x_k, p_k) \to \gamma, x_k \to x, p_k \to +\infty\},$$

therefore, we obtain that $G\tilde{f}(x, p) = \partial(x)_+$.

This also shows that $\tilde{f}(x, p)$ satisfies the gradient consistency.

(IV). By formula (1), there exists a constant κ_0 such that $\rho(x) \leq \kappa_0$ holds for $\forall x \in R$. Then for any $x_1, x_2 \in R$, we obtain that

$$\left\| \nabla_x \tilde{f}(x_1, p) - \nabla_x \tilde{f}(x_2, p) \right\| = \left\| \frac{1}{2} + \int_0^{px_1} \rho(\tau) d\tau - \frac{1}{2} - \int_0^{px_2} \rho(\tau) d\tau \right\|$$

$$= \left\| \int_0^{px_1} \rho(\tau) d\tau - \int_0^{px_2} \rho(\tau) d\tau \right\|$$

$$= \left\| \int_{px_2}^{px_1} \rho(\tau) d\tau \right\| \leq p\kappa_0 \|x_1 - x_2\|.$$

Therefore, $\nabla_x \tilde{f}(x, p)$ is Lipshitz continuous.

3 Discussion on Application of Smoothing Function

The standard SVM problem can be equivalently transformed into an nonsmooth unconstrained optimization problem [7–10] as follows

$$\min_{(\omega, \gamma) \in R^{n+1}} \frac{1}{2} v \left\| (e - D(A\omega - e\gamma))_+ \right\|_2^2 + \frac{1}{2} (\omega^T \omega + \gamma^2) \tag{6}$$

where A is $m \times n$ matrix, $A_i (i = 1, \ldots, m)$ represents the m point in R^n, and $D_{m \times n} = diag(d_1, \ldots, d_m)$, where $d_i = +1$ or -1, and $e = (1, \ldots, 1) \in R^m$.

Due to the non smoothness of the plus function, many scholars put forward the idea of smoothing (6). For instance, the sigmoid function is presented in [4], after that, Xiong et al. proposed a new class of smoothing functions by interpolation function in [5].

When we choose $\rho^{[2]}(\tau) = \frac{2}{(\tau^2 + 4)^{\frac{3}{2}}}$ in (2), then we get the smoothing function of $(x)_+$ is

$$\tilde{f}(x, p) = \frac{1}{2} \left(x + \sqrt{x^2 + \frac{4}{p^2}} \right). \tag{7}$$

We use smoothing function (7) to smooth the objective function of problem (6), so that we can obtain a equivalent smoothing unconstrained optimization problem of problem (6), as follows

$$\min_{x} \tilde{\varphi}_p(x) \tag{8}$$

where $\tilde{\varphi}_p(x) = \frac{1}{2} v \left\| \frac{1}{2} \left(r + \sqrt{r^2 + \frac{4}{p^2}} \right) \right\|_2^2 + \frac{1}{2} (\omega^T \omega + \gamma^2), r = (e - D(A\omega - e\gamma)),$

and $p > 0$ is a smoothing parameter.

Remark: Due to the equivalence of problems (6) and (8), we can use the general method for solving the unconstrained optimization problem to solve the problem (8), then we can get the solution of problem (6).

References

1. Chen, X.J., Zhou, W.J.: Smoothing nonlinear conjugate gradient method for image restoration using nonsmooth nonconvex minimization. SIAM J. Imaging Sci. 3(4), 765–790 (2010)
2. Chen, X.J.: Smoothing methods for nonsmooth, nonconvex minimization. Math. Progr. 134, 71–99 (2012)
3. Chen, C., Mangasarian, O.L.: A class of smoothing functions for nonlinear and mixed complementarity problems. Comput. Optim. Appl. 5(2), 97–138 (1996)
4. Lee, Y.J., Mangasarian, O.L.: A smooth support vector machine for classification. Comput. Optim. Appl. 20(1), 5–21 (2001)
5. Xiong, J.Z., Hu, J.L., Yuan, H.Q., Hu, T.M., Li, G.M.: Research on a new class of smoothing support vector machines. Acta Electr. Sin. 35(2), 366–370 (2007). (in Chinese)
6. Yuan, Y.B., Yan, J., Xu, C.X.: Polynomial smooth support vector machine. Chin. J. Comput. 28(1), 9–17 (2005). (in Chinese)
7. Zheng, J., Lu, B.L.: A support vector machine classifier with automatic confidence and its application to gender classification. Neurocomputing 74(11), 1926–1935 (2011)
8. Zhou, X., Wang, Y., Wang, D.L.: Application of kernel methods in signals modulation classification. J. China Univ. Posts Telecommun. 18(1), 84–90 (2011)
9. Zhao, Z.C.: Combining SVM and CHMM classifiers for porno video recognition. J. China Univ. Posts Telecommun. 19(3), 100–106 (2012)
10. Lin, H.J., Yeh, J.P.: Optimal reduction of solutions for support vector machines. Appl. Math. Comput. 214(2), 329–335 (2009)

Mixed Power Saving Mode for D2D Communications with Wi-Fi Direct

Shuo Xiao[✉] and Wei Li

School of Computer Science and Technology, China University of Mining and Technology,
Xuzhou 221116, China
sxiao@cumt.edu.cn

Abstract. Wi-Fi Direct is a technology that in order to enhance the D2D communications. Nowadays power management is becoming more and more important. The existing two power saving modes are inadequate to meet the power efficiency and reliability requirements of various services. This paper presents a performance analysis of Wi-Fi Direct power saving modes and exploits their problems to design a novel power saving mode—Mixed Power Save Mode (MPSM).

Keywords: MPSM · Power saving · Wi-Fi direct · D2D

1 Introduction

Wi-Fi Direct is a Wi-Fi standard that the device can connect to each other easily, without the need of the wireless Access Point (AP). With the recent advances in Wi-Fi Direct devices, power conservation issues have also risen sharply [2]. The exist two Wi-Fi Direct power management modes are Opportunistic Power Saving Mode (OPSM) and Notice of Absence (NOA) mode. They can not supply enough functionality to respond to different services in the network. Therefore, it is necessary to improve the mode in Wi-Fi Direct to cope with services and applications.

A novel power saving mode is proposed to alleviate the problems of Wi-Fi Direct. The rest of the paper is organized as follows: in Sect. 2, the two power management schemes of Wi-Fi Direct specification are analyzed, and introduce some related work. We analyze the proposed scheme MPSM in Sect. 3. Then in Sect. 4, simulation and analysis are provided. Section 5 gives some conclusions.

2 Related Works

2.1 OPSM and NoA

Wi-Fi Direct devices communicate through the peer-to-peer networks, which are functionally equivalent to the traditional Wi-Fi infrastructure network. When the P2P group is set up, the new devices can be found and joined the group by using active or passive scanning mechanisms. In order to balance the power between GO and P2P client, Wi-Fi Direct defines two power saving modes for P2P GO: OPSM and NoA.

© Springer International Publishing AG 2018
F. Qiao et al. (eds.), *Recent Developments in Mechatronics and Intelligent Robotics*,
Advances in Intelligent Systems and Computing 691, DOI 10.1007/978-3-319-70990-1_35

OPSM can be considered as a P2P GO initiated mode in which the absent/present depends mainly on the client's data transfer. At the start of each beacon, the Client Traffic Window (CTWindow) is configured for a predetermined time shorter than the beacon interval itself. NoA can be used by the P2P GO to save power, regardless of the power status of its related clients. In the case of NoA, the P2P GO defines absence periods with a signaling element included in beacon frames and Probe Responses.

2.2 Related Works

There have some studies on power saving mode. [4] presents a brief overview and the performance on NoA. [5] has tried to use Wi-Fi Direct power saving to enhance the power efficiency. To improve power efficiency, [7, 8] dynamically adjustment the duty cycle. Other study like [9] proposed a technique that uses audio channels. The performance analysis is carried out by using the actual test platform in [10].

3   MPSM: Mixed Power Saving Mode

3.1 Mechanism of MSPM

OPSM is more suitable for small data transmission, and NoA for large data transmission,so that the goal of MPSM is to strengthen the energy management of mixed network. To achieve this goal, we will set up a Back Parameters (*BP*) for P2P client. P2P GO will use OPSM and NoA by mixed them according to client's feedback. The process is shown in Fig. 1. When a client has a small amount of data needs to transmitter, it will set the $BP = 1$, GO will change the power-saving mode to OPSM after receiving the BP. Referring to the analysis above, we know that in a small amount of data,this model has better efficient power saving. When client find that it have a lot of data need to be transmitted, it will reply to P2P GO a $BP = 0$, then P2P GO will change the mode to NoA, the analysis above shows that the large throughput is suitable for this mode. According to this timely feedback from the client, P2P GO can promptly adjust its power saving mode, to achieve the goal of power saving.

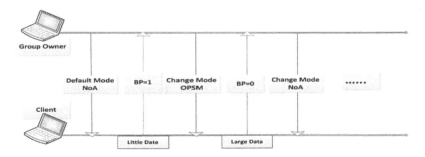

Fig. 1. Mixed power saving mode

3.2 Theoretical Analysis

The MPSM is evaluated through mathematical analysis. We use following time parameters: T_{CTW} is the length of the CTWindow; T is the beacon interval time; t is the cyclicity
data transmission. For OPSM if PM equal to 1, the duty cycle will be expressed as (1).

$$C_{OPSM(PM=1)} = \frac{T_{CTW}}{T} \tag{1}$$

If the PM is set to 0.Then the duty cycle depends on the CTWindow and time t.
Therefore, this situation can be summarized as (2). And the P_{awake} is the probability of
remaining awake.

$$C_{OPSM(PM=0)} = \begin{cases} 100\% & if\ t < T_{CTW} \\ \dfrac{T_{CTW}}{T} + \dfrac{T - T_{CTW}}{T} P_{awake} & if\ t \geq T_{CTW} \end{cases} \tag{2}$$

For NoA, the duty cycle is given as a static value decided by the P2P GO. Thus, the
NoA duty cycle can be described in (3). The T_{NoAA} is active phase of one sleep cycle,
T_{NoAS} is absent period of a sleep cycle.

$$C_{NoA} = \frac{T_{NoAA}}{T_{NoAA+NoAS}} \tag{3}$$

When it has only a small amount of data to be transmitted, MPSM uses OPSM, when
it has large amounts of data to be transmitted, it uses NoA. So we assume that the
probability of using OPSM is P_{OPSM}, and the probability of using NoA is P_{NoA}. The duty
cycle of MPSM can be expressed as (4)

$$C_{MPSM} = \frac{T_{NoAA}}{T_{NoAA+NoAS}} P_{NoA} + C_{OPSM} + P_{OPSM} \tag{4}$$

Notice that, the sum of P_{OPSM} and P_{NoA} is equal to 1, $P_{OPSM} + P_{NoA} = 1$. They both
greater than or equal to 0 and less than or equal to 1. So we can get the inequality (5) as
follow:

$$min\{C_{NoA}, C_{OPSM}\} \leq C_{MPSM} \leq max\{C_{NoA}, C_{OPSM}\} \tag{5}$$

From the above (5) we can see that the value of C_{MPSM} is between the C_{NoA} and the
C_{OPSM}. So that we can be prior to ensure the quality of service, and on this basis, as much
as possible to save power.

4   Analysis and Simulation

Table 1 lists the parameter configurations. The analysis and simulation of the network environment configuration is shown in Table 2. The main factor in the comparison of the performance is duty cycle.

Table 1. Settings of parameter

Unit(ms)	MPSM	OPSM	NoA
T	90	90	90
T_{CTW}	25	25	/
T_{NoAA}, T_{NoAS}	15	/	15

Table 2. Settings of Simulation

Data	Value
Type	CBR
Size	1 Kb
Interval	90 ms-1 ms
Interface	802.11 g
Range	100 m

For the sake of further analyze the proposed MPSM, we model the Wi-Fi Direct energy saving mode which using NS-3 simulator and to evaluate its performance. In the simulation, we are utilizing some channels from 1 to 12 in frequency range of 2.4 GHz. One node is settled as P2P GO, and others nodes are set as the clients. Within a time of 90 s, these nodes generate periodic CBR and 1 K packet sizes and intervals from 90 ms to 1 ms. MPSM is compared to NoA and also to OPSM when the PM configuration is 0 or 1. We use Table 1 with the parameters setting in it to simulate as well.

As the following figures show is that the performance of the energy saving mode when where is one GO and one client. Parameter n represents the number of clients in the network. The distance between the nodes was configured between 1 to 100 meters. In order to observe the effect on energy saving mode we reduce the interval of the data. The simulation result is seeing as Fig. 2. We can see that the trends in performance analysis and simulation results are similar. In addition, as shown in the Fig. 3, due to the mechanism of the OPSM, resulting in an increase in duty cycle with PM = 1 while $n = 2$. The problem with multi-user environment is that if the customer does not transfer data during the CTWindow, the customer will assumes that the GO is sleeping and then does not transmit data during the beacon interval. However, the fact is that the GO may be awake in CTWindow and is receiving packets from another client. This is the reason that the OPSM has a high duty ratio as shown in Fig. 4.

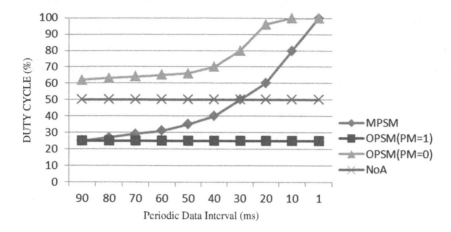

Fig. 2. Duty Cycle (n = 1)

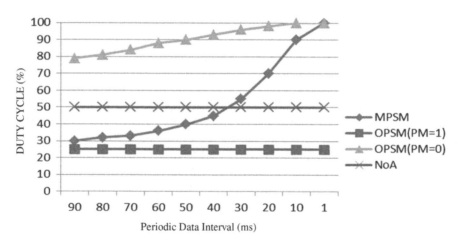

Fig. 3. Duty Cycle (n = 2)

With the number of nodes increased and the data generated by each client in the network, the amount of data generated increases. As showed in Fig. 4, for the OPSM, the duty cycle increased to 100%, because there are too many data processing and GO need to keep awake. When the number of clients is few, MPSM can reduce the duty cycle. And when the network has many clients, it sharply increases in the duty cycle. This is because the MPSM is based on the Quality of Service (QoS). Compared to OPSM and NOA, MPSM is more suitable for the mixed network environment.

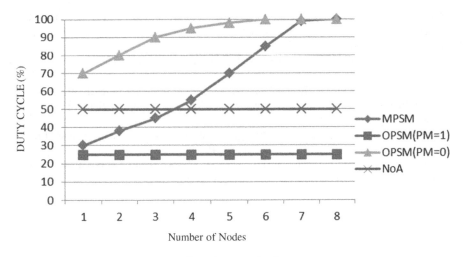

Fig. 4. Duty Cycle

5 The Conclusion

Wi-Fi Direct is promising in the future that it will become the next outstanding D2D communication technology for a variety of services and applications. Based on this, we propose a hybrid model to improve the power saving performance of Wi-Fi Direct. The performance evaluation indicates that the mode we presented can adjust the duty cycle of the data throughput model to guarantee better power efficiency and the reliability of the service. For the future, we can take more complex traffic models and multi-clients into account.

Acknowledgement. This work was supported by the Natural Science Foundation of Jiangsu Province of China under Grant BK20150193 and BK20150201.

References

1. Vergara, E., Nadjm-Tehrani, S.: Watts2Share: energy-aware traffic consolidation. In: IEEE Greencom 2013, August 2013
2. Trifunovic, S., Picu, A., Hossmann, T., Hummel, K.: Slicing the battery pie: fair and efficient energy usage in device-to-device communication via role switching. In: ACM CHANTS 2013, September 2013
3. IEEE P802.11™-2012, Part 11.: Wireless LAN medium access control(MAC) and physical layer (PHY) Specifications, IEEE (2012)
4. Camps-Mur, D., Garcia-Saavedra, A., Serrano, P.: Device to device communications with WiFi direct: overview and experimentation. IEEE Wirel. Commun. Mag. **55**, 2838–2855 (2012)
5. Camps-Mur, D., Perez-Costa, X., Sallent-Ribes, S.: Designing energy efficient access points with Wi-Fi direct. Comput. Netw. **55**, 2838–2855 (2011)

6. Choi, J., Ko, Y.-B., Kim, J.: Enhanced power saving scheme for IEEE 802.11 DCF based wireless networks. In: 8th International Conference on Personal Wireless Communications, pp. 835–840 (2003)
7. Jung, E., Vaidya, N.: Improving IEEE 802.11 power saving mechanism. Wirel. Netw. **14**, 375–391 (2008)
8. Hu, X., Chen, Z., Yang, Z.: Energy-efficient scheduling strategies in IEEE 802.11 wireless LANs. In: International Conference on Computer Science and Automation Engineering, pp. 570–572 (2012)
9. Uddin, M., Nadeem, T.: A2PSM: audio assisted Wi-Fi power saving mechanism for smart devices. In: ACM HotMobile 2013, February 2013
10. Conti, M., Delmastro, F., Minutiello, G., Paris, R.: Experimenting opportunistic networks with WiFi direct. In: IFIP Wireless Days 2013, November 2013

Evidence Combination Based on Singular Value

Dawei Xue[1,2(✉)], Yong Wang[2], Chunlan Yang[1], and Kangkai Gao[2]

[1] School of Electronics and Electrical Engineering,
Bengbu University, Bengbu, China
bbxuedawei@163.com
[2] School of Information Science and Technology,
University of Science and Technology of China, Hefei, China

Abstract. Evidence theory is an efficient tool to fuse multi-source information. However, when Dempster's rule of combination is applied to combine highly conflicting evidences, counter-intuitive results may emerge. In order to resolve such a problem, a new weighted averaging evidence combination method based on singular value was put forward. Minimum singular value can measure the conflict between two evidences more accurately than other common methods. Thus, using singular value to determine the weights of evidences is more reasonable. Based on the weighted averaged evidences and Dempster's rule, the final results are acquired. The numerical example illustrates that the proposed method can fuse highly conflict evidences effectively and has faster convergence speed and better focusing degree than some other existing combination methods.

Keywords: Information fusion · Evidence theory · Evidence combination · Singular value

1 Introduction

Multi-source information is more and more applied in pace with the development of sensor technology. Generally, this information has some uncertainty. Evidence theory [1] is regarded as an effective method to deal with uncertain information, which has been used in many fields such as target recognition [2], fault diagnosis [3], pattern recognition [4], etc. In evidence theory, Dempster's rule is the most common method to combine evidences from different sources. However, it can derive counter-intuitive results when highly conflicting evidences are combined. In order to solve such a problem many methods have been proposed. In general, there are two main types of methodologies. One view is to modify Dempster's rule and another view is to modify evidence source. The first type of methodology imputes the counter-intuitive behaviors to the combination rule. Yager proposed redistributing the conflicting belief assignments to the unknown domain [5]. Lefevre proposed a unified formulation for modified combination rule [6]. However, these methods are complex and destroy the good properties of Dempster's rule such as commutativity and associativity. The second type of methodology imputes the counter-intuitive behaviors to the evidence sources obtained by sensors. Murphy put forth a preprocessing method based on simple

F. Qiao et al. (eds.), *Recent Developments in Mechatronics and Intelligent Robotics*,
Advances in Intelligent Systems and Computing 691, DOI 10.1007/978-3-319-70990-1_36

arithmetic average of the original evidences [7]. It effectively suppresses the influence on the combination results caused by interference evidences and converges faster than Dempster's rule.

We prefer to modify the evidence source. In Murphy's simple averaging approach all the evidence have equal weights. However, it is not always reasonable. Minimum singular value can measure the conflict between two evidences more effectively than traditional methods [8]. If singular value can be used to determine the weights of evidences, better performance can be expected. Thus, a new weighted averaging combination approach is proposed in this paper. Minimum singular value is used to produce the weights. A numerical example illustrates the reasonability and effectiveness of the presented method.

2 Fundamental Theory and Its Deficiency

2.1 Basis of Evidence Theory

Suppose that the frame of discernment (FOD) is Θ whose power set is 2^{Θ}. The basic belief assignment (BBA) $m: 2^{\Theta} \rightarrow [0,1]$ can be defined as follows [1]:

$$\begin{cases} m(\emptyset)= 0 \\ m(A) \geq 0, \sum_{A \subseteq \Theta} m(A)=1, \forall A \subseteq \Theta \end{cases} \tag{1}$$

When $m(A) > 0$, A is called a focal element.

When n independent BBAs are available, the combination result with Dempster's rule of combination is described as follows [1]:

$$m(A) = \begin{cases} 0, & A = \emptyset \\ \dfrac{\sum_{\cap A_j=A} \prod_{i=1}^{n} m_i(A_j)}{1-k}, & A \neq \emptyset \end{cases} \tag{2}$$

where k indicates the degree of conflict among all the BBAs, which is called conflict coefficient, and

$$k = \sum_{\cap A_j=\emptyset} \prod_{i=1}^{n} m_i(A_j) \tag{3}$$

2.2 Deficiency of Dempster's Rule of Combination

When using Dempster's rule to combine the highly conflicting evidences, one can often get counter-intuitive results.

Example 1. Let FOD be $\Theta = \{\omega_1, \omega_2, \omega_3\}$, two BBAs obtained are as follows:

$$m_1(\{\omega_1\}) = 0.99, m_1(\{\omega_3\}) = 0.01; m_2(\{\omega_2\}) = 0.99, m_2(\{\omega_3\}) = 0.01.$$

The final decision by using Dempster's rule of combination to fuse these two BBAs is as: $m(\{\omega_1\}) = 0$, $m(\{\omega_2\}) = 0$, $m(\{\omega_3\}) = 1$. It means that event $\{\omega_3\}$ definitely occurs. Obviously, this combination result is counter-intuitive.

3   A New Evidence Combination Method

Assume that there are n BBAs $m_1, m_2, \ldots, m_n$, the modifying of evidence sources based on weights can be illustrated as:

$$m_{WAE} = \sum_{i=1}^{n} (\beta_i \cdot m_i) \tag{4}$$

where β_i is the weight of m_i and m_{WAE} denotes the weighted averaged BBA of n BBAs of the original evidences. Then, Dempster's rule of combination can be used to fuse m_{WAE} for $n - 1$ times to acquire the final result.

3.1   Evidence Conflict Measure Method Based on Singular Value

It has been verified that conflict coefficient k cannot accurately represent conflict between two evidences. Some new approaches to measure evidence conflict were proposed [9–11], Jousselme distance and Pignistic probability distance are representatives of which.

Assume that m_1 and m_2 are two BBAs defined over FOD Θ, Jousselme distance is define as [9]:

$$d_J(m_1, m_2) = \sqrt{\frac{1}{2}(\boldsymbol{m_1} - \boldsymbol{m_2})^T \cdot \boldsymbol{D} \cdot (\boldsymbol{m_1} - \boldsymbol{m_2})} \tag{5}$$

where $\boldsymbol{m_1}$ and $\boldsymbol{m_2}$ are 2^n column vectors of m_1 and m_2, $\boldsymbol{D}$ is a $2^n \times 2^n$ matrix whose elements is defined as follows [9]:

$$D(A, B) = \frac{|A \cap B|}{|A \cup B|}, \ A, B \subseteq \Theta \tag{6}$$

Assume that the Pignistic probability transitions of m_1 and m_2 are $BetP_{m1}$ and $BetP_{m2}$ respectively. Then, Pignistic probability distance is defined as [10]:

$$difBetP(m_1, m_2) = \max_{A \subseteq \Theta}(|BetP_{m1} - BetP_{m2}|) \tag{7}$$

Example 2. Let FOD be $\Theta = \{\omega_1, \omega_2, \omega_3, \omega_4, \omega_5\}$, and m_1 and m_2 are two BBAs:

$$m_1(\{\omega_1\}) = 0.8, m_1(\{\omega_2, \omega_3, \omega_4, \omega_5\}) = 0.2; m_2(\Theta) = 1.$$

According to Eq. (3), the conflict coefficient $k = 0$, which is not consistent with the practice. According to Eqs. (6) and (7), the results are $d_J = 0.732$ and $difBetP = 0.6$, both of which can represent the conflict reasonably. Are d_J and $difBetP$ are always effective to measure the evidence conflict?

Example 3. Let FOD be $\Theta = \{\omega_1, \omega_2, \omega_3\}$, and m_1, m_2 and m_3 are three BBAs over Θ:

$$m_1(\{\omega_1\}) = 1/3, m_1(\{\omega_1\}) = 1/3, m_1(\{\omega_3\}) = 1/3; m_2(\{\omega_1\}) = 1; m_3(\Theta) = 1.$$

Analyzing these three BBAs intuitively, we can draw the conclusion: $conf(m_2, m_3) > conf(m_1, m_2) > conf(m_1, m_3)$. According to Eqs. (6) and (7), the results are as follows: $d_J(m_1, m_2) = 0.577$, $d_J(m_1, m_3) = 0.577$, $d_J(m_2, m_3) = 0.817$; $difBetP$ $(m_1, m_2) = 0.667$, $difBetP$ $(m_1, m_3) = 0$, $difBetP$ $(m_2, m_3) = 0.667$. Both d_J and $difBetP$ are not corresponding with intuitive analysis.

Suppose that m_1, m_2 are two BBAs over Θ, and F_1 and F_2 are the sets which consist of the focal elements of each BBA. If F is denoted as $F = F_1 \cup F_2 = \{C_1, C_2, \cdots, C_m\}$, the corresponding matrix of two BBAs can be described as [8]:

$$\boldsymbol{M} = \begin{bmatrix} m_1(C_1)\, m_1(C_2) \cdots m_1(C_m) \\ m_2(C_1)\, m_2(C_2) \cdots m_2(C_m) \end{bmatrix} \tag{8}$$

Then, the measure method for evidence conflict based on singular value is defined as [8]:

$$diSV = \min(\sigma(\boldsymbol{M} \cdot \boldsymbol{D})) \tag{9}$$

where $\boldsymbol{D}$ is a $m \times m$ matrix and $\sigma(\boldsymbol{M} \cdot \boldsymbol{D})$ represents singular values of $\boldsymbol{M} \cdot \boldsymbol{D}$. The elements of $\boldsymbol{D}$ can be defined as follows [8]:

$$\boldsymbol{D}(C_i, C_j) = \frac{|C_i \cap C_j|}{|C_i \cup C_j|}, C_i, C_j \in F \tag{10}$$

Consider the three BBAs in example 3 again. Using Eq. (9), one can get the result: $diSV_{13} = 0.667 > diSV_{12} = 0.442 > diSV_{23} = 0.296$. This judgement is corresponding with intuitive analysis. So as to prove the effectivity of the evidence conflict measure method based on singular value further, let's see example 4.

Example 4. Let FOD be $\Theta = \{1, 2, \ldots, 20\}$, and m_1, m_2 are two BBAs over Θ:

$$m_1(A) = 0.8, m_1(\{7\}) = 0.05, m_1(\{2, 3, 4\}) = 0.5, m_1(\Theta) = 0.1; m_2(\{1, 2, 3, 4, 5\}) = 1.$$

A is a subset of Θ, which changes as $\{1\}, \{2\},..., \{1, 2, ..., 20\}$. The conflict between two BBAs with the change of A measured by k, d_J, $difBetP$ and $diSV$ are shown in Fig. 1.

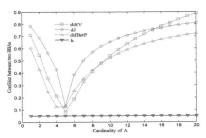

Fig. 1. Changing curves of evidence conflict measured by different methods

As shown in Fig. 1, $diSV$ can represent conflict between two BBAs more precisely than other method. Therefore, using singular value to determine the weights of evidences is more reasonable.

3.2 Weights Determination Based on Singular Value

It is easy to prove that $diSV$ has such properties [8]: $0 \le diSV(m_i, m_j) \le 1$; $diSV(m_i, m_j) = diSV(m_j, m_i)$; $m_i = m_j \Longleftrightarrow diSV(m_i, m_j) = 0$.

The smaller the $diSV$ is, the more similar two evidences are. Then, the degree of similarity between m_i and m_j can be defined as:

$$Sim(m_i, m_j) = 1 - diSV(m_i, m_j) \tag{11}$$

where m_i and m_j are two BBAs defined on FOD Θ.

The degree of similarity between two evidences reflects the degree of support between two evidences. Suppose that $m_1, m_2, ..., m_n$ are n BBAs over FOD Θ. Then, the total degree of support of the evidence can be described as follows:

$$Sup(m_i) = \sum_{j=1, j \neq i}^{n} Sim(m_i, m_j) \tag{12}$$

If one evidence has relatvely high degree of support defined in Eq. (12), it was relatively credible. Then, the degree of credibility of the evidence is as follows:

$$Crd(m_i) = \frac{Sup(m_i)}{\sum_{j=1}^{n} Sup(m_j)} \tag{13}$$

When some evidences have higher degree of credibility, it will be more helpful for make right and solid decisions. Such evidences should be distributed relatively larger weights. Therefore, we use the degree of credibility defined in Eq. (13) as the weights of evidences.

4 Numerical Example

Assume there is a multi-target recognition system, and the FOD is $\Theta = \{\omega_1, \omega_2, \omega_3\}$. There are five evidences acquired by five different sensors, the BBAs of which are as follows:

$$m_1(\{\omega_1\}) = 0, m_1(\{\omega_1\}) = 0.9, m_1(\{\omega_3\}) = 0.1; m_2(\{\omega_1\}) = 0.5, m_2(\{\omega_2\}) = 0.2,$$
$$m_2(\{\omega_3\}) = 0.3; m_3(\{\omega_1\}) = 0.55, m_3(\{\omega_2\}) = 0.1, m_3(\{\omega_3\}) = 0.35; m_4(\{\omega_1\}) = 0.55,$$
$$m_4(\{\omega_2\}) = 0.1, m_4(\{\omega_3\}) = 0.35; m_5(\{\omega_1\}) = 0.55, m_5(\{\omega_2\}) = 0.1, m_5(\{\omega_3\}) = 0.35.$$

The fusion results of several different combination approaches are listed in Table 1. According to Table 1, Dempster's rule cannot fuse the conflicting evidences effectively, which generate counter-intuitive results. Similarly, Yager's rule cannot derive rational results. The combination results of both Murphy's simple averaging approach and weighted averaging method presented in this paper are reasonable. However, the approach presented in this paper has both faster convergence speed and better focusing degree than Murphy's method. It is because that the proposed method can enhance the effect of credible evidences and suppress the effect of incredible evidences simultaneously.

Table 1. Fusion results by using several combination approaches

Method	m_1, m_2	m_1, m_2, m_3	M_1, m_2, m_3, m_4	m_1, m_2, m_3, m_4, m_5
Dempster's rule	$m(\{\omega_1\}) = 0$ $m(\{\omega_2\}) = 0.8571$ $m(\{\omega_3\}) = 0.1429$	$m(\{\omega_1\}) = 0$ $m(\{\omega_2\}) = 0.6316$ $m(\{\omega_3\}) = 0.3684$	$m(\{\omega_1\}) = 0$ $m(\{\omega_2\}) = 0.3288$ $m(\{\omega_3\}) = 0.6712$	$m(\{\omega_1\}) = 0$ $m(\{\omega_2\}) = 0.1228$ $m(\{\omega_3\}) = 0.8772$
Yager's rule	$m(\{\omega_1\}) = 0$ $m(\{\omega_2\}) = 0.1800$ $m(\{\omega_3\}) = 0.0300$ $m(\Theta) = 0.7900$	$m(\{\omega_1\}) = 0$ $m(\{\omega_2\}) = 0.0180$ $m(\{\omega_3\}) = 0.0105$ $m(\Theta) = 0.9715$	$m(\{\omega_1\}) = 0$ $m(\{\omega_2\}) = 0.0018$ $m(\{\omega_3\}) = 0.0037$ $m(\Theta) = 0.9945$	$m(\{\omega_1\}) = 0$ $m(\{\omega_2\}) = 0.0002$ $m(\{\omega_3\}) = 0.0013$ $m(\Theta) = 0.9985$
Murphy's rule	$m(\{\omega_1\}) = 0.1543$ $m(\{\omega_2\}) = 0.7469$ $m(\{\omega_3\}) = 0.0988$	$m(\{\omega_1\}) = 0.3500$ $m(\{\omega_2\}) = 0.5224$ $m(\{\omega_3\}) = 0.1276$	$m(\{\omega_1\}) = 0.6027$ $m(\{\omega_2\}) = 0.2627$ $m(\{\omega_3\}) = 0.1346$	$m(\{\omega_1\}) = 0.7958$ $m(\{\omega_2\}) = 0.0932$ $m(\{\omega_3\}) = 0.1110$
This paper	$m(\{\omega_1\}) = 0.1543$ $m(\{\omega_2\}) = 0.7469$ $m(\{\omega_3\}) = 0.0988$	$m(\{\omega_1\}) = 0.5821$ $m(\{\omega_2\}) = 0.2393$ $m(\{\omega_3\}) = 0.1786$	$m(\{\omega_1\}) = 0.8110$ $m(\{\omega_2\}) = 0.0479$ $m(\{\omega_3\}) = 0.1411$	$m(\{\omega_1\}) = 0.9001$ $m(\{\omega_2\}) = 0.0078$ $m(\{\omega_3\}) = 0.0921$

5 Conclusions

In this paper, a novel weighted averaging combination method based on singular value is proposed. Numerical example illustrates that proposed approach can fuse highly conflict evidences effectively. Compared with some other existing combination method, this method has faster convergence speed and better focusing degree. In our future research, to make the evidence combination method more powerful, more factors will be considered in determining the weights of evidences.

6   Foundation Item

Supported by the National Science Foundation of China (61573332), the Key Program for Domestic and Oversea Visit and Research for Excellent Young and Middle-aged Backbone Talents in University of Anhui Province (gxfxZD2016272).

References

1. Shafer, G.A.: Mathematical theory of evidence. Princeton University Press, Princeton (1976)
2. Geng, T., Lu, G.S., Zhang, A.: Intuitionistic fuzzy evidence combination algorithm for multi-sensor target recognition. Control Decis. **27**(11), 1725–1728 (2012)
3. Xu, X.B., Wang, Y.L., Wen, C.L.: Information-fusion method for fault diagnosis based on reliability evaluation of evidence. Control Theory Appl. **28**(4), 504–510 (2011)
4. Liu, Z.G., Pan, Q., Dezert, J.: A new belief-based K-nearest neighbor classification method. Pattern Recogn. **46**(3), 834–844 (2013)
5. Yager, R.R.: On the Dempster-Shafer framework and new combination rules. Inf. Sci. **41**(2), 93–137 (1987)
6. Lefevre, E., Colot, O., Vannoorenberghe, P.: Belief function combination and conflict management. Inf. Fusion **3**(2), 149–162 (2002)
7. Murphy, C.K.: Combining belief functions when evidence conflicts. Decis. Support Syst. **29**(1), 1–9 (2000)
8. Ke, X.L., Ma, L.Y., Wang, Y.: A new method to measure evidence conflict based on singular value. Acta Electron. Sin. **41**(10), 2109–2112 (2013)
9. Jousselme, A.L., Grenier, D., Bosse, E.: A new distance between two bodies of evidence. Inf. Fusion **2**(2), 91–101 (2001)
10. Smets, P.: Decision making in the TBM: the necessity of the pignistic transformation. Int. J. Approx. Reason. **38**(2), 133–147 (2005)
11. Jiang, W., Peng, J.Y., Deng, Y.: New representation method of evidential conflict. Syst. Eng. Electron. **32**(3), 562–564 (2010)

The New Mechanism of Advancing Hierarchical Mobile IPv6 Handoff

Shilin Fang[✉]

College of Computer Science, Hunan Institute of Science and Technology, Yueyang 414006, Hunan, China
36662628@qq.com

Abstract. Mobile IPv6 protocol is time-consuming in duplicated address detection (DAD) and location update link. On the basis of original HMIPv6 module, this scheme has built a DAD table lookup and an active care of address mechanism by introducing an IP address management module with the help of the expanding neighbor discovery protocol. this scheme as resolved the traditional DAD operation problem, which needs passive waiting for detection result; and it also has reduced handoff delay effectively. In the end, this scheme has built the network topology by using the NS-2 simulation tool, which used to take simulation test to that optimization scheme. It has been proved by the result that this scheme can not only require less time for the handoff delay, but also reduce the package lost rate of mobile node.

Keywords: Mobile IPv6 · Duplication address detection · Neighbor discovery protocol

1 Introduction

In order to resolve mobile IPv6 handoff delay, IETF has put forwards the hierarchical mobile IPv6 which divides the whole network into different domains in a logical way by introducing a new functional entity—mobility anchor point (MAP) and using the idea of "region partition". Each domain is managed by a specific mobile anchor point. And this special entity will act as a temporal home agency for mobile node to transmit the groups between mobile node and correspondent node [1, 2]. Hierarchical mobile IPv6 has adopted two cares of address—on-link care-of address and regional care-of address, to distinguish the micro-mobile in domain and macro-mobile in inter-domain of mobile nodes. When mobile node micro moves into the mobility anchor point domain, changes will only happen to the link care-of address. And mobile node only needs to register the process which updates a link care of address LCoA to mobile anchor point. And it is unnecessary to get registration from delivering binding update BU to home agent and correspondent node. This can prevent the delivering of BU message from being sent to home agent and correspondent node, no matter when the mobile node moves. Thus, the reduced handoff delay has brought broad prospects for the promotion of mobile IPv6 and integration of Internet and communication [3, 4].

© Springer International Publishing AG 2018
F. Qiao et al. (eds.), *Recent Developments in Mechatronics and Intelligent Robotics*,
Advances in Intelligent Systems and Computing 691, DOI 10.1007/978-3-319-70990-1_37

2   The Handoff of Hierarchical Mobile IPv6

2.1   The Handoff Process of Hierarchical Mobile IPv6

Hierarchical mobile IPv6 divides the mobile into micro-mobile and macro-mobile. In micro-mobile, when mobility happens in a MAP domain, it will only change LCoA, not RCoA [5, 6]; while in macro-mobile, when mobile node crosses a MAP domain, RCoA will change. When mobile node moves from an AR to another one, it will receive a routing broadcast of AR. MAP options in the broadcast include many groups of MAP messages. It mainly has MAP Global Unicast Address, priority and their distance from mobile nodes. From those messages, mobile nodes can decide whether to do macro-mobile or not. In micro-mobile, we only need to change the LCoA of mobile node then deliver binding update to MAP; in macro-mobile, it needs to change the RCoA of mobile node and deliver binding update to home agent and correspondent node. There will be a series of handoffs and message delivering in moving.

2.2   The Limitation of Hierarchical Mobile IPv6 Handoff

With the help of MAP, HMIPv6 has achieved the objective of reducing location update times. But there is no obvious improvement in handoff delay especially when MN does inter domain macro-mobile. Owing to the adaptation of MIPv6's DAD working pattern, its handoff performance is not superior to MIPv6 in collocating care-of address. So in order to reduce the duplicate address detection's affection on delay, it is especially important to improve the handoff performance of HMIPv6.

3   The Optimization Scheme of Hierarchical Mobile IPv6 Handoff

Based on the original HMIPv6 pattern, this scheme has solved the problem of traditional DAD operation that requires passive waiting test result; and it also has reduced handoff delay effectively by expanding neighbor discovery protocol and introducing an IP address management module which aims to build a DAD look-up and active generation mechanism of care of address. The extension of neighbor discovery protocol has changed the standard HMIPv6 mobile test process. In this process, access router can do duplicate address test to LCoA by searching neighbor cache address. Whereas the duplicate address test to RCoA is accomplished by searching the address list of IP address management module

3.1   New IAAM Modules

With IAAM, MAP router can not only monitor neighbor notification messages which delivered by neighbor node in the back, but also record the source addresses of messages in address list to show that this address has already been used in MAP domain. On the list, every address entry is unique and has lifetime. IAAM will clean the list periodically to delete the overdue entry. When taking DAD operation on MN's RCoA, MAP router can't deliver neighbor request message to neighbor node. It can only check whether under-detected RCoA is in that address list by using table look-up algorithm. If it is, it will proves there is an address confliction. At this time, IAAM will generate an interface identifier randomly and combine it with the address prefix of MAP router, with the aim to configure a new care-of address, then check the table again. If that care-of address is not on the address list, it will prove that DAD operation is successive and IAAM module will add a correspondent address entry; otherwise, it will start to generate IID continually.

While taking DAD operation on care-of address, MAP router is monitoring the neighbor request message from other nodes in network. In it, MAP router will Check whether the address in the message segment Target Address is the same with care-of address in detection or not. If they are the same, it will prove that the node's delivery of neighbor requesting message is attempt to use that address and do DAD operation. MAP router can't replay that message directly; it can only see that situation as there is a confliction between under-detect care-of address and other addresses.

3.2   Handover Algorithms of Optimization Scheme

Handover algorithm macro-mobility is as follow:

(1)   NAR delivers router advertisement (RA) message to AP.
(2)   MN can receive storage message RA from AP, comparing the prefix of RA message with the prefix existed in cache. If they are different, MN will use stateless auto configuration mechanism to generate an LCoA address.
(3)   MN delivers NRs message to NAR through unicast signal. When NAR receives the message, it would carry out DAD and mobility detection through neighbor cache.
(4)   If address DAD detects successfully, NAR would deliver NRA to MN to bind LCoA address. Otherwise, NAR must find an alternative address in neighbor cache. The alternative address can be changed by choosing from the configured table of router structure, following by inserting that address to neighbor cache as a new entry. MN will configure RCoA address after receiving NRA message.
(5)   MN delivers BU to MAP for registration. When receiving it, MAP extracts RCoA from message and checks the address list in IAAM to take DAD detection. If RCoA has already been there, it will configure an effective RCoA through IAAM and bind RCoA and LCoA for MN. Then it replies a BAck with RCoA options in which the segment RCoA is a new RCoA configured by IAAM. In contrary, if there is no RCoA in IAAM list, MAP will bind RCoA and LCoA directly, then reply a right BAck to MN.
(6)   MN delivers BU binding request to HA and CN.
(7)   HA and CN deliver BAck message to confirm binding, the handoff is finished.

4   Simulation Test and Analysis

To detect the feasibility of our optimized scheme, we have a simulation test. We choose NS-2 simulation platform which is an open network protocol. On this platform, which can be widely applied to a variety of operating systems, simulation can be softly developed under Linux operating system.

4.1 Handoff Delay Analysis

With the application of the same simulation environment and simulation topology, we simulate the HMIPv6 of optimization scheme and the standard HMIPv6, which helps us to get the MN handoff delay under the two environments. We regulate the size of wireless link delay and get the MN handoff delay change in different link delays.

When doing micro-mobile handoff, MN only needs to reconfigure LCoA but not RCoA, and nor binding update home agent and end node. Therefore, the total delay depends on the configuration time of LCoA. In the standardized HMIPv6, MN use standardized DAD detection; in the scheme of this paper, MN can do DAD detection by searching the address of neighbor cache in AR. From the comparison in Fig. 2 we can know, in this scheme, when MN is doing micro-mobile handoff, its handoff delay is far little than that in standard HMIPv6.

When MN doing macro-mobile handoff, it needs to reconfigure RCoA and LCoA, and register the home agent and correspondent node. Thus, the total handoff delay is decided by the configuration time of both RCoA and LCoA. It is similar to the analysis of the comparison of micro-mobile handoff that MN uses standard DAD detection in standard HMIPv6. In the scheme of this paper, MN does DAD detection to LCoA by searching the address of neighbor cache in AR, and does DAD detection to RCoA by searching the address of IP address distributing management module. From the comparison in Fig. 1 we can know, in this scheme, MN handoff test is much lower than standardized HMIPv6. Under the same test conditions, it saves at least 1000 ms handoff delay than standard HMIPv6.

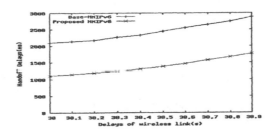

Fig. 1. Handoff delay curve graph of macro-mobile handoff

4.2 The Analysis of Package Lost Rate

There are obvious connections among mobile node's package lost rate, home agent registration and the receiving of care-of address. From Fig. 2 we can know that under HMIPv6 module, mobile node only needs to register new link care-of address to MAP, when it moves in the same MAP domain. And MAP will intercept and transmit all the data packages delivered to mobile nodes. When mobile node moves into different MAP domains, it needs to register new care-of address to home agent. Then the whole data packages will be intercepted and transmitted by home agent, and the already intercepted data packages will be dropped for dealing too late. After mobile nodes' care-of address registering to home agent, the whole data packages which intercepted by home agent, will be delivered to the new care-of address of mobile nodes through tunnel. At the same time, the package lost rate will not change again. This scheme has advanced care-of address's generation by expanding neighbor discovery protocol. Thus, it also has reduced the package lost rata a little.

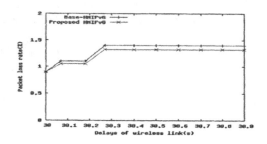

Fig. 2. The package lost rate of mobile node

5 Summary

This paper has mainly discussed the HMIPv6 handoff process on the basis of analyzing mobile IPv6 protocol. Aimed at the characteristic that DAD test takes most of handoff delay, it has put forward an optimization scheme of hierarchical mobile IPv6 handoff by expanding neighbor discovery protocol and adding IP address distribution module in mobile anchor point. And with the testimony of using NS-2 to make simulation test, it has proved this scheme is feasible and valuable.

References

1. Deering, S., Hinden, R.: Internet Protocol, Version 6 Specification. RFC2460, December 1998
2. Johnson, D., Perkins, C., Arkko, J.: Mobility Support in IPv6. RFC3775, June 2004
3. Soliman, H., Castelluccia, C., ElMalki, K., Bellier, L.: Hierarchical Mobile IPv6Mobility Management. RFC4 140, August 2005
4. Perkinsc, C.E., Wangky, K.Y.: Optimized smooth handoffs in mobile IP. In: The Fourth IEEE symposium on Computers and Communications, pp. 340–346 (1999)
5. Peng, X.H., Ma, L., et al.: An improved handoff algorithm for using mobile IPv6 in wireless environment with overlapping areas. In: Proceedings of IEEE 6th CAS Symposium on Emerging Technologies: Mobile and Wireless Conun, Shanghai, vol. 1, pp. 1–4 (2004)
6. Chang, B., Lin, S., Liang, Y.: Minimizing roaming overheads for vertical handoff in heterogeneous wireless mobile networks. In: Proceeding of the 2006 Interactional Conference on Communications and Mobile Computing, NewYork, pp. 957–962 (2006)

Improve Communication Efficiency Between Hearing-Impaired and Hearing People - A Review

Lei Wei[1], Hailing Zhou[1], Junsheng Shi[2(✉)], and Saeid Nahavandi[1]

[1] Institute for Intelligent Systems Research and Innovation (IISRI), Geelong, VIC, Australia
[2] Institute of Color and Image Vision, Yunnan Normal University, Kunming, China
shijs@ynnu.edu.cn

Abstract. Sign languages are one of the most essential communication skills for hearing-impaired people, yet they are not easy to understand for hearing people and this situation has created communication barriers through many aspects of our society. While recruiting a sign language interpreter for each hearing-impaired people is apparently not feasible, improving the communication effectiveness through up-to-date research work in the field of haptics, motion capture and face recognition can be promising and practical. In this paper, we review a number of previous methods in sign language recognition using different approaches, and identify a few techniques that may improve the effectiveness of the communication pipeline between hearing-impaired and hearing people. These techniques can be fit into a comprehensive communication pipeline and serve as a foundation model for more research work between hearing-impaired and hearing people.

Keywords: Sign language · Haptics · Motion capture · Face recognition

1 Introduction

Hearing-impaired people communicate through lip reading, hearing aid and sign languages. Among these approaches, sign languages are the most commonly used communication approaches. There are more than 130 [1] sign languages currently used across different regions in the world, and a typical sign language contains thousands of different signs, each differing in many slight differences on both manual and non-manual features. In contrast to the languages used for hearing people which are communicated through written texts and vocal sound patterns, sign languages combine both manual and non-manual expressions through a series of movements, including hand translation and orientation, finger gestures, upper limbs, torso, head as well as facial expressions.

Although sign languages are crucial among the hearing-impaired people, normal hearing people usually find it hard to understand them and therefore impossible to talk back to the hearing-impaired people, which creates barriers for the communication. This is mainly due to the following reasons:

(1) Sign languages involve facial expressions as well as subtle movements of arms, hands and fingers, which requires extensive training to capture and appreciate.

F. Qiao et al. (eds.), *Recent Developments in Mechatronics and Intelligent Robotics*,
Advances in Intelligent Systems and Computing 691, DOI 10.1007/978-3-319-70990-1_38

(2) Sign languages have their own lexicons and syntaxes, and in many cases, there is no direct one-to-one translation between the two. This is extremely challenging for sign language recognition in continuous manner.

(3) Sign language gestures vary among singers, just as accent and handwriting for different hearing people. This variation creates many challenges for accurate recognition.

(4) Hearing-impaired people who use sign languages can be either left-handed or right handed. While most gestures are simply mirrored, the time line expression is invariable from left to right.

In many situations, a sign language interpreter is required for communications between hearing-impaired and normal hearing people, yet it is not always possible to have an interpreter for a hearing-impaired people. This situation has greatly restrained the communication scope for hearing-impaired people and created barriers for them to involve into the mainstream society.

In this paper, we aim to review several previous methods on sign language recognition and identify their drawbacks. We then discuss the possibility of adopting up-to-date research work and techniques, such as those in the field of haptics, motion capture and face recognition, into the communication pipeline, and potentially solve some of the existing problems.

The rest of the paper is organized as follows: In Sect. 2, we discus related work and try to understand the available approaches for sign language recognition. In Sect. 3, we discuss existing problems in existing work and demonstrate the potential issues need to be solved for more effective communications. In Sect. 4, we discuss these issues in a more generalized communication framework and propose possible solutions for them. Finally, in Sect. 5, we summaries the current status of the art and propose future research directions.

2 Literature Survey

There have been many research work on improving the communication effectiveness between hearing and hearing-impaired people. These previous methods can generally be classified as vision-based and non-vision-based.

Vision-based approaches are natural and direct as they represent how human see sign languages. However, vision-based approaches suffer from issues such as occlusion, and the results can be easily affected by external factors such as lighting and background conditions. Non-vision-based approaches usually adopt external data capture devices such as data gloves [2, 3] and exoskeletons to obtain clear and occlusion-free data, which is crucial for further processing and analysis. However, these external devices can be very expensive, and they usually constraint the movement of the signer and cause the signing data distorted. To date, most research work on sign language recognition is still vision-based.

A few research groups have spent more than a decade on sign language recognition. Specifically, a joint research group across many European countries have accomplished a series of research projects funded by EU IST (ICT after 2007). Their initial project is

VisiCAST between 2000 and 2002, where they focus on SiGML, a XML-based markup language based on HamNoSys notation. They then proceeded to eSIGN project (Essential Sign Language Information on Government Networks), which targeted at providing sign language representations through virtual avatar and help hearing-impaired people explore government networks and accomplish daily tasks where possible. After this, they carried on the jAsigning project, a virtual signing system that adopts virtual avatars to synthesise real-life sign language expression. Their latest effort is the 3-year Dicta project (Sign Language Recognition, Generation and Modelling with Application in Deaf Communication), in which they aim to adopt webcam to capture and recognise the sign language a user signs and sign back through animated avatars. Their framework also enables translations among different sign languages, through the language-independent internal representation. This effort has also led to a project LinguaSign, which is designed for teaching primary school students on new languages through voice and animated signs. Another group lead by Professor Sarkar in University of South Florida, have worked on automatic sign language recognition from videos for the past decade [4–10].

A straightforward approach for sign language recognition would be to build per-gesture classifiers but this approach can easily become inapplicable for large vocabulary tasks. Besides, there is another major difference between sign languages and the languages used by hearing people that should not be neglected: the sub-units in sign languages can be expressed simultaneously through multiple body parts, while languages used by hearing people are usually expressed sequentially. In [11], the authors have extensively surveyed related research state-of-the-art in the field before 2005. In [12], the authors discussed usage of sub-unit feature extraction and consider each sign as an N-dimensional vector, with each component corresponding to a specific constraint, such as location and the motion of a body part. The authors have also proposed classifications for sign language recognition, appearance-based approach and tracking-based approach, where either 2D or 3D tracking can be employed for the latter approach.

Appearance-based approach adopts different types of classifiers to handle various sub-units, such as location, motion and hand arrangement. Specifically, location sub-units look for the location relationship between the person and the detailed motion through spatial grids. Motion sub-units seek the continuous motion details, without considering their locations. Hand arrangement sub-units are similar to motion but only focus on a single frame at one time. Although appearance-based approach has a few advantages such as it is not dependent on tracking results and that it contains information more than just trajectory, its disadvantage should not be neglected: it requires time-consuming data annotation through experts and can be easily fooled by complex background data. In [13], the authors proposed to obtain features of a Bayesian classifier through motion gradient image sequences using PCA. In [14], the authors proposed to skip entire tracking and using volumetric Harr-like features to build classifiers.

On the other hand, tracking approach has maintained its popularity in the past decades due to the advances in computer and imaging technologies. Specifically, 2D tracking refers to those tracking based on 2D cameras without depth sensing, while 3D tracking refers to the tracking involving both 2D and depth information.

2D tracking adopts the tracking results to extract both motion and location information. Specifically, motion sub-units link the tracked planar coordinates with abstract linguistic concepts and derive hand positions through deterministic rules. Local sub-units adopt a similar approach and describe planar coordinates relative to the captured person in the frame. Hand shapes, which is crucial for accurate tracking recognition, are extracted from the Histograms of Gradients (HOG), and segmented, rotated and scaled to match various feature vectors, as discussed in [15].

3D tracking uses similar approach to 2D tracking but adds in depth data for more accurate recognition. Previously, depth data were not directly available. In [16], the authors adopted calibrated light source along with a camera to estimate depth. In [17], the authors constructed depth images through multi-view geometry lit by external lights. In [18], HMD and camera were used together, with the HMD facing towards the user's hands for recognition. Stereo cameras and side/vertical mounted cameras were also used in [19] and [20], respectively.

The latest prosperity of 3D tracking starred with the arrival consumer-level depth tracking device, featuring Microsoft Kinect and other similar devices licensed by Prime-Sense. With both RGB and Depth camera in one device and straightforward calibration, these 3D tracking devices can directly obtain the locations of major human joints and have greatly simplified the 3D tracking procedure. Many techniques have been done using Kinect. In [21], the authors have shown that using the Kinect™ tracking in place of their previous skin based results boosts performance from 20% to 95% on a complex data set of ten number gestures. Similar work has been done in [12, 22–27], where real-time recognition of sign language words can be achieved with higher accuracy.

3   Proposed Method

Nevertheless, Kinect and devices alike merely provides more accurate and interactive input, the detailed processing approaches are still far from being satisfactory, and there are still a number of problems that need to be attended further:

(1) The vocabulary of the sign language through 3D tracking are very limited in certain work. In [28], the authors adopted Kinect for isolated Arabic sign language recognition and their system are able to recognize 4 signs. In [23], the authors declared that their system can recognise 11 words and phrases using Kinect. In [24], the authors have achieved 97% of recognition rate over 25 signs. Some work focuses only on hand gesture, while actual sign language involves the movement of torso, upper limbs, hands, fingers head as well as facial expressions. Results from these methods are insufficient for comprehensive sign language expression. In [12], the author adopted contour tracing algorithm to identify finger gestures and tested on 9 gestures. The recognition rate is between 84% and 99% for single hand gestures and between 90% and 100% for dual hand gestures. Similar work is also discussed in [29].

(2) Even though some other work has extended the number of sign language words, the detailed hand and finger gestures are still difficult to obtain, causing the success rate of recognition relatively low. In [25], the authors claim that their system is able

to recognise 239 Chinese sign language words (out of 4000 words) with an accuracy between 83.51% and 96.32%. However, their approach is based on Euclidean distance measurement of the 3D trajectory of hand tracking and does not take detailed finger gestures into consideration. In [26], the authors have tested 1000 sign language phrases on their Kinect-based system and achieved only 51.5% and 76.12% recognition rate when users are seated and standing.

(3) Many techniques are focusing on recognising split words and phrases, rather than whole sentences. For example, the work in [25] has specifically defined gestures (Raising and putting down hands) to indicate as the start and end gesture of each SL word to be recognized, which is confusing and unnatural for hearing impaired people to use. Besides, their system heavily involves user interaction and it lists multiple candidate choices for each recognised gesture, and asks the user to select the intended one before proceeding to the next recognition. This puts an extra requirement for the hearing-impaired people to understand written language, which is not their preferred language and not all of them know written language. There have been pioneering research work done to solve the issue of continuous recognition. In [30, 31], the authors proposed a matching algorithm to solve the movement epenthesis in sign language recognition, which refers to the transition between signs. These transitional motions belong to neither signs and bring ambiguity to the recognition process. In [32], they proposed the concept of signemes that is robust to coarticulation effects for modelling signs. In [33], they proposed to use iterated conditional models to automatically extract signs from continuous sign language sentences. Although these techniques are crucial for automatic sign language recognition and processing, many of them were based on 2D tracking techniques and they may need a revisit to work more efficiently and accurately with the latest 3D tracking techniques.

With the recent advances of the updated Kinect device, it is possible to also recognise fingers along with other major joints on a human body, which extends positive opportunity for more precise sign language recognition. In [34], the authors have shown that they are able to accurately track all ten human fingers in real time, with various lighting conditions. This, along with the work discussed above, may open new prospects for faster, more robust and more accurate recognition algorithms for automatic sign language recognition.

4 Discussion

Based on the review and analysis presented above, we identified that to be able to improve the communication effectiveness between hearing-impaired and hearing people, the following issues will need to be addressed properly.

Figure 1 describes the generalised communication pipeline between hearing and hearing-impaired people. Left side indicates the required steps from converting speech into understandable sign language and right side indicates the required steps from converting sign language into speech. We list the names of the key modules in the pipelines.

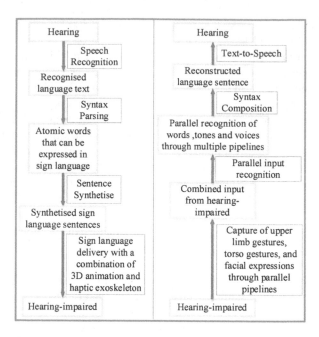

Fig. 1. A generalized communication pipeline between hearing-impaired and hearing people.

Speech recognition, syntax parsing, sentence synthesize, syntax composition and text-to-speech are fields related to natural language processing shown as blue colors, which have been investigated for many years. Recently, the steps shown as red colors are becoming popular, which are related to the up-to-date research on haptics [35], motion capture and face recognition. The details are discussed as following:

(1) Sign language delivery from hearing people to hearing-impaired people through 3D animation and haptic exoskeleton. Since video-based delivery could not fulfil the required accuracy, 3D animations can be used as a visual clue for better immersion, while haptic exoskeleton can help accurately represent the subtle differences between similar words of a sign language. With the introduction of low-cost hand and arm exoskeleton devices such as Novint XIO [36] and Dexmo F2 [37], this can be a very promising direction for the enhancement of sign language delivery.

(2) Capture of the movement of hearing-impaired people, including upper limb gestures, torso gestures and facial expressions. While torso gestures can be acquired rather easily through existing video processing and motion capture techniques, upper limb gestures, especially finger gestures, are more difficult to capture due to accuracy and occlusion issues. Devices such as Leap Motion and Kinect 2 provides stable detection results, yet occlusion issues could not be avoided. Hand and arm exoskeletons on the other hand, are able to provide occlusion free input, which is of extreme importance in capturing clean and unsaturated data before processing similar hand gestures in the next steps.

On the other hand, facial expression is also crucial for the tone and voice of the recognized sentences. Features such as head rotation, including nod, shake, tilt, eye shape and mouth shape, should all be taken into consideration for an appropriate recognition. Existing techniques on this topic can definitely contribute to the correct understanding of the expression from hearing-impaired people.

(3) Recognition of the captured result through parallel input. This basically combines multiple input from different pipelines together with reference to temporal coherence. The same hand gesture may represent totally different meanings with different facial expressions, and these inputs may need to be assigned with weight for each atomic word recognition, while these weights contribute to the final composition of the sentence in the syntax composition stage.

5 Conclusion

In this paper, we reviewed a number of previous work on sign language recognition and identified a few existing issues that adversely affect the effectiveness of communication between hearing and hearing-impaired people. We then identify a few places within the communication procedure which can be improved through the recent advances in haptic exoskeleton, motion capture and face recognition. In the future, we plan to work on the implementation of these proposed improvements and aim for a more effective and intuitive communication pipeline for hearing-impaired people.

References

1. http://en.wikipedia.org/wiki/Sign_language
2. Hernandez-Rebollar, J.L., Lindeman, R.W., Kyriakopoulos, N.: A multi-class pattern recognition system for practical finger spelling translation. In: Proceedings of the IEEE International Conference on Multimodal Interfaces, p. 185 (2002)
3. Kadous, M.W.: Machine recognition of Auslan signs using PowerGloves: "Towards large lexicon recognition of sign language". In: Proceedings of the Workshop: Integration of Gesture in Language and Speech (1996)
4. Nayak, S., Duncan, K., Sarkar, S., Loeding, B.: Finding recurrent patterns from continuous sign language sentences for automated extraction of signs. J. Mach. Learn. Res. 13, 2589–2615 (2012)
5. Nayak, S., Sarkar, S., Loeding, B.: Automated extraction of signs from continuous language sentences using iterated conditional modes. IEEE Trans. Pattern Anal. Mach. Intell. 31(5), 795–810 (2012)
6. Sarkar, S., Loeding, B., Yang, R., Nayak, S., Parashar, A.: Segmentation-robust representations, matching, and modeling for sign language. In: Computer Vision and Pattern Recognition Workshops, pp. 13–19. IEEE Computer Society (2011)
7. Nayak, S., Sarkar, S., Loeding, B.: Automated extraction of signs from continuous sign language sentences using iterated conditional modes. In: IEEE Conference on Computer Vision and Pattern Recognition, June 2009
8. Yang, R., Sarkar, S.: Handling movement epenthesis and hand segmentation ambiguities in continuous sign language recognition using nested dynamic programming. IEEE Trans. Pattern Anal. Mach. Intell. 2009)
9. Yang, R., Sarkar, S.: Coupled grouping and matching for sign and gesture recognition. Comput. Vis. Image Underst. 113(6), 663–681 (2009)

10. Nayak, S., Sarkar, S., Loeding, B.: Distribution-based dimensionality reduction applied to articulated motion recognition. IEEE Trans. Pattern Anal. Mach. Intell. **31**(5) (2009)

11. Ong, S.C.W., Ranganath, S.: Automatic sign language analysis: a survey and the future beyond lexical meaning. IEEE Trans. Pattern Anal. Mach. Intell. **27**(6), 873–891 (2005)

12. Cooper, H., Ong, E.J., Pugeault, N., et al.: Sign language recognition using sub-units. J. Mach. Learn. Res. **13**(1), 2205–2231 (2012)

13. Wong, S.F., Cipolla, R.: Real-time interpretation of hand motions using a sparse Bayesian classifier on motion gradient orientation images. In: Proceedings of the BMVC, Oxford, UK, vol. 1, pp. 379–388 (2005)

14. Cooper, H., Bowden, R.: Sign language recognition using boosted volumetric features. In: Proceedings of the IAPR Conference on Machine Vision Applications, Tokyo, Japan, pp. 359–362 (2007)

15. Buehler, P., Everingham, M., Zisserman, A.: Learning sign language by watching TV (using weakly aligned subtitles). In: Proceedings of the IEEE Computer Society Conference on Computer Vision and Pattern Recognition, pp. 2961–2968 (2009)

16. Segen, J., Kumar, S.: Shadow gestures: "3D hand pose estimation using a single camera". In: Proceedings of the IEEE Computer Society Conference on Computer Vision and Pattern Recognition, vol. 1 (1999)

17. Feris, R., Turk, M., Raskar, R., Tan, K., Ohashi, G.: Exploiting depth discontinuities for vision-based fingerspelling recognition. In: Proceedings of the IEEE Computer Society Conference on Computer Vision and Pattern Recognition: Workshop, vol. 10 (2004)

18. Starner, T., Weaver, J., Pentland, A.: Real-time American sign language recognition using desk and wearable computer based video. IEEE Trans. Pattern Anal. Mach. Intell. **20**(12), 1371–1375 (1998)

19. Munoz-Salinas, R., Medina-Carnicer, R., Madrid-Cuevas, F.J., Carmona-Poyato, A.: Depth silhouettes for gesture recognition. Pattern Recognit. Lett. **29**(3), 319–329 (2008)

20. Vogler, C., Metaxas, D.: ASL recognition based on a coupling between HMMs and 3D motion analysis. In: Proceedings of the ICCV, Bombay, India, pp. 363–369 (1998)

21. Doliotis, P., Stefan, A., Mcmurrough, C., Eckhard, D., Athitsos, V.: Comparing gesture recognition accuracy using color and depth information. In: Proceedings of the Conference on Pervasive Technologies Related to Assistive Environments (PETRA) (2011)

22. Li, Y.: Hand gesture recognition using Kinect. In: Proceedings of the IEEE 3rd International Conference on Software Engineering and Service Science (ICSESS), pp. 196–199 (2012)

23. Li, K.F., Lothrop, K., Gill, E., et al.: A web-based sign language translator using 3d video processing. In: Proceedings of the IEEE 14th International Conference on Network-Based Information Systems (NBiS), pp. 356–361 (2011)

24. Lang, S., Block, M., Rojas, R.: Sign language recognition using kinect. In: Artificial Intelligence and Soft Computing, pp. 394–402. Springer (2012)

25. Chai, X., Li, G., Lin, Y., et al.: Sign language recognition and translation with kinect. In: Proceedings of the IEEE Conference on AFGR (2013)

26. Zafrulla, Z., Brashear, H., Starner, T., et al.: American sign language recognition with the Kinect. In: Proceedings of the ACM 13th International Conference on Multimodal Interfaces, pp. 279–286 (2011)

27. Sun, C., Zhang, T., Bao, B.K., et al.: Discriminative exemplar coding for sign language recognition with Kinect. IEEE Trans. Cybern. **43**(5), 1418–1428 (2013)

28. Ershaed, H., Al-Alali, I., Khasawneh, N., Fraiwan, M.: An Arabic sign language computer interface using the Xbox Kinect. In: Proceedings of the Annual Undergraduate Research Conference on Applied Computing (2011)

29. Raheja, J.L., Chaudhary, A., Singal, K.: Tracking of fingertips and centre of palm using kinect. In: Proceedings of the 3rd IEEE International Conference on Computational Intelligence, Modelling and Simulation, pp. 248–252 (2011)
30. Yang, R., Sarkar, S., Loeding, B.: Enhanced level building algorithm for the movement epenthesis problem in sign language recognition. In: Proceedings of the IEEE Conference on Computer Vision and Pattern Recognition, pp. 1–8 (2007)
31. Yang, R., Sarkar, S., Loeding, B.: Handling movement epenthesis and hand segmentation ambiguities in continuous sign language recognition using nested dynamic programming". IEEE Trans. Pattern Anal. Mach. Intell. **32**(3), 462–477 (2010)
32. Yang, R., Sarkar, S.: Detecting coarticulation in sign language using conditional random fields. In: Proceedings of the IEEE 18th International Conference on Pattern Recognition, pp. 108–112 (2006)
33. Nayak, S., Sarkar, S., Loeding, B.: Automated extraction of signs from continuous sign language sentences using iterated conditional modes. In: Proceedings of the IEEE Conference on Computer Vision and Pattern Recognition, pp. 2583–2590 (2009)
34. http://research.microsoft.com/en-us/collaboration/stories/kinect-sign-language-translator.aspx
35. Jia, D., Bhatti, A., Nahavandi, S., Horan, B.: Human performance measures for interactive haptic-audio-visual interfaces. IEEE Trans. Haptics **6**(1), 46–57 (2012)
36. http://www.novint.com/index.php/novintxio
37. http://www.dextarobotics.com/products/Dexmo

Numerical Analysis of Water Bursting Mechanism and Safety Thickness of Mine Fault

Huitao Wang, Honglu Yang, Shucai Li, Rentai Liu$^{(\boxtimes)}$, Haojie Liu,
Jiwen Bai, Dazhuang Gui, Yanan Liu, and Qichen Jiang

Research Center of Geotechnical and Structural Engineering,
Shandong University, Jinan 250061, Shandong, China
rentailiu@163.com

Abstract. Based on the percolation theory, fluid- solid coupling equation, using COMSOL Multiphysics, the delayed water inrush model is established. The mechanical behavior of the unsaturated seepage stage, the low velocity stable seepage flow stage and the rapid saturated seepage stage are analyzed by Richard equation, Darcy's law and Brinkman equation. It is used to determine the elastoplastic area of geological rock and excavation of the safety thickness. It is calculated that when the safety thickness is 46 m, the safety of the project is guaranteed.

Keywords: COMSOL multiphysics · Delayed water inrush model · Seepage · Safety thickness

1 Introduction

China is a big country of resource demanding. Coal resources have a pivotal position in China's energy security. In China's current energy consumption structure, coal resources account for about 70% of total consumption, a dominant position. In the future, coal resources will continue to be an important part for the rapid development of China's national economy. However, with the development of the country, the demand for energy is increasing and the shallow coal resources have been depleted, so the development of coal mining starts to become the large depth, high ground stress, strong water rich area [1]. Due to the complex and changeable hydrogeological conditions, various geological disasters occur frequently. In non-gas mine, the mine water inrush is ranked first [2]. With the increase of mining depth, the mine water disaster also presents the characteristics of high frequency and strong hazard. The project relies on Wanglou coal [5] mine with deep mine special engineering disaster: the underground water pressure reaches 2.5 Mpa, the water gushing quantity reaches 800 m3/d. It is the typical fault type delayed water inrush [6, 7]. The working face of the project reveals a set of oblique working face in the mining process. In order to ensure the safe production of the mine, the finite element software COMSOL Multiphysics is used to construct the three-dimensional finite element numerical model based on the main

© Springer International Publishing AG 2018
F. Qiao et al. (eds.), *Recent Developments in Mechatronics and Intelligent Robotics*,
Advances in Intelligent Systems and Computing 691, DOI 10.1007/978-3-319-70990-1_39

geological features of the relevant strata and faults, and the mechanism of delayed water inrush and the key technologies are studied.

2 Model Building and Control Equations

2.1 Model Building

The finite element software COMSOL Multiphysics is used to set the model, and the influence of time factor is not considered. According to the main geological characteristics of coal seam related strata and faults, a three-dimensional finite element numerical model is built. The model takes the calculated depth elevation $-650 \sim -900$ m, X direction is 1267 m, Y direction is 524 m.

2.2 Control Equations

2.2.1 Basic Equation of Seepage

In the process of calculation of the model, the body load caused by the gravity of the rock mass is considered in the model. The influence of the groundwater self-weight on the seepage process is considered in the calculation process. The continuous equation of seepage is:

$$\nabla(\rho \overrightarrow{v}) = \frac{\partial(\rho \varphi)}{\partial t} \tag{1}$$

Darcy's equation of motion is:

$$\overrightarrow{v} = \frac{k}{\mu} \nabla \overrightarrow{p} \tag{2}$$

Considering the compressibility of water, the fluid state equation is:

$$\rho = \beta \rho_0 e^{-\beta p} \tag{3}$$

Among them, ρ is the density of water, $\bar{v}$ is the seepage velocity, ϕ is the media porosity, p is the fluid pressure, μ is the dynamic viscosity of water, ρ_0 is the standard density of water, β is the compressibility of water.

2.2.2 Basic Equations of Mechanics

According to Euler's view, the force and displacement generated by the external load of discrete soil particles are transformed into the stress field and displacement field of the surrounding rock in classical elastic mechanics theory. The rock and soil mainly obey the equilibrium equation, the geometric compatibility equation and the constitutive equation. The equilibrium equation of surrounding rock is:

$$\sigma_{ji,j} + F_i = 0 \tag{4}$$

Geometric coordination equation is:

$$\varepsilon_{ij} = \frac{1}{2}\left(u_{i,j} + u_{j,i}\right) \tag{5}$$

Constitutive equation is:

$$\varepsilon_{ij} = \frac{1+v}{E}\sigma_{ij} - \frac{v}{E}\sigma_{kk}\delta_{ij} \tag{6}$$

$$\varepsilon_{kk} = \frac{1-2v}{E}\sigma_{KK} \tag{7}$$

Among them, σ is the stress of rock and soil, F is the additional stress for rock and soil, ε is the deformation of rock and soil, E is the elastic modulus, v is the Poisson's ratio.

2.3 Boundary Conditions and Model Grids

The boundary of the model is the roller support boundary, and the displacement can only occur in the upper and lower directions. The lower boundary is a fixed constraint, and the upper and lower left and right can not be displaced. The model grid uses a very thinner free triangular mesh. At the same time, in order to improve the accuracy of the calculation, encryption grid is used in the vicinity of the fault. The results of model mesh generation are shown in Fig. 1.

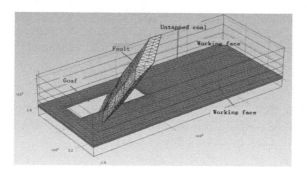

Fig. 1. Three - dimensional model and mesh

3 Phased Simulation

In the simulation process, the fault weakening stage is decomposed into unsaturated seepage stage, low velocity stable saturated seepage stage and rapid saturated seepage stage which are respectively described by the Richard equation, Darcy's law and Brickman equation. A finite element model is presented to describe the process of realization of the internal fault group groundwater from unsaturated to saturated, complete process from low speed flow to fast flowing. From the point of view of fluid flow of porous media, they correspond to the three flow modes of unsaturated seepage, low velocity seepage and rapid seepage, which can be described by Richard equation, Darcy's law and Brinkman equation respectively. Among them, the non-saturated phase calculation and grid partition are shown in Fig. 2.

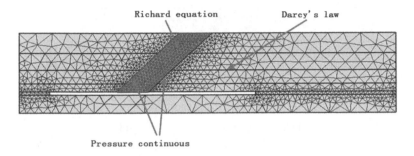

Fig. 2. Calculation of unsaturated stage and grid

After the transition from the fault to the saturation stage, the seepage law of the flow is calculated according to Darcy's law. Due to the low flow velocity in the unsaturated seepage stage, the variation of porosity in this stage is ignored. Therefore, the initial value of the porosity in the fault can be approximated treated as the initial porosity after the fault is disturbed. The free triangular mesh is used to increase the mesh density to improve the computational accuracy in the interior of the fault and at the boundary between the fault and the rock, as shown in Fig. 3.

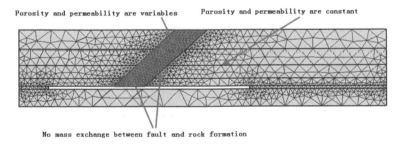

Fig. 3. Low-speed saturation stage and grid

The porosity and permeability obtained in the saturated seepage stage are calculated as the initial value of the stage, and the flow field in the fault and the variation of water inflow in the goaf are calculated. The seepage in the fault is a Brinkman equation, and the surrounding rock mass is Darcy's law, and the pressure continuous condition is set at the boundary of the fault and the surrounding rock. In this Stage, by using the free triangle mesh, the grid density is increased in the interior of the fault and the boundary between the fault and the rock, as shown in Fig. 4. The stage of the fault quickly formed a nearly stable flow field. Flow rate and low-speed saturated seepage stage compared to a substantial increase, then slow growth. When the time reaches 30 days it is basically stabilized.

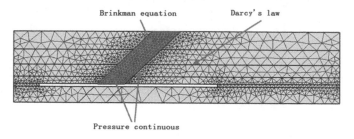

Fig. 4. Rapid seepage stage and grid

4   Simulation Results

The thickness of the coal seam is simulated by the model. When the thickness is 30, 40, 50, 60, 70, 80 m, the outburst coal pillar plastic zone calculation results are shown in Fig. 5. The red area in the figure is the plastic zone and the blue region is the elastic region.

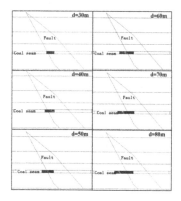

Fig. 5. Plastic zone range

5   Conclusion

Figure 5 shows that when the thickness of coal pillar is 25 or 30 m, the outrigger pillar plastic area extends into the fault inside. Under the influence of mining disturbance, the possibility of fracture initiation and infiltration weakening process are very large. When the thickness of the outburst coal pillar is 35 or 40 m, the boundary of the plastic zone of the outburst coal pillar is near the fault boundary. The fault still has the risk of weakening. When the thickness of the outburst coal pillar is 45 or 46 m, the distance between the boundary of the coal pillar plastic zone and the fault boundary is 2.5 or 4.6 m respectively. Considering the risk of water bursting and coal mining cost, it is concluded that when the thickness of the outburst coal pillar is 46 m, the excavation of the working face will not cause the activation of the fault group.

Acknowledgments. The authors acknowledge the financial support provided by the National Key Research and Development Program of China (No. 2016YFC0801604).

References

1. Huang, P., Han, S.: Assessment by multivariate analysis of groundwater surface water interactions in the Coal-mining Exploring District, China. Earth Sci. Res. J. **20**(1), G1–G8 (2016)
2. Zhaoping, M., Yi, W., Lan, H., Men, W.: Water inrush characteristics of Fangezhuang coalmine field in kailuan and its geological condition analysis of water inrush from coal seam floor. Chin. J. Rock Mech. Eng. **28**(2), 228–237 (2009)
3. Boyang, Z., Haibo, B., Kai, Z.: Study on the mechanism of delayed water inrush of collapse column under the influence of mining. J. China Univ. Min. Technol. **45**(3), 447–454 (2016)
4. Li, W.-P., Liu, Q.-M., Sun, R.-H.: Theoretical and experiment study on vadose conversion of water inrush later occurred from structure broken zone. Coal Sci. Technol. **39**(11), 10–13 (2011)
5. Jiwen, B., Shucai, L., Rentai, L., et al.: Multi-field information monitoring and warning of delayed water bursting in deep rock fault. Chin. J. Rock Mech. Eng. **34**(11), 2327–2336 (2015)
6. Hua, X., Wanga, L., Yinlong, L., Yua, M.: Analysis of insidious fault activation and water inrush from the mining floor. Int. J. Min. Sci. Technol. **24**(4), 477–483 (2014)
7. Zhang, S.-C., Guo, W.-J., Sun, W.-B., et al.: Experimental research on extended activation and water inrush of concealed structure in deep mining. Rock Soil Mech. **36**(11), 3111–3120 (2015)

Numerical Simulation on Seepage Law and Thickness of Reinforcing Ring of Mine Ordovician Aquifer

Huitao Wang, Haojie Liu, Rentai Liu$^{(\boxtimes)}$, Shucai Li, Honglu Yang, Dazhuang Gui, Qichen Jiang, and Yanan Liu

Research Center of Geotechnical and Structural Engineering, Shandong University, Jinan 250061, Shandong, China
rentailiu@163.com

Abstract. Considering that there are very complex hydrogeological conditions, high water level and large quantity of Karst water of Ordovician limestone in Feicheng mining area, and the situation of safety mining of lower coal seams is affected, numerical simulation study of grouting reinforcement of Karst water inrush of Ordovician limestone is carried out. The COMSOL and PFC software were used to study the variation of pressure field, flow field, stress field and water inflow of stope before and after the grouting reconstruction of Ordovician limestone roof. The results show that streamline distribution of flow field becomes very sparse, stress gradient is obviously reduced, and with the increase of depth of grouting, spatial distribution of groundwater pressure field becomes more uniform, water inflow is obviously reduced, which show that water-resisting ability of Ordovician limestone roof has been improved. In addition, it is determined that the optimal depth for grouting reconstruction is about 60 m below Ordovician top.

Keywords: Numerical simulation · Water inrush of ordovician limestone · Grouting reconstruction · Fracture · Darcy's law

1 Introduction

China is a country with large resource demand, and coal resources occupy important strategic position in Chinese energy sector, which will not change in the next 50 years. In recent years, with the increase and expansion of mining depth, mining intensity and mining scale, the calamity of high-pressure confined karstic water of North China type coal field which is an important part of China's coal is increasing day by day, especially the stability of mine roadway formed by water-bearing Ordovician limestone is substandard, and it has a phenomenon of a wide range of gushing water, which make support operation very difficult. The Ordovician limestone fault can easily lead to severe sudden water inflow, which severely threaten to coal mine safety production, so how to control Ordovician limestone water inrush has become one of the key factors to ensure the deep mining safety production.

In consideration of water inrush from coal floor, domestic and foreign scholars had made researches on the topic from the beginning of last century. In 1944 the Hungarian

© Springer International Publishing AG 2018
F. Qiao et al. (eds.), *Recent Developments in Mechatronics and Intelligent Robotics*,
Advances in Intelligent Systems and Computing 691, DOI 10.1007/978-3-319-70990-1_40

scholar M.S. Reibieic [1] first proposed the concept of relatively impermeable layers; In 1988 O. Sammarco [2] found that warning of mine water inrush can be achieved by precursor information of monitor in water bursts such as sudden changes in water levels; According to the layer structure characteristics of floor rock mass, Qian Minggao, Li Liangjie [3] built the key layer of floor water inrush mechanism; Gao Yanfa [4] discussed in detail the mechanism of grouting reinforcement.

At present, study on the mechanism, monitoring and forecasting and management of coal seam floor water inrush has made great progress, however, because of the special geological conditions of Baizhuang coal mine in Feicheng coal field, there are two problems in control of Ordovician limestone water inrush under coal seam floor: In the coal mine, karst fissures are developed, Ordovician limestone water level is high and the water quantity is large and the recharge is also abundant, so drainage for decreasing water pressure is failed; Then with the mining of coal is into the deep areas, floor grouting transformation technology [4] of long-term implementation which is to reinforce No. 5 limestone can't fully meet the safety mining requirements of 9, 10 coal seam. In the light of this situation, this paper adopt a method of grouting transform of Ordovician top on basis of geological condition in Baizhuang coal mine by means of numerical simulation, and study quantitatively grouting effect.

2   Engineering Situation

Baizhuang coal mine is located in the west of Feicheng coal field. It was built in 1970, the mine field is about 4.0 km long, about 2.5 km in slope and an area of 15.8 km^2. The topography of mine is that northeast is higher than southwest, elevation is +71~+125 m. The boundary of all around are faults, and more faults in the field make No. 4 limestone, No. 5 limestone, Ordovician limestone connect or water-bearing stratum and water-resisting layer connect. The Ordovician limestone aquifer which is widely exposed in the mountains around basin is located in the coal strata chassis, and about 800 m thick, and directly accepts the supply of atmospheric precipitation which is rich. In addition, it is the nourishment source of coal measure aquifer.

3   Establishment of Numerical Model

3.1   Generation of Discrete Fracture Network of Ordovician Limestone

The core of problem of gushing water from fractured rock mass is to determine actual crack distribution and effect tendency of grouting transform on crack density, crack aperture, so variation of water inflow can be analyzed before and after grouting reconstruction. First of all we need to use the mathematical statistics theory to establish statistical model of crack distribution. In the calculation of this paper, according to field investigation, assuming that crack size density obeys distribution law of power function:

$$n(l) = -\alpha \cdot l^{-r} \tag{1}$$

Assuming that whole area is a square with a side length of L, the total number of fissures contained in the rock mass can be obtained by integrating above equation to l.

$$n(l_{\min} \leq l \leq l_{\max}) = \alpha\left(\frac{l_2^{1-r} - l_1^{1-r}}{1 - r} \cdot L^3\right) \qquad (2)$$

Where $n(l)$ represents number of cracks per unit area, l represents size of the crack, namely length, α represents scaling index.

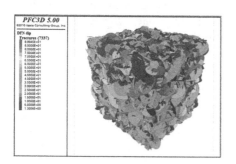 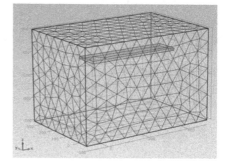

Fig. 1. Numerical model and fracture network generation

Fig. 2. Three-dimensional model mesh generation of stope

3.2 Model and Boundary Conditions

According to the layout of working face, the three-dimensional geometric model of fissure water gushing of Ordovician limestone floor is established, as shown in Fig. 1. Main geometric size of the model and boundary conditions are shown in Fig. 2, in which, the length of mine stope is 400 m, the width is 100 m, influencing depth of the floor transformation is 300 m, bilateralis are coal pillars. Assuming that stope floor is a free-out boundary condition and other boundaries are impervious boundary. In order to ensure the accuracy of calculation, save time and improve calculation efficiency, vicinity of the stope is encrypted and the distance should reduce grid and number of elements in the process of grid segmentation and element generation. The floor below 300 m and surrounding area are approximate replaced by infinite domain.

3.3 Control Equation

During grouting, the flow of slurry is fast in the fracture, but slow in the rock mass. Grout pressure is continuously distributed in the fissures and adjacent rock mass. Flow of the slurry in the rock mass obey to Darcy's law.

$$\rho S \frac{\partial p}{\partial t} - \nabla \cdot (\rho u) = Q_m \qquad (3)$$

$$u = \frac{k}{\mu}(\nabla p + \rho g N) \tag{4}$$

Where p is grout pressure in the pores, S is comprehensive compression coefficient, Q_m is source. is permeability of the rock mass, ρ is slurry density, μ is dynamic viscosity of the slurry, u is Darcy speed of the slurry in the rock, g is gravity acceleration, N is unit vector along direction of gravity.

The flow of the slurry can be calculated in accordance with laminar flow formula in the perforated and highly permeable cracks. However the permeability coefficient of cracks is often affected by filling, roughness and connectivity, making the permeability of fracture low. Therefore flow of slurry should be described by Darcy's law in less permeable and non-penetrating cracks.

In the interior of crack, flow direction of slurry is parallel to crack surface, so the formula (3) and (4) of the Darcy's Law are modified to tangent formula (5), (6) and (7):

$$\rho S_c d_c \frac{\partial p}{\partial t} + \nabla_T \cdot (\rho q_c) = d_c Q_m \tag{5}$$

$$q_c = \frac{k_c}{\mu} d_c (\nabla_T p + \rho g N) \tag{6}$$

$$u_c = \frac{q_c}{d_c} \tag{7}$$

Where q_c is volume flow rate per unit length in the cracks, k_c is permeability of the cracks, d_c is fissure width, ∇_T is gradient operator along fracture tangent plane, u_c is average velocity in the cracks, S_c is comprehensive compressibility coefficient of cracks.

4 Calculation and Analysis of Grouting Reinforcement

It can be seen from Figs. 3 and 4 that the distribution of groundwater pressure field is relatively regular. According to comparison of Figs. 3 and 4, it is found that distribution of groundwater pressure field exhibits anisotropic characteristics under disturbance of water-conducting fissure, and the closer the distance to surface is, the greater the pressure gradient is, the more serious the gushing water is. After grouting transform of Ordovician top, pressure gradient near the free face gradually became smooth, spatial distribution of pressure gradually became uniform, and degree of anisotropy decreased. The results show that after grouting transformation, cracks in the Ordovician limestone are basically blocked and greatly improve water-resisting properties. At the same time, with increase of depth of grouting, the affected area is widening, water-insulating capability of Ordovician top is improved

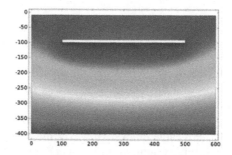

Fig. 3. Change of groundwater pressure field without consideration of fracture distribution

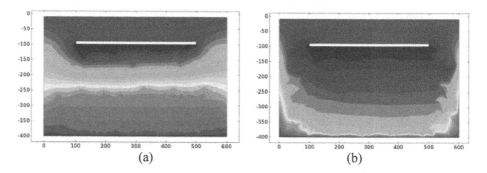

(a)   (b)

Fig. 4. Distribution of pressure field at different depths of grouting

Figures 5 and 6 show the changes of flow field and stress field of the Ordovician limestone before and after grouting transform. It can be seen from the figure that the streamline distribution of flow field is very dense, and connectivity of different fissure units is very strong before grouting transform. At the same time, because of mechanical function of groundwater on the Ordovician limestone, stress gradient inside rock mass is very large, which seriously affects the self-stabilizing ability of rock mass. The coupling between groundwater and rock mass is very significant, which leads to serious water inrush. But after the grouting transform, the boundary condition of the limestone layer is changed, so it limits its field strength and distribution range. It can be seen from Fig. 6b that streamline distribution becomes very sparse in the coal floor and its adjacent area. So grouting material plays a role in blocking the water pressure of Ordovician limestone, this limits the whole flow field and groundwater pressure in the deeper position, avoiding its influence on the coal floor. At the same time, stress gradient of the Ordovician limestone is obviously reduced, and self-stabilizing ability of the rock mass is higher through the grouting transformation.

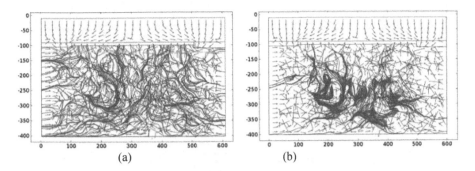

Fig. 5. Flow field distribution of Ordovician limestone before and after grouting

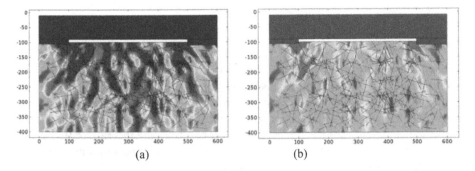

Fig. 6. Stress field distribution of Ordovician limestone before and after grouting

The variation of water inflow of mining stope with depth of grouting can be obtained by parameterizing the different diffusion range of grouting and integrating water inflow in the whole field under different working conditions, as shown in Fig. 7. As can be seen from graph, variation of water inflow with the depth of grouting is nonlinear and the rate of change is gradually reduced and tend to be stabilized. When the depth of grouting is increased to 60 m, water inflow of the whole stope is reduced to about 50 m^3/h, so the effect of grouting on water blocking capacity of Ordovician limestone is significant. With the sequential increase of the depth of grouting, water inflow is further reduced, but the degree of change is reduced. According to practical engineering needs and economic factors, the grouting depth is determined to 60 m, this can effectively control the water inrush and ensure the stability and safety of the of water project.

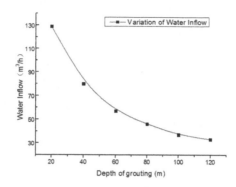

Fig. 7. Variation curve gushing quantity with grouting depth

5   Conclusion

1. The result of numerical simulation shows that after the grouting reconstruction of Ordovician limestone roof, the anisotropy of the distribution of groundwater pressure field decreases, streamline distribution of flow field becomes very sparse, stress gradient is obviously reduced, the fissures in the Ordovician are basically blocked, and self-stabilizing ability of rock mass is higher. With the increase of the depth of grouting, spatial distribution of groundwater pressure field becomes more uniform, and water blocking capacity of roof is higher.
2. Based on the principle of safety, taking the economic benefit as the goal, the optimal depth of grouting transformation of Ordovician limestone roof is 60 m.

Acknowledgments.   The authors acknowledge the financial support provided by the National Key Research and Development Program of China(No.2016YFC0801604)

References

1. Reibieic, M.S.: Hydrofracturing of rock as a method of water, mud and gas inrush hazards in underground coal mining, 4th IMWA, 1, Yugoslavia (1991)
2. Sammarco, O.: Spontaneous inrushes of water in underground mines. Int. J. Mine. Water. **5**(2), 29–42 (1986)
3. Liangjie, L., Minggao, Q., Quan, W.: Relationship between the stability of floor structure and water-inrush from floor. J. China. Univ. Min. Technol. **24**(4), 18–23 (1995)
4. Yanfa, G.A.O., Longqing, S.H.I., Huajun, L.O.U.: Rule of water burst and advantage face of water inrush. China University of Mining and Technology Press, Xuzhou (1999)
5. Jiwen, B., Shucai, L., Rentai, L., Qingsong, Z.: Multi-field information monitoring and warning of delayed water bursting in deep rock fault. Chin. J. Rock. Mech. Eng. **34**(11), 2327–2335 (2015)
6. Longqing, S., Mei, Q., Chao, N., Jin, H.: Feasibility analysis of grouting reinforcement of Ordovician top in Feicheng coalfield. J. Min. Saf. Eng. **32**(3), 356–362 (2015)
7. Chunhu, Z., Xiang, D., Hao, W., Yuan, H.: Numerical simulation for Ordovician limestone water in mine with high pressure. Coal. Min. Technol. **15**(3), 47–49 (2010)

Recommend Algorithm Combined User-user Neighborhood Approach with Latent Factor Model

Xiaojiao Yao[(✉)], Beihai Tan, Chao Hu, Weijun Li, Zhenhao Xu, and Zipei Zhang

Guangdong University of Technology, Guangzhou, China
3176745808@qq.com

Abstract. The item-item neighborhood model became very volatile for current items rapidly replaced, such as online article and news items. And the neighborhood model faces the problem of data sparsity and cold start. The factor model can alleviate data sparseness problem, but it does not take the historical behavior data into consideration. Therefore, in this paper, we proposes a recommendation algorithm based on user-user neighborhood model and latent factor model, which can make accuracy improved significantly and can effectively address the problem of data sparsity. When the number of neighborhoods k increases, the accuracy of the algorithm has improved. The experiment result shows that this approach is correct and feasible.

Keywords: Recommend system · User-user model · Latent factor model

1 Introduction

Collaborative filtering [1] is one of the most widely used approach for building recommend system. And it has been used in industry, food, news, movies and so on [2]. The main advantage of collaborative filtering is that it has no request for their recommend object. The algorithm of collaborative filtering can be divided into neighborhood and model-based methods.

Neighborhood model [3] is divided into user-user neighborhood model and item-item neighborhood model. The main idea of user-user model [4] is to first find the user's neighbor user based on the rate, the item that the neighbor user likes, but the user is not concerned about to recommend to the user. The main idea of item-item model is to first find the neighborhood item based on the rate data, and recommend the similar items to the user when the user is concerned about a item. The basic approach to neighborhood model is to find the neighbor users/items. When the item is updated faster, the item-item model [5] will be very unstable. If we want to predict, but user u is a new user. Then item-item approach will face obvious difficulties. To solve those problem, we choose the user-user model in this paper.

Latent factor model [6], also known as matrix factorization model. Applying SVD [7] to explicit ratings in collaborative filtering raises difficulties due to the high portion of missing values. The traditional SVD [8] factorization method is mainly to fill the missing values and make the rating matrix dense. But there is not the fit approach for computing. Hence, many scholars put forward a series of methods. In 2006, Simon Funk [9] published

© Springer International Publishing AG 2018
F. Qiao et al. (eds.), *Recent Developments in Mechatronics and Intelligent Robotics*,
Advances in Intelligent Systems and Computing 691, DOI 10.1007/978-3-319-70990-1_41

a SVD-based algorithm (Funk-SVD) on the blog, which was called the factor model by Netflix Prize champion Koren. The basic idea is to map both users and items in the high-dimensional vector space into a joint latent space of low-dimensional of f, such that user-item interaction in the same space. And this algorithm can increase prediction accuracy [10].

Although collaborative filtering has many advantage, it still have some problem. Paper [10] proposed that neighborhood model only take their neighborhoods into consideration, it does not consider the all rates, and it faces the problem of sparse data and cold start. Latent factor think about global rates, regardless of historical behavior and neighborhoods. Concerning the flaws of two models, many scholars put forward a series of approaches, such as SVD++ [11]. In this paper, we proposes a algorithm hybrid the user-user neighborhood model and latent factor model, This method alleviates the problem of sparsity, and takes the users' historical behavior into account.

2   User-User Neighborhood Model

User-user model is to calculate the similarity between two users. The method of calculating the similarity of object has been done research by Badrul Sarwar [12]. In this paper, we use the person correlation to calculate the similarity between users. The predicted rating of u for i is as follows:

$$\hat{r}_{ui} = \bar{r} + \frac{\sum_{v \in N_i(u)} w_{uv}(r_{vi} - \bar{r})}{\sum_{v \in N_i(u)} w_{uv}} \tag{1}$$

$N_i(u)$ contains all the items rated by user u that is most similar to item i. The important flaw of this model is computational, especially there are more users than items. Storing all user-user relations is overly expensive and completely impractical. When user is a new user, then the algorithm can not give a recommendation to her. All those flaws disappear by combining neighborhood model with latent factor model.

3   Combining Neighborhood Model with Latent Factor Model

In the hybrid model, the algorithm will add the baseline prediction [13], and the similarity matrix is decomposed into p_u and z_v matrix. This can solve the computational complexity problem of computing user u and user v similarity. When the user u is a new user, it is still can predict the score of item i. The prediction rating of user u for item i is as follows:

$$\hat{r}_{ui} = \mu + b + b + |R(i)|^{-\frac{1}{2}} \sum_{v \in R(i)} (r_{vi} - b_{vi}) p_u^T z_v \tag{2}$$

μ represents the global average of the all rates in the training set. b_u represents the user rating biases. b_i represents the user rating biases. $b_{vi} = b_v + b_i$ represents the biases of user v and item i. $R(i)$ contains all the users who rated item i. The model parameters can be learned by minimizing the following loss function:

$$C(p, q) = \sum_{(u, i) \in Train} \left(r_{ui} - \mu - b_u - b_i - |R(i)|^{-\frac{1}{2}} \sum_{v \in R(i)} (r_{vi} - b_{vi}) p_u z_v \right)^2 + \lambda \left(b_u^2 + b_i^2 + \|p_u\|^2 + \|z_v\|^2 \right) \quad (3)$$

$\lambda(b_u^2 + b_i^2 + \|p_u\|^2 + \|z_v\|^2)$ is added to prevent overfitting. Parameter λ controls the level of regularization. The four parameters p_{uf}, z_{vf}, b_u and b_i need to be advanced along the stochastic gradient descent method, we modify the parameters by moving in the opposite direction of the gradient, so the following recursive formula can be obtained:

$$b_u := b_u + \alpha \left(e_{ui} - \lambda b_u \right) \quad (4)$$

$$b_i := b_i + \alpha \left(e_{ui} - \lambda b_i \right) \quad (5)$$

$$p_{uf} := p_{uf} + \alpha \left(e_{ui} \cdot z_{vf} - \lambda p_{uf} \right) \quad (6)$$

$$z_{vf} := z_{vf} + \alpha \left(e_{ui} \cdot p_{uf} - \lambda z_{vf} \right) \quad (7)$$

α is the learning rate, $e_{ui} = r_{ui} - \hat{r}_{ui}$ represent the predicting rate and the practical rate.

4   Experiment Results

4.1   Datasets and Evaluation Criteria

In this paper, we evaluate the method on the MovieLens [14] data set, MovieLens data set is provided by the GroupLens Research project. MovieLens 100k datasets contain 100,000 ratings by 943 users on 1682 movies, each user at least has rated 20 movies, each film was rated at least three times. The rate of this data set is 1–5, this data set includes the training sets and testing sets. For each datasets, there is a very important feature is that the sparsity of the data set. There are many items does not give a rate in the data set. The formula for computing the sparsity is that (Table 1):

$$sparsity = \frac{all\ rated\ items\ in\ the\ data\ set}{amount\ of\ users \times amount\ of\ items} \quad (8)$$

Table 1. MovieLens datasets

Types	Numbers of users	Number of items	Number of rates	Spareness
MovieLens 100 k	943	1682	100,000	93.7%
MovieLens 1 M	6040	3883	1,000,209	95.7%
MovieLens 10 M	71,567	10,681	1,000,054	98.7%
MovieLens 20 M	138,493	27,278	20,000,263	99.5%

From the above table, the sparse degree of MovieLens 100k is 93.7%, data sparsity is very high.

The RMSE (root mean square error), which is used in Netflix, is used as the basis of the evaluation. The smaller the root mean square error, the better the accuracy of the algorithm.

$$RMSE = \sqrt{\frac{\sum_{u,i \in R_{test}} (r_{ui} - \hat{r}_{ui})^2}{|R_{test}|}} \qquad (9)$$

4.2   Experiment Design and Results

This experiment is based on u1.base as a training set, u1.test as a testing set. In this algorithm, there are four major parameters, the number of implicit features f, the nearest neighborhoods k, the learning rate α and the regularization parameter λ. It is found that the number of implicit features f has little effect on the performance of the algorithm, here there is no benefit in increasing f. So in this experiment, suppose the values of f is 10. While the number of neighbors k has a great influence on the performance of the algorithm. Therefore, we used the following values for predicting: $\alpha = 0.02$, $\lambda = 0.05$, f = 10, the number of iterations is chosen 50 times, the learning rate is reduced by a factor of 0.9 after each iteration, and then observing the change of the accuracy of this algorithm. The result is shown in Fig. 1. By the increasing of neighbors k, the accuracy of the algorithm is increasing, too.

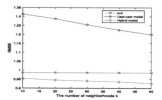

Fig. 1.   Experimental results

It can be seen from Fig. 1 that the hybrid algorithm is more accurate than the user-user model and the latent factor model, and the recommendation can still give recommendations when the data set sparseness is very high, so the hybrid algorithm can address the problem of data sparsity.

In the hybrid algorithm, the learning rate also has an effect on the recommended accuracy, we used the following values for predicting: $\lambda = 0.05$, f = 10, k = 10, the number of iterations is chosen 50 times, the learning rate is reduced by a factor of 0.9 after each iteration. The experimental results are shown in Table 2. The lower the learning rate, the higher the recommended accuracy of the algorithm, but the iterations times increase, too. Therefore, it is necessary to consider the accuracy and running time of the algorithm.

Table 2. The relationship between learning rate and RMSE

Learning rate	0.015	0.02	0.025	0.03
RMSE	0.9440	0.9544	0.9610	0.9656
Iteration times	214	43	25	17

5 Conclusion

In this paper, we have combined user-user neighborhood model with latent factor model, taking into account the global and local information. Experimental results on the Movie-Lens datasets show that the proposed algorithm can improve accuracy and can be effectively recommended under sparse conditions, which can effectively address the problem of data sparsity. And through the experiment, we can observe that implicit feedback has no benefit for this algorithm, we can show that the increasing of neighbors k, and the accuracy of the algorithm is increasing.

Acknowledgment. This work was supported by the National Natural Science Foundation of China (Grant No. 61203117) and Yao is the corresponding author.

References

1. Glodberg, D., Nichols, D., Oki, B.M., Terry, D.: Using collaborative filtering to weave an information tapestry. Commun. ACM **35**, 61–70 (1992)
2. Yao, P., Zou, D., Niu, B.: Collaborative filtering recommender algorithm based on user preferences and project properties Beijing. Comput. Syst. Appl. **24**(7), 15–21 (2015)
3. Ketchantany, W., Derrde, S., Martin, L.: Pearson-based mixture model for color object tracking. Mach. Vis. Appl. **19**, 457–466 (2008)
4. Breese, J.S., Heckerman, D., Kadie, C.: Empirical Analysis of Predictive Algorithms for Collaborative Filtering, pp. 43–52. Morgan Kaufmann Publishers, Wisconsin (1998)
5. Linden, G., Smith, B., York, J.: Amazon.com recommendations item-to-item collaborative filtering. IEEE Internet Comput. **7**(1), 76–80 (2003)
6. Bell, R., Koren, Y., Volinsky, C.: Modeling relationships at multiple scales to improve accuracy of large recommender systems. In: Proceedings of the 13th ACM SIGKDD International Conference on Knowledge Discovery and Data Mining (KDD 2007), New York, pp. 95–104 (2007)
7. Ricci, F., Rokach, L., Shapira, B.: Recommender System Handbooks, pp. 83–84. Springer, New York (2011)
8. Xiang, L.: Recommended System Practice, pp. 184–193. Posts and Telecom Press, Beijing (2012)
9. Simon, F.: http://sifter.org/~simon/journal/20061211.html
10. Koren, Y., Bell, R., Volinsky, C.: Matrix factorization techniques for recommender systems. IEEE Comput. **42**(8), 30–37 (2009)
11. Koren, Y.: Factorization meets the neighborhood: a multifaceted collaborative filtering model. In: Proceedings of ACM SIGKDD Conference, pp. 426–434 (2008)

12. Sarwar, B., Karypis, G., Konstan, J.: Item-based collaborative filtering recommendation algorithms. In: Proceedings of the 10th International Conference on World Wide Web, pp. 285–295. ACM (2001)
13. Koren, Y.: Collaborative filtering with temporal dynamics. In: ACM SIGKDD International Conference on Knowledge Discovery and Data Mining, pp. 447–456 (2009)
14. https://grouplens.org/datasets/movielens/

Numerical Simulation of Mould in Autoclave Process for Fiber Reinforced Composites

Fengyang Bi[1(✉)], Bo Yang[2], Tianguo Jin[3],
Changxi Liu[1], and Xiaohong Wang[1]

[1] School of Mechatronics Engineering,
Heilongjiang Institute of Technology, Harbin, China
bfy952421@163.com
[2] State Key Laboratory of Mechanical Transmission,
Chongqing University, Chongqing, China
[3] School of Mechatronics Engineering,
Harbin Institute of Technology, Harbin, China

Abstract. The heat transfer and deformation processes of the mould in the autoclave process for fiber reinforced composites were analyzed, the simulation methods for mould heat transfer and deformation were built, then the mould heat transfer and deformation processes during autoclave molding were simulated. By the case verification, the simulation model established has high simulation accuracy.

Keywords: Mould structure design parameters · Sensitivities of the mould · Mould stiffness · Mould heat transfer

1 Introduction

Carbon Fiber Reinforced Plastic, or CFRP for short, with the unique advantages over metal materials in specific strength, stiffness, corrosion resistance, dimensional stability, anti fatigue-fracture property, devisable property and the property of being easily overall molded in large formats, are being widely used in the fields of aviation and astronavigation. Particularly in aviation, most of the existing large aircraft tend to the application of CFRP parts. For example, the resin-based composite materials used for Boeing787 amount to 45%, correspondingly reducing the weight by 20% in structure [1, 2]. At present, our large aircraft composite parts are also being developed, and large commercial aircrafts made of domestic composite parts are planned to make the first flight in several years. During the course of making large-format CFRP parts with the format of 2 to 8 m and thickness of 2 to 20 mm, in consideration of their high intensity, forced assembly is forbidden when aircrafts are being assembled. Their shapes which may affect the aerodynamic performance of aircrafts are required to be made in high precision with a small amount of deformation and consistent performance of each part. Autoclave molding technology is the main technology of molding large-format CFRP parts of complex shapes, which, compared with other molding technology, has the advantages of making parts of higher performance and more stable quality.

© Springer International Publishing AG 2018
F. Qiao et al. (eds.), *Recent Developments in Mechatronics and Intelligent Robotics*,
Advances in Intelligent Systems and Computing 691, DOI 10.1007/978-3-319-70990-1_42

Concerning mould design in recent years, scholars at home and abroad mainly guide the mould design by way of establishing numerical models to simulate the physical state of the mold during the autoclave molding process. As for parts deformation control, Kappel and Stefaniak [3–5] made a research on reducing the deformation by adjusting process parameters, and based on experiments, they established a box-shaped tool to forecast part deformation. In heat transfer simulation of mould, Zeng [6], based on experiments, established a semi-analytical model of mould heat transfer, which could simulate mould heat transfer and the interaction between mould and parts which caused warps. Kuppers and Walczyk [7] stimulated and analyzed the heat transfer process of hot-pressing curing molding method which is between soft mould curing and thermal curing.

By analyzing the heat transfer process of the mould in the autoclave and the elastic deformation process of the mould under stress and simulating mould heat transfer analysis and elastic deformation of the mould under stress, this paper selected some design parameters sensitive to the mould stiffness and the results of heat transfer design to establish the optimized model of Metamodel, furtherly to optimize the design of the support structure of the mould.

2   Temperature Difference on Molded Surface and Establishment of Deformation Simulation Model

In the process of autoclave forming and parts forming, the mould is forced to convect with the gas in the autoclave, which can be seen as the thermal load that the mould bears. Vacuumizing the vacuum bag on the molded surface to make pressure difference between the upper surface and lower surface can be seen as the pressure load that the molded surface bears. Under the thermal load and pressure load in the autoclave, the temperature difference on the molded surface and the deformation of the mould produce important effect on the quality of the molded parts. To study the factors and rules of heat transfer and deformation of mould, a method of simulating the process of heat transfer and deformation in the autoclave is established.

In autoclave forming, heating and cooling process are necessary. The former refers to heating the gas forced to flow in the autoclave by an input power which is a resistance heater installed in front of the fan, and the latter refers to cooling the gas in the autoclave by a heat exchanger of cooling medium set in front of the fan. Then the mould is heated and cooled by gas as the medium. As a thermostat keeps controlling the temperature inside the autoclave and the process of heating and cooling, the gas in the autoclave stays flowing. More over, as there is a thermal-protective coating on the autoclave wall, it can be considered that the temperature of the gas in the autoclave is consistent. Heat exchange between the mould and outer medium turns up between the mould and the gas in the autoclave and exists mainly as convection, and the heat inside the mould transfers from the surface to the center by way of thermal conduction. The heating temperature by autoclave forming is mostly below 200 °C, so radiant heat exchange hardly affects the distribution of temperature, thus can be ignored during simulation.

Construction of the simulation method:

(1) Fluid heat exchange area

Suppose that gas in the autoclave can not be compressed during the heat exchange process, so in the fixed Cartesian coordinate system, the three basic physical laws are respectively indicated as:

(1) mass conservation equation (continuity equation)

$$\frac{\partial \rho}{\partial t} + div(\rho u) = 0 \tag{1}$$

ρ as fluid density

(2) energy equation

According to the law of energy conservation, import the Fourier law, then get the following energy equation:

$$\frac{\partial(\rho h)}{\partial t} + \frac{\partial(\rho u h)}{\partial x} + \frac{\partial(\rho v h)}{\partial y} + \frac{\partial(\rho w h)}{\partial z} = -p divU + div(\lambda gradT) + \Phi + S_h \tag{2}$$

λ as fluid thermal conductivity; S_h as inner thermal source of the fluid; h as coefficient relevant to fluid pressure and temperature; Φ as dissipation function. Then the equation can be indicated as:

$$\Phi = \eta \left\{ \begin{array}{l} 2\left[\left(\frac{\partial u}{\partial x}\right)^2 + \left(\frac{\partial v}{\partial y}\right)^2 + \left(\frac{\partial w}{\partial z}\right)^2\right] + \\ \left(\frac{\partial u}{\partial y} + \frac{\partial v}{\partial x}\right)^2 + \left(\frac{\partial u}{\partial z} + \frac{\partial w}{\partial x}\right)^2 + \left(\frac{\partial v}{\partial z} + \frac{\partial w}{\partial y}\right)^2 \end{array} \right\} + \lambda divU \tag{3}$$

(3) momentum equation

Indicated by three dimensional transient equation, Navier–Stokes:

$$\begin{aligned} \frac{\partial(\rho u)}{\partial t} + div(\rho u U) &= div(\eta gradu) + S_u - \frac{\partial p}{\partial x} \\ \frac{\partial(\rho v)}{\partial t} + div(\rho v U) &= div(\eta gradv) + S_v - \frac{\partial p}{\partial y} \\ \frac{\partial(\rho w)}{\partial t} + div(\rho w U) &= div(\eta gradw) + S_w - \frac{\partial p}{\partial z} \end{aligned} \tag{4}$$

ρ as fluid density; p as fluid pressure; η as fluid dynamic viscosity

Generalized source term of momentum equation, indicated as equation:

$$S_u = \frac{\partial}{\partial x}\left(\eta\frac{\partial u}{\partial x}\right) + \frac{\partial}{\partial y}\left(\eta\frac{\partial v}{\partial x}\right) + \frac{\partial}{\partial z}\left(\eta\frac{\partial w}{\partial x}\right) + \frac{\partial}{\partial x}(\lambda divU)$$

$$S_v = \frac{\partial}{\partial x}\left(\eta\frac{\partial u}{\partial y}\right) + \frac{\partial}{\partial y}\left(\eta\frac{\partial v}{\partial y}\right) + \frac{\partial}{\partial z}\left(\eta\frac{\partial w}{\partial y}\right) + \frac{\partial}{\partial y}(\lambda divU) \qquad (5)$$

$$S_w = \frac{\partial}{\partial x}\left(\eta\frac{\partial u}{\partial z}\right) + \frac{\partial}{\partial y}\left(\eta\frac{\partial v}{\partial z}\right) + \frac{\partial}{\partial z}\left(\eta\frac{\partial w}{\partial z}\right) + \frac{\partial}{\partial z}(\lambda divU)$$

S_u, S_v, S_w as the generalized source term of momentum equation
Processing of the fluid dynamic viscosity η: as gas flows turbulently in the autoclave, which accords with zero equation turbulence model, so the viscosity coefficient is thought to be constant and indicated as algebraic equations:

$$\eta_t = \rho f_\mu U_t l_t \qquad (6)$$

η_t as turbulence viscosity coefficient; ρ as fluid density; f_μ as proportionality coefficient; U_t as maximum velocity vector; l_t as fluid mixed length
fluid mixed length is determined by equation:

$$l_t = \left(V_D^{1/3}\right)\Big/7 \qquad (7)$$

V_D as regional volume of fluid
(2) solid heat exchange area
The heat transfer on the molded surface and in the framework is thermal conduction, indicated as equation:

$$\frac{\partial(\rho cT)}{\partial t} = \frac{\partial}{\partial x_j}\left(\lambda\frac{\partial T}{\partial x_j}\right) \qquad (8)$$

ρ as material density of the mould; T as the temperature of the mould; c as the specific heat of the mould
(3) Grid division
It is impossible to get the accurate solution of the mould of complex shapes through analytical method, so the simulation method is adopted in this paper. First, discretize the computational domain into a finite element mesh. Finite element mesh is divided into two categories: structured grid and unstructured grid. The structured grid computing is of high precision and convergence speed, but it is difficult to automatically generate parts of complex shape, so manual intervention is often required, which means that the parts are divided into sub regions manually by the software operator, then a grid division can be realized. As the establishment of the frame mould is based on features, the basic features are exactly equivalent to the sub region division when mapping method is used to divide the structured grid, so it is easier to automatically generate a structured grid

on the basis of setting the basic features as a sub region. Since the gas convects between the autoclave and the mould, the grid division is relatively complex, especially the molded surface. As there is fluid, the heat transfer layer is very thin. And if the unit is too large and the distance between nodes is too far away, non physical simulation of the heat transfer layer and poor simulation precision will be caused. So the mould is moved in the direction of the outer normal by a distance of twice the thickness of the mould wall and it is guaranteed that in the direction of the thickness of the heat transfer layer, there are at least 5 layer units and 6 layer nodes when the grid is divided. As for the convection section of other fluids, unstructured grid is adopted to improve the efficiency of grid division.

3   Applications

The mould deformation simulation is based on the simulation results of the temperature difference of the mould (Fig. 1). The mould deformation is divided into two parts in the autoclave, including Thermal expansion caused by the temperature rise and the elastic deformation caused by the compression of the surface. Suppose that the contact part of the mould and the platform of the autoclave is fixed, superpose the two deformed parts in the simulation, then the results of the mould deformation simulation is shown in Fig. 2. The average thermal expansion coefficient of steel Q235 below 200 °C is $12 \times 10^{-6}/°C$ and in the range of 120 °C, steel Q235 expands 1.4 mm per meter. But due to the mutual restraint of the mould structure, not all the thermal expansion and contraction can be reflected in the macro deformation of the mould. As a result, the maximum deformation of the mould surface is 0.7664 mm, warping upward in the four corners. That is because the temperature of the mould edge is higher than the internal part when the temperature rises.

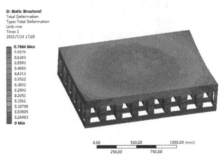

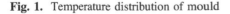

Fig. 1. Temperature distribution of mould   **Fig. 2.** Compressive deformation of mould

4 Conclusion

The heat transfer and deformation processes of the mould in the autoclave process for fiber reinforced composites were analyzed, the simulation methods for mould heat transfer and deformation were built, then the mould heat transfer and deformation processes during autoclave molding were simulated. By the case verification, the simulation model established has high simulation accuracy.

Acknowledgments. This research was supported by the National Natural Science Foundation of China (no. 51575139, 51605057), Natural Science Foundation of Heilongjiang Province (no. E2015027), China Postdoctoral Science Foundation (no. 2016M600721), Special Financial Aid to Post doctor Research Project of Chongqing. (no. Xm2016058), Fundamental and Frontier Research Project of Chongqing (no. cstc2016jcyjA0456), Self-Planned Task of State Key Laboratory of Mechanical Transmission (no. SKLMT-ZZKT-2016Z04), Project of Innovative Talents Foundation of Harbin City (no. 2016RAQXJ075, 2016QN001011, 2016QN017023), Project of Doctor Foundation of Heilongjiang Institute of Technology (no. 2015BJ07). Their support is gratefully acknowledged.

References

1. Shanyi, D.: Advanced composite materials and aerospace engineering. Acta Mater. Compos. Sin. **24**(01), 1–12 (2007)
2. Naibin, Y.: Composite structures for new generation large commercial jet. Acta Aeronaut. Astronaut. Sin. **29**(3), 596–604 (2008)
3. Stefaniak, D., Kappel, E., Spröwitz, T., Hühne, C.: Experimental identification of process parameters inducing warpage of autoclave-processed CFRP parts. Compos. Part A **43**(7), 1081–1091 (2012)
4. Kappel, E., Stefaniak, D., Holzhüter, D., Hühne, C., Sinapius, M.: Manufacturing distortions of a CFRP box-structure-a semi-numerical prediction approach. Compos. Part A **51**(8), 89–98 (2013)
5. Kappel, E., Stefaniak, D., Spröwitz, T., Hühne, C.: A semi-analytical simulation strategy and its application to warpage of autoclave processed CFRP parts. Compos. Part A **42**(12), 1985–1994 (2011)
6. Zeng, X., Raghavan, J.: Role of tool-part interaction in process-induced warpage of autoclave-manufactured composite structures. Compos. Part A **41**(9), 1174–1183 (2010)
7. Kuppers, J., Walczyk, D.: Thermal press curing of advanced thermoset composite laminate parts. Compos. Part A **43**(4), 635–646 (2012)

Analysis for Sensitivities of the Mould Stiffness and Heat Transfer Design Result to the Mould Structure Design Parameters

Fengyang Bi[1(✉)], Bo Yang[2], Tianguo Jin[3], Changxi Liu[1],
and Xiaohong Wang[1]

[1] School of Mechatronics Engineering,
Heilongjiang Institute of Technology, Harbin, China
bfy952421@163.com
[2] State Key Laboratory of Mechanical Transmission,
Chongqing University, Chongqing, China
[3] School of Mechatronics Engineering,
Harbin Institute of Technology, Harbin, China

Abstract. Sensitivities of the mould stiffness and heat transfer design result to the mould structure design parameters were analyzed. Based on the parameters of the mould structure design, the temperature difference of the mould surface should be reduced, the stiffness of the mould be increased, and the mass of the mould should be reduced as much as possible. In the process of designing, proper thickness of molded surface is of most importance, then come proper numbers and sizes of air vent.

Keywords: Mould structure design parameters · Sensitivities of the mould · Mould stiffness · Mould heat transfer

1 Introduction

Resin Matrix Composite, or RMC for short, with the unique advantages over metal materials in specific strength, stiffness, corrosion resistance, dimensional stability, anti fatigue-fracture property, devisable property and the property of being easily overall molded in large formats, are being widely used in the fields of aviation and astronavigation. Particularly in aviation, most of the existing large aircraft tend to the application of RMC parts. For example, the resin-based composite materials used for Boeing787 amount to 45%, correspondingly reducing the weight by 20% in structure [1, 2]. At present, our large aircraft composite parts are also being developed, and large commercial aircrafts made of domestic composite parts are planned to make the first flight in several years. During the course of making large-format RMC parts with the format of 2 to 8 m and thickness of 2 to 20 mm, in consideration of their high intensity, forced assembly is forbidden when aircrafts are being assembled.

Concerning mould design in recent years, scholars at home and abroad mainly guide the mould design by way of establishing numerical models to simulate the physical state of the mold during the autoclave molding process. As for parts

© Springer International Publishing AG 2018
F. Qiao et al. (eds.), *Recent Developments in Mechatronics and Intelligent Robotics*,
Advances in Intelligent Systems and Computing 691, DOI 10.1007/978-3-319-70990-1_43

deformation control, Kappel and Stefaniak [3–5] made a research on reducing the deformation by adjusting process parameters, and based on experiments, they established a box-shaped tool to forecast part deformation. Kuppers and Walczyk [6] stimulated and analyzed the heat transfer process of hot-pressing curing molding method which is between soft mould curing and thermal curing. The research indicated that under the action of the circular flow of gas in the autoclave, whether in heating stage or cooling stage, the temperature distribution of the mould is uneven, and the maximum temperature difference on the surface may reach ten or more degree centigrade.

2 Sensitivity Analysis of Mould Design Parameters

In this section, the design parameters of temperature difference and deformation sensitivity of the molded surface are selected through the single factor and multi level change test. The mould design personnel, according to their experience, select 17 alternative parameters of mould design optimization among more than 40 mould structures and process parameters affecting temperature difference and deformation of the molded surface, and integrate three categories and nine parameters of radial vent, air flow vent and supporting frame into the vent opening ratio (hereinafter referred to as the opening ratio) for combined analysis. It's possible for the design personnel to find out the major parameters, but the sensitivity of the design results is not very clear to them. So in the following paragraphs, there is an analysis of the sensitivity of each factor to the results through the single factor and multi level test.

The subjects in this test includes mould design control parameters ΔT_{up} and ΔT_{down}, and structural factors of the parts like length and width, thickness; process factors of the autoclave, such as heating rate, gas flow rate, pressure and so on; thickness of the molded surface, the air vent opening ratio. Through the single factor and multi level change test, the sensitivities of the factors mentioned above on the temperature difference of the mould are compared to select and optimize the parameters. The level coverage of each factor is fixed in the range of the parameters commonly used in mould design, then the level value is taken evenly in the range of parameter. The selection of factor level value and the test results are shown in Table 1. Effect that the level value of each factor has on temperature difference of the molded surface when they change individually is indicated by first order linear equation. The curve slope signifies the response of temperature difference to the variation of factor level. As different factors have different ranges of parameter values and units, equation coefficients can not be used directly as contrast sensitivity. In consideration of the effect of the value range, unit is ignored and the sensitivity of each factor on the temperature difference of the molded surface is compared by comparing the rates of numerical change of the results caused by the level changes of per percentage within the value range of each factor, then this value is defined as the sensitivity coefficient shown in Eq. 1. SEM as the sensitivity coefficient; k as coefficient of equation (the slope); Δx as the value range of factor parameter.

Table 1. Sensitive degree of single factor change on the mould surface temperature difference

No.	Change on the mould surface temperature difference	SEM
1	$$\Delta T = 0.00144l + 8.29$$	7.056
2	$$\Delta T = 0.00199w + 13.84$$	1.89
3	$$\Delta T = 1.47\delta_c + 13.84$$	11.76
4	$$\Delta T = 0.756\delta_m + 9.358$$	7.56

5	$\Delta T = -3.32 v_{wind} + 25.03$	−11.62
6	$\Delta T = -1.90 p_{autoclave} + 14.01$	−9.5
7	$\Delta T = 9.16 T_{autoclave} + 0.71$	22.9
8	Rectangle: $\Delta T = -8.55 R_{aperture} + 18.22$ Round: $\Delta T = -6.74 R_{aperture} + 15.72$	Rectangle: -2.14 Round:- 1.68

$$SEM = k\Delta x \tag{1}$$

According to the analysis of sensitivity shown in Table 1, the effect that mould structure design parameters have on the temperature difference of the molded surface is: (1) The increases of the length and width of the mould together with the thickness of the molded surface all cause the increase of the temperature difference of the molded surface, among which the thickness of the mould is the most sensitive. (2) The effect that the opening rate of mould air vent has on the temperature difference of the molded surface is: with the increase of the opening rate, the temperature difference of the molded surface will decrease, but not as sensitive as the thickness of the molded surface. (3) The openings in different shapes cause different results. Of the same opening rate value, temperature difference of round opening mould. (4) The increase in the thickness of the parts will increase the temperature difference of the molded surface.

The effect that process parameters has on the temperature difference of the molded surface is: (1) The increase of wind speed in the autoclave can reduce the temperature difference of the molded surface. (2) The increase of pressure in the autoclave can also reduce the temperature difference of the molded surface. (3) The increase of the heating rate of the air in the autoclave can dramatically increase the temperature difference of the molded surface. (4) The process parameters are more sensitive to the temperature difference of the molded surface than the mould structure.

When a mould is being designed, its length and width are determined by the size of the molded part, so they can not serve as the parameters of optimization. The process parameters are mainly determined by the temperature and required time for molding the resin material used in making parts, which is hardly adjustable, so they can neither serve as the parameters of optimization. Parameters which are sensitive to the temperature difference of the molded surface, adjustable during the process of mould design and most suitable for optimizing can be used as the opening rates of the thickness of the molded surface and air vent.

3 Conclusion

Therefore, based on the parameters of the mould structure design, the temperature difference of the molded surface should be reduced, the stiffness of the mould be increased, and the mass of the mould should be reduced as much as possible. In the process of designing, proper thickness of molded surface is of most importance, then come proper numbers and sizes of air vent.

Acknowledgments. This research was supported by the National Natural Science Foundation of China (no. 51575139, 51605057), Natural Science Foundation of Heilongjiang Province (no. E2015027), China Postdoctoral Science Foundation (no. 2016M600721), Special Financial Aid to Post-doctor Research Project of Chongqing (no. Xm2016058), Fundamental and Frontier Research Project of Chongqing (no. cstc2016jcyjA0456), Self-Planned Task of State Key Laboratory of Mechanical Transmission (no. SKLMT-ZZKT-2016Z04), Project of Innovative Talents Foundation of Harbin City (no. 2016RAQXJ075, 2016QN001011,2016QN 017023), Project of Doctor Foundation of Heilongjiang Institute of Technology (no. 2015BJ07). Their support is gratefully acknowledged.

References

1. Du, S.: Advanced composite materials and aerospace engineering. Acta Mater. Compos. Sin. **24**(01), 1–12 (2007)
2. Yang, N.: Composite structures for new generation large commercial jet. Acta Aeronau. Tica Et Astronau. Tica Sin. **29**(3), 596–604 (2008)
3. Stefaniak, E.K., Spröwitz, T., Hühne, C.: Experimental identification of process parameters inducing warpage of autoclave-processed CFRP parts. Compos. Part A **43**(7), 1081–1091 (2012)
4. Kappel, E., Stefaniak, D., Holzhüter, D., Hühne, C., Sinapius, M.: Manufacturing distortions of a CFRP box-structure A semi-numerical prediction approach. Compos. Part A **51**(8), 89–98 (2013)
5. Kappel, E., Stefaniak, D., Spröwitz, T., Hühne, C.: A semi-analytical simulation strategy and its application to warpage of autoclave processed CFRP parts. Compos. Part A **42**(12), 1985–1994 (2011)
6. Zeng, X., Raghavan, J.: Role of tool-part interaction in process-induced warpage of autoclave-manufactured. Compos. Part A Appl. Sci. Manuf. **41**(9), 1174–1183 (2010)

A Low Cost Instrumented Insole to Measure the Center of Pressure During Standing

Xinyao Hu[✉], Jun Zhao, Yonghao Li, Yingting He, and Xingda Qu

Institute of Human Factors and Ergonomics,
Shenzhen University, Shenzhen, China
huxinyao@szu.edu.cn

Abstract. Fall risk assessment can help identify fall-prone individuals and design fall prevention and intervention programs, therefore it is recognized as an effective fall prevention measure. The foot center of pressure (COP) is an important fall risk assessment parameter. This paper presents a novel approach for measuring the COP trajectory by a wearable device based on low cost force sensitive resistors (FSRs). A pilot study was carried out for model development and validation purpose. The results showed that COP trajectory can be estimated accurately. This study can benefit the future clinical application to allow the COP be analyzed in home settings and used as the fall risk assessment parameter for elderly who are prone to fall accidents.

Keywords: Falls · COP trajectory · FSRs · Wearable device · Low cost

1 Introduction

Falls are a major health problem among aged population. Falls often result in serious physical injuries and lead to adverse psychological consequences [1]. To reduce fall accidents among elderly, many fall prevention approaches have been proposed. Fall risk assessment is recognized as an effective fall prevention measure because it can help identify fall-prone individuals and design fall prevention or intervention programs [2]. Specific biomechanical parameters were often selected to identify the risk of falls. They are also known as the fall risk assessment parameters [3]. Postural stability, is among the import fall risk assessment parameters. Many studies have revealed that postural stability generally correlated with the risk of falls [4].

The foot center of pressure (COP) is a parameter that can reflect the postural sway of an individual. Thus, COP and COP related measures have been used to in fall risk assessment. A variety of insole based plantar pressure measurement systems are developed, such as the F-scan measurement system (Tekscan, Inc., MA, USA) and Novel Pedar system (Novel Inc., CA, USA). Those systems can capture temporal and spatial pressure distribution and calculate the COP trajectory accurately. However, these systems in general are still too expensive for personal or daily use.

The recent advancements in microelectronics technologies have allowed the instrumented insoles based on low cost sensors to be used for foot plantar measurement [5]. Force Sensitive Resistors (FSRs) are the most commonly used sensors [6]. FSRs are capable of monitoring the pressure over a fair range of force. And they are very thin,

F. Qiao et al. (eds.), *Recent Developments in Mechatronics and Intelligent Robotics*,
Advances in Intelligent Systems and Computing 691, DOI 10.1007/978-3-319-70990-1_44

therefore they are unobtrusive and won't discomfort the wearer. FSRs have been widely used for measuring the foot plantar pressure.

This paper presents a novel approach for measuring the COP trajectory by low cost FSRs. This approach relies on 12 integrated FSRs, strategically placed on an instrumented insole. A linear regression model was used to weight the sensor output to match the COP trajectory measured by a F-scan System (served as the ground truth). The established model was used to predict the COP trajectory. A pilot study is carried out to develop and validate this model.

2   Method

2.1   System Overview

The instrumented insole consists of 12 integrated FSRs (FSR402, Interlink Electronics, CA, US), an ankle mounted block, and a PC end graphic user interface (GUI) with the COP trajectory estimation algorithm (Fig. 1).

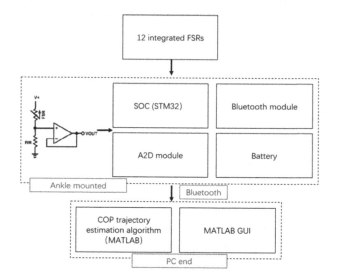

Fig. 1. Block diagram of the instrumented insole

Twelve FSRs were strategically adhered onto an insole made of silicone. The size of the insole is US8. The length, measured from the tip top to the bottom edge, is 270 mm. The width, measured by the distance between the two tangent lines at the left and right edge, is 90 mm. The layout of the 12 FSRs is depicted in Fig. 2.

The sensor placement is similar to what has been suggested by Howell et al. [7]. Twelve sensors cover the important foot plantar pressure distribution area such as great

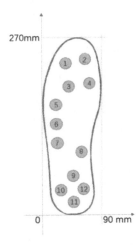

Fig. 2. The FSR layout of the insole

toe, metatarsophalangeal (MTP) joint, arch of the foot, and heel. An insole coordinate system was used to help identify the actual location of each sensor. The x-axis is the tangent line to the bottom edge of the insole. The y-axis is the tangent line to the left

Table 1. The coordinate of each FSR

FSR No.	x-coor	y-coor	FSR No.	x-coor	y-coor
1	30	236	7	9	122
2	60	239	8	49	109
3	37	194	9	42	69
4	62	197	10	14	40
5	10	176	11	37	31
6	9	148	12	52	45

(The No. of each FSR was indicated in Fig. 2, the unit of the coordinate is mm)

edge of the insole. The origin is the intersection between those two lines. The coordinate of each FSR was shown in Table 1

2.2 COP Trajectory Estimation Model

In general, the COP trajectory is calculated based on a linear model that weight the FSR output (voltage) to match the COP trajectory measured simultaneously by a F-scan System (which is recognized as the reference COP trajectory in the present study). To do so, a sensor sheet from F-scan was tailored to make sure it has exactly the same shape of our instrumented insole. It was then adhered underneath the instrumented insole. It was assumed that the plantar COP trajectory should be the same while these two systems (i.e. F-scan sensor and the instrumented insole) were used to collect foot plantar pressure data simultaneously. Therefore, the COP trajectory presented in X and Y coordinate can be modeled as the weighted average of the coordinate of each FSR:

$$X = \frac{\sum_i^{12}(c_i^x x_i V_i)}{\sum_i^{12} V_i} + c_{13}^x \tag{1}$$

$$Y = \frac{\sum_i^{12}(c_i^y y_i V_i)}{\sum_i^{12} V_i} + c_{13}^y \tag{2}$$

Where c_i^x and c_i^y are the weighting coefficients that weight each FSR output. c_{13}^x and c_{13}^y are the coefficients that help adjust the COP trajectory estimation. x_i and y_i are the corresponding FSR coordinate of the instrumented insole, as shown in Table 1. v_i is the voltage output of each FSR. In this model, c_i^x, c_i^y, c_{13}^x and c_{13}^y are the unknown model coefficients that were estimated by an experimental study.

2.3   Experimental Study

As a pilot study, one volunteer was invited to conduct the experiment study. To validate this model, one instrumented insole (for right foot) was used with the F-scan sensor sheet underneath. The volunteer was asked perform 2 standing trials on the insole with his right foot. The data from the first trial was used as the training data. The data was obtained approximately 5 s after the volunteer start to stand on the insole. Data was recorded for 12 s. The COP trajectory was obtained from F-scan sensor. It was used as the input for the linear model to estimate the model coefficient (i.e. c_i^x, c_i^y, c_{13}^x and c_{13}^y) through a least square error approach. The data from the second trial was used as the validation data. The COP trajectory estimated by the instrumented insole by the linear model was compared with that obtained from the F-scan system. The Root Mean Square errors and the correlation coefficients were calculated.

3   Results

Figure 3 shows the comparison of the estimated COP trajectory coordinates (X and Y) and the COP trajectory coordinates measured by the F-scan system with the validation data set. The Root Mean Squared Error (RMSE) between the estimated COP trajectory and the F-scan COP trajectory is 1.74 mm and 2.05 mm for X and Y coordinate, respectively. The correlation coefficients were 0.97 and 0.93 for X and Y coordinate, respectively. The high overall curve similarity between the estimated COP trajectory coordinates and the reference is shown in Fig. 3.

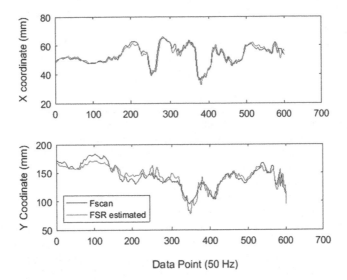

Fig. 3. The estimated X and Y coordinates of COP compared with the reference(F-scan)

4 Discussion

This study demonstrated that foot plantar COP trajectory can be estimated by a low-cost wearable device during quiet standing. In this pilot study, the RMSE between the estimated COP trajectory and the reference measurements indicated that the COP trajectory were estimated accurately. In addition, the correlation coefficient results reveal that the change of the estimated COP trajectory follows closely to the reference measurement. The measurement error is within an acceptable range [8].

A linear regression model was developed to estimate the COP trajectory in a systematic way. The weighting coefficients were determined and used to weight the output of each FSR. Therefore, no calibration procedure was required, unlike the previous studies. As the calibration of FSR remains a tedious and challenging task, our approach offered a more practical solution to implement FSRs directly in relevant applications.

5 Conclusion

The presenting study proposed a novel approach for measuring the COP trajectory by a wearable device based on low cost FSRs. The overall cost of the instrumented insole is within 80 USD. Therefore, it is highly feasible that the low cost instrumented insole will facilitate the COP to be effectively measured and it can be used as the fall risk assessment tool for elderly who are prone to fall accidents.

Acknowledgement. This work was supported in part by Natural Science Foundation of Guangdong Province under Grant number 2015A030313553 and 2016A030310068; Science, Technology and Innovation Committee of Shenzhen City under Grant number JCYJ20160422145322758 and JCYJ20150525092940994.

References

1. Berg, W.P., Alessio, H.M., Mills, E.M., Tong, C.: Circumstances and consequences of falls in independent community-dwelling older adults. Age Ageing **26**(4), 261–268 (1997)
2. Rubenstein, L.Z.: Clinical risk assessment, interventions and services: falls in older people: epidemiology, risk factors and strategies for prevention. Age Ageing **35**(1), 37–41 (2006)
3. Perell, K.L., Nelson, A., Goldman, R.L., Luther, S.L., Prietolewis, N., Rubenstein, L.Z.: Fall risk assessment measures: an analytic review. J. Gerontol. **56**(12), M761–M766 (2001)
4. Kang, H.G., Dingwell, J.B.: A direct comparison of local dynamic stability during unperturbed standing and walking. Exp. Brain Res. **172**(1), 35–48 (2006)
5. Shu, L., Hua, T., Wang, Y., Li, Q.: In-shoe plantar pressure measurement and analysis system based on fabric pressure sensing array. IEEE Trans. Inform. Technol. Biomed. **14**(3), 767–775 (2010). A Publication of the IEEE Engineering in Medicine & Biology Society
6. Chockalingam, N., Healy, A., Naemi, R., Burgesswalker, P., Abdul Razak, A.H., Zayegh, A., et al.: Comments and reply to: foot plantar pressure measurement system: a review. Sensors **12**, 9884–9912 (2013)
7. Howell, A.M., Kobayashi, T., Hayes, H.A., Foreman, K.B., Bamberg, S.J.M.: Kinetic gait analysis using a low-cost insole. IEEE Trans. Biomed. Eng. **60**(12), 3284–3290 (2013)
8. Dyer, P.S., Bamberg, S.J.M.: Instrumented insole vs. force plate: a comparison of center of plantar pressure. In: 2011 Annual International Conference of the IEEE Engineering in Medicine and Biology Society, Boston, pp. 6805–6809 (2011)

A Minimum Distance Cluster Based on Region Growing Method

Kai Zhao[1]([✉]), FengYun Cao[1], AiPing Wang[2], Jia Jing[3], Fengmei Yin[1], and XueJie Yang[1]

[1] School of Computer Science and Technology, Hefei Normal University, Hefei, China
kzhao@aliyun.com
[2] School of Computer Science and Technology, Anhui University, Hefei, China
[3] School of Electronic Science and Applied Physics, Hefei University of Technology, Hefei, China

Abstract. In order to acquire accurately virtual 3D human models, the region of interest (ROI) on medical images is separately segmented out. The proposed method based on region growing algorithms, at first plots adjacent regions around ROIs by rectangles. Then, the seed of each region is detected as centers. To make the method grow at a correct region, the ROI seed is selected which evolves in the direction of eight neighbors under the limit of minimum distance. Finally, depended on the seeds enlargements of ROI, a whole organ are extracted out. In experiments, the segmentation result is compared with an image thresholding method based on minimizing the measures of fuzziness (TMMF). Quantitative evaluation is performed within acceptable limits using volumetric overlap error (VOE) and relative volume difference (RVD).

Keywords: Minimum distance cluster · Image segmentation · Region growing · Image thresholding

1 Introduction

With the development of medical images (CT/MIR), scientists could easily observe the inside of human body with the help of computer. Medical image segmentation shows fundamental value in computer-aided diagnosis, cancer radiation therapy, 3D visualization and other areas of medicine [1–3]. Especially, the virtual 3D human models are often required as the reference of authentic human for scientific simulations. Unfortunately, the medical images are usually so ambiguous that even a specialist could not exactly discriminate his/her interested regions. With the application of the fuzzy theory in the segmentation, different memberships are designed according to different centers on images. The noted algorithm called fuzzy c-means algorithm (FCM) was proposed by Joseph C Dunn in 1973 [4, 5]. With different parameters given beforehand, it can easily segment different regions of images by minimizing a target energy function which is designed according to the weight of distance from the target pixel to the center pixels. The color of the medical image usually shows in grey scales of which the difference can largely separate organs, thus the segmentation by the threshold of grey has been

F. Qiao et al. (eds.), *Recent Developments in Mechatronics and Intelligent Robotics*,
Advances in Intelligent Systems and Computing 691, DOI 10.1007/978-3-319-70990-1_45

developed in many algorithms such as maximum entropy thresholding segmentation [6], the famous FCM and its derivations.

Naturally they are useful during some image segmentation processes, but more often than not, the region of interested (ROI) could not be easily separated independently. As is observed, existing region-based image thresholding methods [7, 8] are unstable for overlapping grey levels between object and background. Given a seeds of ROI, region growing algorithms could solve that segmentation problem in dispersed areas [9, 10]. With good seed selection and the growing criterion, its advantages in medical image segmentation [11] are obvious. Anyway, these methods are in difficulties for the consideration of the position, surrounds characters, grey distribution of the images. As for certain ROI area, it is even difficult to operate the separation wholly and independently.

In this paper, a minimum distance cluster was designed through combing of the region growing method [12]. Based on the grey distribution in histogram and plotted regions on images, the seeds and its neighbors are selected for distances evaluations. Finally, experimental evaluation is performed on examples of medical images, and the grey range of segmentation result to have a comparison with that of IBMFF (image thresholding method based on minimizing the measures of fuzziness [13]). To judge the performance of our proposed approach, the volumetric overlap error (VOE) and relative volume difference (RVD) indexes are calculated.

2   Proposed Method

2.1   The Seed Detection

In our work, the medical images are obtained by a set of DICOM format files [14], from which CT value is read and converted to grey CT images. Most of the CT images binging in low grey difference, have unclear organ borders and are dispersed with noises. In order to reduce the influences, gauss filter is used for the pre-processing. After that, the DICOM files are converted to grey images with the resolution of 512×512, and the grey between 0 and 255.

The seed selection is an important step because an incorrect seed can cause segment errors. In the paper, the features such as recurrence rates, independence of seeds are taken. Supposing the nearest adjacent regions denotes the sets of R_i, the distance from *ROI* to R_i must be smaller than that to any other regions on the image. That is, if the total regions of the image is divided into n, then

$$R_i = \arg \frac{\min}{k = 1, ..n} \{d(\text{ROI}, R_k)\} \tag{1}$$

Given the distance between the two regions is defined as the grey Euclidean distance, the region whose minimum grey is larger than *ROI* is defined as a region of right (*ROR*) and the maximum which is smaller than *ROI* is a region of left (*ROL*). Here, the regions for detections, which are divided into three independent areas expressed as sets of *ROX* (referred as *ROLROL* or *ROR*), agree with the following hypothesis.

$$ROR = \{R_i|R_i > ROI\}, ROL = \{R_i|R_i < ROI\} \tag{2}$$

According to the hypothesis, let the image be the set of I, the relationship of grey sets are derived as

1. $ROR \cap ROI = \Phi, ROL \cap ROI = \Phi$
2. $(ROR \cup ROI \cup ROL) \subset I$
3. $ROR > ROI, ROL < ROI$

The above features tell that *ROR* and *ROL* cannot be so large in that they have the common grey value with *ROI*. *ROX* is the subset of the image with the range of their pixels value is defined.

The detected seed is retrieved from *ROX* based on the above hypothesis. To narrow the area range, a rectangle is plotted on the image according to the pixels occurring frequency (Fig. 1(b)). Hereafter, each pixel is checked whether its eight neighbors have the same grey scale or not. Finally, the detected seed is chosen from the pixels that has the largest number of neighborhoods.

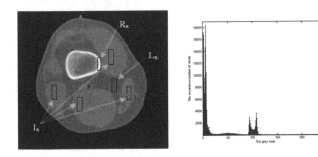

Fig. 1. Sample of the medical image with left (the plotted sub-region of ROL, ROI, and ROR) and right (the grey level histogram).

Take a part of thigh medical image (Fig. 1(a)) for the illustration of *ROX* whose detected sub-regions are marked on the images as *Xx* (referred as *Lx*, *Ix* or *Rx*). The symbol *Ix* by dotted line are the extension of *ROI*, which is used for the afterward combination of different *ROI* areas. According to (1), it is clear to see that *Rx* is not a requirement to the segmentation region of *Ix* by dotted line. But for *Ix* by solid line, *Rx* is a necessary region without which the algorithm would run to the *NOT − ROI* fields.

In the proposed methods, let $h(g)$ denote the occurrence number of pixels at grey of g in the detected sub-region *Xx*, the schema of detecting the seeds proceeds as follows.

Step1. Find the peaks of $h(g)$ and associate it with grey as a vector G. That is,

$$G = \left\{ g \middle| \arg\frac{\max}{g = 1..255}(h(g)) \right\} \tag{3}$$

Step 2. If the members of G are multiple, the average grey of G is calculated to be G. And then G is replaced with any member whose grey is close to $\overline{G}$.

$$G \Leftarrow \left\{ g \,|\, \arg \frac{\min}{g} \left\| G - \overline{G} \right\| \right\} \tag{4}$$

Otherwise go to Step 3.

Step 3. Find the candidate pixels in the detected sub-region Xx which have the same grey as G. The candidate seed sets can be obtained by

$$S = \left\{ (x_i, y_i) \,|\, g(x_i, y_i) = G, (x_i, y_i) \in Xx, i = 1, 2, ..n \right\} \tag{5}$$

Where n is the total number of pixels in Xx, and $g(x_i, y_i)$ is the grey of pixel (x_i, y_i).

Step 4. For each pixel in S, check and record the number of its neighbors who have the same grey as G in eight directions (Table 1). Then the ultimate seed is selected from S, where its neighbor number is maximal.

$$(x_i, y_i) = \mathrm{argmax} \left\{ n(x_i, y_i) \,|\, g(x_{it}, y_{it}) = \mathrm{G} \right\} \tag{6}$$

Where $n(x_i, y_i)$ the neighbor is number of (x_i, y_i), $(x_{it}, y_{it}) \in S$ and $t = 1, 2, ..8$ is the marker of the neighbors' position.

Table 1. Serial numbers of pixels derived from (x_i, y_i).

Left	Middle	Right
(x_{i2}, y_{i2})	(x_{i3}, y_{i3})	(x_{i4}, y_{i4})
(x_{i1}, y_{i1})	(x_i, y_i)	(x_{i5}, y_{i5})
(x_{i8}, y_{i8})	(x_{i7}, y_{i7})	(x_{i6}, y_{i6})

2.2 Distance to the Cluster

The seeds detected from above are aimed at avoiding the similarity between different areas, but just the difference in grey value is not enough. In the distance definition, the distribution of the grey based on the histogram and the grey variance on images are referred as:

$$\mu_{ji} = \sum_{g=g_j}^{g_i} gh(g) / \sum_{g=g_j}^{g_i} h(g) \tag{7}$$

Where μ_{ji} expresses the relativity of the grey distribution, g_i is the grey value of the ultimate seed. Given the number of the pixel j and its neighbor's number n, the grey variance is written as

$$\sigma_j = \sqrt{\frac{\sum_{t=1}^{n} \left(g_t - \frac{1}{n} \sum_{t=1}^{n} g_t \right)^2}{n}} \tag{8}$$

Where σ_j illustrates the grey distribution of the pixel surrounded by its neighbors, and the distance from the pixel j to the cluster i is put forward as

$$d_j(i) = \left|g_j - \mu_{ji}\right| + \left|\sigma_j - \sigma_i\right| \tag{9}$$

So, the cluster indicates the measurement of the distance, it is a set of pixel coordinates on the image.

$$\left\{(x_i, y_i)\right\} = \arg \frac{\min}{j}(d_j(i)) \tag{10}$$

2.3 Minimum Distance Cluster Based on Region Growing (MDRG)

Different regions in the medical image usually demonstrate similar grey, especially on the margins. Even through definite grey range is acquired, *ROI* could not be separated absolutely. Practically, many of the regions in the image are mutually connected. The region growing algorithm thus is an effective way to preserve this connectivity. It is usually good at the region which is self-reliant and contiguous.

As described in Sect. 2, *ROL* and *ROR* are bilateral nearest regions to *ROI*. In this paper, the seed detected from them is obtained as the center of the cluster which is regarded as an initial seed. Combined with the distance cluster, the proposed method calculates the neighbor's distance to *ROI*, *ROL* and *ROR* while growing.

3 Experimental Results

The proposed thigh's CT image is shown in Fig. 2(a), while *Ix*, candidate seed sets, and the ultimate seed is visualized in Fig. 2(b). There the candidate seeds, which occurs in high frequency are illustrated in white color. The candidates with the most neighbors in same grey are elected as ultimate seeds. Sometimes the ultimate seeds are multiple, each of which has the same priority to be taken as an initial seed. As the grey variance is a member in the distance definition, its value is mapped into 0~255 visualized in Fig. 2(c). It reveals that the variances are high on the edges but it tends to be zeroes inside the organ.

In regard to merging different parts by the application of MDRG, seeds in *ROIs* is detected. With the design of the cluster, the MDRG result is shown in Fig. 3(a–c).

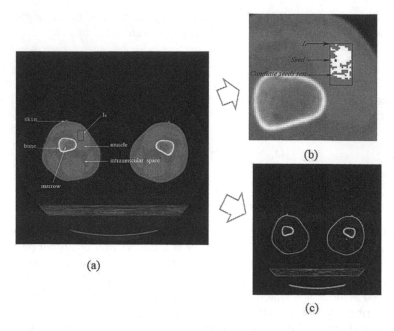

Fig. 2. Seeds detection and variances visualization of a thigh's CT image

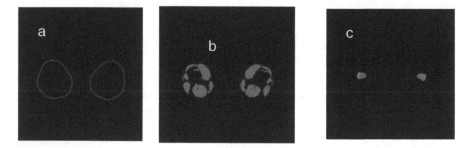

Fig. 3. The segmentation result.

3.1 Quantitative Evaluation of the Segmentation

Thresholding segmentation is one of the most common applications in the extracting *ROI* for image analysis [15]. A better threshold segment is based on the concept of fuzzy set, defining the membership between the grey level and the average grey level of its belonging region. The optimal threshold can be effectively determined by minimizing the measure of fuzziness of an image (TMMF). In order to be compared with TMMF, the seed grey of Xx are divided into $\{g(Lx) \sim g(Ix)\}$ and $\{g(Ix) \sim g(Rx)\}$. Based on the histogram, TMMF are taken to determine the minimal grey value (min-TMMF) in $\{g(Lx) \sim g(Ix)\}$ and the maximul grey value (max-TMMF) in $\{g(Ix) \sim g(Rx)\}$.By preforming the MDRG segmentation based on the histogram (Fig. 4 (right)), the muscles grey range from the minimal grey (Min-MDRG) and the maximal grey (max-MDRG) are obtained. Through changing the plotted area of *Ix*, *Rx* and *Lx* at times, the grey range in different experiments can be illustrated in Fig. 4 (left).The errors of this method are compared with that of TMMF, which have not gone beyond 3.84%.The thresholding segmentation by TMMF is shown in Fig. 4 (middle) which cannot distinguish the edges effectively (compared with Fig. 3(b)). The proposed MDRG method in this paper shows best performance on ROI segmentations

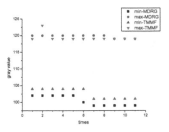
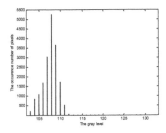

Fig. 4. Comparisons of MDRG and TMMF. Left: grey value result by the two methods. Middle: the thresholding segmentation of muscles by TMMF. Right: The histogram of ROI segmentation of Fig. 3b.

Quantitative evaluation is performed using volumetric overlap error (VOE) and relative volume difference (RVD) [16, 17]. To use the two indexes, Fig. 3 is converted to binary images to compare with those produced by manual segmentation method. For example, the VOE and RVD of Fig. 3(b) are 10.57 and 1.19% respectively, because manual segmentation inevitably includes subjectivity, artificial errors are sometimes tremendous. So from the experiment image, the results are within acceptable limits and the MDRG algorithm proposed in this paper works well. Moreover, different images using the MDRG have also verify this result.

4 Conclusion

Medical image segmentation is a crucial step in image processing field which is a hotspot of research. The region of interest in medical images are often ambiguous so that it brought many difficulties in the segmentation. Based on the variance of pixels and the seed detections, a minimal distance cluster combining with region growing methods is designed. In this paper, adjacent regions are used for restraining the seeds growing. To test its effectiveness, the grey range of the segmentation is compared with the thresholding segmentation methods which are based on minimal segment entropy and fuzzy sets theory. It shows that the positions of the plotted region have little influences on the result. For practical application, minute adjustment of *Xx* that obeys the conditions of the minimal distance cluster makes the proposed algorithm perform better. Finally, quantitative evaluation indexes of VOE and RVD are calculated for the segmentation. The result of the experiments shows that our method is a stable one that operates well on ROI segmentation.

However, some interested areas are composed of many dispersed regions which need more plotted areas for seed detections and adjustments. Besides, medical images are often in low grey difference, so to use this method properly relies on more complex work. In the future the complete automation for performing the proposed methods is more challenging. Even so, the segmentation result of certain region can not only provide accurate data for medical diagnosis and make a treatment plan for radiation therapy, but also establish the foundation for three-dimensional applications.

Acknowledgments. The authors would like to thank the associate editor and the anonymous reviewers for their careful work and valuable suggestions for this paper. This work was supported by Provincial Key Foundation for Excellent Young Talents of Colleges and Universities of Anhui (No.2013SQRL063ZD), Key University Science Research Project of Anhui Province (KJ2017A927 and KJ2017A926) and the National Natural Science Foundation of China under Grant No. 61573022.

References

1. Wu, Y.C., Song, G., Cao, R.F., Wu, A.D., Cheng, M.Y., et al.: Development of accurate/ advanced radiotherapy treatment planning and quality assurance system (ARTS). Chin. Phys. C **32**, 177–182 (2008)
2. Bogovic, J.A., Prince, J.L., Bazin, P.L.: A multiple object geometricde Deformable, model for image segmentation. Comput. Vis. Image Underst. **117**(2), 145–157 (2013)
3. Yaniv, Z., Cleary, K.: Image-guided procedures: a review. Compute. Aided Interventions and Med. Robotics, Imaging Science and Inf. Systems Center, Dpt. of Radiology, Georgetown University Med. Center, Technical report, April 2006
4. Yager, R.R.: On the measures of fuzziness and negation, part 1: membership in the unit interval. Int. J. Gen. Syst. **5**(4), 221–229 (1979)
5. Dunn, J.C.: A fuzzy relative of the ISODATA process and its use in detecting compact well-separated clusters. J. Cybernet. **3**(3), 32–57 (1973)
6. Wu, W.: Multilevel thresholding algorithm for image segmentation based on maximum fuzzy entropy Systems. Eng. Electron. **27**(02), 357–360 (2005). Science Press, China
7. de Portes Albuquerque, M., Esquef, I.A., Gesualdi Mello, A.R.: Image thresholding using tsallis entropy. Pattern Recogn. Lett. **25**, 1059–1065 (2004)
8. Sezgin, M., Tasaltin, R.: A new dichotomization technique to multilevel thresholding devoted to inspection applications. Pattern Recogn. **21**(2), 151–161 (2000)
9. Mostafa, A., Fouad, A., Elfattah, M.A., et al.: CT liver segmentation using artificial bee colony optimisation. Procedia Comput. Sci. **60**, 1622–1630 (2015). Singapore
10. Badakhshannoory, H., Saeedi, P., Qayumi, K.: Liver segmentation based on deformable registration and multi-layer segmentation. In: IEEE International Conference on Image Processing, pp. 26–29. IEEE, New Jersey (2010)
11. Wells, W.M., Grimson, E.L., Kikinis, R., Jolesz, F.A.: Adaptive segmentation of MRI data. IEEE Trans. Med. Imag. **15**, 429–442 (1996)
12. Pohle, R., Toennies, K.D.: Segmentation of medical images using adaptive region growing. In: Proceedings of SPIE Medical Imaging Image Process, vol. 4322, pp. 1337–1346 (2001)
13. Huang, L.K., Wang, M.J.: Image thresholding by minimizing the measures of fuzziness. Pattern Recog. **28**(1), 41–51 (1995)

14. ACR/NEMA, Standards Publication PS3/DICOM 3: Digital Imaging and Communications in Medicine Standard (1993)
15. Sankur, B., Sezgin, M.: A survey over image thresholding techniques and quantitative performance evaluation. J. Electr. Imaging **13**(1), 146–165 (2004)
16. Lu, X.Q., Wu, J.S., Ren, X.Y., et al.: The study and application of the improved region growing algorithm for liver segmentation. OPTIK **125**(09), 2142–2147 (2014)
17. Warfield, S.K., Zou, K.H., Wells, W.M.: Simultaneous truth and performance level estimation (STAPLE): an algorithm for the validation of image segmentation. IEEE Trans. Med. Imag. **23**(7), 903–921 (2004)

The Minimum Perimeter Convex Hull of a Given Set of Disjoint Segments

Yiyang Jia and Bo Jiang[✉]

Dalian Maritime University, Linghai Road 1, Dalian, China
bojiang@dlmu.edu.cn

Abstract. In this paper, we present an algorithm to compute the minimum perimeter convex hull of a given set of disjoint segments, so that each segment is contained in the hull completely or intersects with the boundary of the hull. The problem discussed in this paper is a novel transformation of TSP and MPSP. To solve this problem, we use a contraction strategy to contract the convex hull from a larger one which contains all endpoints of given segments to the direction of a smaller one which only contains some necessary points. We also assess the spatial relationships between outside segments and its convex hull, and add necessary segments into the convex hull successively by finding the shortest path of a sequence of segments. As a result, we present an $O(n^5)$ algorithm for computing the minimum perimeter convex hull of a given set of disjoint segments.

Keywords: Computational geometry · Algorithm · Minimum perimeter convex hull · Contraction strategy

1 Introduction

Over the past two hundred years, the problem of stabbing a set of objects with different types of stabbers has been widely studied. Computing the minimum perimeter convex hull stabbing a given set of disjoint segments is an open problem posed by Löffler and van Kreveld [1]. In [2], an $O(3^m n + n\log n)$ time algorithm has been presented (m is the number of different segment directions). For some special prerequisites, such as parallel segments, Goodrich et al. gave an $O(n\log n)$ time algorithm that decides whether the minimum perimeter convex hull stabbing them exists [3]. Several approximation algorithms are also known as in [4, 5]. Recently, José Miguel Díaz-Báñez et al. proved that if S is a set of pairwise disjoint segments, this problem can be solved in polynomial time and for general segments, this problem is NP-hard [6].

Also, as a classical computational geometry problem, Traveling Salesman problems (TSP) are well known. An original TSP asks for a shortest route to a given set of points, along this path each point must be visited once and only once. In 1979 Garey M.R. and Johnson D.S. [7] proved the NP- Completeness of TSP. A lot of related problems such as the problem of finding simple paths, finding the shortest paths of ordered or partially ordered line segments and so on have been presented and solved in [8, 9].

Motivated by these problems, computing the minimum perimeter convex hull of a given set of disjoint segments is one of the most important extensions, different from

© Springer International Publishing AG 2018
F. Qiao et al. (eds.), *Recent Developments in Mechatronics and Intelligent Robotics,*
Advances in Intelligent Systems and Computing 691, DOI 10.1007/978-3-319-70990-1_46

traditional convex problems and MPSP such as computing the largest and smallest convex hulls for imprecise points [1], computing the minimum perimeter polygon stabbing a given set of segments [6] and so on, it asks for the minimum perimeter convex hull that each segment in the given set either is contained in it or intersects with its boundary. The convex hull can contain only a sub-segment as well as a single endpoint of each given segment. In other words, the convex hull defines a region with a minimum perimeter, each given segment is either completely contained in this region or only has a part in it. We call this convex hull as the minimum perimeter convex hull of a given set of disjoint segments. Its solution can be used to deal with many practical problems such as finding the containing region of some necessary elements with a minimum perimeter, determining some spatial extent limited by some fixed objects and so on. Thus, it is a worthy of thorough research problem.

In this paper, we present an $O(n^5)$ time algorithm to solve this novel problem by using both a contraction strategy and the method of finding the shortest paths of a set of ordered segments, which will be specifically explained in following sections. The contraction strategy can probably be used for other kinds of spatial contraction as well.

2   Preliminaries and Notation

Let Φ denote a set with n disjoint segments in a plane. Each line segment in Φ is denoted by l_i $(0 \le i < n)$. In addition, the set of all endpoints of all the given segments in Φ is denoted by P, each of them is denoted by p_k $(0 \le k < 2n)$.

Our objective is to find the minimum perimeter convex hull of Φ, its interior or boundary contains at least one point of each l_i, and denote it as C_{opt}.

A convex polygon is a simple polygon (not self-intersecting) with the property that each of its internal angles is strictly smaller than π. A convex hull of a given set of points is a convex polygon that contains all the points in the point set. In a similar fashion, C_{opt} is defined as a minimum perimeter convex polygon which satisfies the following two conditions:

(1) Each segment in Φ either is contained in it or intersects with the boundary of it;
(2) It has the minimum perimeter among the convex hulls satisfying condition (1).

In this paper, we define two kinds of convex hulls, the first one denoted by Ω is the convex hull of P; the second one, denoted by Ψ, is a lower bound of C_{opt} which only contains some necessary points in P. Therefore C_{opt} can be found by contracting Ω and expanding Ψ in both directions of Ψ and Ω. We denote the convex hull got in the j-th contraction as Ω_j, and the convex hull got in the k-th expansion as Ψ_k $(j, k \in \mathbb{N}+)$.

3   Construction of the Minimum Perimeter Convex Hull

3.1   Construction of Ω and Ψ

Observation 1. *The convex hull of a set of points R contains the convex hull of a set of points Q if $Q \subseteq R$.*

We first construct the convex hull Ω based on P with *the Graham Scanning Algorithm* [10], as in Fig. 1(the thick segments denote the given segments).

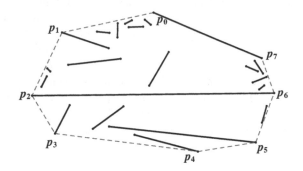

Fig. 1. The convex hull Ω of a given Φ

Lemma 1. Ω *contains* C_{opt}.

Proof. Suppose that P_{opt} is the set of some necessary points on the segments in Φ for building C_{opt}, P_{Ω} is the set of points on the segments in Φ for building Ω, because $P_{opt} \subseteq P_{\Omega}$, by Observation 1, Lemma 1 is hold. □

Then, construct Ψ with the method below:

(1) Denote each segment which has points on the boundary of Ω by l_j ($j \in N^+$) and put them into a set L, as in Fig. 2.

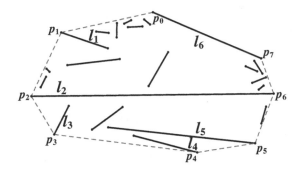

Fig. 2. Denoting segments on the boundary

(2) Divide L into three sets L_0, L_1, L_2:
$L_0 = \{l_j | l_j \in L$ and l_j has only one endpoint on the boundary of $\Omega\}$,
$L_1 = \{l_j | l_j \in L$ and l_j has only two endpoints on the boundary of $\Omega\}$,
$L_2 = \{l_j | l_j \in L$ and the whole segment l_j is on the boundary of $\Omega\}$.
For instance, for the segments in Fig. 2: $L_0 = \{l_1, l_3, l_4, l_5\}$, $L_1 = \{l_2\}$, $L_2 = \{l_6\}$.

(3) If L_0 has less than two line segments, we consider that Ψ does not exist (only the contraction process is need). Otherwise, compute the convex hull of the endpoints

inside Ω of the segments in L_0, then Ψ is obtained by discarding its vertices whose corresponding line segments have parts inside this convex hull.

Lemma 2. *The boundary of C_{opt} must intersect with the line segments in L_0 whose internal endpoints are the vertices of Ψ.*

Proof. As can be seen from the definition of C_{opt}, the line segments in L_0 whose internal endpoints are the vertices of Ψ either intersect with C_{opt} or inside C_{opt}. For the latter, C_{opt} either equals to Ω or intersects with Ω. By Lemma 1, C_{opt} cannot intersect with Ω; by the definition of L_0, C_{opt} cannot equal to Ω, otherwise, if we move a vertex of C_{opt} which is also an endpoint of a segment in L_0 to the direction of the other endpoint of this segment, we can get a convex hull with a smaller perimeter. Thus Lemma 2 is hold.

For instance, we construct Ψ for the line segments in Fig. 1, the inner endpoint of l_5 is discarded (See in Fig. 3).

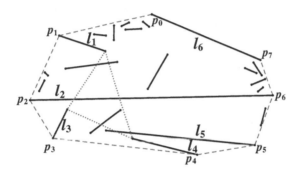

Fig. 3. Ψ of Φ in Fig. 1

3.2 Construction of C_0

C_0 is a precursor of C_{opt}, which is a result of contracting from Ω to Ψ without dealing with the segments in L_2. The strategy for constructing C_0 is as follows:

(1) Create a set of fixed segments F that is initialized to be empty.
(2) At this step, we only process the vertices which are endpoints of the segments in L_0 or in L_1. From any vertex of Ω, contract Ω in the counterclockwise (clockwise) order until no vertex on its boundary can be changed again. For convenience, if the position of a vertex can be changed, we call it movable, otherwise it is called immovable.

For each vertex p_i on the boundary of Ω, we denote its precursor vertex as p_{i-1} and its succeeded vertex as p_{i+1} in the counterclockwise (clockwise) order, and denote the segment connecting p_{i-1} and p_{i+1} as $\overline{p_{i-1}p_{i+1}}$. According to the location of $\overline{p_{i-1}p_{i+1}}$ and a newly defined Ψ_k, it can be divided into the following cases:

Case 1: $\overline{p_{i-1}p_{i+1}}$ does not intersect with $\varPsi_k$. If p_i is moved to the intersection point of $\overline{p_{i-1}p_{i+1}}$ and the given line segment where p_i is located, no given line segment will be outside the new convex hull, then we contract $\varOmega$ by moving p_i to this intersection point.

Case 2: $\overline{p_{i-1}p_{i+1}}$ intersects with $\varPsi_k$. If p_i is moved to the corresponding position on the boundary of $\varPsi_k$, no given line segments will be outside, then we contract $\varOmega$ by moving its boundary between p_{i-1} and p_{i+1} to the corresponding boundary of $\varPsi_k$, see in Fig. 6.

Case 3: $\overline{p_{i-1}p_{i+1}}$ does not intersect with $\varPsi_k$, but if p_i is moved to the intersection point of $\overline{p_{i-1}p_{i+1}}$ and the given line segment where p_i is located, some given line segments will be outside; or $\overline{p_{i-1}p_{i+1}}$ intersects with $\varPsi_k$, if the boundary of $\varOmega$ between p_{i-1} and p_{i+1} is moved to the boundary of $\varPsi_k$, some given line segments will be outside. At this time, we define $\varPsi_j$ as a convex hull of the original vertices of $\varPsi_k$ and all the endpoints of these outside segments, and then contract $\varOmega$ by moving the boundary of $\varOmega$ between p_{i-1} and p_{i+1} to the boundary of $\varPsi_j$, see in Fig. 5.

For each contraction, we can get a new convex hull. Loop the process above until all vertices are immovable, we can get C_0.

(3) Add all the given segments on which the vertices of C_0 located to F.

For C_0, we have the following conclusion:

Lemma 3

(a) *All the segments in F must be accessed by the boundary of C_{opt}.*
(b) *C_{opt} and C_0 access same given segments in the same order.*
(c) *The contracting process costs $O(n^2)$ time at most.*

Proof. For (a), assuming that C_{opt} does not access some of the given line segments on which some vertices of C_0 are located, then these lines segments must be inside C_{opt}, otherwise it will conflict with the definition of C_{opt}. In this case, since C_0 is obtained by a contracting process, so C_0 either contains or intersects with each of the segments in $\varPhi$, thus the perimeter of C_{opt} can be reduced by moving the portion of C_{opt}'s boundary which is outside C_0 to the boundary of C_0, that also conflicts with the definition of C_{opt}.

For (b), since both C_{opt} and C_0 are convex polygons that can be obtained by contracting $\varOmega$, and the line segments accessed by their boundaries are disjoint segments, so the access order does not change during the contracting process. By (a), C_{opt} and C_0 access the same given segments, thus (b) is proved.

For (c), the worst case for the contracting process is that in every j-th contraction $\varPsi_j$ is always a point, as in Fig. 4, C_{opt} is its center point. Assume that $\varPhi$ contains n line segments with same angle and length $|l|$ (for segments with different angles and lengths the time complexity will be reduced).

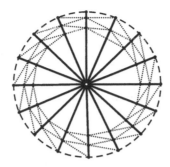

Fig. 4. An example of the worst situation of contraction

We calculate its contraction times. Each time of contraction at least reduces the radius of the convex hull $\cos(2\pi/n)$ times, so after the n^2-th contraction process, the radius of the convex hull is $|l| \times (\cos(2\pi/n))^{\wedge}n^2$. We define a function: $f(x) = (\cos(1/x))^{\wedge}x^2$. When x approaches $+\infty$, $f(x)$ takes the maximum:

$$\lim_{k \to \infty} \left(1 - \frac{2}{k}\right)^k = 1^k \times \left(-\frac{2}{k}\right)^0 + C_k^1 \times 1^{k-1} \times \left(-\frac{2}{k}\right)^1 + C_k^2 \times 1^{k-2} \times \left(-\frac{2}{k}\right)^2 + \cdots < 2/15.$$

That means the radius of the convex hull is $(2/15)^{\wedge}(2\pi) \approx 3.9 \times 10^{-6}$ times of $|l|$, this accuracy is within an acceptable range. Thus at most the contracting process costs $O(n^2)$ time. Therefore, Lemma 3 is proven.

As an example, we deal with Φ in Fig. 1 as follows:

First, for p_1, l_7, l_8, l_9, l_{10} will be excluded from the convex hull if we change p_1 to the intersection point of $\overline{p_0 p_2}$ and l_1, so define Ψ_1 and move the boundary of Ω between p_0 and p_2 to the corresponding boundary of Ψ_1, see in Fig. 5.

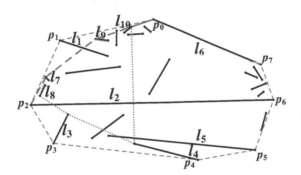

Fig. 5. The first step in contraction of Φ in Fig. 1

Second, move the boundary of Ω (after the first step) between the endpoint of l_7 ($\in \Psi_1$) and p_3 to the boundary of Ψ_1, similarly we deal with p_3 and p_4, as shown in Fig. 6.

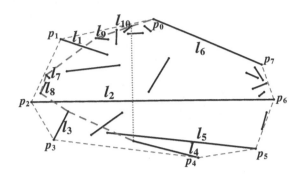

Fig. 6. The second step in contraction of Φ in Fig. 1

Third, the line segment connecting the endpoint of l_4 ($\in \Psi_1$) and p_6 will exclude l_{11} from the new convex hull, so define Ψ_2 and move the boundary of Ω between the endpoint of l_4 ($\in \Psi_1$) and p_6 to the corresponding boundary of Ψ_2, see in Fig. 7. At this point all the vertices of Ω are immovable, and C_0 is obtained (indicated by thick dashed segments).

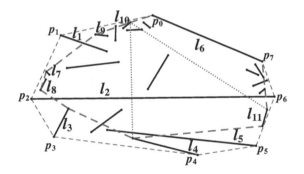

Fig. 7. The result of contraction of Φ in Fig. 1

3.3 Construction of C_{opt}

We construct the convex hull C_{opt} based on C_0 by dealing with the segments in L_2. The constructing method is as follows:

(1) Initialize an access sequence of all the segments in F ordered by the boundary of C_0 in counterclockwise (clockwise), denoted as A, and let the start segment equal to the end segment. For Φ in Fig. 1, $A = \{l_1, l_7, l_8, l_3, l_4, l_{11}, l_6, l_{10}, l_9, l_1\}$.

(2) Compute the shortest path traversing the line segments in sequence A by iterating the method in [9] over a line segment for n times to get a same point both as the starting point and the ending point, the shortest path must be the boundary of a convex hull, we denote this convex hull as C. For example, for Φ in Fig. 1, we first find C as in Fig. 8.

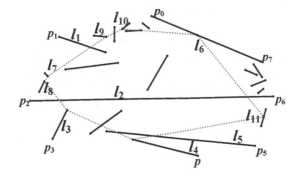

Fig. 8. The initial shortest path of the fixed line segments set F

(3) For each line segment l excluded from C by the shortest path, record the subsequence of A (two segments) which makes it outside the new convex hull, denote it as: l_s, l_t. It can be seen that if l is accessed by the boundary of C_{opt}, then it must be accessed between l_s and l_t. Otherwise there must be right turns on the boundary of C_{opt}, which cannot form a convex hull. Insert l between l_s and l_t. If the right position of l is found, then the shortest path of the new sequence after inserting l will not intersect with itself; otherwise the shortest path of the new sequence after inserting l must intersects with itself since Φ contains only disjoint segments, as in Fig. 9. Sometimes the segment in the wrong position may be accessed two times by the reverse sequence.

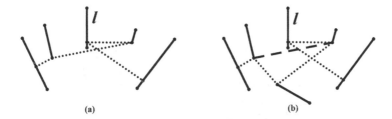

Fig. 9. Inserting segment l in a wrong place

In order to find the right position, we examine all positions between l_s and l_t, if there exist m line segments between l_s and l_t, it can take up to $(m + 1)$ times calculation until the correct position can be founded.

(4) After all the line segments outside C are inserted to sequence A, we get a shortest path. For each of the segments which form right turns on the shortest path, if remove it from A, the new shortest path of A will not make it outside of the area surrounded by the new shortest path, then remove it. Until no segment can be removed, compute the shortest path of the segments in the new sequence A, C_{opt} is obtained.

Lemma 4. *C_{opt} has the shortest boundary length.*

Proof. To prove that C_{opt} has the shortest boundary length, we can simply prove that the line segments accessed by C_{opt} and the order of them cannot be changed. If the order of the segments visited by the boundary of C_{opt} is changed, the order of its corresponding visiting sequence on the shortest path must be changed at the same time, self-intersections must be formed, which is eliminated. Because all the right turns are removed in (4), so the line segments accessed by C_{opt} cannot be changed, Lemma 4 holds.

Lemma 5. *The process of obtaining C_{opt} from C_0 takes up to $O(n^5)$ time.*

Proof. Since at least one line segment is added to the shortest path for each calculation and the operation is performed on up to n line segments. So at most we need to calculate the shortest path for $1 + 2+ ... + (n-1) = n^2/2$ times. Finding a shortest path with same starting and ending point of a set of segments costs at most $O(n^3)$ time [9]. Therefore, the process of obtaining C_{opt} from C_0 takes up to $O(n^5)$ time.

For instance, we deal with the outside given line segments in Fig. 8 as follows: first add l_{12} between l_6 and l_{10}, then add l_{13}, l_{14}, l_{15} between l_{11} and l_6 in order, the new sequence is $l_1, l_7, l_8, l_3, l_4, l_{11}, l_{15}, l_{14}, l_{13}, l_6, l_{12}, l_{10}, l_9, l_1$. If we add l_{16} between l_{11} and l_{15} or between l_{15} and l_{14}, a self-intersection will occur, so these locations are incorrect. Therefore add l_{16} between l_{14} and l_{13}. Since l_{16}, l_{13}, l_6 form a right turn, remove l_{13} and get C_{opt} as in Fig. 10. The final sequence accessed by the boundary of C_{opt} is $l_1, l_7, l_8, l_3, l_4, l_{11}, l_{15}, l_{14}, l_{16}, l_6, l_{12}, l_{10}, l_9, l_1$.

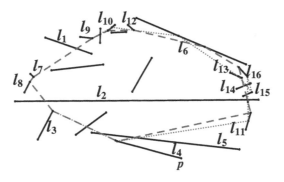

Fig. 10. C_{opt} of Φ in Fig. 1

We give below the pseudo-code of our algorithm MPCH.
Algorithm MPCH:
Input: A set of segments Φ.
Output: The minimum perimeter convex hull C_{opt}.
1 Construct Ω including all the endpoints of the line segments in Φ.
2 Construct $\Psi\Omega$.
2.1 Divide the given line segment involved in Ω's boundary into sets L_0, L_1, L_2.
2.2 Use the inside endpoints of the segments in L_0 to construct Ψ.
3 Construct C_0 by a contracting process from Ω to Ψ.

4 Construct C_{opt}.
4.1 Initialize the sequence A of accessing line segments according to C_0.
4.2 Find C by computing the shortest path of A.
4.3 For each line segment outside C, insert it into its corresponding subsequence.
4.4 Remove the unreasonable segments that form right turns and get C_{opt}.
For the algorithm above, we have Theorem 1.

Theorem 1. *Algorithm MPCH can be done in $O(n^5)$ time.*

Proof. Constructing Ω and Ψ costs $O(n\log n)$ time according to *the Graham Scanning Algorithm* [10]. And constructing C_0 costs at most $O(n^2)$ time, constructing C_{opt} from C_0 takes up to $O(n^5)$ time. Therefore, the entire time complexity of the algorithm MPCH is $O(n^5)$. Theorem 1 is hold.

4 Conclusion

In this paper, we first introduced a problem of computing the minimum perimeter convex hull of a given set of disjoint segments, a given line segment either inside the convex hull or intersects with its boundary. Then we proposed an $O(n^5)$ time solution of this problem. This method can be further used to calculate the shortest path of traversing a set of potentially ordered segments with the help of this convex hull we get. This is an interesting research area and we are now working on it. On the other hand, there are some limits in this paper. For example, this algorithm cannot be applied to computing the minimum perimeter convex hull of a given set of arbitrarily intersecting segments.

Acknowledgements. This work is supported by the General Project of Liaoning Province Natural Science Foundation (No. 20170540147) and Liaoning Province Science and Research (No. L2015105).

References

1. Löffler, M., van Kreveld, M.: Largest and smallest convex hulls for imprecise points. Algorithmica **56**, 35 (2008)
2. Rappaport, D.: Minimum polygon transversals of line segments. Int. J. Comput. Geom. Appl. **5**(3), 243–256 (1995)
3. Goodrich, M.T., Snoeyink, J.: Stabbing parallel segments with a convex polygon. Comput. Vis. Graph. Image Process. **49**(2), 152–170 (1990)
4. Dumitrescu, A., Jiang, M.: Minimum-perimeter intersecting polygons. Algorithmica **63**(3), 602–615 (2012)
5. Hassanzadeh, F., Rappaport, D.: Approximation algorithms for finding a minimum perimeter polygon intersecting a set of line segments. In: Dehne, F., Gavrilova, M., Sack, J.-R., Tóth, C.D. (eds.) WADS 2009, vol. 5664. LNCS, pp. 363–374. Springer, Heidelberg (2009)
6. Díaz-Báñez, J.M., Korman, M., Pérez-Lantero, P., Pilz, A., Seara, C., Silveira, R.I.: New results on stabbing segments with a polygon. Comput. Geom. **48**(1), 14–29 (2015)

7. Garey, M.R., Johnson, D.S.: Computers and Intractability: A Guide to the Theory of NP-Completeness. Freeman, San Francisco (1979)
8. Cheng, Q., Chrobak, M., Sundaram, G.: Computing simple paths among obstacles. Comput. Geom. **16**, 223–233 (2000)
9. Wang, L., Jiang, B., Wei, Q., He, D.: Research on the Algorithm for euclidean shortest paths of visiting line segments in the plane. ICIC Expr. Lett. **8**(6), 1683–1688 (2014)
10. Graham, R.L.: An efficient algorithm for determining the convex hull of a finite planar set. Inf. Process. Lett. **1**(4), 132–133 (1972)

Spatial Effect and Influencing Factors Analysis of Waste-to-Energy Environmental Efficiency in China

Liang Cao, Song Xue$^{(\boxtimes)}$, and Yunhua Zhang

Business School, Hohai University, Nanjing, Jiangsu 211100, China
xuer2399@163.com

Abstract. The accelerating urbanization and improvement of living standards have greatly increased the municipal solid waste (MSW) generation in China. Public private partnership (PPP) widely used in the field of waste-to-energy (WTE) industry can decrease the volume of MSW, generate power and address the serious environmental problems caused by MSW. Then using spatial econometric model to analyze spatial effect and influencing factors of waste-to-energy environment efficiency (WEEE) in China. The results showed that: (1) WEEE of China's 31 provinces is related in space, (2) WEEE with a variety of factors. The high economic level, reasonable changes of structure, good environment infrastructure and specialized environmental regulation can improve WEEE.

Keywords: DEA · Waste-to-energy · Environmental efficiency

1 Introduction

Municipal solid waste (MSW) disposal becomes the worldwide problems while rapid economic growth and massive urbanization have deeply influenced people's lifestyle. The generation rate of Municipal Solid Waste (MSW) in cities goes far beyond the management capability and long term sustainability of cities. Valuable materials are leaking from mass production. In a world where demand and competition for finite and sometimes scarce resources will continue to increase, and pressure on resources is causing greater environmental degradation and fragility [1]. In 1980, the generation of Municipal Solid Waste Removal (MSWR) is 31 million tons in China, meanwhile urbanization rate is 19.4%, thirty-five years later, MSWR increasing to 192 million tons in China, and urbanization rate becomes 56.1%, nearly three times than in 1980 [2].

The current treatment of solid waste mainly in the following six: landfilling, energy recovery, material recovery, incineration, composting, and recycling (UNEP, 2011). Normally landfilling, composting and incineration are the main method for MSW treatment.

A number of WTE incineration plants have been developed in China through PPP \BOT\BOO arrangement, which is regarded as an effective means to attract funds from the private sector to provide public works and services and to improve efficiency in the delivery of such works and services [4]. Public private partnership (PPP) has been widely

© Springer International Publishing AG 2018
F. Qiao et al. (eds.), *Recent Developments in Mechatronics and Intelligent Robotics*,
Advances in Intelligent Systems and Computing 691, DOI 10.1007/978-3-319-70990-1_47

used as an infrastructure delivery model by many governments, including toll highway, freeway, bridge, culvert and WTE plants [3], it was often the proposed response to reduce the cost of waste management and improve service quality [4]. One of the reasons for the popularity of PPP is that governments can simultaneously attract private firms and claim that they are not privatizing. Independently of these political reasons, PPP have the potential to increase efficiency and improve resource allocation [5].

Waste incineration in turning can make waste into treasure, reducing land resource consumption, but also carry undesirable output such as emissions of waste gas, waste water, fly ash. It is necessary to introduce the concept of efficiency into the WTE incineration power generation industry. Through the spatial distribution and time series growth of the waste incineration power plant, the environmental efficiency in different time and different regions can be calculated. In this paper, with DEA [6] (Data Envelopment Analysis). The results of the incineration industry layout and industry management to provide the basis for decision-making.

2 Methodology

DEA is a model approach to the relative efficiency evaluation of decision-making units (DMUs) proposed by American operations scientist Charnes and Copper. DEA is an efficiency evaluation method that evaluates the relative effectiveness of the decision-making units of the same type by developing the frontier of production, based on the concept of relative efficiency [7, 8].

As shown in Fig. 1, there are K, M, N, P, Q, 5 decision-making units, respectively, a and b inputs two resources for production activities, the corresponding output y. It can be seen that M is a decision-making unit that is ineffective and others are in the frontier of production, and are effective decision-making units.

The decision unit N is its projection on the leading edge, which can be represented by a linear combination of P and K. The output of the input and output of the decision unit N is not less than the output of the decision unit M, which indicates that M has been put too much Resources; that is, the decision unit M is invalid relative to the decision unit N, and the decision unit N is technically efficient. DEA is based on this idea, to find the relative efficiency of each decision-making unit, when the decision-making unit is located on the frontier, it means that the relative efficiency value of 1.

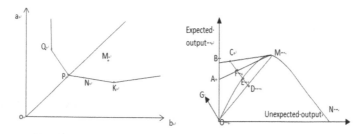

Fig. 1. Directivity distance under disposable condition

A set of production possibilities satisfying the above conditions is expressed as:

$$p^w(x) = \left\{ (y,b) \mid \sum_{j=1}^{n} z_j x_j \le x, \sum_{j=1}^{n} z_j y_j \ge y, \sum_{j=1}^{n} z_j b_j = b, z_j \ge 0 \right\} \tag{1}$$

When the environmental technology satisfies the strong disposable characteristic, its corresponding possible output set is:

$$p^s(x) = \left\{ (y,b) \mid \sum_{j=1}^{n} z_j x_j \le x, \sum_{j=1}^{n} z_j y_j \ge y, \sum_{j=1}^{n} z_j b_j \ge b, z_j \ge 0 \right\} \tag{2}$$

The existing directional distance function and Malmquist-Luenberger index are improved by Ps(x) as shown in (2) to establish the measurement model of WTE. Directional distance function is a generalization of the Shephard output distance function. In order to achieve maximum expected output and minimum undesired output, this paper uses the output-based directional distance function basic form:

$$D(x,y,b;y) = \sup\{\beta : (y,b) + \beta g \in p^s(x)\} \tag{3}$$

Where β is the maximum possible quantity of WTE output growth and contaminant reduction, and g = (gy, gb) is the direction vector of output expansion. g = (y,−b), that is, with the increase in the proportion of expected output.

The existing efficiency measure models like that of joint weakly disposable distance function and SBM model mainly based on the thoughts of environmental efficiency. The strong disposability thought provides a solution for measuring China's MSW environmental efficiency [9]. The index of WEEE can be found in Table 1.

$$\text{WEEE} = \frac{1}{1 + \vec{D}(x,y,b;g)} \tag{4}$$

Notes: "x" is input variable; "y" is the expected output variable; "b" is the unexpected output variable; $D(x,y,b;g) = \sup\{\beta : (y,b) + \beta g \in p^s(x)\}, p^s(x) =$

$$\left\{ (y,b) \mid \sum_{j=1}^{n} z_j x_j \le x, \sum_{j=1}^{n} z_j y_j \ge y, \sum_{j=1}^{n} z_j b_j \le b, z_j \ge 0 \right\}$$

Spatial econometric model

(1) Spatial autocorrelation model

Global Moran's I index describes the cluster state of the distribution of regional economic activities from the whole regional space view. Here is the definition:

$$I = \frac{\sum_{i=1}^{n} \sum_{j \ne i}^{n} W_{ij} z_i z_j}{\sigma^2 \sum_{i=1}^{n} \sum_{j \ne i}^{n} W_{ij}} \tag{5}$$

Notes: "n" is the number of observations; Xi is observation in position "i"; Zi is the standardized transformation of Xi; $z_i = \frac{X_i - \bar{X}}{\sigma}$, $x = \frac{1}{n}\sum_{i=1}^{n} x_i$, $\sigma^2 = \frac{1}{n}\sum_{i=1}^{n}(x_i - \bar{x})^2$.

(2) Spatial lag model

Spatial lag model (SLM) is used to study whether the variables have a spillover effect in the space aspect. The mathematical expression of SLM is as follows:

$$y = \rho Wy + X\beta + \varepsilon \tag{6}$$

Notes: "x" is independent variable; "y" is dependent variable; "β" reflects the effect of "x" on "y"; "ρ" is the spatial autoregressive coefficient; W_y is endogenous variable;

(3) Spatial error model

Spatial error model (SEM) can be used to determine whether an error term is dependent on space level. The definition of SEM is as follows:

$$y = X\beta + \varepsilon \tag{7}$$

Notes: "x" is independent variable; "y" is dependent variable; "β" reflects the effect of "x" on "y"; "ε" represents the random error term vector.

Table 1. Measure index of WEEE

First-grade indexes	Second-grade indexes	Variables interpretation
Input	Investment sum of PPP WTE projects	The cost Including: design, construction, equipment and Land acquisition
	WTE collection	Environmental sanitation funds for each province (covering people, machinery and finance), MSWR
	Operation	Wage costs, Office expenses, Accessories, materials, other consumables
	Conveyance system	Treatment of secondary wastewater, waste gas, waste ash
	Subsidy	WTE disposal fee
Output	Expected output	Power generation (economic benefit)
		Saving land area (social benefit)
	Unexpected output	Emission of GHG (CO_2, SO_2)
		Fly ash Landfilling or treatment

3 Influencing Factors

After comprehensive consideration of the existing research achievements and the characteristics of WTEEE, the indexes in this paper are selected as follows [6] (Table 2): The data above are from annual China Statistical Yearbook, China Rural Statistical Yearbook, China Environmental Yearbook.

Table 2. Regression results of model

Variable	Ordinary least square (OLS)	Spatial lag model (SLM)	Spatial error model (SEM)
CONSTANT	-0.7119*** (-2. 9473)	-0.7566*** (-3.265)	-0.7710*** (-7.1027)
EDL	0.0437 (2.4768)	0.0521* (2.9875)	0.0659** (3.3637)
ESC	0.5822*** (2.6475)	0.5983*** (3.3917)	0.6712*** (5.7768)
CI	0.0071* (1.8012)	0.0075** (2.2238)	0.0069** (2.3521)
WCL	-0.0670 (-1.3450)	-0.0796* (-1.7824)	-0.0785* (-1.7995)
WTEF	0.0121 (0.9185)	0.0107 (0.9816)	0.0136 (1.5101)
ERCOD	-0.2754** (-2.2495)	-0.2910*** (-2.8793)	-0.2569*** (-3.3071)
PE	0.0833** (3.340)	0.0938*** (3.8544)	0.1034*** (3.8772)
R^2	0.7340	0.7425	0.8298
LogL	27.5734	27.8303	30.0695
AIC	-32.9248	-31.2164	-37.9170
SC	-16.9918	-13.6900	-21.9840

Notes: "***", "**", "*" respectively stands for significant level of 1%, 5%, and 10%. The values below OLS are T-Statistics and the values below SLM & SEM are Z-Statistics.

Variable selection and data sources

The efficiency of waste-to-energy environment is influenced by many factors, such as economy, resources, environment, etc. The influence factors here were chose out of two considerations—the existing research papers listed in Table 2 and the characteristics of waste-to-energy environment itself. Based on this, the following influence indexes are selected [10–12].

(1) Economy Development Level (EDL). This index is based on per capita net income.
(2) Economic structure change (ESC). ASCP represents the proportion of poultry industry output in total output value.
(3) City industrialization (CI). The total output value of township enterprises which make up the gross output value is the measure of this indicator.
(4) Wage costs level (WCL). The investment quota of waste infrastructure is the standard.
(5) Environmental regulation (ERCOD&ERNH). Environmental regulation can be reflected in concrete measures like environmental policy making, environmental law enforcement and environmental governance Investment, etc.

(6) WTE disposal fee (WTEF). The PPP incineration plants get subsidy from the governments.
(7) Pollution emissions(PE). Emission of GHG (CO_2, SO_2)

4   Spatial Test

By using global Moran's I method to test the measurement results of waste-to-energy environmental efficiency of China in 2013, the research showed that all the results managed to pass the test under 5% confidence level. The result of this test can support the conclusion that the environmental efficiency of 31 provinces in China is spatially correlated and its distribution has a certain accumulation characteristics.

Comparing the LogL, AIC and SC values between SLM and SEM, it proves that SEM model is better than SLM model. This result shows that the WEEE in China is not only affected by the environment efficiency of neighboring provinces, but also affected by the structural differences between regions, including development level of rural economy, infrastructure, environmental regulation and other factors not included in the model. Since the spatial error coefficient of SEM has passed the 1% significance level test, the existence of significant spatial dependence of waste-to-energy environment efficiency in different provinces is tested.

Fig. 2. WTEEE in China (2005–2015)

5   Conclusion

The result of China's Waste-to-energy environmental efficiency (WEEE) measurement in 2013 can be obtained through formula (1) and Fig. 2 below is the bitmap. The bitmap clearly shows that the spatial distribution characteristics of Waste-to-energy environmental efficiency in China is unbalanced and the distribution pattern can be described to be gradually decreased from east to west, north to south. There are 5 main gathering area: Eastern coastal developed regions; Central regions; Bohai Sea and the Northeast regions; southwest regions; Northwest and Mongolia regions, The average efficiency values of these regions are 0.918, 0.861, 0.719, 0.637, 0.482 in 2005–2015 (Hong Kong, Macao and Taiwan regions not include).

References

1. O'Connor, E.R.: Toronto the green: pollution probe and the rise of the canadian environmental movement. PhD dissertation, University of Western (2010)
2. Menikpura, S.N.M., Sang-Arun, J., Bengtsson, M.: Assessment of environmental and economic performance of Waste-to-Energy facilities in Thai cities. Renew. Energ. **86**, 576 (2016)
3. Sasao, T.: Cost and efficiency of disaster waste disposal: a case study of the Great East Japan Earthquake. Waste Manag. **58**, 3–13 (2016)
4. Chen, Y., Chen, C.: The privatization effect of MSW incineration services by using data envelopment analysis. Waste Manag. **32**(3), 595–602 (2012)
5. Chen, Y., Wu, W.: Constructing an effective prevention mechanism for MSW lifecycle using failure mode and effects analysis. Waste Manag. **46**, 646–652 (2015)
6. Saraei, M.H., Jamshidi, Z., Ahangari, N.: Prioritization of management indexes of municipal solid waste in line with sustainable development: case study: City of Bokan. J. Urban Econ. Manag. **4**(14), 1–17 (2016)
7. Ichinose, D., Yamamoto, M., Yoshida, Y.: Productive efficiency of public and private solid waste logistics and its implications for waste management policy. IATSS Res. **36**(2), 98–105 (2013)
8. Kinnaman, T., Yokoo, H.: Environmental consequences of global Reuse. Am. Econ. Rev. **101**(3), 71–76 (2011)
9. Zhang, T.: Frame work of data envelopment analysis—a model to evaluate the environmental efficiency of China's industrial sectors. Biomed. Environ. Sci. **22**(1), 8–13 (2009)
10. Yang, N., Zhang, H., Chen, M., Shao, L., He, P.: Greenhouse gas emissions from MSW incineration in China: impacts of waste characteristics and energy recovery. Waste Manag. **32**(12), 2552–2560 (2012)
11. Zhao, Y., Christensen, T.H., Lu, W., Wu, H., Wang, H.: Environmental impact assessment of solid waste management in Beijing City, China. Waste Manag. **31**(4), 793–799 (2011)
12. Fruergaard, T., Astrup, T.: Optimal utilization of waste-to-energy in an LCA perspective. Waste Manag. **31**(3), 572–582 (2011)

Dynamic Characteristics of Canard-Controlled Rocket with Deflected Canard

Xinbao Gao[1], Yongchao Chen[1(✉)], Qingwei Guo[2], and Tianpeng Li[1]

[1] Shijiazhuang Mechanical Engineering College, Shijiazhuang, Hebei, China
ycchentgc@126.com
[2] Northwest Institute of Nuclear Technology, Xi'an, China

Abstract. The dynamic characteristics of canard-controlled rocket are analyzed to investigate the stability and controllability with deflected canard. The canard-controlled rocket is regards as a particle and the disturbance motion equations are established by small perturbation method. The transfer functions are respectively derived, where the input is canard deflection angle and the outputs are pitch angle, pitch angular velocity, trajectory inclination angle, trajectory inclination angular velocity and attack angle. The trajectory characteristic points are selected, and the trajectory transitional processes on these points are analyzed by frozen-coefficient method. The results show that with the function of canard step deflection, pitch angular velocity, trajectory inclination angular velocity and attack angle periodical oscillatory damp to a steady value, and pitch angle and trajectory inclination angle are increased with time increased.

Keywords: Rocket · Dynamic characteristics · Small perturbation method · Frozen-coefficient method

1 Introduction

The flight state of canard-controlled rocket will change from one equilibrium state to another when the canard deflected, and this change process is regarded as the dynamic characteristics of rocket [1]. To study the stability and controllability of canard-controlled rocket, its dynamic characteristics are analyzed with the canard step deflected. The disturbance motion equations are established by small perturbation method, the transfer functions are respectively derived, and the dynamic characteristics in tractor characteristic points are analyzed. This study lays the foundation of structure design of this canard-controlled rocket.

2 Disturbance Motion Equations

As shown in function (1), the trajectory disturbance motion equations can be established with the force and moment on canard-controlled rocket by small perturbation method. The detailed establish process is shown in [2, 3].

F. Qiao et al. (eds.), *Recent Developments in Mechatronics and Intelligent Robotics*,
Advances in Intelligent Systems and Computing 691, DOI 10.1007/978-3-319-70990-1_48

$$
\left\{
\begin{aligned}
&\frac{d\Delta x}{dt} = \Delta V \cos\theta + V \sin\theta \Delta\theta; \frac{d\Delta y}{dt} = \Delta V \sin\theta + V \cos\theta \Delta\theta; \frac{d\Delta z}{dt} = -V \cos\theta \Delta\sigma \\
&m\frac{d\Delta V}{dt} = (P^V - R_x^V)\Delta V + (-P\alpha - R_x^\alpha)\Delta\alpha + mg \sin\theta\Delta\theta + F_{cz}\Delta\beta - F_{cy}\Delta\alpha + F_{gx} \\
&mV\frac{d\theta}{dt} = (P^V\alpha + R_y^V)\Delta V + (P + R_y^\alpha)\Delta\alpha + mg \sin\theta\Delta\theta + F_{cy}^\delta\Delta\delta + F_{gy} \\
&-mV\cos\theta\frac{d\Delta\sigma}{dt} = (-P + R_z^\beta)\Delta\beta + (P\alpha + R_y)\Delta\gamma_v + F_{cz}^\delta\Delta\delta + F_{gz} \\
&\frac{d\Delta\gamma}{dt} = \Delta\omega_x - \tan\varphi\Delta\omega_y; \frac{d\Delta\psi}{dt} = \frac{\Delta\omega_y}{\cos\varphi}; \frac{d\Delta\varphi}{dt} = \Delta\omega_z \\
&J_x\frac{d\Delta\omega_x}{dt} = M_x^\beta\Delta\beta + M_x^{\omega_x}\Delta\omega_x + M_x^{\omega_y}\Delta\omega_y + M_x^\delta\Delta\delta + M_{gx} \\
&J_y\frac{d\Delta\omega_y}{dt} = M_y^\alpha\Delta\alpha + M_y^\beta\Delta\beta + M_y^{\omega_y}\Delta\omega_y + M_y^{\omega_z}\Delta\omega_x + M_y^\beta\Delta\dot\beta - (J_{x4} - J_{z4})\omega_x\Delta\omega_z \\
&\qquad -(J_{x4} - J_{z4})\omega_z\Delta\omega_x - J_{z4}\omega_{z4}\frac{d\Delta\gamma}{dt} - J_{z4}\frac{d\gamma}{dt}\Delta\omega_{z4} + M_{cy}^\delta\Delta\delta + M_{gy} \\
&J_z\frac{d\Delta\omega_z}{dt} = M_z^V\Delta V + M_z^\alpha\Delta\alpha + M_z^\alpha\Delta\dot\alpha + M_z^\beta\Delta\beta + M_z^\beta\Delta\dot\beta - (J_{y4} - J_{x4})\omega_y\Delta\omega_x \\
&\qquad -(J_{y4} - J_{x4})\omega_x\Delta\omega_y + J_{y4}\omega_{y4}\frac{d\Delta\gamma}{dt} + J_{y4}\frac{d\gamma}{dt}\Delta\omega_{y4} + M_z^{\omega_z}\Delta\omega_z + M_{cz}^\delta\Delta\delta + M_{gz} \\
&\Delta\alpha = \Delta\varphi - \Delta\theta; \Delta\beta = \Delta\psi - \Delta\sigma; \Delta\gamma_v = \tan\theta\Delta\beta + \frac{\cos\varphi}{\cos\theta}\Delta\gamma
\end{aligned}
\right.
\tag{1}
$$

3   Transfer Function

The Eqs. (1) can be divided into vertical and lateral perturbation motion equations. As the canard-controlled rocket is axisymmetric, the vertical and lateral perturbation motions are the same. Therefore, the vertical perturbation motion is analyzed in this study as an example.

To reduce the study difficulty, the vertical perturbation motion equation is simplified, where the velocity component ΔV is ignored, and only the short cycle perturbation motion is analyzed [4]. In addition, as this rocket is tail-stable, the Magnus and gyroscopic couple are small and can be ignored.

Therefore, the short cycle vertical perturbation motion equation can be writing as:

$$
\left\{
\begin{aligned}
&\frac{d^2\Delta\varphi}{dt^2} - \frac{M_z^{\omega_z}}{J_z}\frac{d\Delta\varphi}{dt} - \frac{M_z^\alpha}{J_z}\Delta\alpha = \frac{M_{cz}^\delta}{J_z}\Delta\delta + \frac{M_{gz}}{J_z} \\
&\frac{d\theta}{dt} - \frac{g\sin\theta}{V}\Delta\theta - \frac{(P + R_y^\alpha)}{mV}\Delta\alpha = \frac{F_{cy}^\delta}{mV}\Delta\delta + \frac{F_{gy}}{mV} \\
&\Delta\alpha + \Delta\theta - \Delta\varphi = 0
\end{aligned}
\right.
\tag{2}
$$

Kinetic coefficients are used in Eq. (2) to simplify the perturbation motion equation. The kinetic coefficients are shown in Table 1.

Table 1. Kinetic coefficient

Kinetic coefficient expression	$a_{22} = \dfrac{M_z^{w_z}}{J_z}$	$a_{24} = \dfrac{M_z^{\alpha}}{J_z}$	$p_2 = \dfrac{M_{cz}^{\delta}}{J_z}$	$a_{26} = \dfrac{1}{J_z}$
Kinetic coefficient expression	$a_{33} = \dfrac{g \sin \theta}{V}$	$a_{34} = \dfrac{C_y^{\alpha} qS + P}{mV}$	$a_{35} = \dfrac{C_{ny}^{\delta} qS}{mV}$	$a_{36} = \dfrac{1}{mV}$

Then the vertical perturbation motion equation can be writing as:

$$
\begin{cases}
\dfrac{d^2 \Delta \varphi}{dt^2} - a_{22} \dfrac{d \Delta \varphi}{dt} - a_{24} \Delta \alpha = a_{25} \Delta \delta + a_{26} M_{gz} \\[2mm]
\dfrac{d\theta}{dt} - a_{33} \Delta \theta - a_{34} \Delta \alpha = a_{35} \Delta \delta + a_{36} F_{gy} \\[2mm]
- \Delta \varphi + \Delta \theta + \Delta \alpha = 0
\end{cases}
\tag{3}
$$

Develop the Eq. (3) by Laplace transforms, and set the initial value as 0. Then we can obtain the algebraic equations of short cycle perturbation motion as shown in Eq. (4).

$$
\begin{bmatrix}
p^2 - a_{22}p & 0 & -a_{24}p \\
0 & p - a_{33} & -a_{34} \\
-1 & 1 & 1
\end{bmatrix}
\cdot
\begin{bmatrix}
\Delta \varphi(p) \\
\Delta \theta(p) \\
\Delta \alpha(p)
\end{bmatrix}
=
\begin{bmatrix}
a_{25} \\
a_{35} \\
0
\end{bmatrix}
\Delta \delta(p) +
\begin{bmatrix}
a_{26} M_{gz}(P) \\
a_{36} F_{gy}(P) \\
0
\end{bmatrix}
\tag{4}
$$

There are two column matrixes at the right of the Eq. (4), they respectively represent the perturbation influence induced by canard deflection and interference effect. So, the solves of $\Delta \varphi(p)$, $\Delta \theta(p)$ and $\Delta \alpha(p)$ are all have two parts. Suppose the canard deflection and interference effect are mutually independent, then their effects on perturbation motion can be respectively solved and linearly combined. To analyze the dynamic characteristics with canard deflection, only the first column matrix at the right of the Eq. (4) should be solved.

Solve the Eq. (4) with Cramer's rule, and we can obtain the transfer function as shown in Eq. (5).

$$
\begin{cases}
\Delta \varphi(p) = \dfrac{a_{25}p + a_{25}(a_{34} - a_{33}) - a_{35}a_{24}}{p^3 + A_1 p^2 + A_2 p + A_3} \Delta \delta(p) \\[3mm]
\Delta \theta(p) = \dfrac{a_{35}p^2 - a_{22}a_{35}p + a_{25}a_{34} - a_{24}a_{35}}{p^3 + A_1 p^2 + A_2 p + A_3} \Delta \delta(p) \\[3mm]
\Delta \alpha(p) = \dfrac{-a_{35}p^2 + (a_{22}a_{35} + a_{25})p - a_{25}a_{33}}{p^3 + A_1 p^2 + A_2 p + A_3} \Delta \delta(p)
\end{cases}
\tag{5}
$$

As the velocity of rocket is high, the kinetic coefficient a_{33} is small and $a_{35} \ll a_{34}$, so a_{33} and a_{35} can be ignored. Factorizes the Eq. (5), we can obtain the transfer function with canonical form as shown in Eq. (6).

$$
\begin{cases}
\Delta\varphi(p)p = \dfrac{K_d(T_{d1}p + 1)}{T_d^2 p^2 + 2T_d\zeta_d p + 1}\Delta\delta(p) \\[3mm]
\Delta\theta(p)p = \dfrac{K_d}{T_d^2 p^2 + 2T_d\zeta_d p + 1}\Delta\delta(p) \\[3mm]
\Delta\alpha(p) = \dfrac{K_\alpha}{T_d^2 p^2 + 2T_d\zeta_d p + 1}\Delta\delta(p)
\end{cases}
\tag{6}
$$

K_d is the transfer coefficient of rocket, T_d is the time constant of rocket, T_{d1} is the aerodynamic time constant of rocket, ξ_d is the damping coefficient of rocket, K_α is the transfer coefficient of attack angle, T_α is the time constant of attack angle.

$$
\begin{cases}
K_d = \dfrac{-a_{25}a_{34}}{a_{22}a_{34} + a_{24}}; T_d = \dfrac{1}{\sqrt{-a_{24} - a_{22}a_{34}}} \\[3mm]
\xi_d = \dfrac{-a_{22} + a_{34}}{2\sqrt{-a_{24} - a_{22}a_{34}}}; T_{d1} = \dfrac{1}{a_{34}} \\[3mm]
K_\alpha = \dfrac{-a_{25}}{a_{22}a_{34} + a_{24}} = K_d T_{d1}
\end{cases}
\tag{7}
$$

4   Dynamic Characteristics Analysis

As the reaction speed of guidance system is higher than change speed of kinetic coefficient for canard-controlled rocket, only the dynamic characteristic at many trajectory characteristic points should be analyzed.

4.1   Kinetic Coefficient at Trajectory Characteristic Point

Flight times and transfer coefficients for trajectory characteristic points are shown in Table 2.

Table 2. Transfer coefficient

Time (s)	K_d	T_d	ζ_d	T_{d1}	K_α
20	0.0545	0.0721	0.0378	2.936	0.160
40	0.0259	0.1033	0.0329	4.670	0.121
60	0.0250	0.0982	0.0363	3.983	0.101

4.2   Dynamic Processes

Solve the Eq. (6) using the dates in Table 2, we can obtain the dynamic processes curves of $\Delta\dot\varphi(t)$, $\Delta\varphi(t)$, $\Delta\dot\theta(t)$, $\Delta\theta(t)$ and $\Delta\alpha(t)$ at each trajectory characteristic point as shown in Figs. 1, 2, 3, 4, 5, and 6.

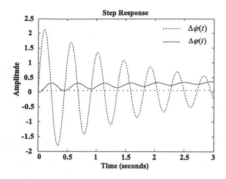

Fig. 1. Dynamic curves of $\Delta\dot{\varphi}(t)$ and $\Delta\varphi(t)$ at 20 s

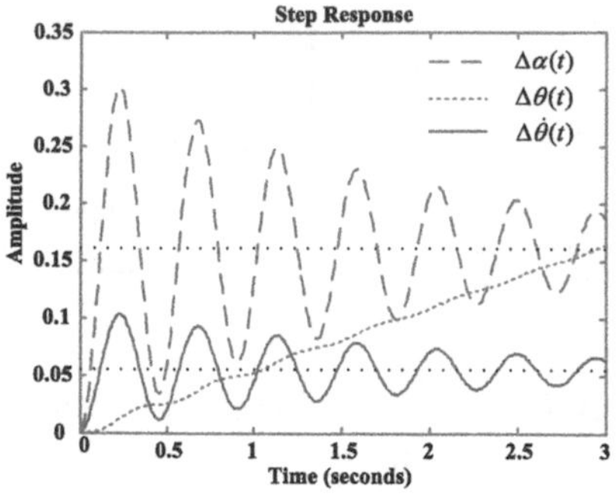

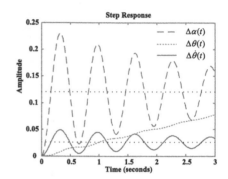

Fig. 2. Dynamic curves of $\Delta\dot{\theta}(t)$, $\Delta\theta(t)$, and $\Delta\alpha(t)$ at 20 s

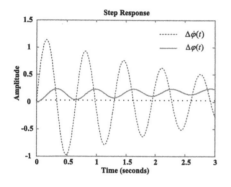

Fig. 3. Dynamic curves of $\Delta\dot{\varphi}(t)$ and $\Delta\varphi(t)$ at 40 s

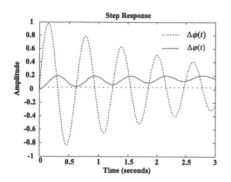

Fig. 4. Dynamic curves of $\Delta\dot{\theta}(t)$, $\Delta\theta(t)$, and $\Delta\alpha(t)$ at 40 s

and $\Delta\alpha(t)$ at 40s

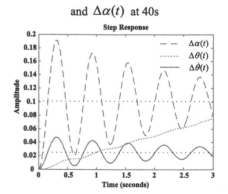

Fig. 5. Dynamic curves of $\Delta\dot{\varphi}(t)$ and $\Delta\varphi(t)$ at 60 s

Fig. 6. Dynamic curves of $\Delta\dot{\theta}(t)$, $\Delta\theta(t)$, and $\Delta\alpha(t)$ at 60 s

From Figs. 1, 3 and 5, it can be seen that the dynamic curves of $\Delta\dot{\varphi}(t)$ oscillatory damp to a steady value and the steady value increased with the rocket transfer coefficient increased, the dynamic curves of $\Delta\varphi(t)$ is oscillatory increased and there are no steady value.

From Figs. 2, 4, and 6, it can be seen that the dynamic curves of $\Delta\dot{\theta}(t)$ and $\Delta\alpha(t)$ oscillatory damp to a steady value, the steady value for $\Delta\dot{\theta}(t)$ increased with the transfer coefficient increased and the steady value for $\Delta\alpha(t)$ increased with the attack angle transfer coefficient, the dynamic curves of $\Delta\theta(t)$ is oscillatory increased and there are no steady value.

In addition, the peak values of $\Delta\dot{\varphi}(t)$ and $\Delta\dot{\theta}(t)$ increased with the transfer coefficient increased, and the peak value of $\Delta\alpha(t)$ increased with the attack angle transfer coefficient.

5 Conclusion

The dynamic characteristic of one canard-controlled rocket is studied when the canard is deflected. The results show that with the function of canard step deflection, pitch angular velocity, trajectory inclination angular velocity and attack angle periodical oscillatory damp to a steady value, and this shows that the damping coefficient of this rocket is small, so, the damping loop should be added in the guidance and control system. In addition, pitch angle and trajectory inclination angle are increased with time increased and there are no steady values, so, the control loop should be added in the guidance and control system.

References

1. Wang, L., Han, J., Chen, Z.: Flight Dynamics of Field Rocket. National Defense Industry Press, Beijing (2015)
2. Guo, Q.: Research on Trajectory Characteristics and Guidance Control Algorithm of Trajectory Corrected Rockets with Canards under Single Channel Control. Shijiazhuang Mechanical Engineering College, Shijiazhuang (2016)
3. Qian, X., Lin, R., Zhao, Y.: Flight Dynamic of Missile. Beijing Institute of Technology Press, Beijing (2015)
4. Guo, Q., Song, W., Song, X.: Study on transfer functions of canard trajectory-corrected rocket with single channel control. J. Ballist. **28**(2), 06 (2016)
5. Zarchan, P.: Tactical and Strategic Missile Guidance. American Institute of Aeronautics and Astronautics, Inc. 1801 Alexander Bell Drive, Reston, Virginia 20191-4344

Teaching Methods of Hydraulic Circuits Based on Hydraulic-Eelectrical Analogy

Yuesong Li[✉]

School of Mechatronics Engineering, Henan University of Science and Technology,
Luoyang Shi, China
liyaosong707@163.com

Abstract. In order to help beginner to learn hydraulic circuits, the analogy teaching method of hydraulic basic circuits is introduced in this paper. Firstly, Analogy of electrical and hydraulic components are introduced. Then the typical hydraulic basic circuits and their corresponding analogy electrical circuits are given respectively. Finally, a complex hydraulic circuits composed of several basic circuits and its corresponding analogy electrical circuit are given. By analogy analysis for their working principle, the complex hydraulic circuit can be better understood by beginners.

Keywords: Hydraulic circuit · Analogy teaching · Electrical circuit · Hydraulic components

1 Introduction

Hydraulic basic circuits are the basis of design and analysis of a hydraulic system, but its learning is difficult for students and beginners [1–3]. The teaching methods of hydraulic basic circuits is always important in the hydraulic teaching research.

The survey shows that students have deeply studied on the electrical engineering course before learning the hydraulic course and they are familiar with electrical engineering. Because the electrical circuits and hydraulic circuits have many similarities in the basic theorem and functions, the analogy method can make the knowledge organize and make learners understand the similarities between the hydraulic and electrical circuits. In order to make the hydraulic beginners quickly grasp the function and usage of hydraulic basic circuits, this paper will introduce a hydraulic teaching method based analogy between hydraulic circuits and electrical circuits.

2 Analogy of Electrical and Hydraulic Typical Components

According to the role of hydraulic components and electrical components, the analogy of components in function can be shown in Table 1 [4, 5].

© Springer International Publishing AG 2018
F. Qiao et al. (eds.), *Recent Developments in Mechatronics and Intelligent Robotics*,
Advances in Intelligent Systems and Computing 691, DOI 10.1007/978-3-319-70990-1_49

Table 1. Analogy of electrical and hydraulic components

Hydraulic cylinder	Hydraulic Motor	Constant current source	Check valve	2/2way directional control valve	3/4 way directional control valve
Linear motor	DC Motor	Hydraulic pump	Diode	Switch	Bipolar switch
Reducing valve	Pilot operated check valve	Accumulator	Adjustable fluid resistance	Fluid resistance	Relief valve
Transformer	Silicon control led	Capacitor	Adjustable electrical resistance	Electrical resistance	Voltage stabilizing diode

3 Basics Hydraulic Circuits and Their Analogy Electrical Circuits

3.1 Speed-Regulating Circuit

Figures 1(a) and 2(a) are common hydraulic throttle speed-regulating circuits, which are composed of relief valve, throttle valve, fixed-displacement pump and hydraulic cylinder [6].

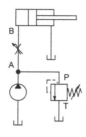
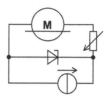

(a)   Inlet throttle speed-regulating circuit   (b)   Analogy electrical circuit

Fig. 1. Inlet throttle speed-regulating circuit and its analogy electrical circuit

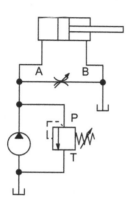
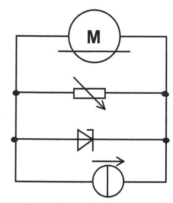

(a)   Bypass throttle speed-regulating circuit   (b)   Analogy electrical circuit

Fig. 2. Bypass throttle speed-regulating circuit and its analogy electrical circuit

In Fig. 1(a), the relief valve is connected in parallel with the pump to stabilize the outlet pressure of the pump. The throttle valve is connected in series between the hydraulic pump and the hydraulic cylinder to control the input flow of the hydarulic cyliner. Then the speed regulation of the hydraulic cylinder is realized by controlling the input flow of the hydraulic cylinder.

It can be seen from Table 1 that the fixed-displacement pump is equivalent to the constant current source and the throttle valve is equivalent to the variable resistance. Hence, the analogy electrical circuit of Fig. 1(a) is Fig. 1(b). In Fig. 1(b), the linear motor speed regulation is realized by adjusting the input current of the linear motor, which is controlled by variable resistance.

In Fig. 2(a), the normally closed relief valve, which is used as a safety valve, is connected in parallel with the pump. The throttle valve is connected with the hydraulic cylinder in parallel, and the input flow of the hydraulic cylinder is controlled by the opening of the throttle valve. Hence, the speed regulation of the hydraulic cylinder is realized by controlling the opening of the throttle valve. The analogy electrical circuit is shown in Fig. 2(b).

3.2 Directional Circuit

Figure 3(a) is a hydraulic directional circuit, which consists of a reversing valve, a fixed-displacement pump pump and a hydraulic cylinder [6].

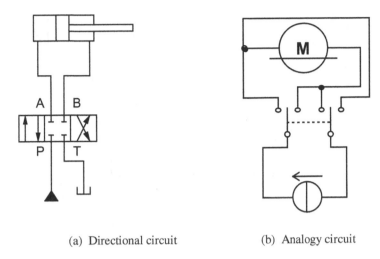

(a) Directional circuit   (b) Analogy circuit

Fig. 3. Directional circuit and its analogy electrical circuit

When the left position of the reversing valve works, the pressure oil flows into the rodless cavity of the hydraulic cylinder and the rod stretches out. When the right position of the reversing valve works, the pressure oil flows into the right cavity of the hydraulic cylinder and the rod returns. The corresponding electrical circuit diagram is shown in Fig. 3(b). The analogy circuit is composed of a constant current source, a bidirectional switch and a linear motor. The motion direction of the linear motor is controlled by the bidirectional switch.

3.3 Unloading Circuits

The function of a pressure venting circuit is to unload the pump and to keep the pump idling when the hydraulic system is during the regular off-working, which it will realize to save the energy.

There are two typical unloading circuits: one is realised by pilot-operated relief valve, as shown in Fig. 4(a), and the other is realised by the neutral position of reversing valve, as shown in Fig. 4(b). In Fig. 4(a), when the pilot control port of pressure relief valve is communicated with the oil tank, the pump outlet pressure is close to zero, the pump unloads; in Fig. 4(b), when the reversing valve is at the neutral position, the hydraulic pump is communicated with the oil tank, the outlet pressure of the pump oil is close to zero, the pump unloads. Figure 5(c) is the analogy electrical unloading circuit. When the switch closes, the output power is near to zero.

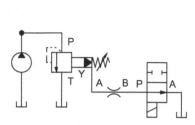
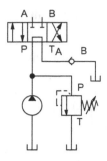

(a) unloading by pilot-operated relief valve   (b) unloading by the neutral of directional valve

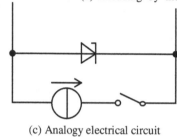

(c) Analogy electrical circuit

Fig. 4. Unloading circuit and its analogy electrical circuit

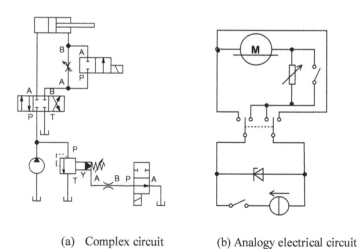

(a)   Complex circuit   (b) Analogy electrical circuit

Fig. 5. A Complex circuit and its analogy electrical circuit

4   A Complex Hydraulic Circuit and Its Analogy Electrical Circuit

Figure 5(a) is a complex hydraulic circuit composed of several basic hydraulic circuits, which consists of a pressure-venting circuit, directional circuit and throttle speed-regulating circuit. It can realize the functions of unloading, reversing, speed-regulation, working feed, quick feed and return.

Figure 5(b) is the corresponding Analogy electrical circuit. Figure 5(b) shows when the switch in parallel with variable resistance is closed, the current and the speed of linear motor both reach maximum value. When the switch in parallel with variable resistance is open, the current and the speed of linear motor change with the variable resistance's value. The motion direction of linear motor can be controlled by bidirectional switch.

5   Conclusion

In this paper, an analogy teaching method of hydraulic basic circuits is introduced. The corresponding analogy electrical circuits of hydraulic basic circuit are given. By introducing the working principles of analogy electrical circuits, the working principles of hydraulic basic circuits becomes easy to be understood.

References

1. Zhu, Y.: Similarity compare between electrical technology and fluid power technology. Mach. Tool Hydraul. **39**(24), 61–79 (2011)
2. Chen, J.H.: Comparison of basic characteristics of electric control and hydraulic control. Electr. Drive Autom. **38**(2), 10–12 (2016)
3. Hu, Y., Huang, Z.: Numerical simulation of flow control valve with bridge hydraulic resistance network. J. Wuhan Univ. Technol. **27**(6), 76–78 (2005)
4. Zheng, H., Wang, P.: Design on flow- resistance network of proportional valves. Fluid Power Transm. Control **2**, 29–30 (2007)
5. Hu, Y.: Fluid Resistance Network Systematics. China Machine PRESS, Beijing (2002)
6. Merrit, H.E.: Hydraulic Control System. Wiley, New York (1967)

Theoretical Modeling and Parameters Simulation of Positioning System

JingShu Wang[✉]

College of Mechanical Engineering, Chongqing University of Technology, Chongqing, China
donot@cqut.edu.cn

Abstract. In order to understand the characteristics of the positioning system, the theoretical model has been developed based on the transmission process. Then the state equation has been calculated. A part of unknown parameters are simulated by the 3D model of Pro/E software. The other unknown parameters are estimated by the system identification method. With all parameters being known, the response of the theoretical model is accordant with the actual position system.

Keywords: Positioning system · Modeling · State equation · System identification

1 Introduction

Precision positioning technique is widely used in precision instruments, machine tools, IC technique, and grating ruling engine [1–3]. In order to reach the high precision, the characteristics of the position system should be researched in detail. Many system modeling and identification methods have been developed to understand position system. Many types of model [4–7], such as ARMA model, ARX model, ARMAX model, and Box-Jenkins model, are applied to describe position system. The system identification is frequently carried out based on black-box model, and various artificial neural networks are employed for function approximation [8–10]. This paper focuses on modeling the actual structure of the position system and the artificial neural network is used to obtain unknown parameters of the system.

The paper is structured as follows. In Sect. 2, the structure of the position system is introduced and simplified. How to simulate a part of unknown parameters of the position system is discussed in Sect. 3. Section 4 presents the system identification to obtain unknown parameters and Sect. 5 explains conclusions and future work to be carried out.

2 Structure and Modeling

Dual stage position system is applied to ensure high precision and long travel of the precision position system. As shown in Fig. 1, the position system is composed of motor, worm gearing, screw-nut pairs, and dual stages, which consist of an inside stage and an outside stage. The outside stage is driven by motor through worm gearing, screw-nut

© Springer International Publishing AG 2018
F. Qiao et al. (eds.), *Recent Developments in Mechatronics and Intelligent Robotics*,
Advances in Intelligent Systems and Computing 691, DOI 10.1007/978-3-319-70990-1_50

pairs. It is designed to obtain long travel, as long as 1000 mm, while the inside stage is designed to get high precision. The inside stage is installed on the outside stage and driven by piezoceramic, the position accuracy is higher than 10 nm.

Fig. 1. The actual structure of the position stage

To understand the characteristics of the position system, the model of the system will be developed. In order to efficiently simplify the system, we assume the connections between the moving components are rigid connections and the main driving parts are flexible bodies. Based on the assumptions, the topological structure of the position system is illustrated in Fig. 2.

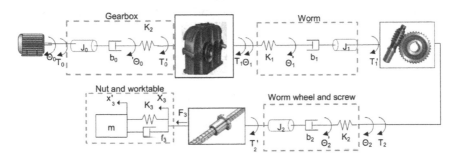

Fig. 2. The topological structure of the position system

The total system is divided into four parts: gearbox, worm, worm wheel and screw, nut and worktable. Firstly, the transmission process of the four parts will be discussed. Secondly, transmission relationship between the four parts) should be researched. Then, the state equation can be established. In the four subsystems, the inputting torque is equal to the torsional moment and the active force is equal to the elastic deformation driving force.

As shown in Fig. 2, for the gearbox subsystem, we can get:

$$T_0 = G_0(\theta_0 - \theta_0') = J_0 \frac{d^2\theta_0'}{dt^2} + f_0 \frac{d\theta_0'}{dt} + T_0' \tag{1}$$

The θ_0 and θ_0' is output angle of motor and input angle of gearbox, respectively. T_0 is output torque of motor while T_0' is the input torque of gearbox. The torsional rigidity, torsional damping, and inertia of motor output shaft is expressed by T_0, f_0, and J_0.

Due to the topological structure of the worm, the worm wheel and screw, the following equations can be obtained.

$$T_1 = G_1(\theta_1 - \theta_1') = J_1 \frac{d^2\theta_1'}{dt^2} + f_1 \frac{d\theta_1'}{dt} + T_1', \ T_2 = G_2(\theta_2 - \theta_2') = J_2 \frac{d^2\theta_2'}{dt^2} + f_2 \frac{d\theta_2'}{dt} + T_2' \tag{2}$$

Where G_1 and G_2 mean the torsional rigidity, f_1 and f_2 mean the torsional damping; J_1 and J_2 mean the inertia of each subsystem. θ_1' is the output angle of worm while T_1' is the output torque of worm. T_2 and T_2' indicate the input torque of worm wheel and the output torque of screw, respectively. θ_2 is the input angle of worm wheel while θ_2' is the output angle of screw.

The transmission process of the nut and worktable is expressed as following equation:

$$F_3 = G_3(y_3 - y_3') = m \frac{d^2 y_3'}{dt^2} + f_3 \frac{dy_3'}{dt} \tag{3}$$

Where F_3 is the driving force of the nut, and y_3 is the displacement of the nut. G_3 and f_3 mean rigidity and damping coefficient of the subsystem, respectively. y_3' indicates displacement of worktable and m for mass of worktable. The relationships between the subsystems can be expressed as follow.

$$T_1 = p_0 T_0'$$

$$T_1' = T_2 \frac{d_1}{D} \tan(\gamma_1 + \rho_v') = N_1 T_2, \ N_1 = \frac{d_1}{D} \tan(\gamma_1 + \rho_v')$$

$$\theta_2 = P_1 \cdot \theta_1'$$

$$T_2' = F_3 \frac{d_2}{2} \tan(\gamma_2 + \rho_v) = N_2 F_3, \ N_2 = \frac{d_2}{2} \tan(\gamma_2 + \rho_v)$$

$$\theta_2' = 2\pi \frac{y_3}{p_2} = P_2 y_3, \ P_2 = \frac{2\pi}{p_2} \tag{4}$$

Where P_1 is the worm gear transmission ratio, p_2 is the lead of screw. d_1 means the diameter of worm while d_2 for pitch diameter of screw.

Therefore, based on all the equations above, the whole transmission process can be described as

$$\ddot{\theta}_1' = \frac{2G_1}{J_1}\theta_0 - \frac{G_1 + N_1G_2P_1}{J_1}\theta_1' - \frac{f_1}{J_1}\dot{\theta}_1' + \frac{N_1G_2}{J_1}\theta_2' + \frac{G_1}{J_1}\Delta_\theta$$

$$\ddot{\theta}_2' = \frac{G_2P_1}{J_2}\theta_1' - \frac{G_2 + N_2G_3/P_2}{J_2}\theta_2' - \frac{f_2}{J_2}\dot{\theta}_2' + \frac{N_2G_3}{J_2}y_3' \qquad (5)$$

$$\ddot{y}_3' = \frac{G_3}{P_2m}\theta_2' - \frac{G_3}{m}y_3' - \frac{f_3}{m}\dot{y}_3'$$

3 State Equation and Parameters Simulation

The state variables of position system consist of six variables as follow $x_1 = \theta_1'$, $x_2 = \dot{\theta}_1'$, $x_3 = \theta_2'$, $x_4 = \dot{\theta}_2'$, $x_5 = y_3'$, $x_6 = \dot{y}_3'$.

The input of state equation is $u = \theta_0$ and the output is y_3'. Then, we can obtain the whole state equation as

$$\begin{bmatrix} \dot{x}_1 \\ \dot{x}_2 \\ \dot{x}_3 \\ \dot{x}_4 \\ \dot{x}_5 \\ \dot{x}_6 \end{bmatrix} = \begin{bmatrix} 0 & 1 & 0 & 0 & 0 & 0 \\ -\dfrac{G_1 + N_1G_2P_1}{J_1} & \dfrac{f_1}{J_1} & \dfrac{N_1G_2}{J_1} & 0 & 0 & 0 \\ 0 & 0 & 0 & 1 & 0 & 0 \\ \dfrac{G_2P_1}{J_2} & 0 & -\dfrac{G_2 + N_2G_3/P_2}{J_2} & \dfrac{f_2}{J_2} & \dfrac{N_2G_3}{J_2} & 0 \\ 0 & 0 & 0 & 0 & 0 & 1 \\ 0 & 0 & \dfrac{G_3}{P_2m} & 0 & -\dfrac{G_3}{m} & -\dfrac{f_3}{m} \end{bmatrix} \begin{bmatrix} x_1 \\ x_2 \\ x_3 \\ x_4 \\ x_5 \\ x_6 \end{bmatrix} + \begin{bmatrix} 0 \\ \dfrac{2G_1}{J_1} \\ 0 \\ 0 \\ 0 \\ 0 \end{bmatrix} u \qquad (6)$$

$$y = \begin{bmatrix} 0 & 0 & 0 & 0 & 1 & 0 \end{bmatrix} x(t)$$

According to physical dimension and material of each part, the inertia of each part is shown in Table 1. The 3D models of each parts have been established by Pro/E, and the rigidity of worktable is simulated to be $G_3 = 4.40335 \times 10^6$ N/m. The torsional stiffness of other parts has also been simulated as Table 2 based on the 3D models. The other parameters of the actual position system are also listed in Table 3.

Table 1. The inertia of each part.

Rotary inertia	Output shaft of motor	Worm	Screw	Worm wheel
Value (kg mm^2)	232.86	232.86	2696.63	117307.3

Table 2. The torsional stiffness of each part.

Torsional stiffness	Output shaft of motor	Worm	Worm wheel and screw
Value (N m/rad)	55478.5	5047.45	37861

Table 3. The value of actual system parameters.

Symbol	m	P_1	P_2	N_1	N_2	p_0
Value	43.28 kg	1/600	2000π	0.0225	2.7146	0.99^3

4 System Identification

After the simulation, there are still three unknown parameters in the position system, f_1, f_2 and f_3. The system identification method is applied to obtain the three parameters. With the signal of the motor being ramp signal, the output of the position system, namely, the displacement of the position stage had been measured.

According to the state equation of the system, the transfer function of the system can be calculated. The system response of ramp signal can also been developed. In combination with the actual response of the position system, a BP neural network is employed to estimate the unknown parameters. The result is $f_1 = 322.2963$, $f_2 = 7.60385$, $f_3 = 3.7454$.

We can get the open-loop transfer function of the position system is

$$H = \frac{327644611423.7918}{(s + 1.229 \times 10^6)(s + 15.66)(s^2 + 0.08875s + 1.017 \times 10^5)(s^2 + 63.37s + 3.155 \times 10^5)} \tag{7}$$

The response of the theoretical model and the actual system is shown in Fig. 3. Obviously, the stabilized displacement of actual position system is less than that of theoretical model. That's because the backlash of the mechanical structure, including the worm and wheel, the screw, and the gearbox. It is accordant with the practical situation.

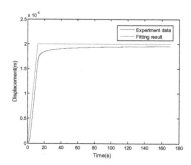

Fig. 3. Response of the theoretical model and the actual system

5 Conclusion

In this paper, a state equation has been implemented to describe the position system based on the transmission process. A part of the unknown parameters are simulated based on the 3D models. The last three unknown parameters are estimated by the system identification method. After obtaining all parameters of the state equation, the responses of the theoretical model and the actual position system had been compared. The result conforms to the actual condition.

Future work will pay more attention to the detailed analysis of the backlash of the position system. Further-more, a more precise model of the position system should been studied to understand more characteristics of the position system.

Acknowledgments. This project is supported by Scientific and Technological Research Program of Chongqing Municipal Education Commission (Grant No. KJ1400936), Foundation and Frontier Research Program of Chongqing Science and Technology Commission (Grant No. cstc2016jcyjA0538).

References

1. Yong, Y.K., Aphale, S.S., Moheimani, S.O.R.: Design, identification, and control of a flexure-based XY, stage for fast nanoscale positioning. IEEE Trans. Nanotechnol. **8**(1), 46–54 (2009)
2. Gao, W., Sato, S., Arai, Y.: A linear-rotary stage for precision positioning. Precis. Eng. **34**(2), 301–306 (2010)
3. Wang, J., Zhu, C., Feng, M.: Thermal error modeling and compensation of long-travel nanopositioning stage. Int. J. Adv. Manuf. Technol. **65**(1), 443–450 (2013)
4. Wang, J., Guo, J., Zhu, C.: System identification of a positioning stage based on theoretical modeling and an ARX model. J. Vib. Shock **32**(13), 66–69 (2013)
5. Bertha, M., Golinval, J.C.: Identification of non-stationary dynamical systems using multivariate ARMA models. Mech. Syst. Signal Process. **88**, 166–179 (2017)
6. Li, Y., Su, Y., Shu, L.: An ARMAX model for forecasting the power output of a grid connected photovoltaic system. Renew. Ener. **66**(6), 78–89 (2014)
7. Magalhães, R.S., et al.: Identification of hybrid ARX–neural network models for three-dimensional simulation of a vibroacoustic system. J. Sound Vib. **330**(21), 5138–5150 (2011)
8. Langdon, G.S., et al.: Fracture of aluminium foam core sacrificial cladding subjected to air-blast loading. Int. J. Impact Eng. **37**(6), 638–651 (2010)
9. Nemat-Nasser, S., et al.: Experimental investigation of energy-absorption characteristics of components of sandwich structures. Int. J. Impact Eng. **34**(6), 1119–1146 (2007)
10. Tagarielli, V.L., Deshpande, V.S., Fleck, N.A.: Prediction of the dynamic response of composite sandwich beams under shock loading. Int. J. Impact Eng **37**(7), 854–864 (2010)

CFD Simulation of Internal Fluid Characteristics of Vehicle Friction Damper

Li Xiang Qin, Dong Zhao[✉], Dong Mei Cai, Yan Zhou,
Zhuang Zhuang Fan, and Kai XiChuan

School of Mechanical Engineering, University of Jinan, Jinan, Shandong, China
me_zhaod@ujn.edu.cn

Abstract. The shock absorber has a very important influence on the stability and smoothness of the vehicle. The fluid characteristics of the shock absorber affect the damping force of the shock absorber directly. In this paper, the vehicle friction damper is studied. Using Solid Works to establish the three-dimensional model of internal flow field of shock absorber, using Gambit software to mesh the model, the variation law of the internal fluid pressure is obtained in the Fluent software based on the dynamic mesh technique and the two-phase flow technique. This paper provides a fast and effective method for the design of friction damper.

Keywords: CFD simulation · Friction damper · Dynamic mesh · Two-phase flow

1 Introduction

The shock absorber is responsible for the vibration attenuation in the automobile suspension system. It affects the vehicle's comfort, handling stability, security and other important performance. At present, there is still a certain gap in the research of vehicle friction damper abroad. In this paper, a new type of vehicle friction damper is taken as the research object, and the characteristics of the internal flow field of the shock absorber are analyzed by using the Fluent software, the internal pressure variation curve of the shock absorber is obtained, which provides a theoretical basis for the design and optimization of the friction damper.

2 Working Principle of Vehicle Friction Damper

The utility model relates to a friction damper for vehicles, which is mainly composed of shell body, an inner cylinder, a friction plate, an actuating rod, a connecting piston, a lower base and a pressurizing air chamber. The structure of the damper is shown in Fig. 1. The outer shell is connected with the vehicle body when the shock absorber works. The shock absorber is arranged inside the inner cylinder, the inner cylinder is pressurized gas piston chamber is divided into two chambers, the lower part of the upper plenum chamber into the oil storage chamber, the actuating rod from the lower part of the oil storage chamber into the internal oil storage room, the actuating rod are connected together by a connecting pin shaft and piston and the outer shell and in the oil storage

F. Qiao et al. (eds.), *Recent Developments in Mechatronics and Intelligent Robotics*,
Advances in Intelligent Systems and Computing 691, DOI 10.1007/978-3-319-70990-1_51

room on the surrounding wall transfer hole oil pressure, circumferential wall outside is sheathed with a rubber body, both sides of the rubber body by the locking sleeve pressing seal, a central screw friction plate, friction plate and screw shell body to form a pair of friction, is the main part of the shock absorber damping force is generated.

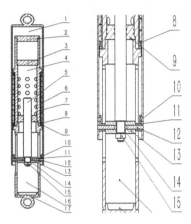

Fig. 1. Schematic diagram of vehicle friction damper

1 shell body, 2 upper cover, 3 pressurized air chamber piston, 4 inner cylinder, 5 actuating rod, 6 rubber body, 7 spiral friction plate, 8 locking sleeve, 9 under cover, 10 short selling, 11 connection block, 12 long pin, 13 connecting piston, 14 sleeve, 15 nut, 16 under the base, 17 under the seat lug.

The oil pressure generated by the internal friction damper of the car is due to the pressurizing chamber filled with pressure gas through the pressurizing chamber and the piston squeeze liquid. Oil pressure by the inner cylinder wall oil pressure transmission hole transfer to the rubber body, the rubber body is squeezed oil pressure to expansion screw extrusion friction plate, friction plate and a spiral diameter occur on the external shell wall extrusion. When the shell body and the spiral friction plate moves will produce friction damping force, the actuator rod end into the internal pressure of oil is lower, and the other end of the shell, the oil pressure in the actuating rod direction constant resistance.

3 Simulation of Shock Absorber Oil

The internal fluid of the shock absorber includes two different kinds of fluid, the compressed gas in the pressurized air chamber and the hydraulic oil in the inner cylinder. Therefore, the multiphase flow technique is used to simulate the two kinds of fluid. When the shock absorber works, the actuating rod moves in the hydraulic oil, which causes the change of the basin boundary.

The displacement of the actuating rod is different from that of the inner cylinder of the shock absorber, which results in the different oil pressure in the inner cylinder.

3.1 Simulation Modeling

There are two kinds of fluid in the basin of the shock absorber, such as gas and liquid, and there are two kinds of liquid in the basin, namely the gas and liquid, and the sketch of the watershed model is shown in Fig. 2. The inner diameter of the inner cylinder of the shock absorber is D, and the diameter of the actuating rod is d. The boundary between the basin and the inner wall of the inner cylinder is wall type, and the boundary of the contact between the basin and the actuating rod is deform, and the boundary of the contact between the basin and the actuating rod is Rigid body.

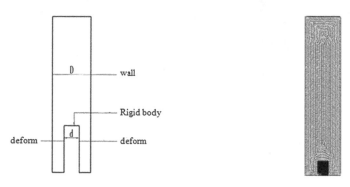

Fig. 2. Sketch map of watershed model   **Fig. 3.** Grid model

The watershed model is divided into Gambit software. Due to the existence of moving boundaries in the basin, the change of the shape of the basin is relatively large, so the triangular mesh of the unstructured grid type is selected to divide the grid, and the grid model is shown in Fig. 3.

3.2 Assumption of Simulation

The watershed model is introduced into the Fluent software to simulate the internal flow field. There are many factors that affect the internal flow field of the shock absorber. In order to carry out effective simulation analysis, the following assumptions are put forward in this paper:

(1) The compressed gas in the gas chamber is an ideal gas;
(2) Hydraulic fluid is Newton fluid and incompressible;
(3) Ignoring the friction between the piston and the inner cylinder;
(4) Ignoring the influence of temperature on the internal flow field.

4 Simulation Results and Analysis

Using Gambit to establish the basin geometry model ($D = 50$ mm, $d = 12$ mm). Import the grid model into the FLUENT software; Open the multiphase flow model, select the standard k-ε equation; The movement of the rod top is defined as the piston model in the software. The speed of the actuating rod is shown in formula 1.

$$v = \frac{L}{2}\omega \sin (\omega t + \varphi) \tag{1}$$

In formula: v- actuating rod velocity, L- actuator amplitude, ω-vibration angular velocity, φ-phase angle.

The height of the upper chamber of the model is 55 mm, and the initial pressure is 3 Mpa. The gas chamber is the first phase fluid as the ideal gas, and the lower part of the air chamber is a second phase hydraulic oil. The two fluid distribution is shown in Fig. 4. The red part is the ideal compression gas, the blue part is the hydraulic oil.

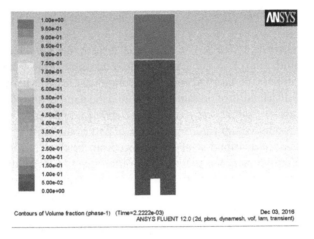

Fig. 4. Distribution of fluid flow in shock absorber

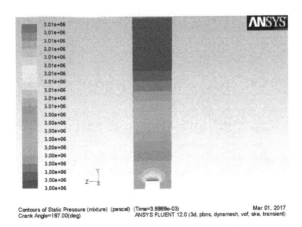

Fig. 5. Fluid pressure nephogram of the actuating rod

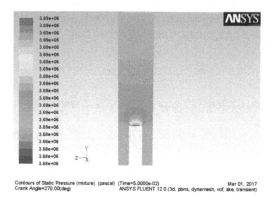

Fig. 6. The pressure distribution of the actuating rod 0.05 s

Through the analysis of the flow field within the damper can change the internal pressure of the fluid damper, which can provide a theoretical basis for the design of shock absorber damper structure and energy performance analysis, in order to facilitate the analysis of the distribution of pressure flow within the damper, the software extracted pressure nephogram of a vibration cycle representative of analysis.

The actuating rod moves upward from the lower part of the inner cylinder, as shown in Fig. 5. The pressure distribution of oil is increasing gradually from the top to the bottom of the inner cylinder, and the pressure difference is 0.014 Mpa. As shown in Fig. 6, when the actuating rod moves up to 40 mm, the pressure in the inner cylinder rises from the initial 3 to 3.69 Mpa. The pressure difference is 0.019 Mpa.

As shown in Fig. 7, the actuating rod moves 0.1 s to reach the maximum displacement of 100 mm, then the pressure of the inner cylinder reaches the maximum pressure of 4.69 Mpa, and the pressure difference in the whole basin is the 0.0016 Mpa. The oil pressure change curve is shown in Fig. 8. Through the analysis of internal hydraulic pressure damper is a cycle of movement, can draw the cylinder pressure within the damper with the actuating rod into the displacement increases the increment is about 1.7 Mpa; the same time inner cylinder oil pressure distribution is uneven but the

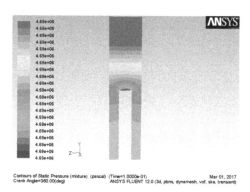

Fig. 7. The pressure distribution of the actuating rod 0.1 s

difference is very small, so the analysis of damping force, average cylinder oil hydraulic pressure is used for the value to calculate the output damping force of shock absorber.

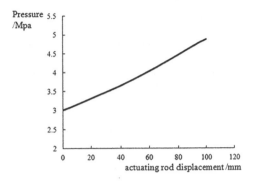

Fig. 8. Variation curve of oil pressure

5   Concluding Remarks

By using fluent finite element analysis software, the dynamic simulation of the fluid inside the shock absorber is carried out. Compared with the traditional methods, the research method is more intuitive and accurate. It provides a fast and effective method for the structural design and parameter optimization of the vehicle friction damper.

References

1. Liu, X., Zhao, D., Liu, H.: Design and mechanical characters study of a self feedback friction damper. Appl. Mech. Mater. **71**, 1724–1727 (2011)
2. Shu, H., Zhang, W., Feng, Y.: Micro-process model of hydraulic shock absorber with abnormal structural noise. J. Central South Univ. Technol. **15**(6), 853–859 (2008)
3. Herr, F., Mallin, T., Lane, J., et al.: A shock absorber model using CFD analysis and EasyS. SAE Pap. **26**(1), 13–22 (1999)
4. Martins, F.P.: Development and validation of a CFD model to investigate the oil flow in a shock absorber. SAE Pap. 143–145 (2005)
5. Duym, S., Stiens, R., Baron, G.V., et al.: Physical modeling of the hysteretic behavior of automotive shock absorbers. SAE Pap. **3**, 45–48 (2005)

Study on Optimization of Impeller Structure Parameters of ESP for Offshore Oilfield Based on ANSYS

Wen-duo He[✉], Qing Zhang, and Jin-xin Zhang

School of Mechanical Engineering, Tianjin University, Tianjin 300350, China
hwd@tju.edu.cn

Abstract. According to the initial design parameters of electric submersible pump and the profile equation of blade, the 3D model of impeller and internal flow passage of electric submersible pump is established in Pro/E. By using the APDL parametric modeling function of ANSYS, the optimal analysis model is established. The design variables, target variables and constraints were selected by referring to the relevant books and empirical formulas, and the structure was optimized by using the ANSYS optimization tool. The parameters before and after optimization of the impeller structure under the same viscosity and the optimization parameters of the different oil viscosity coefficients were compared to obtain the influence of each design variable on the performance of the ESP and the optimum parameters under different viscosity coefficients law, which provides a reference for determining the main design parameters of the electric impeller structure of ESP in offshore oilfield.

Keywords: Electric submersible pump · ANSYS · Optimal design · Impeller structural parameters · Oil viscosity

1 Introduction

With the adjustment of the energy structure, the development of marine resources has been paid more and more attention. In oil and other related industries, the fluid machinery and other equipment have played a pivotal role. As a kind of artificial mining method [1], ESP has the advantages of large flow, high power, strong adaptability, large working pressure, simple process, long service life, significant economic benefits and many other advantages. Because of these advantages of ESP, it has been widely used in the high water content, low gas ratio, low energy and other complex conditions of production wells [2].

Although the ESP has been widely used in offshore oil platform, but its production cost is much higher than the land, mining environment is relatively poor, and all kinds of work are more difficult than on land. For some complex conditions, the oil production still has a great difficulty, especially because of changes in temperature or moisture content caused by changes in oil viscosity. In the process of oil extraction, the viscosity coefficient of crude oil determines the flow performance, and the viscosity of the oil also affects the size of shear stress [3]. There is great practical significance to understand the

© Springer International Publishing AG 2018
F. Qiao et al. (eds.), *Recent Developments in Mechatronics and Intelligent Robotics*,
Advances in Intelligent Systems and Computing 691, DOI 10.1007/978-3-319-70990-1_52

characteristics of the viscosity of the crude oil and the flow law, and to optimize the impeller structure parameters of the ESP according to the size of the viscosity coefficient.

2   Establishment of 3D Model of Impeller and Internal Flow

According to the initial design parameters and the blade line equation, the complete single stage impeller solid model is established by Pro/E software, as shown in Fig. 1, and the impeller blade solid part is removed to generate the internal flow channel model. A single stage impeller 3D solid model as shown in Fig. 2.

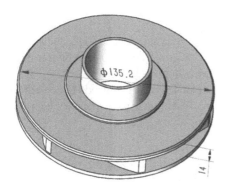

Fig. 1. 3D model of impeller   **Fig. 2.** 3D model of internal flow channel

3   Mathematical Model and Method of Optimization Design

In order to improve the performance of the submersible pump, it is necessary to make some constraints on the parameters with large influence on the impeller structure design, and use it as the constraint condition to optimize the structural parameters of the impeller. The inlet diameter D_1, outlet diameter D_2, inlet angle β_1, outlet angle β_2, outlet width b_1 and the number of blades z have different effects on the final performance of the submersible pump. Therefore, it is necessary to impose the necessary constraint on the optimal constraint equation. Based on the optimization method of finite element software ANSYS Workbench, the flow path model of impeller is modeled and then the simulation and optimization is carried out.

3.1   Optimized Mathematical Model

In general, the objective function and the constraint condition can be expressed as follows:

$$\begin{cases} min f(x) = f(x_1, x_2, \cdots \cdots, x_n) \\ x_{min} \leq x_i \leq x_{max} \quad i = 1, 2, \cdots \cdots, n \\ g_i(x) \leq 0 \end{cases} \quad (1)$$

Where $x = (x_1, x_2, \cdots \cdots, x_n)$ is the design variable; $f(x)$ is the objective function; $g_i(x)$ is a state variable, a constraint function that contains equations and inequality constraints.

3.2 Determination of Constraint Condition and Objective Function

(1) Constraints of design variables

Appropriate choice of design parameters will help save time and reduce the amount of calculation. On the one hand, the value of the design variables cannot be too large, or do not meet the production requirements; on the other hand, if the value range is too small, it may deviate from the optimization point [4]. Therefore, it needs to consult the relevant design books and use the relevant empirical formula to give a reasonable range of design variables.

$$\begin{cases} 0.0631 \leq D_1 \leq 0.0721 \\ 0.1235 \leq D_2 \leq 0.1330 \\ 15° \leq \beta_2 \leq 40° \\ 20° \leq \beta_1 \leq 35° \\ 0.0144 \leq b_2 \leq 0.0231 \\ 3 \leq z \leq 8 \end{cases} \quad (2)$$

Where D_1 is the inlet diameter of the impeller, the outlet diameter of the D_2 impeller, β_2 is the impeller exit angle, β_1 is the impeller inlet angle, b_2 is the impeller exit width, and z is the number of blades.

(2) Constraints for state variables

The state variable constraint is a constraint that determines the structural material to meet certain performance requirements. Such as the allowable stress value of the material:

$$\sigma = max(\sigma_i) \leq [\sigma] \quad (3)$$

Where σ is the maximum equivalent stress on each impeller on the impeller; σ_i is the equivalent stress of the i node; $[\sigma]$ is the allowable stress of the material; i is the node number, $i = 1, 2, 3, \cdots \cdots, N$.

The ESP impeller commonly used materials for high nickel cast iron, containing more than Ni13%, has the mechanical process and good performance [5]. After the heat treatment, the yield strength is 345 MPa, and the safety factor is about 1.5:

$$[\sigma] = \frac{345}{1.5} = 230 \text{ MPa} \tag{4}$$

It can be concluded that the constraint of the state variables is:

$$0 < g_i(x) = \sigma \le 230 \tag{5}$$

Where $g_i(x)$ the stress function for the impeller.

(3) Objective function

The impact of the submersible pump head performance (efficiency and head) are the outlet pressure and outlet speed of the impeller, so choose model data as output target parameters to meet the equivalent stress conditions, so as to reach the maximum value. It can be concluded that the constraint condition of the objective function is:

$$min P_{static} = -f(x_1, x_2, x_3, x_4, x_5, x_6) \tag{6}$$

Where P_{static} on behalf of the export section of the static pressure size. $x_1, x_2, x_3, x_4, x_5, x_6$ represent the design variables D1, D2, β2, β1, b2 and z, respectively.

3.3 Performance Evaluation of ESP

Parameters of electrical submersible pump performance is its work efficiency and head size, formula (7) is the calculation of type ESP single stage head, formula (8) is the calculation of type ESP efficiency.

$$H = (P_2 - P_1) \Big/ \rho g + z_2 - z_1 + \frac{v_2^2 - v_1^2}{2g} \tag{7}$$

$$\eta = \frac{\rho g H Q}{1000 N} \times 100\% \tag{8}$$

Where P_2 is the static pressure at the outlet, MPa; P_1 is the static pressure at the inlet, MPa; ρ is the oil density, kg/m³; V_1 is the initial velocity of the entrance, m/s; V_2 is the speed at the exit, m/s; z_1 import height, m; z_2 export height, m; N is the shaft power, kw.

4   Optimization Calculations and Results Analysis

In this paper, the single-stage flow channel model is chosen as the research object, and the parameters of the single-stage impeller are optimized. Firstly, the optimum parameters of oil viscosity coefficient of 0.01 kg/m-s and density of 850 kg/m³ were analyzed.

The initial sample data in the optimization process is randomly selected by software for 20 groups. After setting the initial sample data and running the analysis, the criterion of the optimization algorithm can be set in the ANSYS optimization module, that is the outlet pressure is the largest and the equivalent stress is the smallest. Different parameters can be set to the priority level, the maximum pressure at the outlet is set to the highest priority and the minimum equivalent stress is set to secondary priority. After updating the operation, three candidate parameter points are determined, as shown in Table 1.

Table 1. Optimized parameter candidate point.

SN	D_1/mm	D_2/mm	β_2/°	β_1/°	b_2/mm	z/个	P_{static}/Mpa	σ/Mpa
1	68.7	126.9	33.6	21.9	16.3	6.1	0.204	136.2
2	67.4	128.0	28.1	34.6	13.8	6.1	0.218	150.7
3	64.2	124.7	33.1	33.1	20.8	4.9	0.190	170.1

4.1 Optimized Performance Cloud Picture

The following figure shows the results of Fluent analysis. Figure 3 is the optimization of the overall pressure of the impeller cloud diagram; Fig. 4 for the optimization of the overall velocity of the impeller cloud diagram.

Fig. 3. Optimized pressure cloud diagram   **Fig. 4.** Optimized velocity cloud diagram

As can be seen from Figs. 3 and 4, the optimized outlet pressure is about 2.04×10^5 Pa and the outlet velocity is 18.9 m/s. Therefore, it can be concluded by the formulas (7) and (8) that the efficiency after optimization reach to 52.7%

Figure 5 shows the deformation of the impeller blade after loading. Figure 6 is the equivalent stress distribution of the impeller blade.

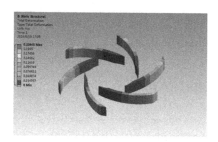
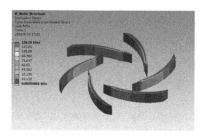

Fig. 5. Deformation cloud diagram   **Fig. 6.** Equivalent stress cloud diagram

It can be seen from the deformation cloud diagram that the maximum deformation of the impeller blades is 0.22 mm. According to the equivalent stress cloud, the maximum equivalent stress value of the impeller blades is 136.15 MPa, which is smaller than the allowable stress.

4.2 Comparison of Parameters Before and After Optimization

The oil viscosity coefficient of 0.1 and 1 kg/m-s were calculated and analyzed respectively.

(1) Comparison of data before and after optimization with a certain viscosity coefficient

Oil viscosity coefficient for the parameter optimization of ESP impeller structure before and after the 0.01 kg/m-s cases are shown in Table 2.

Table 2. Comparison of impeller structure parameters before and after optimization when the viscosity coefficient of oil is 0.01 kg/m-s.

Optimal variable	D_1/mm	D_2/mm	β_2/°	β_1/°	b_2/mm	z/piece	H/m	σ/MPa
Before optimization	70	135.2	35	25	14	5	10	139.79
After optimization	68.7	126.9	33.6	21.9	16.3	6.1	12.3	136.15

Through the comparison of the data in Table 2, it can be seen that the inlet diameter D_1, the outlet diameter D_2, the exit angle β_2 and the inlet angle β_1 of the impeller structure parameter are reduced to some extent, indicating that it is beneficial to improve the performance of the submersible pump when these parameters take a small value; while the export width b_2 is just the opposite, the optimized value is slightly larger than the previous one; the number of leaves is 6.1, which needs to be rounded to 6 in the modeling application.

(2) Comparison of optimization parameters for different oil viscosity coefficients

The data of the optimized parameters of the electric impeller of the ESP under different oil viscosity coefficients (0.01, 0.1, 1 kg/m-s) are shown in Table 3.

Table 3. Contrast of structural parameters of impeller under different oil viscosity coefficient.

Viscosity coefficient	D_1/mm	D_2/mm	β_2/°	β_1/°	b_2/mm	z/piece
0.01	68.7	126.9	33.6	21.9	16.3	6.1
0.1	67.4	126.1	34.1	21.4	17.3	5.6
1	66.5	125.1	35.0	20.8	18.8	5.9

By comparing the data in Table 3, it can be seen that the calculated impeller optimization parameters are different under different oil viscosity coefficients. With the viscosity coefficient of the oil increases, the values of the impeller inlet diameter D_1, the outlet diameter D_2, and the inlet angle β_1 are slightly reduced, which shows that the three parameters of a smaller value can help improve the performance of ESP. While the export angle β_2 and the outlet width b_2 is just the opposite, with the increase of oil viscosity coefficient, the number of these parameters will be slightly larger. For the number of impeller blades, the optimization value of the three kinds of oil viscosity has little difference, taking the number of 6 after the round.

5   Conclusion

In this paper, the optimal mathematical model is established by choosing the design variables, target variables and constraints using the relevant empirical formula, and the optimization analysis of the structure is carried out by ANSYS optimization tool. We can draw the following conclusions:

(1)   The diameter of the inlet diameter of the submersible pump and the decrease of the outlet diameter will reduce the viscosity of the oil to a certain extent, which will help to reduce the energy loss and improve the efficiency of the ESP. The increase in export angle and entrance angle leads to a slight reduction in efficiency and pumping head. The number of leaves is too small, which makes the relative length of the runner becomes smaller. On the contrary, it will cause the flow channel is too narrow, increasing friction and causing energy loss.

(2)   Under the condition of large oil viscosity, the lower value of the inlet diameter, outlet diameter and inlet angle will improve the performance of the submersible pump. The export angle and the width of the outlet is just the opposite, the larger value of these two parameters will improve the performance of the submersible pump; for the number of leaves of the impeller, the three oil viscosity optimization values are basically the same.

References

1.   Liu, J., Wang, L., Ma, C.: Reliability analysis of the dead impeller of electrical submersible pump. Technol. Superv. Pet. Ind. **10**, 008 (2005)
2.   Chen, S., Wang, Z.C.: Numerical simulation of scouring erosion characteristics for electric submersible pump impeller. Adv. Mater. Res. **749**, 535–539 (2013)
3.   Qi, X., Turnquist, N., Ghasripoor, F.: Advanced electric submersible pump design tool for geothermal applications. Trans. Geotherm. Resour. Counc. **36**, 543–548 (2012)
4.   Sukhanov, A., Amro, M., Abramovich, B.: Analyses of operating electric submersible pumps (ESPs) of different manufacturers-case study: Western Siberia. Oil Gas Eur. Mag. **41**(4), 202–204 (2015)
5.   Hakeem, A.A., Elserougi, A.A., Abdelkhalik, A.S., et al.: Performance evaluation of a transformerless multiphase electric submersible pump system. J. Eng. (2015)

Research on a Portable Heat Dissipation System for Cable Splicing

Zhang Hao[1], Li Yan[1], Zhang Yong[2], Liu Han[2], and Liang Kun[3(✉)]

[1] State Grid Shandong Electric Power Company Economic and Technological Research Institute,
Shandong 250021, China
[2] State Grid Shandong Electric Power Company, Shandong 250001, China
[3] College of Computer Science and Information Engineering,
Tianjin University of Science & Technology, Tianjin 300222, China
liangkun@tust.edu.cn

Abstract. Wireless sensor network (WSN) containing thousands of tiny and low-power nodes can be used to monitor environment. An energy-efficient and reliable wireless communication architecture is usually required. In this paper, we propose a novel energy-efficient multi-agent based architecture (EEMA), which is based on a clustering algorithm and multi-agent system to reduce the redundant messages and filter most error messages. EEMA consists of three important agents: classification agent, error agent and filter agent which divide messages processing into three different phases separately and achieve different functions. The simulation results show that EEMA can outperform LEACH and ECDG. It can distinctly prolong the lifetime of the system and improve the system reliability.

Keywords: Agent · Wireless sensor network · Energy efficient

1 Introduction

Through all aspects of power communication, communication cable is to support the construction of a strong smart grid and an important public transport channel [1]. n the power communication, through the cable long-distance transmission signal has been a wide range of promotion and application. Therefore, the fiber cable usually carries a large amount of information and covers a wide range, once the large number of cable failure, it leads to large area signal output disruption. It needs to rapidly response the technical staff and restores the fiber optic cable in the shortest time. However, in the process of welding fiber optic cable, especially high temperature conditions in the summer, the fiber optic cable splicer has to face downtime due to the high temperature, which greatly affects the cable repair time, causes great inconvenience to people's production and life. Therefore, it is urgent to provide a heat dissipation system for fiber optic cable welding, effectively eliminate the heat generated during the operation of the cable splicer, so that the temperature around it to maintain a reasonable space to ensure that the long process of welding cable in the continuity and stability.

© Springer International Publishing AG 2018
F. Qiao et al. (eds.), *Recent Developments in Mechatronics and Intelligent Robotics*,
Advances in Intelligent Systems and Computing 691, DOI 10.1007/978-3-319-70990-1_53

Optical fiber splicer is the common connection equipment to the fiber cable, often in the downtime course of high temperature. The main reason is that the cable welding itself is a large number of heat release process, the optical fiber splicer releases the heat which cannot be discharged in time, it will start the automatic protection mechanism, and causes downtime [2]. As the optical fiber splicer nominally works during 0–40 °C temperature, the downtime situation has occurred frequently especially in the base of the outdoor temperature of 35 °C or more. Table 1 shows the operation and maintenance records, in June 2015 in Shandong province, China.

Table 1. Welding machine operation

Place	Times of splicing	Temperature (>35°C) days	Times of downtime
Dongying	13	12	8
Binzhou	15	13	9
Weihai	15	12	7
Linyi	22	19	17

At present, there are generally two measures for the downtime: first, stop the construction and cool the splicer into the air-conditioned place or shade; second, from other sites to transport another splicer. Two splicers are alternately welded. In the actual construction process, because the splicer is expensive, the construction company configuration is less, and often there will be more on-site construction of the situation, regardless of which of the above measures will seriously affect the fiber repair time, resulting in delays in the progress of the project. Therefore, the cable repair work is without delay, it is very necessary to reduce the burden of construction team and improve work efficiency for the power communication network [3].

This paper is to design a heat dissipation system for fiber optic cable splicer, and improve the heat dissipation during the splicing process, and reduce the number of downtime. Through the actual measurement and experiment, we present the optimal design combined with the actual needs of the cable welding site work application validation.

2   System Mode

Combined with the actual work situation, how to analyze the scene in the environment, we designed a low cost, easy assembly of the jacket heat sink. The radiator according to the realization function is divided into six parts: chassis module, cooling module, control module, temperature acquisition module, display module, power module and so on. According to the combination of the various functions, cable welding machine cooling system function structure shown in Fig. 1.

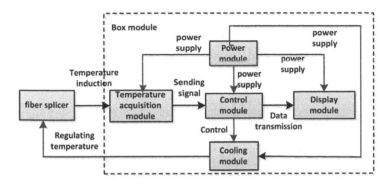

Fig. 1. System mode

When the splicer works, the ambient temperature of the fuselage is gradually increased. When the value reaches a certain value, the temperature acquisition module sends a signal. The control module sends the instruction to the cooling module. The cooling module adjusts the fan speed according to the instruction and takes away excess heat to control Welding machine temperature. The same process, the temperature decreases after the fan speed decreases until the stop, saving power [4].

2.1 Cooling Module

The cooling module is located on the inside of the system, and its structure, layout and size affect the cooling effect of the whole system.

(1) the size of the fan

The fan should be able to play a cooling effect, and the noise is small, does not interfere with the welding work, so that the operator can work comfortably. Fan noise and air volume and its size, the current market, the existing mainstream fan 7 cm fan and 12 cm fan two. Through the comparison test, the impact of two different sizes of fans on the welding operation and applicability analysis, The final choice of small air volume and low noise 7 cm fan, the fan is less than the fuse at least 6.7 cm, welding machine and the fan does not overlap, the applicability of better, will not affect the construction staff welding operations (Fig. 2).

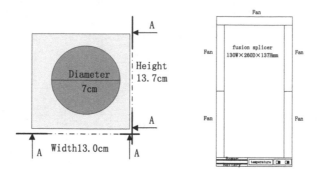

Fig. 2. Fan size and position diagram

(2) the number of fans

The fan should be able to cover the welding machine to the maximum extent possible. Ventilation and cooling with fan from the ventilation or external forced ventilation in two forms. Air as a cooling medium, in the box to do three-dimensional steady turbulent flow. The three-dimensional steady-state turbulence flow of the fluid follows the following control equation.

Continuous equation:

$$\frac{\partial \rho}{\partial t} + \frac{\partial}{\partial x_1}(\rho u_1) = 0 \tag{1}$$

where: ρ represents the fluid density; u1 the velocity of the fluid velocity along the i direction.

Momentum equation:

$$\frac{\partial}{\partial t}(\rho u_1) + \frac{\partial}{\partial x_j}(\rho u_i i_j) = -\frac{\partial p}{\partial x_i} + \frac{\partial \tau_{ij}}{\partial x_j} + \rho g_1 + F_1 \tag{2}$$

where p is the static pressure, τ_{ij} the stress vector, ρg_i is the gravitational component in the i direction, and Fi is the other energy item caused by resistance and energy.

$$\tau_{ij} = \mu \left(\frac{\partial u_1}{\partial x_j} + \frac{\partial u_j}{\partial x_i} \right) - \frac{2}{3} \mu \frac{\partial u_1}{\partial x_1} \delta_{ij} \tag{3}$$

In the above formula, μ is the molecular viscosity; the second term on the right is the result of volume expansion.

2.2 Control Module

Using the controller to achieve accurate and reliable control of the fan, take up a small space, economical and reasonable. At present, the most common control methods include logic circuit control and single-chip control in two ways [5]. Through the test the accuracy of the test method for the production of simple logic control circuit and single-chip circuit, respectively, control a LED lights, relatively bright off the number of times to compare the accuracy of control. After testing, single-chip logic circuit can achieve more precise control, to achieve simple and accurate control, small size, reasonable price design requirements.

We choose the microcontroller chip STC89C52, in order to achieve the timing of temperature collection, different temperature fan at different speeds and other functions, the need for some control in the program requires more precise control set timing/delay to achieve [6]. When programming with a single-chip microcomputer, the timer can be used to simulate the temperature acquisition every 10 s and set the speed (800 rpm) according to the different temperature (35 °C). The CPU occupancy rate and time delay are calculated continuously 30 times Timing temperature acquisition, different temperature fan under different speed and other functions

2.3 Temperature Acquisition Module

The temperature acquisition module requires accurate detection of temperature. Accurate detection of temperature, according to GB-T15768-1995 standard requirements, the error ≤ 5%. Because at the same temperature, the value of the two components output can be detected by the two components of the temperature [7]. Therefore, the accuracy of the two elements to detect the temperature of the experiment, the test temperature can be set at 40 °C, the temperature test error interval [38.4 °C, 41.6 °C]. The experimental results are shown in Figs. 3 and 4.

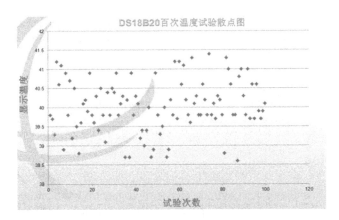

Fig. 3. Temperature test of DS18B20

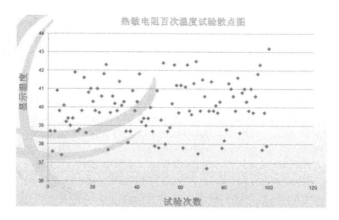

Fig. 4. Temperature test of thermistor

According to the scatter plot, it was found that there were 12 transcendental reference lines in 100 tests with an error range greater than ± 5%. Therefore, the conclusion: DS18B20 chip temperature detection error is maintained at less than 5%, the error is smaller, more stable than the thermistor accurate.

2.4 Power Module

Power module uses lithium battery for the cooling system power supply. Currently users usually adopt the polymer lithium battery or 18650 lithium battery. By carrying out a simple charge and discharge circuit, respectively, we used polymer lithium battery and 18650 battery for four charge and discharge experiments. Meanwhile, we charge the battery with a capacity of 12 V for 20 h using a power supply with a output voltage of 12 V. The constant discharge is carried out at a current of 0.5 A through a 6 W bulb, and the discharge time is calculated by observing the lamp discharge and recording the discharge time. Experiments show in the discharge efficiency of polymer lithium batteries higher than 18650 lithium batteries.

3   Conclusion

Fiber optic cable splicing is a common means of connecting optical fiber. We put the designed splicer cooling system into applications in the four areas. The results show that the splicer machine has been running normally, through the temperature sensor based on single-chip test, splicer machine surface temperature between 37°~43° wandering, no downtime. The cable welding heat dissipation system to the greatest extent to meet the production site communication operation and maintenance requirements, effectively improve the summer cable welding efficiency, shorten the outdoor working hours. Its wide applicability, versatility, not only the power system, in the entire communications industry has a broad application prospects.

References

1. Bible, K.W., Nikles, T.: Optical fiber strand and splice sleeve retention tool for use during fiber optic cable splicing (2016)
2. Wesson, L.N.: Field repair options for fiber optic cable; fusion splicing, mechanical splicing, and field termination. In: 2010 IEEE sAvionics Fiber-Optics and Photonics Technology Conference (AVFOP), pp. 55–56 (2010)
3. Fukuda, M., Ito, K.: Heat treatment device for optical fiber reinforcing member, optical fiber fusion splicer provided with same heat treatment device, and method for heat treating optical fiber reinforcing member (2016)
4. Chen, D.Z.: Optical-fiber mechanical splicer using heat-shrink ferrule (2012)
5. Smith, A.V., Smith, J.J.: Mode instability in high power fiber amplifiers. Opt. Express **19**(11), 10180–10192 (2011)
6. Roth, J.M., Parenti, R., Michael, S., et al.: Observations of power-in-fiber statistics in two recent free-space communication link experiments. In: Applications of Lasers for Sensing and Free Space Communications. LSMB3 (2010)
7. Yue, W.U.: Research on the development of power fiber communication technology. Technological Development of Enterprise (2011)

Optimization and Simulation of Key Parameters of Gas - Type Stirling Engine Burner

Wang Qin[✉], Chen Hua, Tian Junwei, Jiang Zhiyi, and He Xiping

School of Mechanical and Electrical Engineering,
Xi'an Technological University, Xi'an 710021, China
wq992514@163.com

Abstract. This paper proposed a structural design scheme of a gas-type Stirling engine burner. According to the characteristics of Stirling engine heating, the more stable the hot side output temperature is, the better Stirling engine is. In order to ensure the feasibility of the design, a model of flat flame burner is established, carried out simulations by FLUENT software, the temperature field of different cross-section is obtained. On this basis, the model of annular burner is established and simulated on the temperature field. Changed the number and location of the gas inlet. The temperature field simulation is carried out under this condition, and the burner structure can be realized to achieve uniform heating. The simulation results show that the theoretical value of temperature uniformity on the hot side of the Stirling engine can reach up to 93%, which can meet the heating requirements of Stirling engine.

Keywords: Gas generator · Temperature field · Stirling engine · Temperature uniformity

1 Introduction

Stirling engine, also known as the heat engine, is an external combustion engine, that is, rely on external heat source to heat the working fluid that sealed in the machine, making the closed loop, to promote the piston work. Stirling has no special requirements for external combustion. As long as the temperature of the external heat source is higher than the temperature of the working fluid in the machine. Therefore, the heating method is also flexible: both fossil fuel, solar energy, and biomass can be used, meanwhile, industrial waste heat with a certain temperature can be also used as heat source. In addition, Stirling engine also has the advantages of high efficiency, low noise, good operating characteristics, low vibration, reliable operation and low maintenance cost. Nowadays, the petrochemical energy crisis is more and more serious, the use of clean energy as a heating source for power generation is increasingly valued, and become a hot topic of national research.

Abroad has matured Stirling engine technology and applied to reality. China's development of the Stirling engine power generation is still mainly in the 1–5 KW level. At present, one large domestic state-owned enterprises has designed a 20 KW

© Springer International Publishing AG 2018
F. Qiao et al. (eds.), *Recent Developments in Mechatronics and Intelligent Robotics*,
Advances in Intelligent Systems and Computing 691, DOI 10.1007/978-3-319-70990-1_54

power generation Stirling engine, is still using electric heating more, because the heating power is not enough, and the temperature uniformity is poor, making the engine difficult to work in the rated power generation [1]. This paper presents a gas-fired furnace method for the Stirling engine hot side of the heating, the burner's design is the key factor to ensure that Stirling engine power generation efficiency, the hot end of the engine is a 350 mm diameter ring end face, the working gas is helium, and the temperature field formed by the heaters from the burner is the focus of this paper.

At present, the main gas burners are: regenerative nozzle burner, flat flame burner (convection heat transfer), ring flat flame burner (convection, radiation mixed heat transfer) these three one. Regenerative nozzle burner heating device is mainly composed of regenerator, burner (nozzle), valve and other components, the commutation time of the reversing valve has a very complicated effect on the performance of the radiant tube, and the control commutation interval is the key of the regenerative burner [1]. Flat flame burner burn flame is a disc-shaped thin layer of flame. The thin layer of flame of this disk is the centrifugal effect of the rotary air, which makes the air flow through a flaring expansion of the mouth of the trumpet to result in [2]. Flat flame burners are generally have simple structure, its maintenance and operation are very convenient; adjustment range is wide, the fuel and air pressure changes have a strong adaptability [3]. Through the experimental study of a variety of flat flame burner, it was found that there was a problem of flame "black heart" (i.e., the center temperature of the burner brick was low), and the heating end of the Stirling engine was just in the middle area. This paper intends to use ring flat flame burner, it is the traditional flat flame burner's structural improvements, in order to achieve a higher uniform flame.

2   The Structure and Modeling of the Burner

2.1   Using High-Quality Flat Flame Burner

The burner uses a special burner body structure to achieve the purpose of flat flame. Burner body structure shown in Fig. 1. In order to ensure the reliability of combustion, flat flame burner instrument design with fire detection sensor. Burner design temperature resistance up to 1600 °C. The burner is designed to withstand temperatures of 1600 °C. The burner is composed of a main gas channel (11) and a main air channel (12). The gas passes through the main channel of the middle part of the burner and enters the combustion area through 14, 16, 17 and 18 holes, the air enters the combustion zone through the 15-hole from the outside of the burner. Air and gas in the combustion area to achieve premixed, ignition and combustion. The cooling air passes through the radial wind to block the flame, does not form a long straight flame, but forms a uniform flat flame in the combustion area.

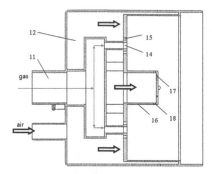

Fig. 1. Schematic diagram of the burner structure

2.2 Burner Modeling, Simulation and Optimization Design

First in GAMBIT in accordance with the design parameters to establish three-dimensional model. By analyzing the gas path of the burner, the simplified model of the burner is obtained. According to the verification results of the burner in this paper, the air - fuel ratio is 10:1, the air flow is 40 m ^ 3/h and the gas flow is 4 m ^ 3/h.

The traditional burner is modeled in GAMBIT. Assuming that the gas inlet diameter is 20 mm, 10 air inlets are distributed around, with a diameter of 20 mm. The rated air-fuel ratio is achieved here by adjusting the inlet speed of air and gas. The model is shown in Fig. 2.

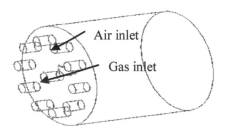

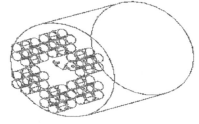

Fig. 2. Traditional burner modeling   **Fig. 3.** Three-dimensional modeling of ring flat flame burner

For the shape of Stirling engine hot side, this paper considers the inlet valve design as a ring, which can maximum fit objects that to be heated. As shown in Fig. 3, a three-dimensional model of a ring-shaped flames burner was established. In order to facilitate simulation analysis, the model was assumed to be 20 mm, and in total of 6; each gas inlets around the cloth are 6 air inlets. Here, the burned air-fuel ratio is adjusted by setting the inlet speed of air and gas. Due to the complexity of the combustion, in this paper, combustion simulation is performed in FLUENT6.3 to verify the rationality of improvement.

Then, divide the three-dimensional model diagram into grid, and define the type of gas inlet, air inlet and outlet in Gambit. Because of the complexity of this geometric structure, in the grid parameter setting, it is using unstructured tetrahedral mesh, to ensure the accuracy of grid division. Import into the unstructured grid that carried out by the pre-processing software GAMBIT. Set the number of surface meshes to 10, and set the import and export border type PRESSURE-OUTLET, import boundary type VELOCITY-INLET.

Import the built-in burner model and the grids that has been divided into FLUENT6.3 solver for the numerical solution. Numerical simulations were performed using the FLUENT solver. In this case, the boundary type is set as the wall boundary; the initial condition is 0; the convergence of the variables is set to 10–5; the iteration step is set to 1000 steps and the 900 steps are converged. In the text box that on the right of Number of Iterations, enter 1000, that is iteration 1000 steps. Click to start the simulation calculation.

First of all, do simulation calculation towards the traditional burner temperature field, this is a list of temperature distributions of the section surfaces of the burner ring at the inlet surface distance of 0.1, 0.2, 0.3 and 0.4 m. As shown in Fig. 4(a–d).

The simulation results show that when the distance from the gas inlet surface is 0.4 m, the flame temperature uniformity is the best, the temperature difference should not more than 500 °C, the flame temperature is the highest and the theoretical maximum temperature is 2000 °C.

At the same time, the temperature profile of the cross section of the burner axis could be obtained, as shown in Fig. 5. It can be seen from the figure, near the right end of the flame, the temperature is highest, this time, at this time, Stirling engine hot end achieve the best heating effect.

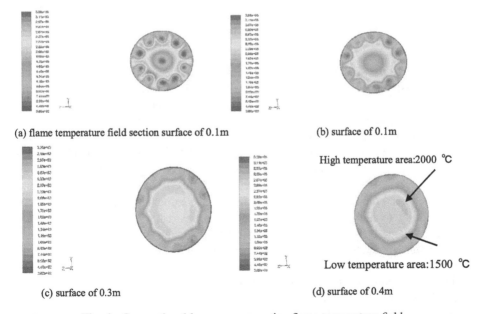

(a) flame temperature field section surface of 0.1m   (b) surface of 0.1m

(c) surface of 0.3m   (d) surface of 0.4m

High temperature area:2000 °C

Low temperature area:1500 °C

Fig. 4. Conventional burner cross section flame temperature field

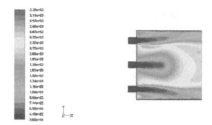

Fig. 5. Burner axis cross section flame temperature field

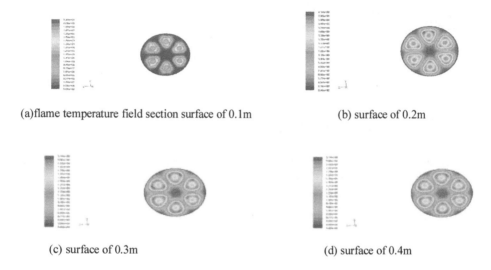

(a)flame temperature field section surface of 0.1m   (b) surface of 0.2m

(c) surface of 0.3m   (d) surface of 0.4m

Fig. 6. Annular flat flame burner section temperature field

Next, make the simulation towards the temperature field of the toroidal burner in Fig. 3. The list of temperature distributions of the section surfaces of the burner ring at the inlet surface distance of 0.1, 0.2, 0.3 and 0.4 m is also be given. As shown in Fig. 6(a–d).

The flame size on the target cross section is shown in Fig. 7(a). The temperature distribution on the cross section of the burner axis is shown in Fig. 7(b). It can be seen from the figure, the temperature near the right end of the flame reach the highest, this time, the top of the Stirling engine heating the best effect. At the same time, there are some problems: 1. The resulting flame shape is not fully consistent with the ring 2. Some part of the area temperature is too low, there are also some high temperature point.

According to the simulation results above, compared with the traditional flame burner, flat flame burner temperature uniformity advantage is clear and stand out, and the optimum temperature range of the burner is 0.4 m away from the burner, where a higher temperature and uniformity can be obtained. Therefore, in the design of Stirling engine heating equipment, should try to ensure that the distance between the engine hot side and the burner is 0.4 meters. In order to get a more optimized structure, this paper changes the structure on the basis of the ring burner and continues the simulation.

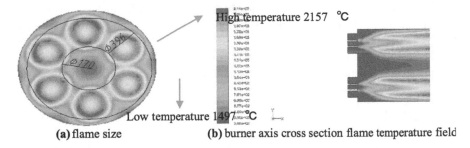

(a) flame size   **(b)** burner axis cross section flame temperature field

Fig. 7. (a) Flame size, (b) burner axis cross section flame temperature field

Assumptions: the gas imports increased to twelve gas imports evenly distributed, the pipe diameter of 20 mm; each gas imports around 8 evenly distributed air imports, diameter is 20 mm. Likewise, the rated air-fuel ratio is achieved by adjusting the inlet speed of air and gas. The burner used for the simulation is consists of a high temperature resistance stainless steel head and a high temperature resistance stainless steel cylinder.

After the simulation calculation, here only lists the temperature field at the distance of 0.4 meters from the burner, as shown in Fig. 8.

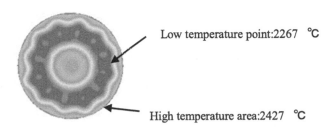

Low temperature point:2267 °C

High temperature area:2427 °C

Fig. 8. Annular flat flame burner for the second design of each section of the flame shape

After optimization, the maximum temperature can reach 2427 °C, the minimum temperature on the torus can reach 2267 °C, the absolute value of temperature is further improved, and the temperature uniformity of the heating end can be improved up to 93%. Could meet the requirements of 20 KW class Stirling engine heating.

3   Conclusion

In this paper, for the characteristics of the gas-type Stirling engine, based on the study of the three burners, through structural modeling and temperature field simulation of the traditional flat flame burner, ring burner, could get these following conclusions:

Ring burner relative to the flat flame burner, can get a higher heating temperature and better temperature heating temperature uniformity;

The best heating distance of the toroidal burner is 0.4 m away from the burner, which can ensure the compact structure, and can also get a more uniform temperature field;

For the annular burner, when the number of gas inlet is increase to 12, a higher heating temperature and a more uniform heating temperature can be obtained.

4   Outlook

1. The practicality of the heater needs to be verified in further combustion tests;
2. The temperature field of the heater can reach the maximum temperature of 2427 °C during simulation, this is the theoretical temperature, in the actual combustion or on the furnace structure design, how to ensure the temperature also need further testing, especially the furnace insulation structure design;
3. In order to achieve a better combustion effect and ensure the complete combustion of gas, during the test, air and gas intake ratio, gas and air premixed problem, should also be fully considered in later furnace structure design.

Funds. The department of project in Shanxi province (NO. 2016GY-175).

References

1. Martin. Development and optimization of a new type of regenerative (HTAC) radiant tube heating device
2. Zheng, Z., Gao, J.: Fast convection heating technology and its application. China Iron Steel Conf. Proc. 238
3. Zhanzeng, L., Cuijiao, D., Zhaoping, Z.: Study on the technology of flat flame burner. Iron Steel Res. **12**(6), 123–147 (2001)
4. de Ojeda, W., Zoldak, P., Espinosa, R., et al.: Development of a fuel injection strategy for partially premixed compression ignition combustion. SAE **1**, 15–27 (2009)
5. Cai, J., Zhou, S., Sun, X.: Design of high temperature burner. Coal Technol. **1**(1), 30–178 (2011)
6. Cuijiao, D., Yanghu, J., Zhaoping, Z., et al.: Experimental study on several burners. Metall. Energy **07**(4), 23–41 (2004)

Research on Attitude Estimation of Quadrotor Based on Improved Complementary Filter

Guo Jiahui[✉]

Academy of PLA, Hefei, Anhui, China
2268539112@qq.com

Abstract. With the development of Micro-Electro-Mechanical System (MEMS), inertial measurement devices can be used on quadrotor. However, because of the problem of the technology, MEMS devices has the problem of being susceptible to interference and data drift. So it hard to use its data to calculate the angle directly. For this problem, this paper present an algorithm of improved complementary filter. This algorithm combines the data of accelerometers, gyroscopes and magnetometers, and also introduces a Fuzzy PI Controller to correct the error between the sensors, it can adjust the parameters of the PI Controller online according to the motion status of the quadrotor. Verified by experiment, this algorithm has the advantages of small computation, fast speed and high precision.

Keywords: Quadrotor · Inertial sensor · Improved complementary filter · Fuzzy PI controller

1 Introduction

In recent years, quadrotor won the wide attention at domestic and foreign because of its vertical takeoff and landing, flexible control and stable flight. It has been widely used in the field of aerial photography, search and rescue. In order to keep the quadrotor flying smoothly, it is necessary to estimate the attitude of the aircraft in real time. But a single MEMS sensor can not provide accurate information for long time because of its low accuracy and data drift. Therefore, the data of accelerometer, gyroscope and magnetometer must be combined.

Aiming at the problem of attitude estimation of quadrotor, This paper present a improved complementary filter algorithm, which fuse the three-axis acceleration, angular velocity and geomagnetic data. In order to suppress the noise of the sensor better, a Fuzzy PI Controller is introduced, which can adjust the parameters of the PI controller according to the motion state of the quadrotor. The improved algorithm offsets the deficiency of the traditional complementary filter algorithm and improves the accuracy of the attitude estimation.

© Springer International Publishing AG 2018
F. Qiao et al. (eds.), *Recent Developments in Mechatronics and Intelligent Robotics*,
Advances in Intelligent Systems and Computing 691, DOI 10.1007/978-3-319-70990-1_55

2 Conversion of Spatial Coordinate Systems

In order to control the flight of quadrotor, it is necessary to calculate the Euler angles of the quadrotor in real time, namely the roll angle (θ), the pitch angle (λ) and the yaw angle (φ). The Euler angle is also the angular relationship between the navigation coordinate system ($X_nO_nY_n$) and the carrier coordinate system ($X_bO_bY_b$). The angular relationship between the two coordinate systems is shown in Fig. 1.

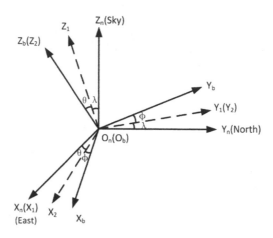

Fig. 1. The angular relationship between the two coordinate systems

The angular relationship between the two coordinate systems can be expressed by the direction cosine matrix(DCM).we can obtain the DCM of the navigation coordinate system to the carrier coordinate system by Euler angles. The DCM can be expressed as formula (1)

$$R_n^b = \begin{pmatrix} \cos\phi\cos\lambda & \cos\phi\sin\lambda\sin\theta - \sin\phi\cos\theta & \sin\phi\sin\theta + \cos\phi\sin\lambda\cos\theta \\ \sin\phi\cos\lambda & \cos\phi\cos\theta + \sin\phi\sin\lambda\sin\theta & \sin\phi\sin\lambda\cos\theta - \cos\phi\sin\theta \\ -\sin\lambda & \cos\lambda\sin\theta & \cos\lambda\cos\theta \end{pmatrix} \tag{1}$$

However, using Euler angle to estimate attitude will appear "Gimbal Lock", the angle of calculation can not exceed 90 degrees. In order to solve this problem, we can use the quaternion to express the attitude of the aircraft. The quaternion Q is a variable of four degrees of freedom and it can be expressed as formula (2).

$$Q = q_0 + i\vec{q}_1 + j\vec{q}_2 + k\vec{q}_3 \tag{2}$$

In the formula (2), q_0 is the scalar part of the quaternion; $\vec{q} = [q_1, q_2, q_3]^T$ is the vector part of the quaternion; i, j, k represents the unit vector of the coordinate axes in the carrier coordinate system. According to literature [1], The DCM can be expressed in quaternion form as Formula (3).

$$R_n^b = \begin{bmatrix} q_0^2 + q_1^2 - q_2^2 - q_3^2 & 2(q_1 q_2 - q_0 q_3) & 2(q_0 q_2 + q_1 q_3) \\ 2(q_0 q_3 + q_1 q_2) & q_0^2 - q_1^2 + q_2^2 - q_3^2 & 2(q_2 q_3 - q_0 q_1) \\ 2(q_1 q_3 - q_0 q_2) & 2(q_0 q_1 + q_2 q_3) & q_0^2 - q_1^2 - q_2^2 + q_3^2 \end{bmatrix} \tag{3}$$

Compared with the method of Euler angle, the method of quaternion avoids the "Gimbal Lock" and the operation of trigonometric functions. So it can also improves the speed of attitude estimation. Because of the these advantages of quaternion, it has been widely used in engineering practice.

3   The Algorithm of Improved Complementary Filter

In order to eliminating the error of drift and the interference of the noise of the inertial sensor, we can use traditional complementary filter to combine the data of senors.

Here, the roll angle θ is taken as an example. Assuming that the transfer function of the low-pass filter is $G_1(S)$, the transfer function of the high-pass filter is $G_2(S)$, and $G_1(S) + G_2(S) = 1$, According to the characteristics of accelerometer and gyroscope, we known that the estimated value of accelerometer $\theta_a(s) = \theta(s) + U_H(s)$ and the estimated value of gyroscope $\theta_\omega(s) = \theta(s) + U_L(s)$. Of which, $U_H(s)$ is the high frequency noise of accelerometer and $U_L(s)$ is the low frequency error of the gyroscope. $\theta(s)$ is the true value of angle. The principle of complementary filter can be expressed as formula (4).

$$\begin{aligned} \hat{\theta}(s) &= \theta_a(s)G_1(s) + \theta_\omega(s)G_2(s) = \theta_a(s)\frac{K}{S+K} + \theta_\omega(s)\frac{S}{S+K} \\ &= \theta(s) + U_H(s)\frac{K}{S+K} + U_L(s)\frac{S}{S+K} \approx \theta(s) \end{aligned} \tag{4}$$

The senor's error can be eliminated by formula (4). But the parameters K of the filter is difficult to choose and its value will be affected by the motive state of the aircraft. So the actual filtering effect is not good as expected. To solve the problem, this paper proposes a improved method of complementary filter based on Fuzzy PI Controller. The Fuzzy PI Controller is used to correct the error between the senors and use the data corrected to compensate the data of the gyroscope. K_p represents the weight of the accelerometer data and K_i can effectively suppress the gyroscope's error of drift. The Fuzzy PI Controller can effectively filter out the high-frequency and low-frequency noise. And the parameters of the PI controller can be adjusted online according to the flight status of the quadrotor, which achieve the purpose of parameter adaptation and improves the accuracy of attitude estimation. The process of attitude estimation is shown in Fig. 2.

When the quadrotor is in the state of static or hover, the Fuzzy PI Controller can appropriately increase the weight of the accelerometer data. On the contrary, the Fuzzy PI Controller will decrease the weight of the accelerometer data when the aircraft is in the state of motion. The angular velocity ω is used to express the state of motion in this paper. The structure of Fuzzy PI Controller is shown in Fig. 3.

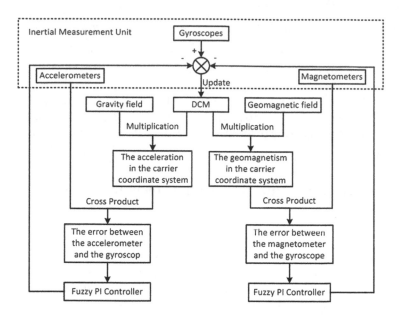

Fig. 2. The process of attitude estimation

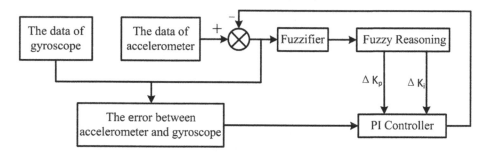

Fig. 3. The structure of Fuzzy PI Controller

The initial parameter of K_p of the Fuzzy PI controller is chosen to be 6, and the Ki is chosen to be 0.005, which the effect is better when the quadrotor is in the state of static. According to the range of the gyroscope, the universe of discuss of angular velocity ω is selected as [0,1000], the ΔK_p's universe of discuss is selected as [−1,0] and the ΔK_i's universe of discuss is selected as [−0.002,0]. Considering the speed and accuracy of attitude estimation, ω, ΔK_p and ΔK_i can be divided into four levels, each level corresponds to a fuzzy set. We can use neglarge (NL), negmedium (NM), negsmall (NS), zero (ZO), possmall (PS), posmedium (PM), poslarge(PL) to describe the variables of the system. So the fuzzy set of ω can be expressed as {ZO,PS,PM,PL}. The fuzzy set of ΔK_p and ΔK_i can be expressed as {NL,NM,NS,ZO}. In order to reduce the amount of calculation, the form of trapezoidal membership function are selected as the membership function of variables. The membership function of ω, ΔK_p and ΔK_i is shown in Figs. 4, 5, 6.

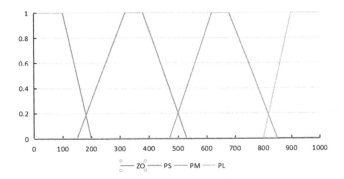

Fig. 4. The membership function of ω

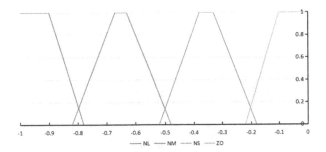

Fig. 5. The membership function of ΔK_p

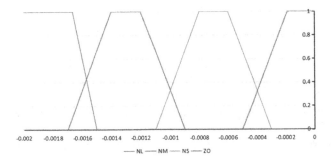

Fig. 6. The membership function of ΔK_i

For Fuzzy PI Controller, it only outputs the adjustment of the two parameters. Which eventually affects the accuracy of the attitude estimation is the actual parameters of the PI Controller. Therefore, according to the effect of each parameter of PI Controller to establish the fuzzy rules. When the quadrotor moves violently, larger $\triangle$Kp and $\triangle$Ki will be selected to decrease the wight of the accelerometer data; While the aircraft moves smoothly, smaller $\triangle$Kp and $\triangle$Ki will be selected. The fuzzy rules are shown in Table 1.

Table 1. Fuzzy Rules

ω	ZO	PS	PM	PB
$\triangle$Kp	ZO	NS	NM	NL
$\triangle$Ki	ZO	NS	NM	NL

4   Experimental Verification

In order to verify the accuracy and feasibility of the improved algorithm. The control board of quadrotor is built based on STM32. There are three-axis accelerometer, three-axis gyroscope and three-axis magnetometer on the board. MPU6050 and HMC5883 is selected as inertial measurement unit. The MCU reads the data through the I2C bus, and send back data to the host computer through the serial port. The hardware of control board is shown in Fig. 7.

Fig. 7. The hardware of control board

According to the hardware platform to write program, Set the frequency of data acquisition and attitude estimation is 500 Hz. In order to simulate the situation of aircraft in flight as much as possible, we fixed the control board on the turntable and made the turntable spin around the X axis of board. The angle of rotation is limited between −45 degrees and 45 degrees, and the speed of rotation will be faster and faster. The acceleration, angular velocity and geomagnetic data collected by experiment are shown in Fig. 8.

On the basis of the experimental data, take the roll angle as an example, the algorithm of traditional complementary filter and improved complementary filter are used to deal with the data. The results are shown in Fig. 9.

We can see the result of calculation from the Fig. 9, when the turntable is in a quiescent state, the accuracy of the two algorithms is not much different. However, with the increase of angular velocity, the angle calculated by the algorithm of traditional complementary filter begins to deviate from the true value because of its fixed parameters. Especially when the turntable is suddenly turning in the opposite direction. The degree calculated by the traditional algorithm is only about 35 degree, this is much smaller than the true angle of the turntable. While the algorithm of improved complementary filter

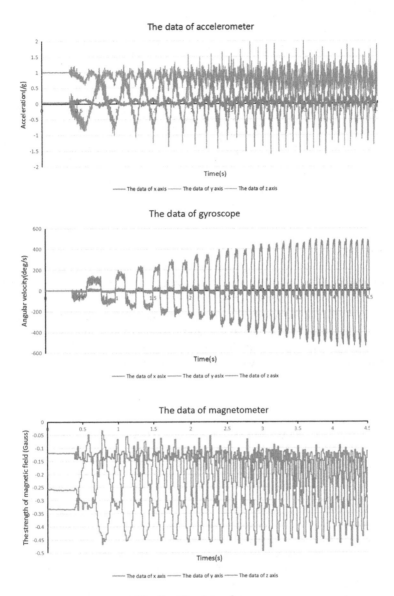

Fig. 8. The data of senors

can adjust the parameters online according to the state of aircraft. So, it can get good accuracy of calculation either in the conditions of movement or quiescency.

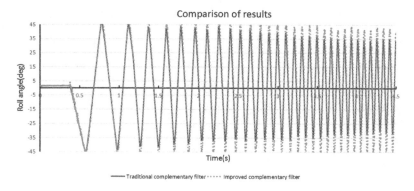

Fig. 9. The comparison of results

5   Conclusion

Aiming at the difficulty of the attitude estimation of quadrotor. This paper improve the algorithm of traditional complementary filter and the Fuzzy PI controller is introduced to correct the error between the senors. Proved by experiment, the accuracy of the improved algorithm is higher than the traditional one and can meet the requirements of the quadrotor's attitude control. Meanwhile, the hardware based on STM32 has been built. It provides a prerequisite for the study of the algorithms of quadrotor's attitude control.

References

1. Chen, M., Xie, Y., Chen, Y.: Attitude estimation of MEMS based on improved quaternion complementary filter. J. Elect. Meas. Instrum. **29**(9), 1391–1397 (2015)
2. Zihao, X., Zhou, Z., Sun, L.: Attitude measure system based on KF. Elect.Meas. Technol. **37**(1), 1–4 (2014)
3. Zhao, H., Wang, Z.: MEMS senors based on attitude measurement system using UKF. Chin J Senors Actuators **4**(5), 643–646 (2011)
4. Yibo, L., Qiling, X., Jianda H.: Modeling and PID Control of Tethered Unmanned Quadrotor Helicopter Mechatronic, Sciences, Electric, Engineering, and, Computer (MEC). In: Proceedings 2013 International Conference on.IEEE, 2013 pp. 58–262
5. Ruicai, J.: Attitude estimation algorithm for low cost MEMS based on quaternion EKF. Chin. J. Sensors Actuators **27**(1), 90–95 (2007)
6. Yibo, F., Shengxi, L.: Adaptive compensation method for heading error of, electronic. Magnetic compass. Chin J. Sci Instrum **35**(11), 2607–2614 (2014)
7. Jingqiu, S., You Youpeng, F., Zhongyun, F.: Attitude estimation based on conjugate gradient and complementary filter. Chin J Sensors Actuators **27**(4), 524–528 (2014)
8. Ronghui, Z., Hongguang, J., Tao, C., et al.: Attitude solution for strapdown inertial navigation system based on quaternion algorithm. Opt. Precis. Eng. **6**(10), 963–1970 (2008)
9. Qing, Y.: Inertial navigation. Science Press, Beijing (2006)
10. Marins, J.L., Yun, X., Bachmann, E.R., et al.: An extended Kalman filter for quaterion-based orientation estimation using MARG sensors. Intelligent Robots and Systems, 2001. In: Proceedings of 2001 IEEE/RSJ International Conference on IEEE, 2001,4: 2003–2011

Optimization of Vehicle Aerodynamic Drag Based on EGO

Chenguang Lai, Chengping Yan$^{(\boxtimes)}$, Boqi Ren, Bao Lv, and Yuting Zhou

Chongqing University of Technology, Chongqing 400054, China
yanchengping@foxmail.com

Abstract. In this paper, the improved EGO global optimization algorithm, based on the Kriging response surface and EI function, was used to complete the aerodynamic drag reduction in the design space of a vehicle combined with data mining technology. The EGO algorithm can usually achieve the global optimum with minimum function evaluations. Data mining technologies provide a method to uncover the influence mechanisms of design variables on aerodynamic drag and to analyze the relationship between variables. Aerodynamic drag of the optimal design is 1.56% lower than that of the original model. The data mining results show that the engine hood inclination and the tail angle play a leading role in the vehicle's aerodynamic drag, and the hood inclination has the greatest impact. The method can efficiently solve the computationally expensive black box problems such as vehicle aerodynamic design optimization.

Keywords: Global optimization · Kriging model · EI function · Drag reduction · Data mining

1 Introduction

Aerodynamic optimization is a typical nonlinear computationally expensive and black box problem. Since they usually yield better designs and time-saving, design methods that combine modern optimization algorithms and advanced numerical simulation techniques are more and more used [1, 2]. Modern optimization algorithms such as evolutionary algorithms and genetic algorithms have good global convergence in solving nonlinear multi-modal problems [3, 4]. The biggest obstacle to their use in vehicle aerodynamic optimization is that they need too much time and resources caused by excessive calls of evaluation operator and CFD calculation.

Approximate model is used to establish the mapping relation between input and output instead of CFD calculations. Considered both the optimal value of the response surface and its uncertainty, an efficient global optimization (EGO) algorithm was proposed by Jones et al. [5], which can obtain the global optimal solution with few function evaluations. In addition, Simpson et al. pointed out that data mining technologies can effectively interpret data feature in the design space [6].

On the basis of EGO algorithm, this paper improved the sampling method and the optimization method of EI function. The new algorithm uses very few sample points to complete the global optimization of the aerodynamic drag reduction of a vehicle.

© Springer International Publishing AG 2018

F. Qiao et al. (eds.), *Recent Developments in Mechatronics and Intelligent Robotics*,
Advances in Intelligent Systems and Computing 691, DOI 10.1007/978-3-319-70990-1_56

The data mining technique analysis of variance (ANOVA) [7] is used to study the influence mechanisms of the design variables on aerodynamic drag.

2   EGO Global Optimization Method

Based on the Kriging response surface and EI function, EGO algorithm can get the global optimal solution with less evaluation times than the other optimization methods. Fitting the Kriging response surface and the optimization method of EI function are the key technologies and they have significant influences on the efficiency and accuracy of the algorithm.

2.1   Kriging Response Surface

Since its high accuracy and can provide the predicted value and its error of the predicted position at the same time, Kriging model is very suitable for global optimization algorithm. In this paper, Kriging response surface was used to establish the mapping relation between input and output instead of CFD calculations.

Suppose that Y is the real value of the initial sample set S = {X(1), X(2), ..., X(n)}, the unknown objective function value y(X(i)) for X(i) can be predicted as:

$$y(X^{(i)}) = \mu + \epsilon(X^{(i)}) \tag{1}$$

Where X denotes the m-dimensional vector; μ denotes the mean of the stochastic process, which is taken as a constant; $\epsilon(X^{(i)})$ denotes the error term, which follows the normal distribution i.e. $\epsilon(X^{(i)}) \sim N(0, \sigma^2)$, σ^2 denotes the process variance of kriging model.

Kriging method assumes that $\epsilon(X^{(i)})$ and $\epsilon(X^{(j)})$ are spatially correlated. Let θ_h denotes the correlation coefficient, and gives $p_h = 2$, the correlation defined as:

$$corr\left[\epsilon(X^{(i)}), \epsilon(X^{(j)})\right] = exp\left[-\sum\nolimits_{h=1}^{m} \theta_h \left|x_h^{(i)} - x_h^{j}\right|^{p_h}\right] \tag{2}$$

Usually we call the above two expressions DACE stochastic process models. The m + 2 unknowns ($\mu, \sigma^2, \theta_1, ..., \theta_m$) in the model are valued by the maximum likelihood function. Let R denotes the correlation matrix composed of the correlation functions of each sample point in training sample set S, and r_i denotes the correlation vector between the unknown point X(i) and the sample points in S. The best linear unbiased estimation of the objective function y(X^{(i)}) is:

$$y(X^{(i)}) = \mu + r_i R^{-1}(y - l\mu) \tag{3}$$

2.2 EI-Based Infill Criterion

The EGO method uses the "expected improvement" (EI) [5] as the quality factor. The search strategy (infill criterion) here is adding the point with maximum EI (maxEI) values to the training sample set and re-iteration. Suppose that the predictor y(x) follows the normal distribution, i.e. $Y \sim N(\hat{y}, s^2)$, $\hat{y}$ denotes the predicted value of the DACE model at x, s denotes the standard deviation, the expectation that the predicted value is less than the current optimal value f_{min} can be defined as:

$$E[I(X)] = \left(f_{min} - \hat{y}\right)\Phi\left(\frac{f_{min} - \hat{y}}{s}\right) + s\varphi\left(\frac{f_{min} - \hat{y}}{s}\right) \tag{4}$$

Where $\varphi(\cdot)$ and $\Phi(\cdot)$ denote the density function and the accumulation function of normal distribution respectively. Because expression considers both the predicted value and its error of the predicted position at the same time, so the EI function can balance local and global search. The EI function has strong robustness, and greatly reduces the number of iterations required for optimization.

3   Aerodynamic Drag Reduction Optimization and Data Mining

3.1   Original Model and Design Variables

The original model is 1:1 model of a real vehicle (see in Fig. 1). Four important parameters in the Y-axis direction are selected as design variables. Firstly, commercial software HyperMesh was used to generate the mesh on the surface of the vehicle. Then, the four control volumes (see in Fig. 1) were used to control the parametric deformations of the four variables (see in Fig. 2), respectively.

Fig. 1. Original model and the control volumes

It should be noted that the 4 design variables, respectively, stand for the (engine) hood inclination, rear window inclination, trunk height and the tail angle. Since the difficulty of real vehicle parameterization, parameters x_1, x_2, x_3, x_4 were used to quantify the deformation of the 4 variables. Table 1 shows the range of the 4 parameters, and all the parameters of the original model were defined as 0.

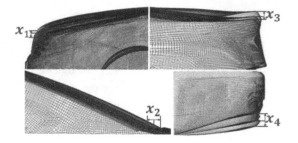

Fig. 2. Four variables and their deformations

Table 1. Range of the parameters

Parameter	Interval (mm)
x_1	$(-50,30)$
x_2	$(-30,30)$
x_3	$(-10,20)$
x_4	$(-40,60)$

3.2 Optimization Process

The objective function value (aerodynamic drag) is calculated by CFD using Commercial software Fluent. Aerodynamic drag of the original vehicle is 319.45 N.

After the design variables are determined, the optimization process is as follows:

Step 1: The initial samples set S are constructed by a sample strategy named uniform design (UD) [8], and the real function values are evaluated by CFD.
Step 2: A kriging response surface is fitted by the training samples set S.
Step 3: Determine whether the response surface meets the accuracy requirements and the final stop criterion or not. If satisfied, the algorithm stops and outputs the optimal design; otherwise the point with maxEI will be added to the training samples set S, and return to step 2.

It should be noted that the final stop criterion for this optimization is that the relative errors of the three successive predicted optimal values are less than 0.1%. Unlike traditional EGO algorithm, the initial designs were constructed using a programmed UD code and a self-programming code of self-adaptive differential evolution named SMODE [9] was used to search the point of maxEI.

3.3 Cross Validation of the Response Surface

In this paper, 21 initial sample points were used to fit the kriging response surface in the beginning. Then 8 other points with maxEI were added to the training samples set S in order to meet the accuracy requirements and the final stop criterion. 29 sample points were used fit the response surface at last.

The accuracy of response surface was diagnosed by a procedure, called "cross validation" [5] (see in Fig. 3). All the standardized cross-validated residuals are in the interval [−3, +3], which means that confidence value of the DACE model can reach 99.7% and the model is valid. The fitted response surface meets the accuracy requirements and can be used for further analysis.

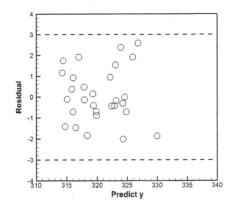

Fig. 3. Cross validation of the DACE model: standardized cross-validated residuals versus cross-validated prediction

3.4 Optimization Result and Data Mining

Table 2 shows the optimal design and its real drag calculated by CFD. Compared with the original model, the drag of optimal design decreased by 1.56%.

Table 2. Details of the optimal design

x_1 (mm)	x_2 (mm)	x_3 (mm)	x_4 (mm)	Predicted drag	Real drag	Prediction error
−33.70	16.31	−10.00	−40.00	314.09 N	314.46 N	0.12%

ANOVA is carried out to investigate the interactions between the parameters and objective function in a quantitative way and intuitively reflect the sensitivities of the objective function to each variable (see in Fig. 4).

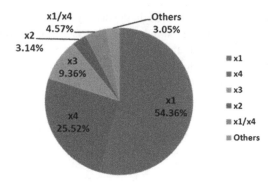

Fig. 4. ANOVA results

There should be 4 main effect and 6 joint effect caused by the 4 parameters. Here, only the effects with its variance proportion $\frac{\delta_i^2}{\delta^2}$ and $\frac{\delta_{ij}^2}{\delta^2}$ larger than 1% are listed in Fig. 4. The variance proportion of x_1 reached 54.36% and it has the greatest effect on the objective function. Except for x_1/x_4, variance proportions of other joint effects are less than 1% and their sum is only 3.05%. The sum variance proportion of x_1, x_4 and x_1/x_4 reached 84.45%. Thus, it can be found that the engine hood inclination (x_1) and the tail angle (x_4) play a leading role in the vehicle's aerodynamic drag.

4   Conclusion

(1) The improved EGO algorithm was used to complete the 4-dimensional drag reduction optimization of a vehicle and only 29 sample points were used to fit the response surface. It means that the improved EGO algorithm can achieve the global optimum with minimum function evaluations.
(2) The aerodynamic drag of the optimal design in the design space is 314.46 N, which is 1.56% less than that of the original model.
(3) Data mining results show that the engine hood inclination and the tail upturn angle play a leading role in the vehicle's aerodynamic drag, and the hood inclination has the greatest impact.

To summarize, the method, which combine the improved EGO global optimization algorithm and data mining technology, can efficiently solve the computationally expensive black box problems such as automotive aerodynamic design optimization.

Acknowledgments. This research was supported by Key Technology Innovation Projects of Key Industries in Chongqing (cstc2015zdcy-ztzx60011) and Graduate Innovation Fund of Chongqing University of Technology (YCX2016109). Part of the work was carried out under the Collaborative Research Project of the Institute of Fluid Science, Tohoku University, Japan.

References

1. Xiong, J., Liu, F., Mcbean, I.: Three-dimensional aerodynamic design optimization of a turbine blade by using an adjoint method. J. Turbomach. **133**(1), 651–664 (2009)
2. Guo, Z.D., Song, L.M., Li, J., et al.: Meta model-based global design optimization and exploration method. Tuijin Jishu J. Propuls. Technol. **36**(2), 207–216 (2015)
3. Doorly, D.J., Peiró, J.: Supervised parallel genetic algorithms in aerodynamic optimisation. In: Artificial Neural Nets and Genetic Algorithms, pp. 229–233. Springer, Vienna (2006)
4. Ahn, C.S., Kim, K.Y.: Aerodynamic design optimization of a compressor rotor with Navier–Stokes analysis. Proc. Inst. Mech. Eng. Part A J. Power Energy **217**(2), 179–183 (2003)
5. Jones, D.R., Schonlau, M., Welch, W.J.: Efficient global optimization of expensive black-box functions. J. Global Optim. **13**(4), 455–492 (1998)
6. Simpson, T.W., Toropov, V., Balabanov, V., et al.: Design and analysis of computer experiments in multidisciplinary design optimization: a review of how far we have come or not. AIAA 2008-5802
7. Schonlau, M., Welch, W.J.: Screening the Input Variables to a Computer Model via Analysis of Variance and Visualization Screening, pp. 308–327. Springer, New York (2006)
8. Fang, K.T., Winker, P., Zhang, Y.: Uniform design: theory and application. Technometrics **42**(3), 237–248 (2000)
9. Song, L.M., Luo, C., Li, J.: Automated multi-objective and multidisciplinary design optimization of a transonic turbine stages. J. Power Energy **226**(2), 262–276 (2012)

Comprehensive Evaluation Research of Campus Network Based on Rough Set

Xia Ouyang and Jiaming Zhong[(✉)]

School of Economics and Management, Xiang Nan University,
Chenzhou, Hunan, China
jmzhongcn@163.com

Abstract. This paper proposes comprehensive evaluation method of campus network based on rough set theory, to make the evaluation of campus network more scientific and objective. At last, examples have proved the practicality and feasibility of this method.

Keywords: Rough set · Campus network · Comprehensive evaluation

1 Introduction

Comprehensive evaluation refers to evaluate matters in an all- around way and from all sides by making use of several indexes, which is an important method for cognition [1]. Foreign and domestic scholars have conducted a large amount of research work on multi- objective comprehensive evaluation method, and thus have proposed many methods such as balanced scorecard [2], gray synthesized evaluation methods [3], fuzzy synthesized evaluation methods [4], analytic hierarchy process [5], artificial neural network [6] and data envelopment analysis method [7]. But these above mentioned methods have some limitations in terms of weight settings and determination of one- factor evaluation matrix [8]. According to object class ability of rough set theory, by making use of knowledge dependence and importance measurement method for attribute, this paper offers a multi-index comprehensive evaluation method that completely driven by data. This method has overcome subjectivity and fuzziness of the above mentioned evaluation method, featuring sound self- study ability.

2 Basic Concept of Rough Set and its Correlation Theory

2.1 Knowledge Expression System [9]

Definition 1. Quadruple $S = (U, A, f, V)$ is named as one knowledge expression system, Where U indicates nonempty finite set of objects, i.e. $U = \{x_1, x_2, \ldots, x_m\}$, which is called domain of discourse; A denotes nonempty finite set, i.e., $A = \{a_1, a_2, \ldots, a_n\}$, and each $a_i (i \leq n)$ is called an property; $V = \underset{a \in A}{\cup} V_a$, V_a is the range of property a; $f : U \times A \rightarrow V$ is an knowledge expression function, which endows an knowledge expression value to each property of all objects, i.e., $\forall a \in A, x \in U, f(x, a) \in V_a$.

F. Qiao et al. (eds.), *Recent Developments in Mechatronics and Intelligent Robotics*,
Advances in Intelligent Systems and Computing 691, DOI 10.1007/978-3-319-70990-1_57

$\forall P \subseteq A$, non-differentiable relationships ind(P) of property set P is defined as:

$$ind(P) = \{(x,y) \in U \times U | \forall a \in P, f(x,a) = f(y,a)\} \tag{1}$$

if $(\alpha, \beta) \in ind(P)$, then α and β are non-differentiable to P.

2.2 Attribute Reduction [9]

Definition 2. Suppose $S = (U, A, f, V)$ be one knowledge expression system, $\alpha \in A$, if $ind(A - \{\alpha\}) = ind(A)$, then α is unnecessary in A; otherwise α is necessary in A.

Unnecessary property is redundant in knowledge expression system, if which is deleted from the knowledge expression system, it will not change the class ability of knowledge expression system. On the contrary, if α necessary property is deleted from knowledge expression system, then the class ability of knowledge expression system must be changed.

Definition 3. Suppose $S = (U, A, f, V)$ be an knowledge expression system, if $\forall a \in A$ is necessary in A, then property set A is independent; otherwise A is relevant.

For relative property set, unnecessary property are included, which can be reduced.

Definition 4. Suppose $S = (U, A, f, V)$ be an knowledge expression system, the set that combined by all required property in A is called as the core of property set A, which can be written as $core(A)$.

2.3 Knowledge's Dependence and Attribute's Importance

Definition 5 Suppose P and Q be relation of equivalence in U, P positive region of Q is represented as $pos_P(Q)$, that is

$$pos_P(Q) = \bigcup_{X \in U/Q} \underline{P}X \tag{2}$$

The degree of dependency between Q and P is

$$\gamma_P(Q) = \frac{|pos_P(Q)|}{|U|}, \quad \text{wherein} \quad 0 \le \gamma_P(Q) \le 1 \tag{3}$$

3 Comprehensive Evaluation Model of Campus Network Based on Rough Set

3.1 Determination of Index Weight

Definition 6 [10] Suppose $S = (U, A, f, V)$ be an knowledge expression system, $A = \{a_1, a_2, \cdots, a_n\}$, the importance of property $a_i \in A$ in A is $r(a_i)$, then weight of $a_i \in A$ is defined as:

$$\omega(a_i) = \frac{r(a_i)}{\sum\limits_{i=1}^{n} r(a_i)} \tag{4}$$

3.2 Evaluation Value Formula [11]

Let $\omega(A_i)$ be preliminary index after knowledge expression system being reduced, and $\omega(a_{ij})$ the secondary index weigh ('n' and 'm_i' (i = 1,2,...,n) are numbers of preliminary index and secondary index after being reduced), v_{ij} is the value of its object under attribute 'a_{ij}',

$$V_1 = (\omega(a_{11}), \omega(a_{12}), \cdots, \omega(a_{1i_1})) \begin{pmatrix} v_{11} \\ v_{12} \\ \vdots \\ v_{1i_1} \end{pmatrix} \tag{5}$$

$$V_n = (\omega(a_{n1}), \omega(a_{n2}), \cdots, \omega(a_{ni_n})) \begin{pmatrix} v_{n1} \\ v_{n2} \\ \vdots \\ v_{ni_n} \end{pmatrix} \tag{6}$$

then evaluation value of every object can be applied to the following formula:

$$V = (V_1, V_2, \cdots, V_n) \begin{pmatrix} \omega(A_1) \\ \omega(A_2) \\ \vdots \\ \omega(A_n) \end{pmatrix} \tag{7}$$

3.3 Basic Thinking and Steps

Construction of campus network comprehensive evaluation model of the basic ideas and steps can be summarized as follows [10, 11]:

(1) the analysis of all factors affecting the campus network, establish campus network index system; (2) to build a knowledge expression system; (3) the reduction of knowledge expression system index system; (4) reduction of properties; (5) calculating the weight of the index; (6) object evaluation.

4 Imitate Models

Evaluation index system of campus network that built by utilizing references [12] is showed as Table 1.

Table 1. Campus network evaluation of index system in College and University

Content		Result			
Primary indicator	Secondary indicator	Excellent	Good	Ordinary	Poor
Index item	Index item				
Information infrastructure (I_1)	Per capita possession amount of computer product (i_{11})				
	Computer network proportion (i_{12})				
	Network performance level (i_{13})				
	Information safety insurance (i_{14})				
Application system (I_2)	Construction of office and business application system (i_{21})				
	Construction of information resource and basic data (i_{22})				
	Construction of information public (i_{23})				
	Portal website construction (i_{24})				
Cost and income (I_3)	Software, hardware, personnel training, and operation input (i_{31})				
	Campus network and management efficiency enhancement (i_{32})				
	Enhancement of scientific decision level (i_{33})				
	Satisfaction degree of teachers, students, and the society (i_{34})				
Safety system (I_4)	Views and talent environment (i_{41})				
	Information planning and enforcing standard (i_{42})				
	Standards application (i_{43})				
	Security safeguard and technical support (i_{44})				

4.1 Simplification of Rough Set Theory to Index System

(1) Construct knowledge expression system S

At first, regard all the secondary indicators in Table 1 as attributes set of knowledge expression system, that is $A = \{I1, I2, I3, I4\} = \{\{i_{11}, i_{12}, i_{13}, i_{14}\}, \{i_{21}, i_{22}, i_{23}, i_{24}\}, \{i_{31}, i_{32}, i_{33}, i_{34}\}, \{i_{41}, i_{42}, i_{43}, i_{44}\}\}$. Individual Campus networks to be evaluated are regarded as object set of the system. In the evaluation, the result accords with each secondary index, or the attribute value, is measured by the level of "excellent", "good", "ordinary" and "poor", which can be replaced by "4", "3", "2", "1" respectively, thus forming an knowledge expression system S as Table 2.

Table 2. Preliminary knowledge expression system S

U	i_{11}	i_{12}	i_{13}	i_{14}	i_{21}	i_{22}	i_{23}	i_{24}	i_{31}	i_{32}	i_{33}	i_{34}	i_{41}	i_{42}	i_{43}	i_{44}
x_1	3	2	4	3	3	2	4	4	1	4	2	3	2	4	3	3
x_2	4	3	2	1	1	3	2	2	1	2	3	4	1	2	1	4
x_3	2	2	3	2	2	2	1	3	3	3	2	2	3	3	2	2
x_4	3	2	4	3	3	2	4	4	1	4	2	3	3	4	3	3
x_5	2	3	2	1	1	3	2	2	1	2	3	2	1	2	1	2
x_6	2	4	4	2	2	4	1	4	3	4	4	2	3	4	2	2
x_7	4	1	2	3	3	1	3	2	2	2	1	4	2	2	3	4
x_8	4	1	2	3	3	1	2	2	2	2	1	4	2	2	3	4
x_9	2	4	3	2	2	4	1	3	3	3	4	2	3	3	2	2

(2) Deletion of Index System for Preliminary knowledge expression system S

With regard to knowledge expression system $S = (U, A, f, V)$, if the attribute value of evaluation objects that accord with attribute $x, y \in A$ is the same, then attribute x and y are regarded that they have the same resolving power, so only one need to be reserved. After deleting correlative lines, knowledge expression system can be primarily simplified to curtail corresponding primary index system. The preserved secondary indexes are: i_{11}, i_{12}, i_{21}, i_{23}, i_{31}, i_{32}, i_{41}.

(3) Attribute Reduction

According to indiscernible relations, we can get that

$$U/ind(R) = \{\{x_1\}, \{x_2\}, \{x_3\}, \{x_4\}, \{x_5\}, \{x_6\}, \{x_7\}, \{x_8\}, \{x_9\}\},$$
$$U/ind(R - \{i_{11}\}) = \{\{x_1\}, \{x_2\}, \{x_3, x_5\}, \{x_4\}, \{x_6\}, \{x_7\}, \{x_8\}, \{x_9\}\},$$
$$U/ind(R - \{i_{12}\}) = \{\{x_1\}, \{x_2, x_9\}, \{x_3\}, \{x_4\}, \{x_5\}, \{x_6\}, \{x_7\}, \{x_8\}\},$$
$$U/ind(R - \{i_{21}\}) = \{\{x_1\}, \{x_2\}, \{x_3\}, \{x_4\}, \{x_5\}, \{x_6\}, \{x_7\}, \{x_8\}, \{x_9\}\},$$
$$U/ind(R - \{i_{23}\}) = \{\{x_1\}, \{x_2\}, \{x_3\}, \{x_4\}, \{x_5\}, \{x_6\}, \{x_7, x_8\}, \{x_9\}\},$$
$$U/ind(R - \{i_{31}\}) = \{\{x_1\}, \{x_2\}, \{x_3\}, \{x_4\}, \{x_5\}, \{x_6\}, \{x_7\}, \{x_8\}, \{x_9\}\},$$
$$U/ind(R - \{i_{32}\}) = \{\{x_1\}, \{x_2\}, \{x_3\}, \{x_4\}, \{x_5\}, \{x_6, x_9\}, \{x_7\}, \{x_8\}\},$$
$$U/ind(R - \{i_{41}\}) = \{\{x_1, x_4\}, \{x_2\}, \{x_3\}, \{x_5\}, \{x_6\}, \{x_7\}, \{x_8\}, \{x_9\}\}.$$

It can be seen that attribute a_{21} and a_{31} are unnecessary, while attribute $i_{11}, i_{12}, i_{23}, i_{32}$ and a_{41} are necessary. Therefore, the core of knowledge expression system S is $core(A) = \{i_{11}, i_{12}, i_{21}, i_{32}, i_{41}\}$, and there are to reductions: $R_1 = \{i_{11}, i_{12}, i_{21}, i_{23}, i_{32}, i_{41}\}$ and $R_2 = \{i_{11}, i_{12}, i_{23}, i_{31}, i_{32}, i_{41}\}$. Only R_1 is considered in this paper, the corresponding Table 2 of knowledge expression system S can be further deleted as Table 3, whose corresponding primary and secondary index set are $\{I_1, I_2, I_3, I_4\} = \{\{i_{11}, i_{12}\}, \{i_{21}, i_{23}\}, \{i_{31}\}, \{i_{41}\}\}$.

Table 3. Knowledge expression system after final reduction

	i_{11}	i_{12}	i_{21}	i_{23}	i_{31}	i_{41}
x_1	3	2	3	4	4	3
x_2	2	2	2	1	2	3
x_3	4	3	1	2	3	1
x_4	3	2	3	4	4	2
x_5	2	3	1	2	2	1
x_6	2	4	2	1	4	3
x_7	4	1	3	3	2	2
x_8	4	1	3	2	2	2
x_9	2	4	2	1	3	3

(4) Calculation of Index Weight

$$U/ind(I_1) = \{\{x_1, x_4\}, \{x_2\}, \{x_3\}, \{x_5\}, \{x_6, x_9\}, \{x_7, x_8\}\},$$
$$U/ind(I_1 - \{i_{11}\}) = \{\{x_1, x_2, x_4\}, \{x_3, x_5\}, \{x_6, x_9\}, \{x_7, x_8\}\},$$
$$U/ind(I_1 - \{i_{12}\}) = \{\{x_1, x_4\}, \{x_2, x_5, x_6, x_9\}, \{x_3, x_7, x_8\}\},$$
$$U/ind(I_2) = \{\{x_1, x_4\}, \{x_2, x_6, x_9\}, \{x_3, x_5\}, \{x_7\}, \{x_8\}\},$$

$$U/ind(I_2 - \{i_{21}\}) = \{\{x_1, x_4\}, \{x_2, x_6, x_9\}, \{x_3, x_5, x_8\}, \{x_7\}\},$$
$$U/ind(I_2 - \{i_{23}\}) = \{\{x_1, x_4, x_7, x_8\}, \{x_2, x_6, x_9\}, \{x_3, x_5\}\},$$

$$\omega(i_{11}) = 0.35, \omega(i_{12}) = 0.65, \omega(i_{21}) = 0.31, \omega(i_{23}) = 0.69, \omega(i_{32}) = 1, \omega(i_{41}) = 1.$$

$$U/ind(A) = \{\{x_1\}, \{x_2\}, \{x_3\}, \{x_4\}, \{x_5\}, \{x_6\}, \{x_7\}, \{x_8\}, \{x_9\}\},$$
$$U/ind(A - \{I_1\}) = \{\{x_1\}, \{x_2, x_9\}, \{x_3, x_5\}, \{x_4\}, \{x_6\}, \{x_7\}, \{x_8\}\},$$

$$U/ind(A - \{I_2\}) = \{\{x_1\}, \{x_2\}, \{x_3\}, \{x_4\}, \{x_5\}, \{x_6\}, \{x_7, x_8\}, \{x_9\}\},$$
$$U/ind(A - \{I_1\}) = \{\{x_1\}, \{x_2, x_9\}, \{x_3, x_5\}, \{x_4\}, \{x_6\}, \{x_7\}, \{x_8\}\},$$
$$U/ind(A - \{I_1\}) = \{\{x_1\}, \{x_2, x_9\}, \{x_3, x_5\}, \{x_4\}, \{x_6\}, \{x_7\}, \{x_8\}\}.$$

$$\omega(I_1) = 0.3, \omega(I_2) = 0.3, \omega(I_3) = 0.2, \omega(I_4) = 0.2.$$

(5) According to Formula (3, 4),

$$V = 3.765,$$

Therefore, this campus network is good.

5 Conclusions

To determine objective weight by rough set, Based on this, evaluation value formula of object can be formed, thus the comprehensive evaluation model of objects can be obtained. It can be said that the algorithm and the model have strict theoretical basis and logic deduction, thus making the evaluation to campus network more scientific and objective. Examples prove that evaluation is feasible and efficient, which offers brand new method and means to evaluation. This thought is also adaptable to the comprehensive evaluation in other fields.

Acknowledgment. Thanks for being supported by the Science-Technology Planning Program of Hunan Province (Study on key technology and model structuring of related data for wisdom city data space 2014SK3229); Social Science Program of Hunan Province (Nos: 13YBB205 and 16YBA329); Natural Sciences Fund Project of Hunan Province (No. 2017JJ2241).

References

1. Guo, Y.J.: Methods and Application of Comprehensive Evaluation Theory. Science Press, Beijing (2007)
2. Zhang, X.J.: ERP performance evaluation based on balanced score card. J. Huazhong Univ. Sci. Technol. (Nat. Sci. Ed.) **31**(5), 31–33 (2003)
3. Deng, J.L.: Gray Prediction and Decision. Huazhong University Press, Wuhan (2002)
4. Wang, P.Z.: Fuzzy Set Theory and Application. Shanghai Science and Technology Publishing House, Shanghai (1983)
5. Li, X.P.: On application of analytic hierarchy process in knowledge expression system evaluation. J. Inf. **6**, 70–75
6. Wang, Z.J.: Multiple target comprehensive evaluation method and application based on complex object system of bp neural network. Microcomput. Syst. **16**(1), 25–31 (1995)
7. Guo, J.F., YangDeli, D.L.: General methods of data envelopment analysis. J. Dalian Univ. Technol. **38**(2), 236–241 (1998)
8. Huang, G.M., Zhang, W.: Study on comprehensive evaluation method based on rough set. Comput. Eng. Appl. **2**, 36–38 (2004)
9. Pawlak, Z.: Rough set theory and its applications to data analysis. Cybern. Syst. **29**(1), 661–688 (1998)
10. He, W., Zhong, J., Liu, Z.: Research on customer value' data mining model based on rough set. In: 2009 International Conference on E-Business and Information System Security, 2009, pp. 490–494
11. Zhong, J., Li, D.: Application research on rough set and analytic hierarchy process in fuzzy comprehensive evaluation. In: 2008 IEEE International Symposium on IT in Medicine and Education, 2008, 12, pp. 86–89
12. He, W.H., Liu, Z.H., Zhong, J.M.: On comprehensive evaluation of educational websites based on rough set. Comput. Appl. Softw. **26**(5), 57–61 (2009)

Angle Cosine Method for Judgment Matrix Similarity in Group AHP

Yanling Li[1(✉)], Qifei Yan[2], and Qian Miao[1]

[1] Xi'an Hong Qing High Tech Research Institute, Xi'an, China
lyling998@163.com
[2] Xi'an Communication Institute, Xi'an, China

Abstract. Aim at the problem that the measure of judgment matrix similarity is easily affected by 1–9 division, from the perspective of "relative" similarity, four judgment matrix similarity measurement methods were proposed based on angle cosine: method one, by comparing all the experts judgment matrix measurement information, the method based on angle cosine is proposed for judgment matrix relative similarity measurement, to reduce the influence of 1–9 scale; method two at the base of method one, by introducing gray relation degree, each expert judgment matrix similarity relative value is calculated based on comparative meaning; method three at the base of method one, using D-S composition rule to fuse the measurement information of row and column of judgment matrix; method four using D-S composition rule to fuse the measurement information based on angle cosine and the measurement information based on Euclidean distance. In the example analysis, based on the consistency test index CR, the result rationality of the four method is analyzed, the analysis shows that the method three that based on multi-information fusing or the method four that based on different measurement method fusing gets result more reasonable.

Keywords: Group AHP · Judgment matrix similarity · Angle cosine · D-S fusing rule

1 Introduction

Group AHP method [1] is a group decision method that multiple experts with knowledge and experience in related field participate. How to aggregate the expert personal judgment matrix in group AHP method is an important problem. Currently, many judgment matrix aggregation methods are proposed [1–3], mainly two factors are considered in experts weighting: one is the expert personal judgment matrix consistency degree, the other is judgment matrix compatibility. Both of them need to measure the degree of similarity between the matrices. In using AHP method, the element of judgment matrix is always described by 1–9 scale [4], the similarity measurement method always used, such as Euclidean distance, angle cosine, entropy is uncertainly fit for judgment matrix measurement. So the judgment matrix similarity measurement method is a problem worth to research.

© Springer International Publishing AG 2018
F. Qiao et al. (eds.), *Recent Developments in Mechatronics and Intelligent Robotics*,
Advances in Intelligent Systems and Computing 691, DOI 10.1007/978-3-319-70990-1_58

Aiming at the problem above, at the background of judgment matrix consistency degree, this article importantly researched angle cosine using method in judgment matrix similarity measurement.

2 Problem Describe

Set l experts make up the decision group $E = \{e_1, e_2, \ldots, e_l\}$, the kth expert judgment matrix is $A^k = [a_{ij}^k]_{n \times n}$. Where, a_{ij}^k is the relative weight of index i relative to index j, which is given by e_k, n is the number of index, element a_{ij}^k is described with 1–9 scale.

Set the sorting vector of A^k got by AHP method is $W^k = (\omega_1^k, \omega_2^k, \cdots, \omega_n^k)$, set $B^k = [b_{ij}^k]_{n \times n}, b_{ij}^k = \frac{\omega_i^k}{\omega_j^k}$, B^k is called corresponding fully consistent judgment matrix of A^k [2].

Because of the complexity of object world and the variety of human knowledge, judgment matrix given by experts hardly has fully consistency. Generally agreed that the higher consistency degree of expert personal judgment matrix is, the higher consistency level of this expert mind. So, the consistency degree of expert personal judgment matrix is always regards as an import factor to weight expert [5, 6], which can be measured by the similarity between A^k and B^k, the higher the similarity is between A^k and B^k, the higher consistency degree of A^k is, and the higher the expert weight is.

3 Similarity Measurement Method of Judgment Matrix Based on Angle Cosine

3.1 Judgment Matrix Relative Similarity Measurement Method Based on Angle Cosine (Method One)

Angle cosine is a method always used to measure the similarity between vector, it is widely used in group AHP [1, 5, 6]. The similarity calculated by angle cosine actually can be regard as absolute similarity. To reduce the influence of 1–9 scale, the relative similarity is used to measure the similarity between A^k and B^k in this article, here the "relative" mainly refers to compare to all the expert judgment matrix measure situation, like formula (1):

$$\bar{\alpha}_i^k = \frac{\sum\limits_{j=1}^{n} a_{ij}^k \cdot b_{ij}^k}{\sqrt{\sum\limits_{j=1}^{n} (a_{ij}^k)^2} \sqrt{\sum\limits_{j=1}^{n} (b_{ij}^k)^2}} \bigg/ \sum\limits_{k=1}^{l} \frac{\sum\limits_{j=1}^{n} a_{ij}^k \cdot b_{ij}^k}{\sqrt{\sum\limits_{j=1}^{n} (a_{ij}^k)^2} \sqrt{\sum\limits_{j=1}^{n} (b_{ij}^k)^2}} \tag{1}$$

A^k represents the similarity of each row between A^k and B^k in a relative sense. Specific method is as follows:

Step 1. Calculate the sorting vector $W^k = (\omega_1^k, \omega_2^k, \cdots, \omega_n^k)$ of A^k;
Step 2. Construct consistency judgment matrix B^k of each expert;
Step 3. Calculate the $\bar{\alpha}_i^k$ between each row of A^k and B^k by formula (1);
Step 4. Calculate the relative similarity $\cos A^k$ between A^k and B^k by formula (2):

$$\cos A^k = \frac{\sum\limits_{i=1}^{n} \bar{\alpha}_i^k}{\sum\limits_{k=1}^{l}\sum\limits_{i=1}^{n} \bar{\alpha}_i^k} \quad (k = 1, 2, \cdots, l) \tag{2}$$

3.2 Judgment Matrix Similarity Measurement Method Based on Angle Cosine and Grey Correlation Degree (Method Two)

The grey correlation degree can reflect the relative size of the correlation value between the comparison sequence and the reference sequence [7, 8]. To reduce the influence of 1–9 scale, in this article, using grey correlation degree to measure the relative similarity degree between A^k and B^k. Specific method is as follows:
Steps 1–3 are same as the Steps 1–3 in method one;

Step 4. Construct comparison sequence $X_k = \{\bar{\alpha}_1^k, \bar{\alpha}_2^k, \ldots, \bar{\alpha}_n^k\}$ $(k = 1, 2, \ldots, l)$ and reference sequence $X_0 = \{\bar{\alpha}_1^+, \bar{\alpha}_2^+, \ldots, \bar{\alpha}_n^+\}$, among it $\bar{\alpha}_i^+ = \max\{\bar{\alpha}_i^k\}$ $(i = 1, 2, \ldots, n)$;
Step 5. Using the grey comprehensive correlation degree model [7] to calculate correlation degree $\gamma(X_0, X_k)$;
Step 6. Calculating the similarity degree γA^k between A^k and B^k with formula (3):

$$\gamma(A^k) = \frac{\gamma(X_0, X_k)}{\sum\limits_{k=1}^{l} \gamma(X_0, X_k)} \quad (k = 1, 2, \ldots, l) \tag{3}$$

3.3 Similarity Measurement Method of Judgment Matrix Based on D-S Evidence Theory

In method one and method two, the basic information they used is the relative sense angle cosine value of each row. In fact, each column of the judgment matrix also reflects the consistency level of experts on a certain index, if the angle cosine value of each row is fused with angle cosine value of each column, the overall similarity degree between A^k and B^k can be more comprehensively reflected. In addition, there are many methods to measure the degree of similarity or difference between vectors, Euclidean distance and angle cosine are classical measurement methods, but these methods focus on different points, if the judgment matrix similarity obtained by each method is fused, it should be more objective than the results obtained by one method.

The D-S evidence theory has the D-S synthesis rule for synthesizing a variety of information, therefore, it is regarded as an effective information fusion method [7, 8]. In this article, D-S synthesis rule is used in information fusing through two ways:

(1) Judgment matrix similarity measurement method based on fusing of measurement information of row and column (Method three)

The main idea of this method is: set $\Theta = \{e_1, e_2, \ldots, e_l\}$ as identification framework, construct mass function using relative similarity degree of A^k and B^k by row and column as evidence, then use D-S synthesis rule for information fusing, obtain the final similarity degree between A^k and B^k. Specific steps are as follows:

Steps 1–2 are same as the Steps 1–2 in method one;

Step 3. Calculate the $\bar{\alpha}_i^k$ of each row, calculate $\bar{\eta}_i^k$ of each column:

$$
\bar{\eta}_j^k = \frac{\sum\limits_{i=1}^{n} a_{ij}^k \cdot b_{ij}^k}{\sqrt{\sum\limits_{i=1}^{n} (a_{ij}^k)^2} \sqrt{\sum\limits_{i=1}^{n} (b_{ij}^k)^2}} \Bigg/ \sum\limits_{k=1}^{l} \frac{\sum\limits_{i=1}^{n} a_{ij}^k \cdot b_{ij}^k}{\sqrt{\sum\limits_{i=1}^{n} (a_{ij}^k)^2} \sqrt{\sum\limits_{i=1}^{n} (b_{ij}^k)^2}} \tag{4}
$$

Step 4. Calculate $\cos A^k$ by row and calculate $\cos' A^k$ by column:

$$
\cos' A^k = \sum\limits_{j=1}^{n} \bar{\eta}_j^k \Bigg/ \sum\limits_{k=1}^{l} \sum\limits_{j=1}^{n} \bar{\eta}_j^k \quad (k = 1, 2, \cdots, l) \tag{5}
$$

Step 5. Based on $\cos A^k$ and $\cos' A^k$, construct mass functions $m_{hang}(e_k)$ and $m_{lie}(e_k)$, and use D-S synthesis rule to obtain similarity degree of A^k and B^k based on row and column information fusing, as formula (6) shows:

$$
A_{hl}^k = \left(m_{hang} \oplus m_{lie}\right)(e_k)(k = 1, 2, \ldots, l) \tag{6}
$$

where, $m_{hang}(e_k) = \cos A^k$, $m_{lie}(e_k) = \cos' A^k$.

(2) Expert judgment matrix similarity measurement method based on measurement information fusing of distance and angle cosine (Method four)

Similar to method three, use relative similarity degree of A^k and B^k that obtained from Euclidean distance and angle cosine as evidence to construct mass function. Specific steps are as follows:

Steps 1–2 are same as Steps 1–2 in method one;

Step 3. Calculate $\bar{\alpha}_i^k$, calculate $\bar{d}_i^k$ based on distance measurement by formula (7):

$$
\bar{d}_i^k = \sqrt{\sum\limits_{j=1}^{n} (a_{ij}^k - b_{ij}^k)^2} \Bigg/ \sum\limits_{i=1}^{n} \sqrt{\sum\limits_{j=1}^{n} (a_{ij}^k - b_{ij}^k)^2} \tag{7}
$$

Step 4. Calculate $\cos A^k$, and calculate the $disA^k$ based on distance measurement:

$$disA^k = \sum_{i=1}^{n} \bar{d}_i^k \Bigg/ \sum_{k=1}^{l}\sum_{i=1}^{n} \bar{d}_i^k \quad (k = 1, 2, \ldots, l) \tag{8}$$

Step 5. Based on $\cos A^k$ and $disA^k$, construct mass function $m_c(e_k)$ and $m_d(e_k)$, and use D-S synthesis rule to obtain the similarity degree between A^k and B^k:

$$A_{cd}^k = (m_c \oplus m_d)(e_k)(k = 1, 2, \ldots, l) \tag{9}$$

where, $m_c(e_k) = \cos A^k, m_d(e_k) = (1 - disA^k) \Big/ \sum_{i=1}^{n}(1 - disA^k).$

4 Example Analysis

Data of this example is from [9]. Set there are four experts participate the decision making, the judgment matrix shows below:

$$A^1 = \begin{bmatrix} 1 & 1/5 & 1/7 & 1/5 & 1/7 & 3 \\ 5 & 1 & 1/5 & 1/3 & 1/5 & 5 \\ 7 & 5 & 1 & 3 & 1/3 & 5 \\ 5 & 3 & 1/3 & 1 & 1/5 & 5 \\ 7 & 5 & 3 & 5 & 1 & 9 \\ 1/3 & 1/5 & 1/5 & 1/5 & 1/9 & 1 \end{bmatrix} \quad A^2 = \begin{bmatrix} 1 & 1/3 & 1/3 & 1 & 1/7 & 1/9 \\ 3 & 1 & 1 & 1 & 1/5 & 1/7 \\ 3 & 1 & 1 & 1 & 1/5 & 1/7 \\ 1 & 1 & 1 & 1 & 1/3 & 1/3 \\ 7 & 5 & 5 & 3 & 1 & 1 \\ 9 & 7 & 7 & 3 & 1 & 1 \end{bmatrix}$$

$$A^3 = \begin{bmatrix} 1 & 1/5 & 1/7 & 1/9 & 1 & 1/3 \\ 5 & 1 & 1/3 & 1/3 & 5 & 7 \\ 7 & 3 & 1 & 1 & 5 & 7 \\ 9 & 3 & 1 & 1 & 5 & 9 \\ 1 & 1/5 & 1/5 & 1/5 & 1 & 3 \\ 3 & 1/7 & 1/7 & 1/9 & 1/3 & 1 \end{bmatrix} \quad A^4 = \begin{bmatrix} 1 & 3 & 1/7 & 1/3 & 1 & 5 \\ 1/3 & 1 & 1/5 & 1/5 & 1/3 & 1/7 \\ 7 & 5 & 1 & 1/3 & 1/5 & 1/3 \\ 3 & 5 & 3 & 1 & 1/3 & 1/3 \\ 1 & 3 & 5 & 3 & 1 & 1 \\ 5 & 7 & 3 & 3 & 1 & 1 \end{bmatrix}$$

The similarity degree between A^k and B^k calculated by four methods as Fig. 1. It is obvious that the results obtained by various methods are different, according to the sorting of similarity degree value, method one is same as method two, they all get the order of A^2, A^1, A^3, A^4, however method three is same as method four, they all get the order of A^2, A^3, A^1, A^4, the difference between two sorting result shows in the order of A^1, A^3.

In AHP method, CR is a widely recognized consistency test index [1]. If $CR < 0.1$, the judgment matrix is considered with a satisfied consistency, with the increasing of CR, the inconsistency degree of judgment matrix increase. In this example, the CR value of each expert judgment matrix is $CR_1 = 0.107$, $CR_2 = 0.0414$, $CR_3 = 0.0759$, $CR_4 = 0.6107$.

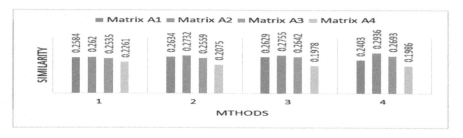

Fig. 1. Result of judgment matrix similarity calculated by four methods

It can be seen that, according to the consistency testing index CR, there is a satisfied consistency in A^2, A^3. So, according to the CR value, the consistency degree order of four expert judgment matrix is A^2, A^3, A^1, A^4. That means the similarity order of four expert judgment matrix should be A^2, A^3, A^1, A^4, it can be seen that result calculated by method three and method four is more reasonable than result calculated by method one and method two.

5   Conclusion

This article researched the method of angle cosine measurement of similarity of judgment matrix in group AHP, aiming at the problem that judgment matrix similarity degree measurement is easily affected by 1–9 scale, this article proposed four judgment matrix similarity measurement methods based on angle cosine, in the point of "relative" similarity.

According to the example, if analysis from the point of consistency test index CR, the result of method three based on measurement information fusing of row and column and the result of method four based on information fusing of angle cosine and Euclidean distance is reasonable, it indicates that the result based on multiple information fusing or result based on multiple method fusing is more objective and more comprehensive.

References

1. Ji, J.: Compatibility test in group decision making based on included angle cosine. J. Chongqing Inst. Technol. (Nat. Sci.) **22**(6), 141–143 (2008)
2. Yue-jin, L.V., Xin-rong, G.U.O.: An effective aggregation method for the AHP judgement matrix in group decision-making. Syst. Eng. Theory Pract. **7**, 132–135 (2007)
3. Lin, C.: Study on Theories and Methods of the Group Preference Aggregation Based on the Judgment Matrices. University of Electronics Science and Technology of China, Chengdu (2015)
4. Saaty, T.L.: The Analytic Hierarchy Process. MoCraw-Hill, New York (1980)
5. Wang, J., Chen, Z.: Synthetic capability evaluation of the gun weapon system based on AHP group decision. Fire Control Command Control **37**(10), 182–184 (2012)

6. He, L., Wang, L., Zhang, L.: A method for determining the experts' weights of multi-attribute group decision-making based on clustering analysis. Oper. Res. Manag. Sci. **23**(6), 65–72 (2014)
7. Li, Y., Zhu, Y., Wu, J.: Research on expert weighting method based on D-S evidence theory. In: IAEAC (2017)
8. Yi, S., Haifeng, H.: Adaptive algorithm for adjusting weights of decision-makers and attributes. Autom. Inf. Eng. **34**(2), 1–5 (2013)
9. Li, Y., Wu, J., Zhu, Y.: A method for determining expert 's weight based on the consistency of judgment matrix. Comput. Modern. Unpublished

Based on the Plastic Screw Holes Research into the Causes of the Fracture

Junjie Lv$^{(\boxtimes)}$

School of Electronic Information Engineering, Wuhan Polytechnic College,
Wuhan, Hubei, People's Republic of China
593316640@qq.com

Abstract. Plastic cracking is fatal flaw, affect the normal use of the related to many factors. This article selects a typical "screw holes rupture" case analysis, found the cause of broken plastic screw hole for its own material has changed, the new material mechanical strength decreased, intolerance to assembly the external force and fracture. At the same time, material control proposal is presented.

Keywords: Plastic parts · Screw holes · Fracture · Failure case

1 Introduction

Engineering plastics because of their excellent characteristics, such as high strength, heat resistance, impact resistance, anti-aging and is widely used in industrial parts and all kinds of shell manufacturing. But in the process of manufacture or use, often because of mold plastic products, molding process and plastic materials, environmental stress factors such as induced cracking failure, thus affect the normal use of the products, serious and even cause the entire finished product scrap. These factors, the cracking failure caused by plastic materials change products accounted for a larger proportion, especially those who lack of quality control of products for raw materials processing and manufacturing enterprise, this kind of failure phenomena are more likely to happen. Individual components change on the one hand, because of the plastic material type changes, even often unable to identify from the outside, on the other hand for outsourcing processing enterprises, outsourcing factories often because of the interest relations with cheap materials or add recycling, pack material in raw materials, the material some performance degradation [1] cause cracking failure of plastic products. Based on a typical plastic screw holes fracture cases, through a series of tests to analysis work, discusses the cause of screw holes fracture and improving Suggestions.

© Springer International Publishing AG 2018
F. Qiao et al. (eds.), *Recent Developments in Mechatronics and Intelligent Robotics*,
Advances in Intelligent Systems and Computing 691, DOI 10.1007/978-3-319-70990-1_59

2   Case Analysis

2.1   Background

In this article for PC plastic material, the fracture in screw holes position, part of products of the fracture occurred in the process of assembly, part assemble into finished products in the client to use after a period of time after fracture. In order to analyze the fracture cause of screw holes, this paper collected has occurred fracture failure of product and use good product.

2.2   The Analysis Process

Failure of plastic parts and good parts appearance observation, found the failure are in screw holes position fracture, fracture surface is smooth, no obvious plastic deformation, fracture did not see the secondary cracks around ryohin keikaku screw holes parts in good condition. Failure product appearance FIG. See Fig. 1a, good product appearance diagram as shown in the Fig. 1b.

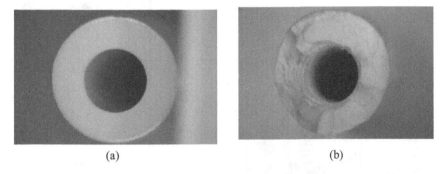

(a)   (b)

Fig. 1.   Failure parts and good parts appearance figure

Respectively screw holes on the failure fracture and good screw holes and scanning electron microscopy (sem) and energy spectrum analysis (SEM&EDS), the results showed that the inner wall of the screw holes for the fracture source area, the central extended end near the outer wall fracture, fracture part not seen obvious inclusions are not detected abnormal elements. By the failure of product and good test position detection and element composition, the failure parts and good parts were detected with carbon (C) and oxygen (O), silicon (Si), titanium (Ti) elements, but two samples of silicon (Si), titanium (Ti) elements have obvious difference, according to its composition may be different. Results images as shown in the Figs. 2 to 3.

402   J. Lv

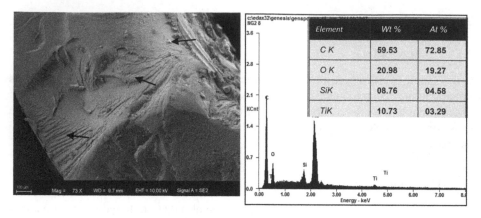

Fig. 2. Screw holes fracture SEM&EDS results failure

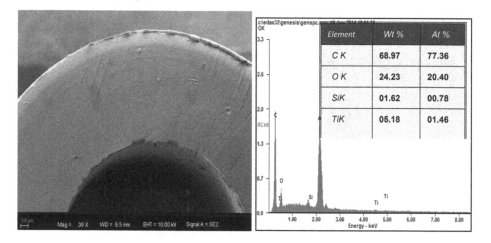

Fig. 3. Ryohin keikaku screw holes end SEM&EDS results

For screw holes position of failure and what goods plastic material consistency analysis (ir, DSC, TGA). By infrared analysis result, the two plastic organic main ingredients are based on PC, but its components are different; DSC results show that the glass transition temperature difference is more than 5 °C; TGA results showed that the failure is tasted the initial decomposition temperature of plastic is markedly lower than the good parts of plastic initial decomposition temperature, a difference of more than 20 °C. The above results that the failure parts and good parts of plastic material component is different. Results images as shown in the Figs. 4, 5 and 6.

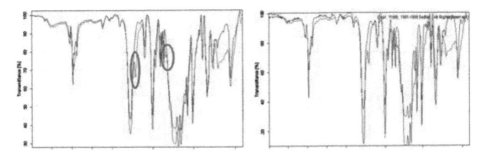

Fig. 4. Failure products, good quality plastic and infrared spectra of PC (red box display spectrogram differences)

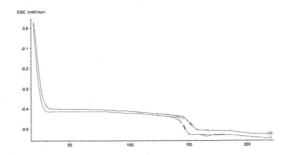

Fig. 5. Failure parts and pieces of plastic ryohin keikaku DSC curves

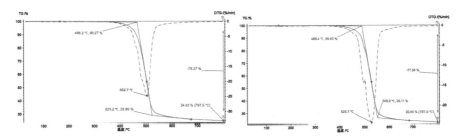

Fig. 6. Plastic failure of product and good TGA curves

Apply failure and good whole pieces of plastic tensile test to confirm its mechanical performance difference, by the test result, the failure product of maximum tensile force average 1.23 KN, good parts of maximum tensile force average 1.56 KN, apparently failure goods the maximum tensile force of a plastic fine products that are lower than 330 N, shows that the mechanical properties of plastic failure product than the good parts.

3 Conclusion

This case, the plastic screw hole fracture reason is because the plastic failure product use material with good product material composition, failure mechanical strength is relatively low, can't stand assembly screws exert force and fracture. Investigate the root cause is lack of manufacturer for material quality control, failed to timely found in incoming material changed, cause subsequent product cracking of serious problems.

Through this case also can give enlightenment on factory and other enterprises, is the first step of all products and equipment quality guarantee is the reliability of raw materials, the enterprises should be a number of ways for the incoming implement effective quality control, so as to improve the reliability of the product. Such as: complete before put into mass production in product design, establishing reasonable technical scheme for multiple vendors in material selection, scheme affect the reliability of the material quality should be considered all measures of performance, applicability, process involving a variety of materials with material should also be considered when using compatibility. In addition, should also consider the price, supply and quality assurance ability of supplier, finally determine the reasonable raw material suppliers. Second in daily production, should also be selected key indicators for each batch of incoming batch consistency inspection, to ensure the quality of different batches of product consistency, this case USES the material of consistency (ir, DSC, TGA) test for organic non-metallic materials is a kind of effective batch inspection method. Believe that through these control method can greatly reduce the risk of product reliability problems.

References

1. Jian, P.: Material certification and consistency control analysis technology. China Build. Mater. Sci. Technol., S1 (2010)
2. The national standard GB/T 1040.2 - determination of tensile properties - part 2 - plastic molding and extrusion plastic test conditions
3. The national standard GB/T 19466.2 plastic differential scanning calorimetry (DSC), part 2: determination of the glass transition temperature
4. The international standard ISO 11358 plastic polymer thermogravimetric analysis (TG) of general principle
5. Jang, J.-S.R.: ANFIS adaptive-network-based fuzzy inference system. IEEE Trans. Syst. Man Cybern. 23(3), 665–685 (1993)
6. Shouman, M., Turner, T., Stocker, R.: Applying k-nearest neighbour in diagnosing heart disease patients. Int. J. Inf. Educ. Technol. 2(3), 220–223 (2012)
7. Mandal, D., Chattopadhyay, I.M., Mishra, S.: A low cost non-invasive digital signal processor based (TMS320C6713) heart diagnosis system. In: 1st Int'l Conf. on Recent Advances in Information Technology I RAIT-2012
8. Perera, I.S, Muthalif, F.A, Selvarathnam, M.: Automated diagnosis of cardiac abnormalities using heart sounds. In: 2013 IEEE Point-of-Care Healthcare Technologies (PHT) Bangalore, India, 16 – 18 January, 2013

The Methodology of Multi-valued Dynamic Uncertain Causality Graph for Intelligent Selection of Excellent Teachers

Kun Qiu$^{(\boxtimes)}$ and Qin Zhang

School of Computer Science and Engineering,
Beihang University, Beijing, China
qiukunn@sina.com

Abstract. This article uses the multi-valued dynamic uncertain causality graph (M-DUCG) to solve the problem of evaluating or selecting teachers in intelligent way. One example is used to illustrate, junior high school wonder select second grade teachers go further education in the form of merit selection. Model is built according to the relationship between characters by principal without any historical data. There are cycles in this model, how to solve them is the advantage of DUCG will be show. The inference parameters are given depend on statistics and experience. With the modelling and reasoning, our method can rank the teachers according to high praise and present a satisfactory conclusion.

Keywords: Intelligent system · Cycle model · Inference · Evaluation

1 Introduction

Rural education has always been the focus of Chinese government, select qualified and eliminate unqualified teachers is the basic requirement. Pursuing the quality of rural education make rural teachers contend fiercely, the promotion and salary increase no longer rely on seniority but actual performance. With the implementation of the competition mechanism, the rural teachers' evaluations have become a key problem. However, it is not easy to recognize a good teacher because there is also no consensus on this topic [1, 2]. Some education institutions only focus on the effectives of teachers in accomplishing high student learning outcomes [1, 3]. But this practice should follow a hypothesis that behavior of teachers and teachers' pedagogical [4] have a directly impact on students' achievement. Moreover, the Chinese government ban merely using graduation rates and test scores to evaluate teachers. So judge only from the results of teachers' professional conduct value is not suitable.

It is feasible to make the graduation rates or test scores as an element among plenty of to operate, other elements include [5] interpersonal, pedagogic competence, subject knowledge and methodological competence and organizational competence and so on. But, there is no unified standard to integrate these factors, therefore, the problem of evaluation we might as well use uncertainty probability to solve, generally, this kind of problem will be accompanied by the causation. Professor Zhang Qin have proposed M-DUCG [6] theory has become an important method to solve this type of problem.

F. Qiao et al. (eds.), *Recent Developments in Mechatronics and Intelligent Robotics*,
Advances in Intelligent Systems and Computing 691, DOI 10.1007/978-3-319-70990-1_60

Representing knowledge in compact way and be able to cope with static directed cycle are the two obvious advantages.

A junior high school is forced to rely on experience to select three outstanding teachers go further education from eleven persons who conduct students in grade two. Because of limitation, admit the best one is the only way. Students can know teacher directly, teachers can evaluate other teacher directly, too. The headmaster's son named Lin, who is also students, can give perspective to teachers while teachers can give Lin assessments, too. The headmaster evaluate teacher through directly contact in workplace and also depend on Lin. According to the above description, we established a causal influence relation model between each role as shown in Fig. 1.

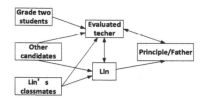

Fig. 1. The relationship among all characters

2 M-DUCG Presentation

In M-DUCG, Uncertain causal relationship is described by virtual causal random event, we say the virtual event happened if cause result consequence, otherwise it is not happen. The existence of causal relationship is also uncertain, using parameter "r" to express, only has relative meaning. We use capital letters 'X', 'B', 'A' as variables or events and lowercase letters 'a' and 'b' represent probability of events 'A' and 'B', subscript of variables or events is letter 'i' who indexes the variable. 'X', whose shape is circular, represents the events or variables of result or effect, it can also be reasons events or variables; 'B', whose shape is rectangle, represents the events or variables of basic or root, it can only be an independent reasons event or variable; 'A' represents the linkage events or variables and is drawn as red arrow or virtual red arrow who is defined as conditional linkage events. Multi-valued can describe the reason of each state independently and the effect of parents to child is weighted average. Parents variable only impact the child's true state is S-DUCG which the expression of relationship is "Logic or". Parent variables affect more than two states of child is M-DUCG. Summarizing a general formula of M-DUCG expression below:

$$X_{nk} = \sum_i \left(r_{n;i}/r_n \right) \sum_{j_i} A_{nk;ij_i} V_{ij_i} \tag{1}$$

'r_n' in the denominator is the sum of all '$r_{n;i}$' parameter. 'V' represents the parent variables include variable 'B', 'X'.

3 M-DUCG Modeling and Parameters

We have identified that using M-DUCG to model the relationship as Fig. 1. Variable X is used to represent principal, Lin and evaluated teacher because there are mutual effects between them. For the evaluation of other teachers, other students and Lin's classmates are external factors and we defined as basic event using variable B. The preliminary M-DUCG as shown in the Fig. 2.

We give each character a variable name, principal is defined X_1, Lin is X_2, evaluated teacher is X_3, define the external factor of affecting evaluated teacher to B_4 and define the external factor of affecting Lin to B_5. Here to explain, there are three external factors impact the X_3 originally, we combine classmates into the other students when talking about evaluated teacher as a result that only two factors left. Moreover, from man's intuition, students are more suitable to evaluate teachers because competition and envy between teachers' mutual evaluation, while teachers' advice is more accurate to students due to there may be "brothers" between students. Accordingly, we adopt an empirical formula to assign two external influencing factors, the formula

$$X = 0.6Y + 0.4Z \qquad (2)$$

is applied to evaluate teachers and

$$X = 0.4Y + 0.6Z \qquad (3)$$

is applied to evaluate Lin. Y represents the evaluation from students and Z represents the evaluation from teachers. Link event is an expression of each arrow. Variable $A_{3;1}$ represents that X_1 influences X_3 and $A_{1;3}$ represents that X_3 influence X_1. An intact M-DUCG model is shown Fig. 3.

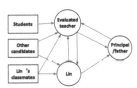
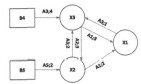

Fig. 2. The preliminary M-DUCG   **Fig. 3.** The intact M-DUCG

In order to quantify the link event, we need define more the one state for each variable, so, the link event become a state matrix [7]. Variable B_4 and B_5 is defined binary variable include negative comment and positive comment, variable X_2 and X_3 also define binary one include behave well and behave bad. Variable X_1 is defined three-phase variable, include good evaluation, bad evaluation and medium evaluation. Therefore, link event $A_{3;1}$ is two row and three column matrix, $A_{1;3}$ and $A_{1;2}$ are three row and two column matrix.

Inference of M-DUCG includes logical operation and probability calculation. Logical operation has been described above only left probability calculation. There are

Parameter "a", "b" and "r" in progress of probability calculation of M-DUCG. Parameter "a", who is the probability matrix of link event, quantifies the relationship between each state of parent nodes and each state of child node. Parameter "b" is the probability of variable B. Parameter r, who is the weight coefficient, denotes influence degree that parent nodes impact child node. We get the parameter "b" using statistical method. There are a total of 460 students, 11 teachers in grade two and 42 persons in Lin's class. The quantity of students' questionnaire that Conform to the requirements for evaluating teachers is 348 and evaluating teachers is 11. The statistical result in Table 1.

From Table 1 we can calculate the positive ratio by students are A = 0.5, B = 0.575, C = 0.848, D = 0.894, E = 0.443, F = 0.546, G = 0.629, H = 0.592, I = 0.103, J = 0.221, K = 0.859, L = 0.825, Lin = 0.769, and positive ratio by teachers are A = 0.455, B = 0.364, C = 0.273, D = 0.545, E = 0.455, F = 1, G = 0.636, H = 0.091, I = 1, J = 0.182, K = 0.364, L = 0.455, Lin = 0.667. According to empirical formula (2) and (2) we can calculate the positive ratio from external factors as follows: A = 0.482, B = 0.491, C = 0.618, D = 0.754, E = 0.448, F = 0.728, G = 0.632, H = 0.392, I = 0.462, J = 0.205, K = 0.661, L = 0.677, Lin = 0.708.

Table 1.

Candidate's name	Evaluation of teacher		Evaluation of student		Candidate's name	Evaluation of teacher		Evaluation of student	
	Positive number	Negative number	Positive number	Negative number		Positive number	Negative number	Positive number	Negative number
A	5	6	174	174	H	1	10	206	142
B	4	7	200	148	I	11	0	36	312
C	3	8	295	53	J	2	9	77	271
D	6	5	311	37	K	4	7	299	49
E	5	6	154	194	L	5	6	287	61
F	11	0	190	158	Lin	8	4	30	9
G	7	4	219	129					

So, the parameters $b_{4,1}$ is the corresponding positive ratio of teacher, and parameter $b_{5,1}$ is the positive ratio of Lin. The parameter r, which is provided by principal, is defined weighted coefficient. For instance, the two coefficient $r_{1;2} = 0.5$ and $r_{1;3} = 0.5$ mean Lin's performance and evaluated teacher's performance have the equal effect to principal's decision. $r_{3;1} = 0.3$, $r_{3;2} = 0.2$, $r_{3;4} = 0.5$ show that principal's evaluation take up 30% for evaluated teacher's performance, Lin's evaluation take up 20% and external factors take up 50% and so on. In the same way, we use parameter $r_{2;3} = 0.6$ and $r_{2;5} = 0.4$ to evaluate Lin. The parameters are shown in Appendix A.

4 Logical Expression and Reasoning Calculation

It has already been said the most important feature of *M*-DUCG is logical expression whose purpose is to express logical relationship between variables. Put parameters into logical expression to calculate the probability after simplification. Sub *M*-DUCGs are extracted from Fig. 3 to support getting logical expression, as the Fig. 4.

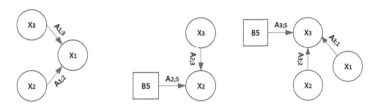

Fig. 4. Three sub *M*-DUCGs

Our purpose is to eliminate all variable *X* in the right of equation, leaving only variable *A* and variable *B*. With the operation going, the parameter r may be changed. In the progress of unfold, we would use event operations of probability theory including absorptive law, commutative law, associative law, distributive law and empty set operation.

We get the most simplify equation:

$$X_1 = (0.5/1.6)A_{1;3}A_{3;4}B_4 + (0.3/1.6)A_{1;3}A_{3;2}A_{2;5}B_5 + (0.5/1.7)A_{1;2}A_{2;5}B_5 + (0.35/1.7)A_{1;2}A_{2;3}A_{3;4}B_4$$
$$X_2 = (1/1.7)A_{2;5}B_5 + (0.7/1.7)A_{2;3}A_{3;4}B_4$$
$$X_3 = (1/2.6)A_{3;4}B_4 + (1/2.6)A_{3;1}A_{1;2}A_{2;5}\ B_5 + (0.6/2.6)A_{3;2}A_{2;5}B_5$$

Combining with parameters of the Appendix A, the teachers' favorable rates, also the $Pr(X_{3,1})$ are listed in the Table 2.

Table 2.

Candidate's name	A	B	C	D	E	F	G	H	I	I	K	L
positive ratio	0.504	0.525	0.622	0.738	0.478	0.691	0.648	0.423	0.493	0.290	0.644	0.653

5 Conclusion

Through analyze the problem of evaluating teacher we select *M*-DUCG theory to model, use empirical method to get parameters, calculate the positive ratio, rank candidates of evaluating teacher and Come to the conclusion which the principal want to get. As regards this problem, it is accurate and flexible to use *M*-DUCG for evaluating.

Appendix A

Candidate name	Parameter a	Parameter b	Parameter r
A	$a_{3,4}=\begin{bmatrix}1&0\\0&1\end{bmatrix}$ $a_{2,5}=\begin{bmatrix}1&0\\0&1\end{bmatrix}$ $a_{3,2}=\begin{bmatrix}0.6&0.3\\0.4&0.7\end{bmatrix}$ $a_{2,3}=\begin{bmatrix}0.8&0.1\\0.2&0.9\end{bmatrix}$ $a_{1,2}=\begin{bmatrix}0.5&0.2\\0.4&0.3\\0.1&0.5\end{bmatrix}$ $a_{3,1}=\begin{bmatrix}0.7&0.6&0.2\\0.3&0.4&0.8\end{bmatrix}$ $a_{1,3}=\begin{bmatrix}0.4&0.1\\0.4&0.3\\0.2&0.6\end{bmatrix}$	$b_4=\begin{bmatrix}0.518\\0.482\end{bmatrix}$ $b_3=\begin{bmatrix}0.292\\0.708\end{bmatrix}$	$r_{1,2}=0.5$ $r_{1,3}=0.5$ $r_{2,3}=0.6$ $r_{2,4}=0.4$ $r_{3,1}=0.3$ $r_{3,2}=0.2$ $r_{3,4}=0.5$
B	$a_{3,4}=\begin{bmatrix}1&0\\0&1\end{bmatrix}$ $a_{2,5}=\begin{bmatrix}1&0\\0&1\end{bmatrix}$ $a_{3,2}=\begin{bmatrix}0.5&0.3\\0.5&0.7\end{bmatrix}$ $a_{2,3}=\begin{bmatrix}0.2&0.1\\0.8&0.9\end{bmatrix}$ $a_{1,2}=\begin{bmatrix}0.3&0.2\\0.4&0.3\\0.3&0.5\end{bmatrix}$ $a_{3,1}=\begin{bmatrix}0.4&0.6&0.2\\0.6&0.4&0.8\end{bmatrix}$ $a_{1,3}=\begin{bmatrix}0.4&0.1\\0.2&0.3\\0.4&0.6\end{bmatrix}$	$b_4=\begin{bmatrix}0.509\\0.491\end{bmatrix}$ $b_3=\begin{bmatrix}0.292\\0.708\end{bmatrix}$	
C	$a_{3,4}=\begin{bmatrix}1&0\\0&1\end{bmatrix}$ $a_{2,5}=\begin{bmatrix}1&0\\0&1\end{bmatrix}$ $a_{3,2}=\begin{bmatrix}0.7&0.3\\0.3&0.7\end{bmatrix}$ $a_{2,3}=\begin{bmatrix}0.1&0.1\\0.9&0.9\end{bmatrix}$ $a_{1,2}=\begin{bmatrix}0.2&0.2\\0.1&0.3\\0.7&0.1\end{bmatrix}$ $a_{3,1}=\begin{bmatrix}0.2&0.6&0.2\\0.8&0.4&0.8\end{bmatrix}$ $a_{1,3}=\begin{bmatrix}0.4&0.1\\0.5&0.3\\0.1&0.6\end{bmatrix}$	$b_4=\begin{bmatrix}0.382\\0.618\end{bmatrix}$ $b_3=\begin{bmatrix}0.292\\0.708\end{bmatrix}$	
D	$a_{3,4}=\begin{bmatrix}1&0\\0&1\end{bmatrix}$ $a_{2,5}=\begin{bmatrix}1&0\\0&1\end{bmatrix}$ $a_{3,2}=\begin{bmatrix}0.7&0.2\\0.3&0.8\end{bmatrix}$ $a_{2,3}=\begin{bmatrix}0.1&0.3\\0.9&0.5\end{bmatrix}$ $a_{1,2}=\begin{bmatrix}0.2&0.6\\0.1&0.3\\0.7&0.1\end{bmatrix}$ $a_{3,1}=\begin{bmatrix}0.2&0.5&0.2\\0.8&0.5&0.8\end{bmatrix}$ $a_{1,3}=\begin{bmatrix}0.4&0.1\\0.5&0.1\\0.1&0.8\end{bmatrix}$	$b_4=\begin{bmatrix}0.246\\0.754\end{bmatrix}$ $b_3=\begin{bmatrix}0.292\\0.708\end{bmatrix}$	
E	$a_{3,4}=\begin{bmatrix}1&0\\0&1\end{bmatrix}$ $a_{1,3}=\begin{bmatrix}1&0\\0&1\end{bmatrix}$ $a_{3,2}=\begin{bmatrix}0.6&0.3\\0.4&0.7\end{bmatrix}$ $a_{2,3}=\begin{bmatrix}0.8&0.1\\0.2&0.9\end{bmatrix}$ $a_{1,2}=\begin{bmatrix}0.5&0.2\\0.4&0.3\\0.1&0.5\end{bmatrix}$ $a_{3,1}=\begin{bmatrix}0.7&0.6&0.2\\0.3&0.4&0.8\end{bmatrix}$ $a_{1,3}=\begin{bmatrix}0.4&0.1\\0.4&0.3\\0.2&0.6\end{bmatrix}$	$b_4=\begin{bmatrix}0.542\\0.448\end{bmatrix}$ $b_3=\begin{bmatrix}0.292\\0.708\end{bmatrix}$	
F	$a_{3,4}=\begin{bmatrix}1&0\\0&1\end{bmatrix}$ $a_{2,5}=\begin{bmatrix}1&0\\0&1\end{bmatrix}$ $a_{3,2}=\begin{bmatrix}0.6&0.3\\0.4&0.7\end{bmatrix}$ $a_{2,3}=\begin{bmatrix}0.8&0.1\\0.2&0.9\end{bmatrix}$ $a_{1,2}=\begin{bmatrix}0.5&0.2\\0.4&0.3\\0.1&0.5\end{bmatrix}$ $a_{3,1}=\begin{bmatrix}0.7&0.6&0.2\\0.3&0.4&0.8\end{bmatrix}$ $a_{1,3}=\begin{bmatrix}0.4&0.1\\0.4&0.3\\0.2&0.6\end{bmatrix}$	$b_4=\begin{bmatrix}0.272\\0.728\end{bmatrix}$ $b_3=\begin{bmatrix}0.292\\0.708\end{bmatrix}$	
G	$a_{3,4}=\begin{bmatrix}1&0\\0&1\end{bmatrix}$ $a_{2,5}=\begin{bmatrix}1&0\\0&1\end{bmatrix}$ $a_{3,2}=\begin{bmatrix}0.3&0.3\\0.7&0.7\end{bmatrix}$ $a_{2,3}=\begin{bmatrix}0.8&0.2\\0.2&0.8\end{bmatrix}$ $a_{1,2}=\begin{bmatrix}0.5&0.3\\0.4&0.3\\0.1&0.4\end{bmatrix}$ $a_{3,1}=\begin{bmatrix}0.3&0.4&0.2\\0.7&0.6&0.8\end{bmatrix}$ $a_{1,3}=\begin{bmatrix}0.4&0.3\\0.2&0.1\\0.4&0.6\end{bmatrix}$	$b_4=\begin{bmatrix}0.368\\0.632\end{bmatrix}$ $b_3=\begin{bmatrix}0.292\\0.708\end{bmatrix}$	
H	$a_{3,4}=\begin{bmatrix}1&0\\0&1\end{bmatrix}$ $a_{2,5}=\begin{bmatrix}1&0\\0&1\end{bmatrix}$ $a_{3,2}=\begin{bmatrix}0.3&0.6\\0.7&0.4\end{bmatrix}$ $a_{2,3}=\begin{bmatrix}0.1&0.2\\0.9&0.8\end{bmatrix}$ $a_{1,2}=\begin{bmatrix}0.5&0.2\\0.1&0.2\\0.4&0.6\end{bmatrix}$ $a_{3,1}=\begin{bmatrix}0.6&0.6&0.2\\0.4&0.4&0.8\end{bmatrix}$ $a_{1,3}=\begin{bmatrix}0.4&0.2\\0.4&0.3\\0.2&0.5\end{bmatrix}$	$b_4=\begin{bmatrix}0.608\\0.392\end{bmatrix}$ $b_3=\begin{bmatrix}0.292\\0.708\end{bmatrix}$	
I	$a_{3,4}=\begin{bmatrix}1&0\\0&1\end{bmatrix}$ $a_{2,5}=\begin{bmatrix}1&0\\0&1\end{bmatrix}$ $a_{3,2}=\begin{bmatrix}0.6&0.3\\0.4&0.7\end{bmatrix}$ $a_{2,3}=\begin{bmatrix}0.8&0.1\\0.2&0.9\end{bmatrix}$ $a_{1,2}=\begin{bmatrix}0.5&0.2\\0.3&0.3\\0.2&0.5\end{bmatrix}$ $a_{3,1}=\begin{bmatrix}0.7&0.5&0.2\\0.3&0.5&0.8\end{bmatrix}$ $a_{1,3}=\begin{bmatrix}0.4&0.1\\0.3&0.4\\0.3&0.5\end{bmatrix}$	$b_4=\begin{bmatrix}0.538\\0.462\end{bmatrix}$ $b_3=\begin{bmatrix}0.292\\0.708\end{bmatrix}$	
J	$a_{3,4}=\begin{bmatrix}1&0\\0&1\end{bmatrix}$ $a_{2,5}=\begin{bmatrix}1&0\\0&1\end{bmatrix}$ $a_{3,2}=\begin{bmatrix}0.4&0.3\\0.6&0.7\end{bmatrix}$ $a_{2,3}=\begin{bmatrix}0.3&0.1\\0.7&0.9\end{bmatrix}$ $a_{1,2}=\begin{bmatrix}0.5&0.2\\0.4&0.4\\0.1&0.4\end{bmatrix}$ $a_{3,1}=\begin{bmatrix}0.7&0.8&0.2\\0.3&0.2&0.8\end{bmatrix}$ $a_{1,3}=\begin{bmatrix}0.5&0.1\\0.3&0.2\\0.2&0.7\end{bmatrix}$	$b_4=\begin{bmatrix}0.795\\0.205\end{bmatrix}$ $b_3=\begin{bmatrix}0.292\\0.708\end{bmatrix}$	
K	$a_{3,4}=\begin{bmatrix}1&0\\0&1\end{bmatrix}$ $a_{2,5}=\begin{bmatrix}1&0\\0&1\end{bmatrix}$ $a_{3,2}=\begin{bmatrix}0.6&0.3\\0.4&0.7\end{bmatrix}$ $a_{2,3}=\begin{bmatrix}0.6&0.1\\0.4&0.9\end{bmatrix}$ $a_{1,2}=\begin{bmatrix}0.5&0.2\\0.4&0.3\\0.1&0.4\end{bmatrix}$ $a_{3,1}=\begin{bmatrix}0.7&0.5&0.2\\0.3&0.5&0.8\end{bmatrix}$ $a_{1,3}=\begin{bmatrix}0.4&0.1\\0.4&0.2\\0.2&0.7\end{bmatrix}$	$b_4=\begin{bmatrix}0.339\\0.661\end{bmatrix}$ $b_3=\begin{bmatrix}0.292\\0.708\end{bmatrix}$	
L	$a_{3,4}=\begin{bmatrix}1&0\\0&1\end{bmatrix}$ $a_{2,5}=\begin{bmatrix}1&0\\0&1\end{bmatrix}$ $a_{3,2}=\begin{bmatrix}0.6&0.3\\0.4&0.7\end{bmatrix}$ $a_{2,3}=\begin{bmatrix}0.8&0.1\\0.2&0.9\end{bmatrix}$ $a_{1,2}=\begin{bmatrix}0.5&0.2\\0.4&0.3\\0.1&0.5\end{bmatrix}$ $a_{3,1}=\begin{bmatrix}0.7&0.6&0.2\\0.3&0.4&0.8\end{bmatrix}$ $a_{1,3}=\begin{bmatrix}0.4&0.1\\0.4&0.3\\0.2&0.6\end{bmatrix}$	$b_4=\begin{bmatrix}0.323\\0.677\end{bmatrix}$ $b_3=\begin{bmatrix}0.292\\0.708\end{bmatrix}$	

References

1. Bakx, A.: Development and evaluation of a summative assessment program for senior teacher competence. Stud. Educ. Eval. **40**, 50–62 (2014)
2. Berliner, D.: Learning about and learning from expert teachers. Int. J. Educ. Res. **35**, 463–482 (2001)
3. Mangiante, E.M.S.: Teachers matter: measures of teacher effectiveness in low-income minority schools. Educ. Assess. Eval. Account. **23**(1), 41–63 (2011)
4. Kleickmann, T., Richter, D., Kunter, M., Elsner, J., Besser, M., Krauss, S., et al.: Teachers' content knowledge and pedagogical content knowledge. The role of structural differences in teacher education. J. Teach. Educ. **64**(1), 90–106 (2013)
5. Snoek, M., Clouder, C., De Ganck, J., Klonari, K., Lorist, P., Lukasova, H., et al.: Teacher quality in Europe: comparing formal descriptions. Paper presented at the ATEE conference 2009, Mallorca (2009)

6. Zhang, Q.: Dynamic uncertain causality graph for knowledge representation and reasoning: discrete DAG cases. J. Comput. Sci. Technol. **27**(1), 1–23 (2012)
7. Zhang, Q., Dong, C., Cui, Y., Yang, Z.: Dynamic uncertain causality graph for knowledge representation and probabilistic reasoning: statistics base, matrix and fault diagnosis. IEEE Trans. Neural Netw. Learn. Syst. **25**(4), 645–663 (2014)

An Action-Based Constraint Satisfaction Algorithm for Planning Problems

Xiao Jiang[✉], Yuting Zhao, Rui Xu, and Wenming Xu

School of Aerospace Engineering, Beijing Institute of Technology, Beijing, China
jiangxiaotwn@hotmail.com

Abstract. In recent years, the complex and growing constraints in the planning domain have made constraint satisfaction technology a hotspot in the automatic planning field. In the procedure of planning, the action processing plays a vital role. However, diverse but conflicting action relation is difficult to solve in a typical constraint satisfaction method. This paper we code the planning actions as constraints to make the classic planning model a domain-dependent constraint satisfaction problem and let the domain-specific action constraints help guide the search. Based on the model, a dynamic constraint set and constraint process method is proposed to solve the conflicts between the specific action constraints and general constraints in constraint satisfaction problem. The simulation experiments show that the proposed algorithm, with specific constraints, can effectively reduce the planning time and achieves a better performance over other constraint programmed planners.

Keywords: Planning · Constraint satisfaction · Domain dependent · Action constraint

1 Introduction

Constraint satisfaction technique, which is the primary means in the field of artificial intelligence, is widely used because of its strong pruning ability and high processing efficiency [1]. In recent years, the complex and growing constraints in the planning domain highlights the importance of constraint handling, so the introduction of CSP technology into the automatic planning community has been widely studied [2]. The benefit to do so is obvious. Firstly the researchers in planning field can directly use the developed CSP algorithms with a constraint-programmed planner. Secondly, planning and CSP focus on the different artificial intelligent method: CSP focuses on inference while planning focuses on search [3]. This has provided a new research method for the planning field after applying the CSP technique.

For this reason, constraint satisfaction technique has become a hotspot in planning field. In 1996, Kautz and Selman first used constraint programming method to solve planning problems. They estimated the length of a planning problem during a preprocess procedure and set this estimated value as a fixed bound, which transfers the planning problem into an NP hard problem. Therefore, the resulting problem can then be solved by CSP, which is an NP-Complete formalism. Based on this theory, more research on

© Springer International Publishing AG 2018
F. Qiao et al. (eds.), *Recent Developments in Mechatronics and Intelligent Robotics*,
Advances in Intelligent Systems and Computing 691, DOI 10.1007/978-3-319-70990-1_61

constraint-programmed planning has been made focused on the coding of the translation and the extensions of CSP to address the complex planning domains such as timelines [4], possibilities [5], and conditional effect [6].

However, the method to process actions in a constraint-programmed planner is still a technical difficulty. Different from typical CSP constraints, which must be all satisfied when a solution is found, the action relations are usually mutex. In this paper our objective is to code the action as constraints and develop a domain dependent, action-based algorithm. A dynamic constraint set and constraint process method is proposed to solve the conflicts between the specific action constraints and general CSP constraints and use the domain-specific action constraints help guide the search.

The paper is structured as follows: section two provides the background on AI planning and CSP and analyses the difficulty of coding a planning into a CSP. Section 3 introduces the proposed algorithm to design a dynamic action constraint set and solve the mutex action problem the we run comprehensive experiments in Sect. 4 to verify the validity of our algorithm. And finally, the conclusion is provided in Sect. 5.

2   Background

Planning is a combinatorial problem that seeks to achieve a given set of states using a set of predictable operations in sequence [7]. The target of planning is to organize an action sequence to transfer a planning problem from the initial state (also denoted as S_0) to the goal state (S_w), which includes every element of the goal set G.

Definition 1 (AI planning). A planning problem can be expressed as a triple $\Pi = \langle A, I, G \rangle$ which includes:

- An initial state set $I = \{i_1, i_2, \cdots, i_l\}$.
- An action set $A = \{a_1, a_2, \cdots, a_m\}$ such that every element in the set is a tuple $\langle pre(a), eff(a) \rangle$, where $pre(a)$ denotes the precondition of the action a and $eff(a)$ denotes the effects.
- A goal set $G = \{g_1, g_2, \cdots, g_t\}$ that denotes the target state of the planning.

Constraint programming (CP) is a powerful paradigm for solving combinatorial problems and is successfully applied in many domains such as scheduling, vehicle routing, configuration, and bioinformatics [8].

Definition 2 (CSP). A constraint satisfaction problem can also be expressed as a triple which includes:

- A variable set $X = \{x_1, x_2, \cdots, x_n\}$.
- A domain set $D = \{D_1, D_2, \cdots, D_n\}$ such that for every element x_i in the set there is a domain D_i.
- A constraint set $C = \{c_1, c_2, \cdots, c_k\}$ such that every element in the set defines a predicate which denotes a relation over a particular subset of variables.

In all the difficulties of transferring planning problems to CSPs, the first problem is that the size of a planning result is unknown before the search starts, whereas the size of a CSP is static. A typical solution is to set a fixed bound k beforehand and based on this bound translate the planning problem into a CSP with k levels. Another problem is that the action relations are different from the traditional constraints in CSP. When a solution of the CSP has been found, all the constraints must be satisfied. However, the situation is different in a constraint-programmed planning problem. For a simple example, one cannot pick up and drop a ball at the same time. In the next section, a dynamic set of constraints is proposed to solve this problem.

3   Proposed Method

Variable ordering is a core part for CSP search algorithm. A good choice of variable will reduce the amount of backtracking. Usually, a variable that is most likely to fail, in other words, the variable bound by the most constraints, will be select first. After the variable selection, we need to handle the action constraints mutex problem. Here, we construct a dynamic set of action constraints with multiple layers.

The rules for constructing this set include the following:

1. The layer number of the constraint equals the selected variable's time step, which means that the layer number starts from one.
2. Action constraints in the same layer cannot be mutually exclusive.
3. Newly added constraints cannot change values of other assigned variable, which will make other constraints no longer satisfied.
4. Action constraints in the same layer can be satisfied simultaneously, which denotes the concurrency of actions.

Based on the above rules, the pseudocode is shown in Algorithm 1.

Because a variable at different time steps always has the same data structure, in the third phase of the heuristic, a driving variable will always be selected before its response variable. Therefore, the layer number of the constraints always equals the time step of the driving variable, thus avoiding confusion in the constraint hierarchy.

In the construction process, the function *isconsistent( )* is the core step in determining whether an action constraint is added to the set or dropped. To do so, the first step is to check whether it is consistent with other constraints of the same layer. Apparently, if two action constraints have the same driving variable but have different values, they are mutually exclusive. When two driving variables have the same value, we check whether they are a changeable variable.

Algorithm 1. The algorithm for constructing the dynamic set

1: **for** every constraint involved *current_variable* **do**
2:     **if** layer(*current_variable*.TimeStep) == Null **then**
3:         $C_i.layer = current\_variable.TimeStep$
4:     **else**
5:         **if**   isconsistent(layer(*current_variable.TimeStep*), C_i) **then**
6:             $C_i.layer = current\_variable.TimeStep$
7:         **else**
                return 0
8:         **end if**
9:     **end if**
10: **end for**

Finally, we must ensure that newly added constraints cannot change the values of other assigned variables, or it will make other constraints no longer satisfied. Thus, the response variables of the checked constraint must be unassigned. The pseudocode is shown in Algorithm 2.

Algorithm 2. Function isconsistent()

1: **for** every constraint in layer(n) **do**
2:     **if** $var \in C_j$ **then**
3:         **if** var.value in C_i != var.value in C_j **then**
4:             return 0
5:         **else**
6:             **if** situation 3 in table 3 no longer satisfied **then**
7:                 return 0
8:             **else**
9:                 **if** response variable of var has been assigned **then**
10:                     return 0
11:                 **else**
                        return 1
12:                 **end if**
13:             **end if**
14:         **end if**
15:     **end if**
16: **end for**
---

4   Experimental Results

The proposed algorithm was written in C/C++. To evaluate the performance, we compared it experimentally with anther constraint based planner GP-CSP. In the comparison experiments, these two planners will test the domains from previous IPCs including BlockWorld, Gripper, Logistics and Mystery. Experiments were performed on a system with an Intel i5-2430 2.4 GHz CPU and 4 GB of RAM.

From the comparison data for different domains, we can see that with more actions in the domain, the proposed algorithm can gain more domain information to accelerate the search process. For example, in domain BlockWorld with only one action, the proposed algorithm performance is worse than GP-CSP. However in domain Logistic which is with the most actions, the efficiency of the proposed algorithm is obvious (Figs. 1 and 2).

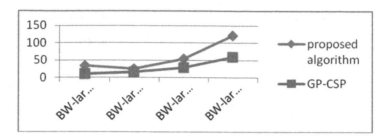

Fig. 1. Comparison experiments of BlockWorld

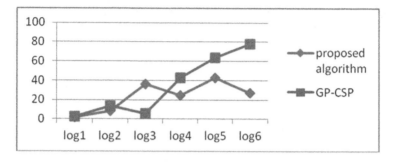

Fig. 2. Comparison experiments of logistics

The result can also be verified by the comparison experiments of domain Gripper, and Mystery. In the experiment results of Gripper and Mystery instances, we can see the proposed algorithm is more efficient than GP-CSP planner, and with an expansion of the scale of the problem, this advantage becomes more obvious with domain-specific action constraints and a means of specific processing to help guide the search. It has been proven that the action constraints greatly improve the efficiency of planning and that the key to their success has been their ability to prune the search space (Figs. 3 and 4).

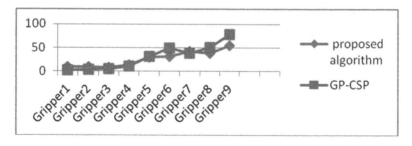

Fig. 3. Comparison experiments of Gripper

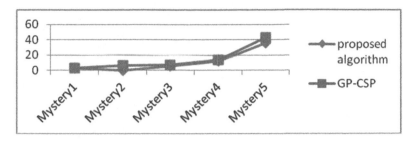

Fig. 4. Comparison experiments of Mystery

5   Conclusion

Based on the characteristics of planning, this paper proposed a domain dependent action-based constraint satisfaction algorithm. With action constraints, it is shown that a higher efficiency can be achieved compared to standard domain-independent planner like GP-CSP. These encouraging results suggest many directions for future work. For example, the time attributes of actions and variables can be more in-depth study. In future research, we will incorporate time constraints and resource constraints into the planning model to achieve a stronger connection with real world planning.

References

1. Roman, B.: Constraint processing. Acta Autom. Sin. **17**(2), 687 (2007)
2. Lozano-Pérez, T., Kaelbling, L. P.: A constraint-based method for solving sequential manipulation planning problems. In: International Conference on Intelligent Robots and Systems, pp. 3684–3691. IEEE (2014)
3. Veksler, M., Strichman, O.: Learning general constraints. Arti. Intell. **238**, 135–153 (2016)
4. Verfaillie, G.: How to model planning and scheduling problems using constraint networks on timelines. Knowl. Eng. Rev. **25**(3), 319–336 (2010)
5. Achlioptas, D., Hassani, S.H., Macris, N., et al: Bounds for random constraint satisfaction problems via spatial coupling. In: Twenty-Seventh ACM-SIAM Symposium on Discrete Algorithms, pp. 469–479. ACM (2016)
6. Rintanen, J.: Heuristics for planning with SAT. In: International Conference on Principles and Practice of Constraint Programming, pp. 414–428. Springer (2010)
7. Jorg, H.: FF: the fast-forward planning system. AI Mag. **22**(3), 57–62 (2014)
8. Edward, T.: Foundations of Constraint Satisfaction: The Classic Text, BoD–Books on Demand (2014)

Automation and Control

Adaptive Control with Parameters Compensator for Aircraft Boundary Control System

Qian Kun[⊠], Liu Kai, Yan Hao, and Liao Kaijun

The First Aeronautical Institute of Air Force, Xinyang, Henan, China
qiankun_0306@163.com

Abstract. This paper proposed a method of adaptive controller with a lead-lag parameters compensator and applied to a kind of aircraft boundary control system (ABCS). This method ensures optimization of fuzzy logic control function, and guarantees on-line adaptation of domain in aircraft boundary control system, which can make a great progress in dynamic characteristics and adaptabilities of flight control system. A proposed hybrid adaptation algorithm has been tested in the analogy flight practice speed control system of an induction motor drive. The digital simulation results demonstrated that it could improve transient properties in flight boundary control system, and provide well disturbance rejection in spite of its simplicity.

Keywords: Aircraft · Adaptive control · Flight control system · Boundary control · Reference model · Fuzzy logic compensation · Servo system

1 Introduction

Aircraft boundary control system, on the basis of full authority digital electronic controller, has been an important developing direction in flight control system. Contrast to conventional centralized control system, flight distributed control system is different essentially in control construction, control method and control realization. It is decreased greatly in weight and improved high in performance and reliability.

For CAN bus characteristics in flight control system application, such as long length and complicated signal types, this paper lucubrated signal transmission delay led to CAN bus length and node. Relation among CAN bus length, nodes, signal type and data stream is reasoned out, and it solved CAN bus network data collision problem led to data steam overload [1–3]. This paper proposed originally a mathematical model for a linear distributed network control system with delays in forward and feedback channels and solved efficiently time lag for distributed real-time control system. It established important theory basic for CAN bus applying to flight control system and realized great breakthrough in CAN bus research.

© Springer International Publishing AG 2018
F. Qiao et al. (eds.), *Recent Developments in Mechatronics and Intelligent Robotics*,
Advances in Intelligent Systems and Computing 691, DOI 10.1007/978-3-319-70990-1_62

2 Adaptive System Structure

There are many kinds of tasks, such as hard real-time tasks, soft real-time tasks, non-time tasks, and so on in ABCS. Among them, hard real-time tasks must be finished before their deadlines, otherwise the catastrophes occur with the second-order transfer function:

$$G_p(s) = \frac{Y(s)}{U(s)} \approx \frac{K_A}{T_A^2 \cdot s^2 + 2\xi \cdot T_A \cdot s + 1} = G_A(s) \tag{1}$$

where $G_p(s)$ is reference model and $G_A(s)$ is its approximation.

Therefore, fault-tolerance must be offered for ABCS. The stability of interval polynomial matrices family is researched. By the method of linear matrix inequality, this problem is demonstrated to be equivalent to a fuzzy logic control problem.

The condition that the task must meet when they are schedulable and the method for setting robust fuzzy controller with parametric uncertainties are investigated. Here we use a second-order reference model to aircraft boundary control system:

$$G_M(s) = \frac{Y_M(s)}{U_R(s)} = \frac{1}{\frac{T_M}{K_M} \cdot s^2 + \frac{1}{K_M} \cdot s + 1} \tag{2}$$

where $U_R(s)$ and $Y_M(s)$ are input and output of reference model. Based on Fuzzy model, a present algorithm is constructed in consideration of parametric uncertainties that can guarantee the stability in aircraft boundary control system in Fig. 1.

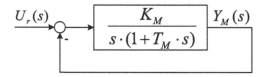

Fig. 1. The structure of 2nd-order reference model

For the first time, we proposed general design for aircraft boundary control system based on control area network bus. For system hardware, research was developed deeply in bus design, central processing unit function design, intelligent unit design, distributed power system design and core circuit module design. Then, we constructed self-object simulation system and realized the simulation for ABCS. Detailed hardware circuit schemes of smart temperature sensors, smart rotate speed sensor, smart automatic adjust-angle transfer sensor and intelligent actuator were finished for the first time.

For the same multi redundant smart sensor, we discussed a data fusion method based on optimized distributed decision. This method can prevent efficiently algorithm loss when the data are compressed and improve performance for the decision system. For different smart sensors, we discussed a sensor data fusion algorithm based on matrix singular value decomposition. The sensor data fusion algorithm can compensate thermocouple's inertia delay. It provide theory basis for practical engineering in future and have important value [4]. The configuration of the online self-tuning aircraft boundary control system is shown in Fig. 2. Flight control compensator function can describe as follows:

$$G_f(s) = K_f \cdot \frac{T_f \cdot s + 1}{T_M \cdot s + 1} \tag{3}$$

where the function defines two adaptive coefficients, gain constant K_f and delay coefficient T_f.

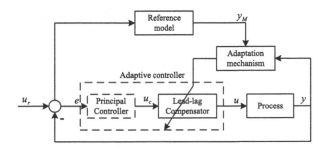

Fig. 2. The configuration of aircraft boundary control system

3   The Algorithm Design

A control law of parameters compensator for aircraft boundary control system can describe as follows [5]:

$$\begin{aligned} K_f^{new} &= K_f^{old} \cdot (1 + \gamma_K \cdot \Delta K_f) \\ T_f^{new} &= T_f^{old} \cdot (1 + \gamma_T \cdot \Delta T_f) \end{aligned} \tag{4}$$

where we define that γ_K and γ_T are adaptive parameters, ΔK_f and ΔT_f are gain constant and delay coefficient respectively.

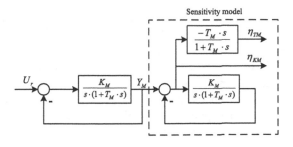

Fig. 3. The configuration of the second order reference model

For the design of CPU logic algorithm for aircraft boundary control system, a new delay parameters compensator algorithm with fuzzy logic control was proposed and applied to flight control system. A dynamic neural network was to identify the boundary control system on-line. The control signal was then calculated iteratively according to the reference model and the identified system [6]. The second order configuration of the reference model is shown in Fig. 3.

4   Digital Simulation

The proposed parameters compensator algorithm has been tested in aircraft boundary control system of an induction motor drive with a fuzzy logic controller. The configuration of the parameters compensator algorithm is shown in Fig. 4.

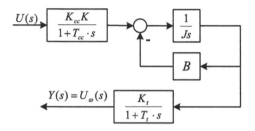

Fig. 4. Configuration of the parameters compensator algorithm

The fuzzy logic parameters of aircraft boundary control system are as follows: $K_{cc} = 0.8$ A/V – current compensator coefficient, $T_{cc} = 0.06$ ms – current compensator delay constant, $K = 0.75$ – motor speed gain, $J = 0.0015$ kg·m^2 – motor of inertia, $B = 0.0004$ Nms – slip friction constant, $Kt = 0.05$ V·s – rotation motor gain and $T_t = 2.5$ ms – rotation motor delay constant.

Digital simulation shows in Fig. 5 for response of induction motor drive speed control system. Form Fig. 6, we can see that an initial delay error $e_M = y_M - y$ is decreased in aircraft boundary control system of an induction motor drive.

In digital simulation, we can see in Figs. 7 and 8. The values of fuzzy logic compensator parameters converge to their steady-state values in boundary control system.

In the next experiment, we constructed self-object simulation system and realized the simulation for ABCS based on CAN bus. In engineering, in order to simulation fault diagnosis for smart sensors and validate the reliability in Fig. 9. We designed integrated detection instrument based on CAN bus for certain flight control system and solved CO sensors fault diagnosis. At the same time, for vibration sensor, we discussed two new fault diagnosis methods in Figs. 9 and 10, based on vibration parameter nephogram and based on stopping process vibration nephogram for components of flight control system.

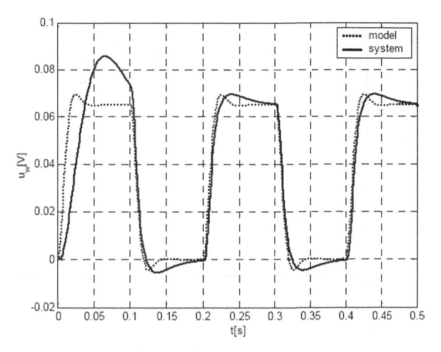

Fig. 5. ABCS responses for J = 5·Jn

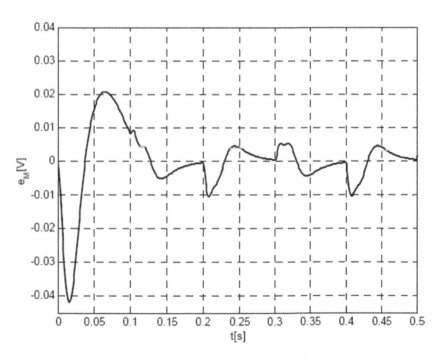

Fig. 6. Motor tracking error for J = 5·Jn

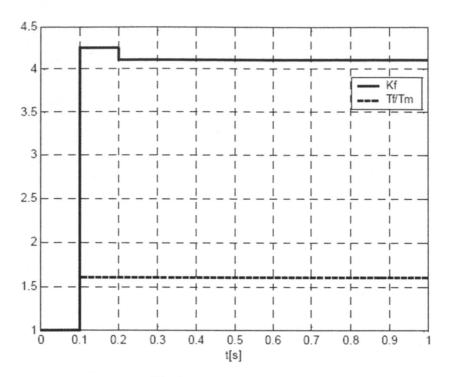

Fig. 7. K_f and T_f for $J = 5 \cdot Jn$

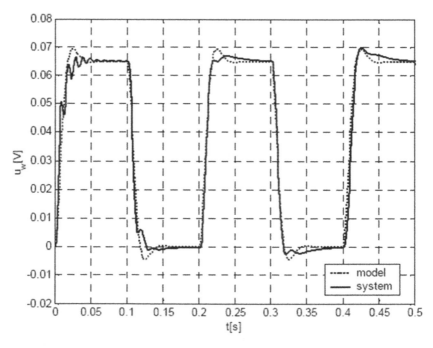

Fig. 8. ABCS responses for $5 \cdot K_{on}$

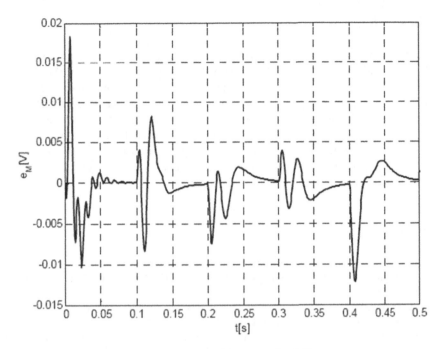

Fig. 9. Motor tracking error for $5 \cdot K_{on}$

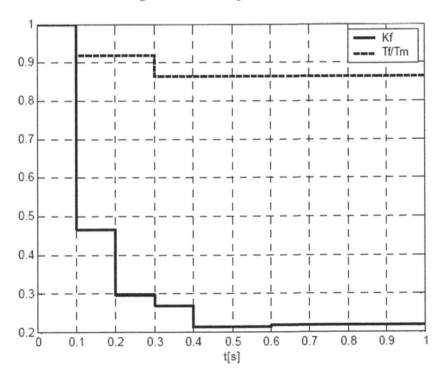

Fig. 10. K_f and T_f for $5 \cdot K_{on}$

5 Conclusion

This paper proposed a method of adaptive controller with a lead-lag parameters compensator and applied to a kind of aircraft boundary control system (ABCS). This method ensures optimization of fuzzy logic control function, and guarantees on-line adaptation of domain in aircraft boundary control system, which can make a great progress in dynamic characteristics and adaptabilities of flight control system. A proposed hybrid adaptation algorithm has been tested in the analogy flight practice speed control system of an induction motor drive. The digital simulation results demonstrated that it could improve transient properties in flight boundary control system, and provide well disturbance rejection in spite of its simplicity.

References

1. Qian, K., Liao, K., Liu, K.: Identification of complex systems using multiple-model adaptive estimation algorithms. In: Proceedings of International Conference on EME 2015, Wuhan, China, vol. 1, pp. 870–876 (2015)
2. Fan, Y., Wang, W., Jiang, X.: Generalized fuzzy hyperbolic model adaptive control design for a class of nonlinear systems. Control Decis. **31**(3), 417–422 (2016)
3. Qian, K., Xie, S., Hu, J.: Aerial pneumatic servo system based on modified feedback linearization control. Control Theory Appl. **23**(3), 465–467 (2005)
4. Gao, X., Ren, X.: Prescribed performance adaptive control for hysteresis hammerstein system. Trans. Beijing Inst. Technol. **36**(4), 412–416 (2016)
5. Li, D., Wu, B., Zhang, Y.: Robust adaptive control for spacecraft attitude tracking with unknown friction torque of reaction wheel. J. Astronaut. **37**(2), 175–181 (2016)
6. Qian, K., Pang, X., Li, B.: Neural network identifier with iterative learning for turbofan engine rotor speed control system. In: Proceedings of CIS 2008, Chengdu, China, vol. 1, pp. 94–98 (2008)

Complete Sets of Aquaculture Automation Equipment and Their Monitoring Cloud Platform

Jiayi Huang[1,2(✉)], Xiangbao Meng[1,2], Qiubo Xie[1,2], Wanyun Chen[2], and Shuyang Wang[2]

[1] Guangdong Institute of Modern Agricultural Equipment, Guangzhou 510630, China
Rachel_jiayihuang@foxmail.com
[2] Guangzhou Jiankun Network Technology Development Co. Ltd., Guangzhou 510630, China

Abstract. For the development of modern aquaculture in Guangdong, a set of intelligent breeding equipment have been researched which consisted of unmanned cruise boat for fish pond or lake, automatic aquaculture plant for indoor, buoy floating with wind and waves or on-demand voyages for marine pasture and their supporting sensors, as well as a remote service cloud platform for centralized online scheduling and monitoring these equipment. Through demonstration application in several enterprises, it was proved that these products can improve aquaculture efficiency, management level and ecological environment by the use of lower late farming inputs.

Keywords: Unmanned cruise boat · Automatic aquaculture plant · Buoy floating · Sensor · Cloud platform

1 Introduction

China is a major production and consumption of aquatic products, which aquaculture accounts is about 60% of the world. With the rapid development of industry and agriculture, water pollution and the shortage of quality resources have been becoming more and more serious that led to the quality of aquatic products can not be effectively protected. In order to improve the ecological environment, promoting aquaculture from the traditional "extensive" to "precision" intelligent farming transformation has become a very important issue of promoting the aquaculture development to a healthy and efficient direction. Thus, it is very necessary to achieve real-time online monitoring water and ecological status, while achieve linkage of aquaculture equipment (aerobics, feeding and fishing machine, etc.) with the change of water quality and the growth demand of aquatic products in [1–4].

The purpose of this paper is to use industrial automatic control technology, combining with agricultural machinery and agronomy, to develop complete sets of aquaculture automation equipment and their remote centralized control cloud platform for three different aquaculture scenarios, namely fish pond or lake culture, and indoor industry culture marine farming.

© Springer International Publishing AG 2018
F. Qiao et al. (eds.), *Recent Developments in Mechatronics and Intelligent Robotics*,
Advances in Intelligent Systems and Computing 691, DOI 10.1007/978-3-319-70990-1_63

2   Method

2.1   Hardware Development

For fish pond or lake culture, prerequisite of scientific aquaculture is to obtain real-time and fast access to water key parameters, usually including temperature (T), pH, dissolved oxygen (DO), oxidation-reduction potential (ORP) and electronic conduction (EC) in [5, 6]. However, most domestic farmers estimated the quality of water still through their own experience, or used artificial sampling analysis that is time-consuming and laborious. Although some companies used multiple sets of fixed sensor groups to obtain water quality at different locations and different depths in breeding area, but it obviously increased business operating costs of purchase and maintenance. Therefore, it is difficult to popularize.

For solving this problem, a kind of equipment was developed, consists of unmanned automatic cruise boat, environmental and ecological monitoring devices and remote service platform. It achieved a real-time online monitoring and precise control of field devices with high efficiency and low cost. By using the technology of automatic heading speed control, navigation and anti-collision, the unmanned boat realized the function of automatic cruise. Then, the boat carried GPS compass, control device, propeller propulsion plant, speed sensors and a self-made multi-functional device (one controller with different probes) to obtain ecological characteristics and various parameters at each specified location, including T, pH, DO, ORP, GPS and video. Detection parameters can be expanded according to the actual needs. According to the monitoring results, the switch of the aerator, feeding machine or other equipment could be opened or closed automatically. The structure of the boat was shown in Fig. 1.

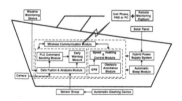

Fig. 1. Hardware schematic diagram of unmanned automatic cruise boat

In order to further simplify the control system of the unmanned boat, left and right propellers (also known as propulsion) were used to control the boat's course and speed (see in [7–10]) instead of a rudder (see in [11]). The overall flow chart of the automatic control system of the rudderless ship is shown in Fig. 2.

Compared with the conventional steering control system, it has several advantages: small turning radius, short response time. What's more, it is simple in structure and operation, and easy to achieve high-precision control in contrast with the rotary course control system.

In order to achieve a large-scale and high-density breeding of valuable varieties, more and more farmers or enterprises choose industry culture, which biggest advantage is the ability to precisely provide suitable aquaculture water for different types of fish

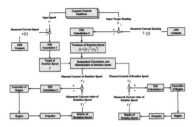

Fig. 2. Overall flow chart of the automatic control system

or other aquatic products based on their growth needs. High-density aquaculture is a general choice for factory farming, which key technology is how to effectively improve the efficiency of water purification with lower costs, and find or even predict the abnormal situation of aquatic products in time. To this purpose, a set of factory automation breeding equipment was designed, as shown in Fig. 3.

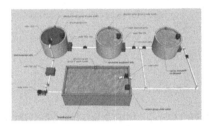

Fig. 3. Design 3D drawing of factory automation aquaculture and water treatment

In each water treatment sink and breeding pool, according to the requirements of water purification and aquaculture objects, the appropriate sensor group was installed, such as sensors of T, pH, DO, ORP and EC are normal configuration. For water treatment, TD, COD and BOD are sometime needed, but their on-line detection equipment is quite expensive. Users can choose on demand or through the sampling method to increase the number of the detection parameters. Then, the data of sensors are transmitted to the monitoring platform and the centralized controller of the devices in real time, such as water pump, electronic valve, etc. Based on the detection result and the parameter threshold, the centralized controller automatically turns on or off the switch of water pump, electronic valve or water treatment equipment. Clearly, stable and reliable sensor and professional expert library is the key to determining the correctness of device control.

The focus of marine farming is to monitor environment and ecology in real time, predict its change trend and achieve efficient management of fishery information. For the fist two requirements, a buoy is usually selected. Continued power supply, high safety and resistance to wind are the core of a buoy's normal work. Therefore, a new buoy with two movement modes (floating with the wind and traffic by designated routes) that can adjust the angle of the solar panel according to the longitude was designed.

2.2 Software Design

Software design includes two parts, equipment software and cloud service platform. The platform in the structures is divided into three levels: IaaS (Infrastructure as a Services), PaaS (Platform as a Services) and SaaS (Software as a Services), as shown in Fig. 4.

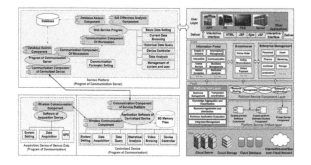

Fig. 4. Architecture diagram of cloud services platform on aquaculture management

3 Results and Discussion

3.1 Fish Pond or Lake Culture

Figure 5 gives the outside and inside of the boat. Based on the moving boat, the number of monitoring sensor group could be obviously reduced and effectively improve the detection accuracy of the sensors. Furthermore, information fusion, statistical analysis and configuration control were used to analyze the obtained data and precisely manage field aquaculture equipment. The result shows that the integrated monitoring equipment can meet the needs of large-scale and factory-style aquaculture, and promote the wider application of precision agriculture equipment in ordinary farmers.

(a) outside of the boat   (b) inside of the boat

Fig. 5. Pictures of the unmanned boat in field

3.2   Indoor Factory Culture

In order to avoid affecting aquaculture production, a support device is designed, so that the sensor group and other equipment installed on the support device can be free to translate and change the depth of immersion. The design and installation of the sensor group in field are shown below (Fig. 6).

(a) design of the sensor group   (b) installation of the sensor group in field

Fig. 6. Design and installation picture of the sensor group in field

3.3   Marine Farming

Figure 7 gives the outside and inside of the buoy, which function is similar to the unmanned boat as shown in Fig. 5, thus it is not repeated here.

(a) outside of the buoy   (b) inside of the buoy

Fig. 7. Pictures of the buoy in field

3.4   Cloud Platform

Users can access the cloud service platform through cell phone, PAD or PC and achieve centralized management of multiple farms. It could help a staff to discover problems in time and accurately lock its location then precisely control field equipment by PLC based on the measurement of sensor group and built-in threshold, to provide a proper condition to aquatic growth. App interfaces of the cloud service platform are shown in Fig. 8.

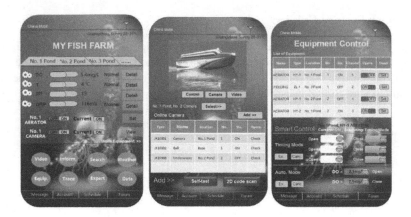

Fig. 8. App interfaces of the cloud service platform

4 Conclusion

For the three aquaculture scenarios, this study provides a complete set of intelligent equipment for environmental and ecological monitor, their cloud service platform as well as personalized industry solutions. Through demonstration application in several enterprises, it can improve aquaculture efficiency effectively, ensure the quality and safety of aquatic products, reduce breeding risk and artificial input, and provide a method for environmental sustainable development. It is particular good to meet efficient management needs of a comprehensive aquaculture enterprise.

Acknowledgments. This work was supported by Pearl River S&T Nova Program of Guangzhou (2014J2200091), Science and Technology Program of Guangzhou (201604046004) and 2016 Provincial Modern Agricultural Science and Technology Innovation Alliance Construction Program (2016LM2147). In addition, I am indebted to a talented workmate, Mr. Rongbin Xian, who helped us draw 3D figure.

References

1. Shi, B., Zhao, D., Liu, X., et al.: Intelligent monitoring system for industrialized aquaculture based on wireless sensor network. Trans. CASE **27**(09), 136–140 (2011)
2. Shi, B., Zhao, D., Liu, X., et al.: Design of intelligent monitoring system for aquaculture. Trans. China. Soc. Agric. Mach. **42**(09), 191–196 (2011)
3. Li, H., Liu, X., Li, J., et al.: Aquiculture remote monitoring system based on IOT Android platform. Trans. CSAE **29**(13), 175–181 (2013)
4. Li, L., Shi, W.: Design of dynamic monitoring system for seawater factory culture. Jiangsu Agric. Sci. **40**(03), 364–367 (2012)
5. Li, D., Fu, Z.: Design and development of intellectual aquacultural information system. Trans. CSAE **16**(04), 135–138 (2000)
6. Ma, C., Zhao, D., Qin, Y., et al.: Intelligent monitoring and control for aquiculture process based on fieldbus. Trans. China. Soc. Agric. Mach. **38**(8), 113–115 (2007)

7. Cui, J., Ma, T., Yang, S., et al.: Optimized analysis of high-speed USV intelligence propulsion system. Ship Sci. Technol. **9**(1), 69–73 (2009)
8. Shanghai Maritime University. A ship propulsion system and navigation control method: China, 200810034634.3, 14 January 2009
9. Zhu, D., Zhang, B., Wang, Y., et al.: Full - swing pod propulsion hydrodynamic performance test. Ship Eng. **35**(S2), 58–61 (2013)
10. Islam, M.F., MacNeill, A., Veitch, B., et al.: Effects of geometry variations on the performance of podded propulsors. In: Maritime Technology Conference & Expo and Ship Production Symposium, Florida, USA (2007)
11. Chen, G.: Research on Ship Heading Control Based on Fuzzy Sliding Mode Variable Structure Control. Jiangsu University of Science and Technology, Zhengjiang (2010)

Application of Genetic Algorithm and Neural Network in Ship's Heading PID Tracking Control

Jianming Sun, Renqiang Wang[✉], Kang Yu, Keyin Miao, and Hua Deng

Jiangsu Maritime Institute, Navigation College, Nanjing 211170, China
wangrenqiang2009@126.com

Abstract. An intelligent PID control algorithm is designed for ship's course tracking control. As known to all, the deficiencies of traditional PID control is that its parameters cannot be adjusted in real time, and thus control effect is not perfect. To this end, Genetic Algorithm is used to optimize PID parameters in this paper in order to achieve feedback control and to ensure system stability. CMAC neural network is used to achieve feed-forward nonlinear control, and to restrain internal and external disturbance for ensuring satisfied control accuracy and response speed of system. The result of simulations shows that the controller devised in the paper has superior performance and strong robustness.

Keywords: Intelligent PID control · Genetic Algorithms · Neural network · Cerebellar Model Articulation Controller · Ship course controller

1 Introduction

Ships in the sea will be influenced by the wind, waves and other strong interference. By the mean time, ship movement has time-delay and non-linear characteristics. The traditional law of PID was unable to track the direction of the ship correctly. The Intelligent Control Theory, which was born in the 60 s of last century, will provide effective solutions for non-linearity of ship motion and uncertainty control. To Combine PID Control that after GA (Genetic Algorithm) optimized and CMAC (Cerebellar Model Articulation Controller) neural networks together organically, one could design a new ship navigation controller with three functions, which could shorten the voyage and reduce resistance and energy consumption as well as protect the steering gear. The overall design process of the controller is to upgrade the conventional PID manual control to intelligent PID control, that is to use the GA optimize and integrate PID parameter automatically, and to avoid cumbersome manual off-line parameter controlling adjustment, thereby, to achieve feedback controlling and stable system. CMAC neural networks are used to carry out feed-forward non-liner controlling and to suppress the non-liner interference from inside and outside the ship as well to control the complicated, dynamic and unstable system, for ensuring the controlling precision and responding speed of the system.

© Springer International Publishing AG 2018
F. Qiao et al. (eds.), *Recent Developments in Mechatronics and Intelligent Robotics*,
Advances in Intelligent Systems and Computing 691, DOI 10.1007/978-3-319-70990-1_64

2   Ship Response Mathematical Model

In 1950s, Japanese scholars Nomoto established the type of ship response mathematical model based on the perspective of controlling theory. The mathematical expression of this model is:

$$\ddot{\psi} + \frac{K}{T}H(\psi) = \frac{K}{T}\left(\delta + \delta_D\right) \tag{1}$$

In this expression, T represents following index, K represents turning index, $H(\dot{\psi}) = \alpha\dot{\psi} + \beta\dot{\psi}^3$, α and β are parameters of nonlinear system, δ is rudder angle, δ_D is external disturbance equivalent rudder angle.

3   Design of Steering Controller Based on PID and CMAC

3.1   Feed-Forward CMAC Neural Network Algorithm

Blended controlling system is the organic fusion of PID optimized controller and CMAC controller, it can realize stable and fast controlling of the whole controlled system. CMAC gets on feed-forward controlling, to create inverse dynamic model of the controlled object; PID optimization gets on feedback controlling, to value the performance of next level CMAC, and thereby, it can ensure the stability of the system and suppress disturbance. CMAC uses the type of mentor learning algorithm, when every controlling cycle ends, the CMAC function calculator will calculate the corresponding output value un(k), comparing with the total controlling input value u(k), then carry out the process of correcting the weight and learning, with the purpose of minimizing the difference between u(k) and un(k) to a minimum. The controlling algorithm of the blended controlling system is,

$$u_n(k) = \sum_{i=1}^{c} w_i a_i \tag{2}$$

$$u(k) = u_n(k) + u_p(k) \tag{3}$$

In this expression, a_i represents binary selection vector, c represents CMAC the generalization parameters of the network and up(k) PID the output value of the optimizer, adjustment index of CMAC is,

$$\begin{cases} E(k) = \dfrac{1}{2}(u(k) - u_n(k))^2 \cdot \dfrac{a_i}{c} \\[2mm] \Delta w(k) = \eta\dfrac{u(k) - u_n(k)}{c}a_i = \eta\dfrac{u_p(k)}{c}a_i \\[2mm] w(k) = w(k-1) + \Delta w(k) + \alpha(w(k) - w(k-1)) \end{cases} \tag{4}$$

In the expression, η represents network learning rate $\eta \in (0, 1)$, α represents Inertia quantity, $\alpha \in (0, 1)$.

3.2 PID Controlling Algorithm

PID controlling algorithm utilizes proportion-differential-integration controlling algorithm to automatic steering instrument for ships, the yaw signal after electronic circuits treatment will give feedback to the controlling device of steering gear, and thereby, to achieve automatic steering. The mathematical expression of this controlling algorithm is:

$$u = K_p \cdot e + K_i \cdot \int edt + K_d \cdot \frac{de}{dt} \qquad (5)$$

In this expression, deviation $e = rin - yout$ (e represents heading deviation, rin represents standard course, $yout$ represents the actual heading), u represents the controlling steering signal, K_p represents Proportional gain, K_i represents Integral gain, K_d represents differential gain.

3.3 PID Control Parameters Online Optimization

The parameters of PID will be optimized by GA, and the optimization includes the following parts.

(1) Coding

The GA cannot deal with the solution data in the solution space directly, therefore, we have to prepare several Initial solutions for the GA, then to form an initial population, which is the first generation of the subsequent evolution. Because the individuals in the initial population was generated randomly, it has common fitness.

(2) Construct fitness function

The fitness function is the interface of GA and optimization, and it is also the evaluation link of the system performance. Every individual who has been broken down into its own representative parameters through translation process, are substituted into the optimized fitness function, thereby, to find the indicators of system performance. This article uses the following economic performance index of ship's handling:

$$\begin{cases} J = \dfrac{1}{N} \sum_{n=1}^{N} \left(\lambda_1 \cdot e_n^2 + \lambda_2 \cdot \delta_{rn}^2 + \lambda_3 \cdot \dot{\delta}_n^2 + \lambda_4 \cdot r_n^2 \right) \\ e_{\min} \le e_n \le e_{\max} \\ -35° \le \delta_r \le 35° \\ -3°/s \le \dot{\delta} \le 3°/s \end{cases} \qquad (6)$$

In the expression, e represents course deviation, its range was defined according to the navigation environment, setting up according to the actual situation, δ_r represents actual rudder angle, $\dot{\delta}$ represents rudder speed, r represents heading rate, N represents iteration, J represents performance index, λ_1, λ_2, $\lambda3$ and λ_4 represent weighting coefficient. Weighting coefficient λ_1, λ_2, λ_4 should been selected according to sea state. When the sea state is getting worse, λ_1 values will become larger, λ_2 values will become smaller, and λ_3 is smaller value.

(3) Copy

The purpose of copy operation is to select excellent individuals from the current group, so that they will have the opportunity to become the parent generation. The criteria for judging good or bad is their own fitness value. Obviously, this operation following the principles of survival of the fittest were eliminated, and is the higher the fitness of the individual, the higher the probability of being selected. The specific method is multiplying each fitness value by a random number, and then choose the maximum product value of two chromosomes as the male parent used to breed individuals.

(4) Alternate

Alternate would been realized in two following steps in the match pool containing gene groups waiting for matching. The new genes generated by copying will be randomly paired. Selecting intersection randomly, and crossing breeding matching genes, we will have a new pair of genes. The specific process as follows,

Assume the length of genetic character is l, randomly select an integer value between the range of $[1, l - 1]$ as cross dot, and change all characters after the place of k, now there are two new bit strings. Multi-point crossover will be in the same way. Crossover operator is the most important genetic operation in algorithm, because it can get a new generation of individuals.

(5) Variation

To avoid missing important information in the process of replication and crossover, we should join mutation operator in the system of artificial inheritance. Mutation operator is quite Subtle genetic manipulation, it needs used with copying, crossing properly, and the purpose of that is excavating the diversity of individual in the group, overcoming the drawback of been lost in the local optimal solution possibly.

3.4   Steering Control Flow Chart Based on PID and CMAC

According to Fig. 1, the yaw information sending from the input signal *rin* and output signal *yout*, to compare the yaw information from the detector, genetic algorithm device would calculate the amount of yaw of the ship, angular velocity of vectoring, left and right symmetry in real time, and send PID linear parameter used of correction conforming the characteristic of Instant yaw motion of the ship. PID would calculate Steering time, frequency and the rudder angle size according to the parameter mentioned above, then send steering command signal to servo, which is optimizing PID linear parameter automatically through genetic algorithm, to ensure of the stability of the system and feedback controlling. Based on the PID optimization algorithm mentioned above, after quantification of *rin* is performed, and sending into the address map at the same time, articulation controller will perceive the messy nonlinear interference from the inside and outside the ship, and will produce additional steering command signal sending to servo, to get on feed-forward nonlinear control and suppresses nonlinear interference, therefore, it can ensure the control precision and responding speed of the system.

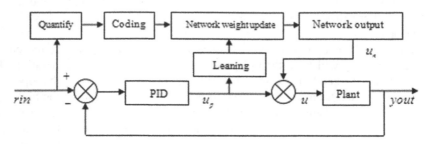

Fig. 1. Mixed controller based on PID and CMAC

4   Research of Simulation

Effectiveness of control algorithm is verified through simulation of training vessel "YUKUN". It is known that the parameters [10] of the ship's nonlinear system is $K = 0.24$, $T = 206.79$, $\alpha = 1$ and $\beta = 30$.

4.1   Course Changes Experiment

Conduct experiment of tracking square wave, set the cycle as 100 s, the change of course as 30°, the experimental results will be shown in Fig. 2.

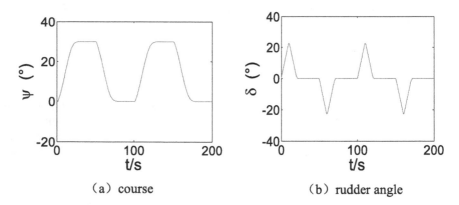

(a) course   (b) rudder angle

Fig. 2. The output of course angle and rudder angle

As can be seen from Fig. 2, in the case of desired track changes, the new controller can do better in steering and easing the helm counter steering and output the right rudder angle according to ship characteristics and sea condition at the right time, and also conduct without overshoot tracking with faster speed.

4.2 Interference Experiment

Setting the protocol initial heading as 000, desired track as 030, propose maneuvering ship model parameters K and T perturb at 40%, apply amplitude as 0.1° and make white noise interference to the course, the experiment results are shown in Fig. 3.

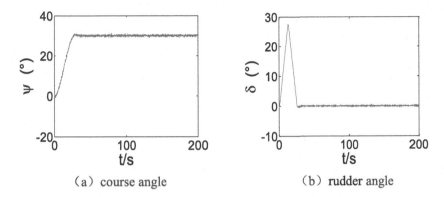

(a) course angle   (b) rudder angle

Fig. 3. The output of course angle and rudder angle

As can be seen from Fig. 3, when the ship model parameters occur perturbation with interference from the outside world, the course output from the new controller has smaller fluctuation around expect course, and has smaller shock amplitude of rudder controlling, thereby, it is illustrated that the anti-interference ability of the new controller is stronger.

5 Conclusion

This article from the mathematical model of Nomoto ship motion point of view, considering the influences of changes of ship parameters and outside interference has designed a algorithm called adaptive control of ship's course change by combining genetic algorithm and CMAC together, and the design flow is also given.

The simulation result shows that the algorithm of adaptive control of ship's course change designed by this article has ideal controlling effect and strong robustness, and it also has a wide range of practical application value in the field of upgraded version of autopilot, maintenance of rudder equipment and economic and fast navigation.

Acknowledgments. This work is supported by Foundation of Jiangsu Maritime Institute under Grant No. XR1501 and No. 2015KJZD-01.

References

1. Liu, J.: Intelligent Control, 2nd edn. Peking University Press, Beijing (2012). pp. 30–32
2. Liu, J.: Advanced PID Control and MATLAB Simulation, 2nd edn. Electronic Industry Press, Beijing (2012). pp. 44–46
3. Zhu, J., Zou, L., et al.: CMAC neural network PID control. Control Syst. **7**, 12–13 (2005)
4. Jia, X., Yang, Y.: Ship Motion Mathematical Model. Dalian Maritime University Press, Liaoning (1999). pp. 52–55
5. Dai, Y.: Ship course control to optimize the fuzzy controller based on genetic algorithm. Ship Ocean Eng. **6**, 22–23 (2009)
6. Song, Y., Du, H.: Control designed based on CMAC and parallel PID controller design and application. Control Syst. **3**, 31–32 (2005)
7. Zhang, W., et al.: Parallel control study based on the improved CMAC and PID. Comput. Meas. Control **13**(12), 6–7 (2005)
8. Zhang, L., Zhao, Q., et al.: CMAC neural network simulation of submarine maneuvers. Ship Mech. **4**, 10–11 (2009)
9. Jin, H., Yu, B.: CMAC and PID composite control of fin stabilizer. Ship Sci. Technol. **1**, 7–8 (2009)
10. Zhang, X., Jia, X.: Ship Motion Control. National Defense Industry Press, Beijing (2006). pp. 60–64

Rapid and Automated Analysis of Portable Flow Cytometer Data

Xianwen Wang[1], Peng Li[2], Jianggang Wu[1], Qingyong Zhang[1], Xiaoliang Li[1], Zhiqiang Yang[1], and Li Li[1(✉)]

[1] Institute for Drug and Instrument Control of Directorate of Medical Services, Logistics Support Department of Central Military Commission, Beijing, China
liwa4675@163.com
[2] Institute of Health Service and Medical Information, Academy of Military Medical Sciences, Beijing, China

Abstract. The advent of mass portable cytometry has lead to an unprecedented increase demand for the automated platform of data analysis. To provide a practical method applied to portable devices, we propose a rapid and accurate approach. This approach, based on K-means, initializes the number of clustering using kernel density estimation and optimizes calculation efficiency with k-d tree. After merging by a two-segment line regression algorithm, the clustering groups closest to the true populations can be achieved. Two different experiments proved the method we proposed would provide a rapid and accurate analysis of the multidimensional data of portable flow cytometers.

Keywords: Portable flow cytometer · Clustering analysis · K-means · k-d tree

1 Introduction

Flow cytometry (FCM) is a technology that can be applied to analyze the physical, chemical and biological properties of biological cells. The principle of FCM is that a micron laser beam is used to excite cells that are aligned by hydrodynamic forces while simultaneously collecting and recording multi-angle scattered light signals and multi-wavelength fluorescence signals. Typically, scattered light and fluorescence signal induced by a single cell are recorded in the form of an event, and all events together into a complete FCM data. With the development of the technology in recent years, portable flow cytometer have been developed and are now widespread use [1–4]. Unlike the traditional universal flow cytometer, these devices often has a characteristic of emphasis on accurate and rapid data analysis. The traditional analysis of FCM data is projected each event into two or three dimensional domain and gating according to the biological background knowledge in some dimension until find the target cells. This analysis often is error-prone, non-reproducible, nonstandardized and time-consuming, therefore, a rapid and automated platform has become an argent demand for the devices [5–9].

Based on the characteristics of FCM data, several attempts have been made to automate the gating process [10–12]. Lo et al. [13] proposed an approach based on *t* mixture models with a Box-Cox transformation, transforming the non-normal distribution data into a nearly normal distribution data, and improve the capacity of

F. Qiao et al. (eds.), *Recent Developments in Mechatronics and Intelligent Robotics*, Advances in Intelligent Systems and Computing 691, DOI 10.1007/978-3-319-70990-1_65

holding outliers. Based on the study of Kenneth, Finak et al. [14] has used an algorithm that merge the mixture components to develop the model's ability to fit the data. Sorensen et al. [15] developed an automated analysis pipeline for datasets with large event counts in high dimensions, sensitive to identify rare cell types even when next to large populations. Rebhahn et al. [16] put forward a scalable weighted iterative clustering technique to the results analyzed by Gaussian mixture models, which present a good fit to the subsamples of low-density that near the high-density groups. Wang et al. [4] proposed a flexible, statistical model-based clustering approach for identifying cell populations in flow cytometry data, which mimic the manual gating process, employs a finite mixture model with a density function of skew t distribution. Compared to the mixture models, non-probabilistic method is another method of analyzing FCM data. Zare et al. [17] have used a spectral clustering algorithm to find cell populations based on representative subsampling of the data; however, this method can potentially decrease the quality of gating because some biological information can be lost during the sampling process. Sugar and Sealfon [18] developed a new, unsupervised density contour clustering algorithm that is based on percolation theory and is very efficient for the analysis of large data sets. Ge and Sealfon [19] combined the finite mixture model and histogram spatial exploration, which by exhaustively searching all local peaks of density function and cluster the associated local peak. However, the smoothing Gaussian mixture model scale variables need a custom parameters, therefore, it is a semi-automatic analysis method in actual use. Malek et al. [20] proposed a method by automating a predefined manual gating approach. The algorithm is based on a sequential bivariate gating approach that generates a set of predefined cell populations. It chooses the best cut-off for individual markers using characteristics of the density distribution. However, the new-generation algorithms cannot be applied directly to high-dimensional data.

With the rapid development of the portable devices, the fast and accurate data analysis platform becomes an urgent demand [21, 22]. K-means is the earliest algorithm that used to analysis of FCM data and its obvious advantages is the high computational efficient [23]. While, the adoption of K-means has been restricted, because it requires identify the number of clustering in advance, and it is limited to modelling spherical cell populations. To provide a data analysis method application in portable devices, we have developed an approach that based on the K-means and directly analyzes multidimensional FCM data via a similar subpopulations-merged algorithm. Because the approach are no human setting initialization parameters, no shape limitation, and rapid and automated analysis of multidimensional data, it can be well applied in portable devices.

2   Materials and Methods

2.1   Initial Number of Clusters

For K-means method, K values need to be determined at the beginning. However, it may not be set up manually in the process of a large number of data analysis. Here we uses the ISODATA algorithm, which finds the scope of the K value firstly, and then combine and do filtering of the values. A method for determining the initial value is treating in each event of FCM data as a group, namely $K = n$, then constantly merge groups according to a certain rule. However, for FCM data always contain tens of thousands or millions of events, this way is obviously inefficient. Therefore, we introduced the kernel density estimation.

Kernel density estimation is a non-parametric estimation method, which is characterized by mixed density form of each sample that is assigned a mixture component density, i.e. $X_1, X_2 \ldots X_n$ is d-dimension independent and identically distributed samples, and the density function is f, then the kernel density estimation $\hat{f}$ can be expressed as

$$\hat{f}(x; H) = n^{-1} |H|^{-\frac{1}{2}} \sum_{i=1}^{n} K \left\{ H^{-\frac{1}{2}} (x - X_i) \right\} \tag{1}$$

Among them, K is the kernel function. H is $d \times d$ positive definite symmetric bandwidth matrix. Usually, the Gaussian probability density function is chosen as the density function $K\{x\}$, that is, $K\{x\} = (2\pi)^{-d/2} \exp(-\frac{1}{2} x^T x)$, while the bandwidth H is determined by method proposed by Duong and Hazelton [24]. For (1), calculate the second derivative and the curvature estimation of $\hat{f}$ can be obtained:

$$\nabla^{(2)} \hat{f}(x; H) = n^{-1} |H|^{-\frac{1}{2}} \sum_{i=1}^{n} H^{\frac{1}{2}} \nabla^{(2)} K \{ H^{-\frac{1}{2}} (x - X_i) \} H^{-\frac{1}{2}} \tag{2}$$

Due to the change of density function gradient can reflect the peak of the density function, as a result, take each dimension of the FCM data into type (2) respectively and adopt linear segmentation variable grid method [25], the number of each dimension significantly negative curvature in the region can be obtained quickly. Because each dimension data will be partial or completely ($d = 1$) reflects the number of groups that may be exist in the whole data. Therefore, the rank of K value is:

$$K = [\min(k_j), \sum_{j=1}^{d} k_j] \quad (j = 1, \ldots, d) \tag{3}$$

Among them, $\sum_{j=1}^{d} k_j$ as the initial K value of K-means algorithm, $\min(k_j)$ as the minimum number of merged groups.

2.2   K-means Algorithm

2.2.1   Initial Centers of Clusters

In general, K-means algorithm is based on Lloyd, which shortcomings is no certain selection method to determine the initial clustering center, usually a random selection was used in actual. Because the initial clustering center influences the clustering results very much, and bad initial centers often leads to a local minimum and reduce the convergence rate. Therefore, we optimized the selection of the initial clustering center, which is based on Arthur and Vassilvitskii [26].

The basic idea of Arthur is to make the distance between initial clustering centers as far as possible. Assuming that $x_1, x_2, \ldots, x_n$ are the samples to be classified, $x_i = (x_{i1}, \ldots, x_{id})$ are d dimensional vectors, and c_S is the clustering center of group S, then the choice of the initial center according to the following steps:

(1) Select x_i randomly in the samples as the first cluster center c_1;
(2) Calculate minimum distance of all samples to first $k - 1$ ($k = 2, \ldots, K$) clustering center:

$$d_i^2 = \min_{\{S=1,\ldots,k-1\}} ||x_i - c_s||^2, i = 1, \ldots, n \qquad (4)$$

(3) Choose x_i with the probability of $d_i^2 / \sum_{j=1}^{n} d_j^2$ as to be the k-th clustering center c_k.

The optimization of the initial clustering center selection can improve the repeatability of clustering results and reduce the possibility of the results appear to local optimal solution.

2.2.2   Optimization of Algorithm Efficiency

Since FCM data often contain tens of thousands or even millions of events, and each event often contain more than one dimension of information (such as the forward scatter, side light scatter, the first fluorescence channel, the second fluorescent channel…), therefore, computation time of K-means is often longer in actual. Accordingly, we put forward to use k-d tree algorithm to optimize K-means' calculation efficiency. Each iteration of K-means algorithm includes two main processes:

(1) Calculate the distance between each value x and the center set C, and associate it to its nearest center c_i;
(2) Update clustering center set C according to all the x value of each center linked.

The optimization is for above processes. Different from the conventional k-d tree construction, we build the k-d tree of each node that has a super rectangle, which has a boundary of d length and other two boundaries of vector $h^{\max}$ and $h^{\min}$, besides, all data points associated with the node are included in the super rectangle. For each non-leaf node of the tree, the node includes the segmentation dimension d and segmentation value v, the subtree of non-leaf node l (or r) represents the super rectangle h_l (or h_r) and it is included in h.

When a center set C (the initial center set of k-means) is given, the cost time taken to determine the c_i to be the nearest neighbor center is about $O(d)$ if there exist a c_i and the distance between c_i to the each point of h is less than the distance from $c_i = C \backslash \{c_i\}$ to h. As a result, the method we proposed will update center set C at a faster rate and improve the efficiency. An iterative process of this method to update center set as follows:

(1) If h is a leaf node, find the recent center c_i of all points within the h, and update the corresponding points.
(2) Otherwise, calculate the distance from h to set C, if there is only one center nearest to c_i, then update the center point set. Otherwise put algorithm genetic to subtree of h, and continue to go back to step (1).

2.3 Merging the K-means Results

2.3.1 Proximity Measure

According to type (3), the real number of groups is within the scope of the initial value of K, so clustering results need to be merged and optimized. Suppose $x_1 = (x_{11}, \ldots, x_{1d})$, $x_2 = (x_{21}, \ldots, x_{2d})$ are two events in FCM data, usually, Euclidean distance is used to estimate the similarity degree of two events. However, given two groups $X = (x_1, \ldots x_M)$ and $Y = (y_1, \ldots, y_N)$, to determine whether the x_i belong to group Y, not only need to consider the distance between x_i and Y, but also the distribution of Y. For two samples sets (x and y) of same distribution and covariance matrix of $\sum$, the similarity measure between the two sets could use Mahalanobis distance, namely, $D(x,y) = \sqrt{(x-y)^T \Sigma^{-1}(x-y)}$. In the case that two clustering groups X and Y are two unknown distribution sample sets, if the X and Y truly belong to one group, then, the cells in the two groups should be subject to same or similar distributions. Accordingly, we can use a distance similar to Mahalanobis distance to estimate the similarity between X and Y, which defined as follow:

$$D(X,Y) = \sqrt{(\overline{X} - \overline{Y}) \Sigma_X^{-1} (\overline{X} - \overline{Y})^T} \tag{5}$$

Among them, $\overline{X}$ is the center of the groups X, $\overline{Y}$ is the center of cluster Y, and Σ_X is the covariance matrix of X. Since the Σ_Y^{-1} could be used to replace Σ_X^{-1} when calculation, so $D(X, Y)$ may has different values, here we take the smaller value.

2.3.2 Optimization Results

As long as the distance between two subgroups is small enough, the two groups should to be merged. After several rounds of merging, the distance between any two of rest subgroups would be more large, otherwise, it will not stop until the number of groups equal to the $min(k_j)$ ($j = 1, \ldots, d$). As each time we merged two groups which has the minimum distance, therefore, the minimum distance before merged must be bigger than the minimum distance after merged. When the number of groups is equal to the actual real number, and we continue to merge groups X and Y, the minimum distance will suddenly become very large (as the merging groups X and Y are different groups in fact). So, if we put the clustering number k and its corresponding minimum distance D_k to two dimensional space in the merging process, a mutations point $P_K = (K, D_K)$ would appear at the real number of groups, by looking for the points P_K, we could get the best clustering results.

To recognize point P_K, we proposes using a two linear regression fitting methods, the specific implementation process is: suppose $m = min(k_j)$ ($j = 1, \ldots, d$), $n = \sum_{j=1}^{d} k_j$, and i represent a variables from $m + 1$ to $n - 1$, namely that $i = (m + 1, ..., n - 1)$, so all points$(P_m,..,P_n)$ are divided into two parts $(P_m, ..., P_i)$ and $(P_i, ..., P_n)$ by P_i. Do regression fitting on each part points respectively, and the results are $\hat{y}_1$ and $\hat{y}_2$. The best clustering number K is equal to the i that can minimum the sum of the residual sum of squares of $\hat{y}_1$ and $\hat{y}_2$, that is

$$K = \arg\min(\sum_{j=m}^{i} (\hat{y}_{1(j)} - D_j)^2 + \sum_{j=i}^{n} (\hat{y}_{2(j)} - D_j)^2) \tag{7}$$

Among them, $\hat{y}_{1(j)}$ and $\hat{y}_{2(j)}$ represent the value of $\hat{y}_1$ and $\hat{y}_2$ at j. The merging result of K value corresponding to is the best clustering results.

3   Results

To demonstrate the performance of proposed automated clustering method, FCM data from peripheral blood samples test were used. The data contains 29000 cells, and three biomarkers, namely CD3, CD8 and CD45. The experiment is designed to find the CD8 + T lymphocyte subpopulations. Firstly, the lymphocyte subpopulations were identified by analyzing the plot of CD45 and SSC scatter, then the data that corresponding to lymphocyte cells was projected into CD3 and CD8 dimensional spaces to find the CD8 + T lymphocytes. Figure 1(a) shows the analysis results gating performed by FloMax software. From the results, we can see that the sample includes four cell groups, R1 to R4 respectively represent the lymphocyte subgroup, monocyte subsets, granulocyte subgroup and dead cells. Figure 1(b) is the analysis result by quadrant method, where cells in Q2 are the CD8 + T lymphocytes.

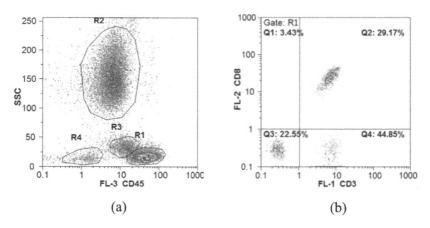

(a)   (b)

Fig. 1. The analysis process and result of expert

Figure 1 shows that the artificial analysis need to put the FCM data into a plot and gate according to the biological background knowledge in some dimension, and then analyze the clustering group in other dimension until finally find the target cells.

Compared to the manual analysis, automatic analysis method not only can be carried out in accordance with the way that based on two-dimensional sequential clustering analysis [4, 13, 14, 20], but also directly find the target cells at once. To validate the method proposed for direct analysis of multi-dimensional data, the method of Sam-SPECTR [17], Misty Mountain [18] and flowPeaks [19] are applied here. The artificial successive two-dimensional projection method, like the way of expert data analysis, also can be used, here no further analysis and discussion. Figure 2 shows that the clustering results of above four analysis methods in SSC and CD45 dimension projection, Fig. 2(a)–(d) are the result of SamSPECTRAL, Misty Mountain, flowPeaks and method proposed in this paper. From the results, we can see that the results of the method we proposed are accordance with the results of the artificial method, while a part of lymphocyte subgroup analyzed by Misty Mountain and flowPeaks are mis-judgment. At the same time, because of the influence of little outliers, analysis results by SamSPECTRAL has a large misclassification rate.

The clustering results of Fig. 2(d) are projected to other dimensions, and the results are shown in Fig. 3. Figure 3(a) shows that the final clustering results in FSC and SSC

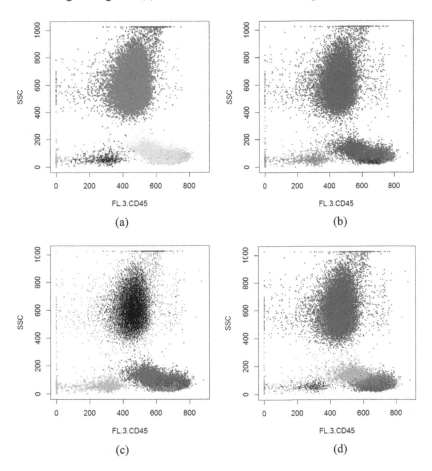

Fig. 2. Results comparison of four methods

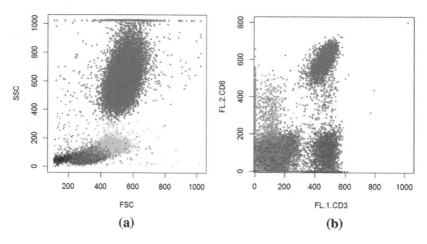

Fig. 3. Results projected in other dimensions

dimensions, which are consistent with Fig. 2(d). Cells on the right top part of Fig. 3(b) is the CD8 + T lymphocyte subgroup.

To assess the performance of above four methods, the misclassification rate (MR) against the benchmark expert clustering, which is taken as being the 'true' class membership, was calculated. Note that the dead cells must be first removed. As shown in Table 1, the outlier is clustering based on the Mahalanobis distance from the particle to the cluster center.

The computing time were recorded at the same time, which is shown in Table 1. According to the results in the table, the time of SamSPECTRAL analysis has exceeded the time of manual analysis (expert cost about 60 s). Although the Misty Mountain is superior to the manual analysis method, but compared to the flowPeaks, it is slow. Obviously, the method we designed is the best in computing time.

Table 1. MR and computing time of four methods

	Methods			
	SamSPECTRAL	Misty Mountain	flowPeaks	Our method
MR	0.1387	0.1125	0.1084	0.0892
Time (s)	132	18	4	3

To further verify the optimal effectiveness of the algorithm, anther FCM data, from the experiment that is to get the proportion of live and dead bacteria in yeast fall, was analyzed. The data contain 10200 cells and two markers, namely, FL1-FDA, FL2-PI, in addition to the FSC and SSC. The first step of manual analysis is to exclude few impurities through FSC and SSC scatter, and find the yeast group, then analyze the yeast groups that removed the impurities in FL1-FDA and FL2-PI dimension, and find out absolute quantity and proportion of the various states of yeast group. As shown in

Fig. 4, the groups gating by experts is using the FloMax software. Because the samples only contain yeast bacteria, so there is only one group distribution in the plot of FSC and SSC, as shown in Fig. 4(a). Figure 4(b) shows the gating results projected in FL1 and FL2 dimensions, the cells of Q1 area represents dead bacteria, the quantity is few, accounts for about 0.95% of the total. The cells of Q2 area represents the bacteria between dead and alive (the membrane is broken, PI has entered the cell), the cells of Q3 area represents the cellular debris, and the cells of Q4 area represents the living bacterium.

We analyzed the data of experiment two by the method proposed and methods based on mixture mode: flowMerge [14], immunoClust [15], SWIFT [16]. According to the results, besides the immunoClust, other methods have similar results (the MR comparison in Table 2), but compared to the results of flowMerge and SWIFT, the clustering results of method we designed are much closer to the expert analysis results.

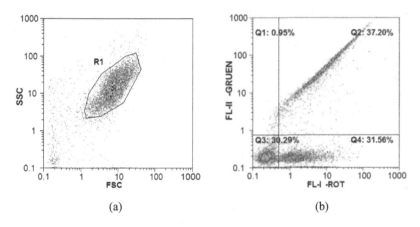

(a)   (b)

Fig. 4. The analysis process and result of expert

Compared to the clustering methods based on probability density, the method designed not only has an advantage in the accuracy, but also in computational effi ciency. The computing time of above four method are shown in Table 2. Obviously, the analysis time of method we proposed is far less than other clustering methods and manual (expert cost about 27 s).

Table 2. MR and computing time of four methods

	Method			
	flowMerge	immunoClust	SWIFT	Our method
MR	0.1106	0.1947	0.0922	0.0784
Time (s)	124	42	20	1

4 Discussion

At present, automatic analysis method for FCM data can mainly divided into two categories: one is the mixture model methods based on probability calculation, which belongs to a kind of "soft" clustering method [13–16]. The principle is to assume that each event/cells in FCM data are in accordance with a certain probability distribution (e.g., t distribution), and uses a mixture models to fit the groups. Since the method is based on probability theory, it can estimate the probability of every event that belongs to the groups and has a low misjudgment rate of analysis of the data that are elliptical symmetrical distribution, but its obvious defect is the computational complexity. Experiment 2 adopted the methods based on mixture models. The results show that the Gaussian distribution is more sensitive to outliers, so the error of fitting data by immunoClust is bigger (MR = 0.1947). However, due to the increase of tail weight to regulation parameters, the robustness of method based on t mixture models significantly increased (MR = 0.1106). Different with the methods based on mixture models, the "hard" clustering method classify the particles in data directly according to a certain rule, rather than to estimate the probability of each event belongs to groups. The method we designed belongs to the class. The main advantage of the method is the high computational efficiency, and the ability to analyze concave data. In experiment 1, we compared the results of four "hard" clustering methods. According to the Fig. 2(a), SamSPECTRAL is more sensitive to outliers. When using the direct analysis of multidimensional FCM data, the misjudgment rate of SamSPECTRAL increased significantly (MR = 0.1387). Because of the using of the matrix method to calculate, SamSPECTRAL takes a long time (132 s), while the Misty Mountain and flowPeaks is relatively fast, and they are obviously better than the methods based on the mixture models.

K-means algorithm is first applied to analysis of FCM data, but because it is very difficult to determinate the initial K value, this method has not been widely used in the analysis of FCM data. Hamerly and Elkan [27] extended the K - means methods by the BIC rule, and the results showed that this method has a poor repeatability. Kaufman and Rousseeuw [28] adopted different K values to clustering data, and then combined with the results of different K value to get the final classification, however, the model parameters need to be set in advance. In the method we proposed, the K value range is determined by the kernel density estimation, the advantage is that it can rapidly and accurately find the peak point in the data. Due to the calculation of kernel density estimation of multidimensional data is very complex, the method estimates the numbers of groups in each dimension of data, gets the initial values of K using the peak number of all dimensions, and merges the redundant clustering to obtain the optimal result.

K-means is widely used in pattern recognition field because of the simply and efficient computing, however, with the increase of the events and the number of dimensions, its computation time increases. For optimization of K-means computing efficiency, there are two strategies. The first is to use a former recursive information to reduce the data quantity of the next distance calculation [29]. The second is to use an appropriate data structure to store all the objects, which can make a given object to find its neighbor quickly [30]. Based on the characteristics of FCM data, we put forward k-d tree optimization method that belongs to the later. When the events of FCM data is low, there is little difference between optimized method and non-optimized method in computing time. However, when the events of data up to ten thousand, the computational efficiency difference is obvious. In experiment 1, to analyze the FCM data, optimized method takes 3 s while the non-optimized method takes 5 s.

5 Conclusion

We have developed a rapid and automated method for identifying cell populations in flow cytometry. This method, based on K-means algorithm, uses kernel density estimation for initializing number of clustering and k-d tree for optimization efficiency, which allows it to be the best method in terms of accuracy and efficiency in portable devices.

References

1. Boutrus, S., Greiner, C., Hwu, D., et al.: Portable two-color in vivo flow cytometer for real-time detection of fluorescently-labeled circulating cells. J. Biomed. Opt. **12**(2), 020507 (2007)
2. Grafton, M.G., Zordan, M.D., Chuang, H.-S., et al.: Portable microfluidic cytometer for whole blood cell analysis. In: SPIE Proceedings, vol. 7593, pp. 1–8 (2010)
3. Friedman, B.: Counting white blood cells with a portable flow cytometry device. Clinical Lab Industry News, 11 April 2013
4. Xianwen, W., Feng, C., Zhi, C., et al.: Automated gating of portable cytometer data based on skew t mixture models. J. Mech. Med. Biol. **15**(3), 1550033 (2015)
5. Chester, C., Maecker, H.T.: Algorithmic tools for mining high-dimensional cytometry data. J. Immunol. **195**(3), 773–779 (2015)
6. Pedreira, C.E., Costa, E.S., Lecrevisse, Q., et al.: Overview of clinical flow cytometry data analysis: recent advances and future challenges. Trends Biotechnol. **31**(7), 415–425 (2013)
7. Verschoor, C.P., Lelic, A., Bramson, J.L., et al.: An introduction to automated flow cytometry gating tools and their implementation. Hypothesis Theory **6**, 380 (2015)
8. Gouttefangeas, C., Chan, C., Attig, S., et al.: Data analysis as a source of variability of the HLA-peptide multimer assay: from manual gating to automated recognition of cell clusters. Cancer Immunol. Immunother. **64**(5), 585–598 (2015)
9. Lee, S.X., McLachlan, G.J., Pyne, S.: Modeling of inter-sample variation in flow cytometric data with the joint clustering and matching procedure. Cytom. A **89**(1), 30–43 (2016)
10. Aghaeepour, N., Finak, G., Hoos, H., et al.: Critical assessment of automated flow cytometry data analysis techniques. Nat. Methods **10**(3), 228–238 (2013)
11. Chen, X., Hasan, M., Libri, V., et al.: Automated flow cytometric analysis across large numbers of samples and cell types. Clin. Immunol. **157**(2), 249–260 (2015)
12. Hasan, M., Beitz, B., Rouilly, V., et al.: Semi-automated and standardized cytometric procedures for multi-panel and multi-parametric whole blood immunophenotyping. Clin. Immunol. **157**(2), 261–276 (2015)
13. Lo, K., Brinkman, R.R., Gottardo, R.: Automated gating of flow cytometry data via robust model-based clustering. Cytom. A **73**(4), 321–332 (2008)

14. Finak, G., Bashashati, A., Brinkman, R., et al.: Merging mixture components for cell population identification in flow cytometry. Adv. Bioinform. **2009**, 247646 (2009)
15. Sorensen, T., Baumgart, S., Durek, P., et al.: immunoClust - an automated analysis pipeline for the identification of immunophenotypic signatures in high-dimensional cytometric datasets. Cytom. Part A **87**(7), 603–615 (2015)
16. Rebhahn, J.A., Roumanes, D.R., Qi, Y., et al.: Competitive SWIFT cluster templates enhance detection of aging changes. Cytom. Part A **89**(1), 59–70 (2016)
17. Zare, H., Shooshtari, P., Gupta, A., et al.: Data reduction for spectral clustering to analyze high throughput flow cytometry data. BMC Bioinform. **11**, 403–419 (2010)
18. Sugar, I.P., Sealfon, S.C.: Misty Mountain clustering: application to fast unsupervised flow cytometry gating. BMC Bioinform. **11**, 502 (2010)
19. Ge, Y., Sealfon, S.C.: FlowPeaks: a fast unsupervised clustering for flow cytometry data via K-means and density peak finding. Bioinformatics **28**(15), 2052–2058 (2012)
20. Malek, M., Taqhiyar, M.J., Chong, L., et al.: Misty Mountain: reproducing manual gating of flow cytometry data by automated density-based cell population identification. Bioinformatics **31**(4), 606–607 (2015)
21. Mair, F., Hartmann, F.J., Mrdjen, D., et al.: The end of gating? An introduction to automated analysis of high dimensional cytometry data. Eur. J. Immunol. **46**(1), 34–43 (2016)
22. Lujan, E., Zunder, E.R., Nq, Y.H., et al.: Early reprogramming regulators identified by prospective isolation and mass cytometry. Nature **521**(7552), 352–356 (2015)
23. Murphy, R.F.: Automated identification of subpopulations in flow cytometric list mode data using cluster analysis. Cytometry **6**(4), 302–309 (1985)
24. Duong, T., Hazelton, M.L.: Plug-in bandwidth matrices for bivariate kernel density estimation. Nonparametric Stat. **15**(1), 17–30 (2003)
25. Duong, T.: ks: Kernel density estimation and kernel discriminant analysis for multivariate data in R. J. Stat. Softw. **21**(7), 1–16 (2007)
26. Arthur, D., Vassilvitskii, S.: k-means+: the advantages of careful seeding. In: Proceedings of the Eighteenth Annual ACM-SIAM Symposium on Discrete Algorithms, pp. 1027–1035 (2007)
27. Hamerly, G., Elkan, C.: Learning the K in k-means. Adv. Neural. Inf. Process. Syst. **17**, 281–288 (2004)
28. Kaufman, L., Rousseeuw, P.: Finding groups in data: an introduction to cluster analysis. Wiley, New York (1990)
29. Cooper, L.A., Kong, J., Gutman, D.A., et al.: Novel genotype-phenotype associations in human cancers enabled by advanced molecular platforms and computational analysis of whole slide images. Lab. Invest. **95**(4), 366–376 (2015)
30. Chen, Y.J., Chen, S.C., Wu, J.L.: A hybrid vector quantization combining a tree structure and a Voronoi diagram. Math. Probl. Eng. **2014**, 1–6 (2014)

Control of Lower Extremity Exo-skeleton Based on Admittance Method

Wenhao Chen$^{(\boxtimes)}$, Shengli Song, Husheng Fan, and Ming Zhao

PLA University of Science and Technology, Nanjing, Jiangsu, China
743202615@qq.com

Abstract. To reduce the tracking error between the exo-skeleton and the human body, a control algorithm based on the admittance principle is proposed. The algorithm draws on the admittance characteristics of velocity and force in the process of moving objects. A man-machine coupling dynamics model is established and a reasonable control function is designed to shield the admittance characteristics of the exo-skeleton. The stability of the control algorithm is analyzed and the admittance parameters are optimized. The simulation results show that the tracking error between the exo-skeleton controlled by the optimized admittance and the calf of the human is 0.01°. The optimized admittance control can achieve more accurate motion tracking.

Keywords: Exo-skeleton · Admittance control · Trajectory tracking

1 Introduction

The exo-skeleton is a mechanical device that mimics the physiological structure of the human body, which can assist wearer in movement or augment physical abilities. Therefore, exo-skeleton in the military, medical, disaster relief areas have broad application prospects [1].

In view of the complex exo-skeleton system, researchers around the world put forward their own control strategies. BLEEX (Berkeley Lower Extremity Exo-skeleton) is an exo-skeleton developed for military use that increases the weight-bearing ability of a soldier [2]. However, the control effect is limited by accuracy of the inverse dynamics model [3]. HAL (Hybrid Assistive Limb) is an exo-skeleton developed for aging problems that can assist older people walk, climb stairs and move items [4]. HAL can obtain more realistic human movement intention, but there are some defects. For instances, there is noise in the measurement data and the EMG signals are vulnerable to environmental [5].

In recent years, the admittance control method has been proposed on the basis of impedance control, and the man-machine coordination movement has been successfully realized by changing the admittance of the man-machine system. Nagarajan et al. for hip exo-skeleton (only provide torque to human hip), by improving the human hip joint admittance, improve the thigh movement of the agility [6]. Liu et al. for knee skeleton, use admittance principle to shield exo-skeleton itself admittance, the results

F. Qiao et al. (eds.), *Recent Developments in Mechatronics and Intelligent Robotics*,
Advances in Intelligent Systems and Computing 691, DOI 10.1007/978-3-319-70990-1_66

show that the exo-skeleton can reduce tracking error and coupling torque [7], but do not analyze stability of system.

In order to further improve the coordination of man-machine movement, based on the knee exo-skeleton, analyze the stability of the system, and optimize the admittance parameters. The man-machine coupling dynamics model is established, and the simulation is carried out.

2 Man-Machine Coupling Dynamics Mode

When analyze the dynamic model of man-machine, the man-machine is usually treated as a rigid connection [2, 8]. However, there is always a coupling torque between the human body and the exo-skeleton, which has an important influence on the stability of the system. This paper is based on knee exo-skeleton, as shown in Fig. 1. The connection between the human calf and the exo-skeleton is bandaged. The dynamics model of the isolated human calf is given by

$$I_h \ddot{\theta}_h(t) + b_h \dot{\theta}_h(t) + k_h \theta_h(t) = \tau_h(t) \tag{1}$$

where I_h, b_h, k_h are the moment of inertia, joint damping coefficient and joint stiffness coefficient of the human calf respectively; $_h(t)$ is the knee joint angle trajectory and $\tau_h(t)$ is the net muscle torque acting on the joint.

The dynamic model of the isolated exo-skeleton is given by

$$I_e \ddot{\theta}_e(t) + b_e \dot{\theta}_e(t) + k_e \theta_e(t) = \tau_e(t) \tag{2}$$

where I_e, b_e, k_e are the moment of inertia, joint damping coefficient and joint stiffness coefficient of the exo-skeleton respectively: $_e(t)$ is exo-skeleton arm's angle trajectory and $\tau_e(t)$ is the actuator torque.

The human - machine coupling dynamics model are given by

$$I_h \ddot{\theta}_h(t) + b_h \dot{\theta}_h(t) + k_h \theta_h(t) = \tau_h(t) - \tau_c(t) \tag{3}$$

$$I_e \ddot{\theta}_e(t) + b_e \dot{\theta}_e(t) + k_e \theta_e(t) = \tau_e(t) + \tau_c(t) \tag{4}$$

where $\tau_c(t)$ is the coupling torque representing the interaction between exo-skeleton and the human calf.

The coupling torque model is mostly simplified as a spring model [2, 8], ignoring the damping effect. This paper will give full consideration to the damping effect, the coupling torque model is given by

$$\tau_c(t) = b_c(\dot{\theta}_h(t) - \dot{\theta}_e(t)) + k_c((\theta_h(t) - \theta_e(t)) \tag{5}$$

where b_c is the damping coefficient and k_c is the coefficient.

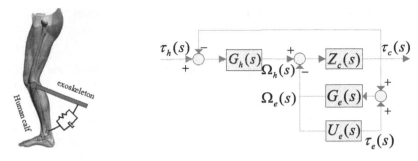

Fig. 1. Man-Machine coupling model   **Fig. 2.** Man-Machine system with admittance control

3   Admittance Control

From (1) to (3), their transfer functions are given by (6) to (8)

$$G_h(s) = \frac{\Omega_h(s)}{\tau_h(s)} = \frac{s}{I_h s^2 + b_h s + k_h} \tag{6}$$

$$G_e(s) = \frac{\Omega_e(s)}{\tau_e(s)} = \frac{s}{I_e s^2 + b_e s + k_e} \tag{7}$$

$$Z_c(s) = \frac{\tau_c(s)}{\Omega_c(s)} = \frac{b_c(s) + k_c}{s} \tag{8}$$

The closed-loop coupled man-machine system is represented by the block diagram of Fig. 2. The open-loop transfer function $G_{hc}(s)$ of the system is given by

$$G_{hc}(s) = \frac{I_e s^2 + b_e s + k_e - sU_e(s)(b_c(s) + k_c)}{(I_e s^2 + b_e s + k_e - sU_e(s) + b_c(s) + k_c)(I_h s^2 + b_h s + k_h)} \tag{9}$$

The feedback control $U_e(s)$ is given by

$$U_e(s) = aI_e s + bb_e s + \frac{ck_e}{s} \tag{10}$$

where a, b and c are control parameters.

From Fig. 2, we can also get

$$G_{eu}(s) = \frac{\Omega_e(s)}{\tau_c(s)} = \frac{s}{(1-a)I_e s^2 + (1-b)b_e s + (1-c)k_e} \tag{11}$$

When the parameter is greater than 0, the exo-skeleton's admittance characteristics can be offset by the control function $U_e(s)$, which makes the exo-skeleton be equivalent to a small mass. Thereby reducing the coupling with the human calf, can play a lower tracking error and coupling torque.

4 Stability Analysis and Parameter Optimization

4.1 Stability Analysis

In order to make the system have better stability, the design system is the minimum phase system, that is, the open-loop transfer function $G_{hc}(s)$ has no pole and no zero in the right half of the complex plane. Substituting Eq. (10) into Eq. (9), it can be shown that $G_{hc}(s)$ has the following equivalent form:

$$G_{hc}(s) = \frac{k_e((1-a)I_e s^2 + (1-b)b_e s + (1-c))(b_c(s) + k_c)}{((1-a)I_e s^2 + ((1-b)b_e + b_c)s + (1-c)k_e + k_c)(I_h s^2 + b_h s + k_h)} \quad (12)$$

For the parameters I_h, b_h, k_h, b_c and k_c are greater than 0, a, b and c can be obtained by:

$$1-a > 0; 1-b > 0; (1-c)k_e + k_c > 0; 1-c > 0; (1-b)b_e + b_c > 0 \quad (13)$$

For the parameters b_e and k_e are both greater than 0, (15) reduces to

$$a < 1; b < 1; c < 1 \quad (14)$$

In addition, the system gain margin $K_g(dB)$ and phase margin $\gamma(^\circ)$ should also be met:

$$k_g > 1; \gamma > 0 \quad (15)$$

4.2 Closed-Loop Coupled Man-Machine System

The optimization of the parameters can reduce the tracking error and the coupling torque. From Fig. 2(d), the closed-loop admittance $G(s)$ is given by

$$G(s) = \frac{G_{hc}(s)}{1 + G_{hc}(s)} \quad (16)$$

When the human calf swing at the angular velocity ω, the module of the $G(j\omega)$ should be the smallest. As the human calf can swing at different angular velocities, so set the objective function:

$$f = \int_0^{\omega_m} |G(j\omega)| d\omega \quad (17)$$

where ω_m is the maximum angular velocity of the human calf.

The parameters used in this paper are shown in Table 1. The parameters about human are set for reference [7] and the remaining parameters are design parameters.

Table 1. parameters of man-machine system

Parameters	Value	Parameters	Value	Parameters	Value	Parameters	Value
I_h	0.220 kg·m²	ω_m	11 rad/s	b_e	0.06 N·m (rad/s)	k_e	2.77 N·m/rad
k_h	11.843 N·m/rad	I_e	0.120 kg·m²	b_c	9.45 N·m (rad/s)	k_c	2000 N·m/rad
b_h	1.714 N·m (rad/s)						

Optimizations were performed in matlab. The initial values of the three parameters a, b and c are set to 0, the step is 0.01, the final value is 1, search the three parameter values which makes the objective function f minimum, and satisfy the Eqs. (15). Optimization results are $a = 0.89$, $b = 0.91$, $c = 0.95$.

5   Simulation

In the case of experimental testers who do not wear exo-skeleton, keep the thighs static, the calf free to swing, action capture system of Noitom Technology co,. Ltd was used to records angular trajectory of knee, the 6 sensors wearing on calf as shown in the red oval box in Fig. 3), which can record information of hip, knee and ankle, recorded angular trajectory curve of knee as shown in Fig. 4(a) green solid line.

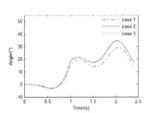

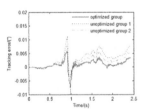

Fig. 3. Experimental site

Fig. 4. Comparison of human calf swing

Fig. 5. Tracking performance

The net muscle torque of the experimental testers can be obtained by using Eq. (1). The situations that wear exo-skeleton without power (the exo-skeleton is completely passive) and wear exo-skeleton adopts the admittance control (the exo-skeleton is in active control and the optimized admittance parameters are adopted), the obtained net muscle torque as input, are simulated. The simulation results are shown in Figs. 4 and 5. Figure 4 (a) shows that the phase of the calf swing in the case 1 (wearing exo-skeleton without power) is significantly lagging behind that in the case 2 (the calf is free to swing without exo-skeleton), and the amplitude is also smaller, indicating that wearing exo-skeleton without power, exo-skeleton is a burden for human body. And the calf's swing in the case 3(adopts the admittance control) is very close to that in the case 2. In order to verify the effectiveness of the optimization algorithm, the unoptimized admittance parameters group 1 ($a = 0.8$, $b = 0.91$, $c = 0.85$) and group 2 ($a = 0.85$,

$b = 0.92$, $c = 0.90$) were selected randomly. Figure 5 shows that the tracking error adopted the optimized parameters is smaller than that adopted unoptimized parameters. And the peak of tracking error adopted unoptimized parameters is about 0.5°. The validity of the optimization algorithm is verified.

6 Conclusion

The admittance control method of exo-skeleton was studied in this paper. Firstly, the man-machine coupling dynamics model is established, and then a reasonable admittance control strategy is designed. Then, the stability of the control algorithm is analyzed and the admittance parameters are optimized. Finally, the validity of the admittance control is verified by simulation. The simulation results show that the admittance control can shield the mechanical admittance of the exo-skeleton itself, thus reducing the tracking error and the coupling torque of the exo-skeleton. The optimized admittance control can achieve more accurate motion tracking.

References

1. Zhou, J., Zhang, A., Mo, X., et al.: Hydraulic system design of hydraulic-driven load exo-skeleton robot. Mach. Tool Hydraul. **21**, 30–34 (2016)
2. Ozkul, F., Erol, D.: Upper-extremity rehabilitation robot RehabRoby: methodology, design, usability and validation. Int. J. Adv. Robot. Syst. **10**(12), 101 (2013)
3. Steger, R., Kim, S.H., Kazerooni, H.: Control scheme and networked control architecture for the Berkeley lower extremity exo-skeleton (BLEEX). In: IEEE International Conference on Robotics and Automation, pp. 3469–3476. IEEE (2006)
4. Zhao, G., He, L., Li, X., et al.: Support vector machine based sensitivity amplification control for a lower extremity exo-skeleton. Comput. Meas. Control **09**, 211–214 (2016)
5. Gopura, R.A.R.C., Kiguchi, K., Li,Y.: SUEFUL-7: a 7DOF upper-limb exo-skeleton robot with muscle-model-oriented EMG-based control. In: IEEE/RSJ International Conference on Intelligent Robots and Systems, pp. 1126–1131. IEEE (2009)
6. Nagarajan, U., Aguirre-Ollinger, G., Goswami, A.: Integral admittance shaping: a unified framework for active exo-skeleton control. Robot. Auton. Syst. **75**, 310–324 (2015)
7. Liu, D., Tang, Z., Pei, Z.: Swing motion control of lower extremity exo-skeleton based on admittance method. J. Beijing Univ. Aeronaut. Astronaut. **41**(6), 1019–1025 (2015)
8. Tang, Z., Tan, Z., Pei, Z.: Design and dynamic analysis of lower extremity exo-skeleton. J. Syst. Simul. **25**(6), 202–208 (2013)

A Two-Loop Control for All-Electric Tank Gun System Based on Disturbance Observer and SMVSC

Hong-yan Wang$^{(\boxtimes)}$, Li-hui Wang, and Jie Ma

Department of Information Engineering,
Academy of Armored Force Engineering, Beijing, China
zxywhy2000@sina.com

Abstract. Considering the whole impact of the external disturbance and uncertainty of inner parameters, a two-loop control approach is proposed, the disturbance observer (DOB) as inner loop and the sliding mode variable structure control (SMVSC) as outer loop. Through analysis the robustness and performance of the DOB system in the name of two-degree-freedom, DOB compensates large perturbation and uncertainties by designing an adequate low pass filter, it can also reduce the chattering of SMVSC. The two-loop control can enhance the All-electric tank gun control system speed tracking performance, robustness and adaptive capacity. Computer simulations show that the tracking accuracy is much better and the control system has strong adaptability and robustness.

Keywords: AC all-electrical tank gun control system · Disturbance observer (DOB) · Sliding mode variable structure control (SMVSC) · Robustness

1 Introduction

Tank servo system with variable load, no-load and full load conditions and poor working environment conditions, load changes in a large range of parameters has strong uncertainty; exist friction and clearance nonlinearity and external environment interference in the system, which will influence the control performance of the servo system. In order to adapt to the development of tank weapon system with high accuracy, the compensation and suppression of jamming becomes more and more urgent. At the same time there is friction and clearance nonlinearity and external disturbance in the system, which will influence the control performance of the servo system In order to adapt to the development of high precision tank weapon system, interference suppression and compensation become increasingly urgent.

In recent years, many scholars have used the model reference adaptive control [1], robust control [2], ADRC control [3] or combined intelligent control [4] method to suppress the disturbance in the system, and enhance the robust performance of the system, but these methods are usually more complicated.

© Springer International Publishing AG 2018
F. Qiao et al. (eds.), *Recent Developments in Mechatronics and Intelligent Robotics*,
Advances in Intelligent Systems and Computing 691, DOI 10.1007/978-3-319-70990-1_67

In this paper, the equivalent interference observed by the disturbance observer is used to compensate the disturbance, at the same time, the gain of sliding mode switching is lowered, and the chattering phenomenon of SMVSC is reduced. The Lyapunov function is used to construct the variable structure robust controller to compensate the disturbance error, and further improve the robustness and tracking performance of the system.

2 Mathematical Model of Gun Servo System

The servo system control system includes a turret, reducer and motor, its composition diagram is shown in Fig. 1. The dynamic equation is expressed as:

$$\ddot{\theta} = -A_P\dot{\theta} + B_P u + C_P M_d \tag{1}$$

Among them: $A_p = \frac{C_e C_m}{JR}$, $B_p = \frac{KC_m}{JR\ i}$, $C_p = \frac{1}{Ji}$, θ is the turret angle; ω is the motor angular velocity; u is the control input of the outer ring angle of the gyroscope; K is the amplification coefficient; J is the moment of inertia; C_e is the back EMF coefficient of the motor; C_m is the torque coefficient of the motor; R is the resistance of reduction gear ratio; i is the reduction ratio; M_L is the external disturbance of the system; M_d is all disturbance torques; M_d includes all disturbance torques.

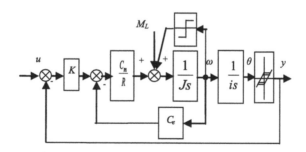

Fig. 1. Dynamic model of tank servo system

3 Design of Disturbance Observer

The disturbance observer is shown in Fig. 2 [6], where u, M_d, ξ, y respectively denote input, interference, noise and output. $\hat{M}_d$ is the interference estimation. $P_n(s)$ is the nominal model of the system. $Q(s)$ is used to implement $P_n^{-1}(s)$ of low-pass filter.

As can be seen from Fig. 2 the relationship between input and output are as follows:

$$y = G_{uy}(s)u + G_{M_dy}(s)M_d + G_{\xi y}(s)\xi \tag{2}$$

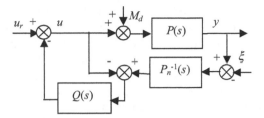

Fig. 2. Structure of disturbance observer

$$G_{uy}(s) = \frac{P(s)P_n(s)}{P_n(s) + Q(s)[P(s) - P_n(s)]} \quad G_{M_{d}y}(s) = \frac{P(s)P_n(s)(1 - Q(s))}{P_n(s) + Q(s)[P(s) - P_n(s)]}$$
$$G_{\xi y}(s) = \frac{P(s)Q(s)}{P_n(s) + Q(s)[P(s) - P_n(s)]} \tag{3}$$

If $Q(s) \approx 1$, formula is: $G_{uy}(s) \approx P_n(s)$ $G_{M_{d}y}(s) \approx 0$ $G_{\xi y}(s) \approx 1$. It is shown that the disturbance observer makes the system $P(s)$ the same as $P_n(s)$, and the interference is suppressed. If $Q(s) \approx 0$, Formula (3) is: $G_{uy}(s) \approx P(s)$ $G_{M_{d}y}(s) \approx P(s)$ $G_{\xi y}(s) \approx 0$, it can be seen that the high frequency noise is attenuated.

For the two order system, it is assumed that the Q filter has the following form:

$$Q_{nm}(s) = \frac{\sum\limits_{i=0}^{m} a_{ni}(\tau s)^i}{(\tau s + 1)^n} \tag{4}$$

τ is the filter time constant, $a_{ni} = n!/((n-i)!i!)$ is a binomial coefficient, m, n respectively for the numerator and denominator of the order. For the two order system, meet the $n - m \geq 2$. The nominal model is assumed to have the following form:

$$P_n(s) = \frac{k}{s^2 + 2\zeta\omega_n s + \omega_n^2} \tag{5}$$

We use the method of Coprime factorization to robustness analysis. In the name of two-degree-freedom, by analysis the robustness, the interference suppression performance and the effect of sensor noise of the DOB system, we can get the following results as Figs. 3, 4 and 5. The detailed relationship can be found in [7, 8].

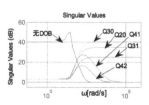

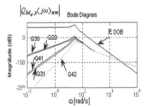

Fig. 3. The Robustness of DOB system

Fig. 4. Interference suppression performance of DOB system

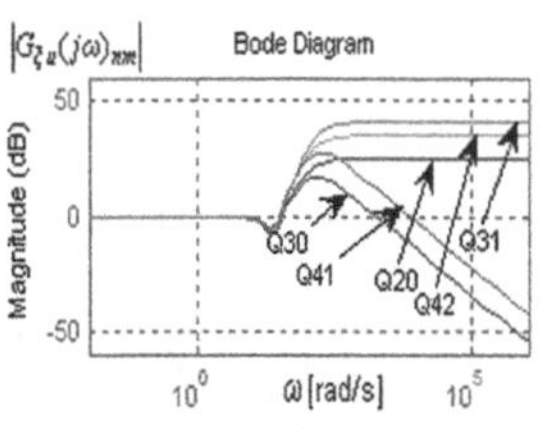

Fig. 5. Effect of DOB noise

Figure 4 compares the robustness of second-order systems with different $Q(s)$ can get the following conclusion: the smaller the relative order of $Q(s)$, the better the robustness; the greater the denominator order, the better the robustness; from Fig. 5 can get: in the low frequency, the greater the molecular order, the better the interference suppression performance; the time constant τ determines the effective frequency range of interference suppression. The time constant τ is small, means that the wide frequency range of interference suppression; from Fig. 6 can get: the relative order of Q (s) is larger, the better the performance of the sensor noise suppression, the smaller the time constant, the more sensitive the sensor noise to the DOB system.

4   Design of SMVSC Controller

The disturbance observer can effectively eliminate the interference, but the dynamic characteristic of the observer is actually a low pass filter, the bandwidth can not be set too wide, and then there are errors in the actual application, which will limit the performance of the system. In this paper, a sliding mode variable structure controller is used to compensate the error caused by the disturbance observer by SMVSC, The performance of the disturbance observer is further enhanced, and the desired trajectory of the system can also be tracked. In addition, due to the compensation effect of the inner loop disturbance observer, the gain of sliding mode switching can be reduced, and the chattering phenomenon of SMVSC can be reduced. The system structure is shown in Fig. 6.

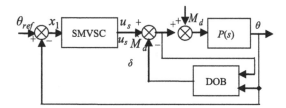

Fig. 6. Sliding mode control structure with DOB

The position command signal is θ_{ref}, define the error and its derivative:

$$x_1 = \theta_{ref} - \theta \tag{6}$$

$$x_2 = \dot{x}_1 = \dot{\theta}_{ref} - \dot{\theta} \tag{7}$$

Considering the formula (1), the error state equation is obtained:

$$\begin{bmatrix} \dot{x}_1 \\ \dot{x}_2 \end{bmatrix} = \begin{bmatrix} 0 & 1 \\ 0 & -A_p \end{bmatrix} \begin{bmatrix} x_1 \\ x_2 \end{bmatrix} + \begin{bmatrix} 0 \\ -B_p \end{bmatrix} u + \begin{bmatrix} 0 \\ -1 \end{bmatrix} \left(M_d - \ddot{\theta}_{ref} - A_{pn}\dot{\theta}_{ref} \right) \tag{8}$$

Select the switch function as:

$$s = cx_1 + x_2 \tag{9}$$

Sliding mode controller is designed as:

$$u_s = \psi x_1 + K_f \text{sgn}(s) \tag{10}$$

where

$$\psi = \begin{cases} \alpha & \text{if } sx_1 > 0 \\ \beta & \text{if } sx_1 < 0 \end{cases} \tag{11}$$

After adopted the DOB, the whole control function is:

$$u = B_{pn}^{-1}(u_s - \hat{M}_d + \ddot{\theta}_{ref} + A_{pn}\dot{\theta}_{ref}) \tag{12}$$

Theorem. For the system (8), DOB is used as inner loop to detect the complex interference of the system, SMVSC is used as the outer loop, the switching function is selected as formula (9), sliding mode control law is designed as formula (10), the control law is adopted as formula (12). To meet the formula (13), the closed-loop system is globally uniformly asymptotically stable.

$$A_{pn} \geq c, \quad \alpha \geq c(A_{pn} - c), \quad \beta \leq c(A_{pn} - c), \quad K_f > |\tilde{\Omega}|_{\max} \tag{13}$$

Proof. Structure the following Lyapunov function:

$$V = \frac{1}{2}s^2 \tag{14}$$

$$\begin{aligned}
\dot{V} &= s\dot{s} = (c\dot{x}_1 + \dot{x}_2)s = (cx_2 - A_{pn}x_2 - B_{pn}u - \Omega + \ddot{\theta}_{ref} + A_{pn}\dot{\theta}_{ref})s \\
&\quad -((c - A_{pn})x_2 - \psi x_1 - K_f \text{sgn}(s) + \hat{\Omega} - \Omega)s \\
&= (c - A_{pn})s^2 - [c(c - A_{pn}) + \psi]sx_1 - K_f|s| + \tilde{\Omega}s
\end{aligned}$$

The formula (13) into the upper, have $\dot{V} \leq 0$, it is shown that the closed-loop system is globally uniformly asymptotically stable.

5 Simulation Results

System simulation parameters: $R = 0.4\ \Omega$, $i = 1650$, $C_m = 0.25$ N m/A, $C_e = 0.25$ V/(rad/s), $J = 0.0067$ kgm^2, $K = 200$, disturbance torque $M_d = 1.5\sin(2t) + 8.5$ N.m. After comprehensive analysis and comparison, the low pass filter Q(s) of DOB is selected as: $Q_{31} = \frac{3\tau s + 1}{(\tau s + 1)^3}$, $\zeta = 0.5$, $\omega_n = 23.3209$. The simulation is divided into the following situations, as shown in Table 1:

Table 1. Simulation parameters

Reference input	Parameter (J)	Simulation map name	
		θ	u
$\theta_{\mathrm{ref}} = \sin(2t)$	$J = 1.5Jn$	Figure 7a	Figure 7b
$\theta_{\mathrm{ref}} = \sin(2t)$	$J = Jn$	Figure 8a	Figure 8b
$\theta_{\mathrm{ref}} = \sin(2t)$	$J = 0.5Jn$	Figure 9a	Figure 9b

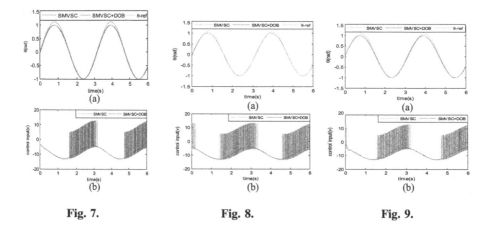

Fig. 7.   **Fig. 8.**   **Fig. 9.**

The simulation results show that the designed controller has a good control effect on external disturbance and system parameter variation.

6   Conclusion

In this paper, a sliding mode variable structure controller based on disturbance observer is proposed for the nonlinear characteristics, parameter uncertainties and unknown disturbances in the tank servo system. Disturbance observer and sliding mode variable structure control complement each other, which can overcome the influence of complex interference, it has a strong practical value.

References

1. Zhang, Z.: Application of model reference adaptive control in artillery. J. Gun Launch Control **4**, 32–37 (1992)
2. Wang, Q., Dong, R., Chen, J.: Application of robust control method of uncertain system in fire control system. Fire Control Command Control **24**(2), 6–9 (1999)
3. Chai, T., Chen, J.: Active disturbance rejection control algorithm for tank stabilizer. Fire Control Command Control **29**(4), 23–25 (2004)

4. Chen, K.: Nonlinear PID control of artillery hydraulic servo system based on neural network. Hydraul. Pneum. **3**, 12–14 (2005)
5. Sun, C., Chen, J.: Sliding mode variable structure control of tank stabilizer based on optimization. Acta Armamentarii **22**(1), 15–18 (2001)
6. Tesfaye, A., Lee, H.S., Tomizuka, M.: Sensitivity optimization approach to design of a disturbance observer in digital motion control systems. IEEE/ASME Trans. Mech. **5**(3), 32–38 (2000)
7. Lu, Y.: Advanced control for tape transport. Carnegie Mellon University, Pittsburgh (2002)
8. Wang, H., Wang, Q.: Sliding mode variable structure control of tank servo system based on disturbance observer. J. Syst. Simul. **21**(14), 4487–4491 (2009)

Robot Vision

Image Fusion Study by Inversible System of Neural Network

Cao Jian[1($\boxtimes$)], Li Gang[1], Zhao Sen[2], and Yan Cong[1]

[1] Air Force Engineering University, Xi'an 710051, Shaanxi Province, People's Republic of China
cao_jian1972@163.com
[2] Jiangnan Design and Research Institute of Electrics and Machinery, Guiyang 550009, Guizhou Province, People's Republic of China

Abstract. This paper discusses arithmetic of inversible neural network. Condensation algorithm is based on the sampling factor for conditional density propagation method, combined with the dynamic model of learning, can complete robust motion tracking. This technology was first widely used in military guidance, introduced into the monitoring system, tracking objects in particular to people or other relatively slow speed of the target, thus tracking method is more simple than the former. Current tracking methods can generally be divided into match tracking and motion tracking. Based on match of track algorithm main has regional match track, and profile match track and the features match track, regional match track of basic thought is put advance extraction of movement regional as match template, set a match metric, and in stay match figure as move template search target. Although the amount of calculation is decreased, but cannot solve the inherent defects profiles matching the basic idea is to extract the contour of objects and to track profile properties. Region matching pursuit algorithm by contour matching benefits can be achieved at a very good match, and the complexity is not high, is a focus of current research.

Keywords: Arithmetic · Offline training · Inversible system · Attitude control · Mutation

1 Introduction

Time points method on dynamic changes scene very effective, but general situation cannot is good to extraction out movement target of all features pixel; background there method can get movement objects is full of features data, but is on by light or other external of dynamic scene changes by raised of background changes very sensitive; light flow method can in camera movement effective of detection movement target, however this method calculation complex degrees is high, does not apply for requirements real-time sex is strong of real-time monitoring occasions; background statistics model classification can since adapted of select threshold value, However the specified threshold are too simple to be completely Adaptive complexity is too high [1, 2]. The shortcomings of traditional motion detection technology, the trend of recent years, there are two: one is to improve on traditional methods. For example was proposed has based on Shi domain difference figure as edge and gray figure as edge strong correlation of algorithm [3, 4]. Fusion has both of movement target edge of precise extraction, for background there

© Springer International Publishing AG 2018
F. Qiao et al. (eds.), *Recent Developments in Mechatronics and Intelligent Robotics*,
Advances in Intelligent Systems and Computing 691, DOI 10.1007/978-3-319-70990-1_68

method on movement scene changes sensitive of defects, was proposed dynamic background update of method, also many scholars also in research reduced light flow field calculation volume of method makes its can applies real-time occasions, these research on traditional movement detection technology of further perfect very useful; another a trend is research new of movement detection method [5].

For example, taking into account the traditional motion detection technology totally ignore the color information for the image, it has been suggested based on color image segmentation method and achieved very good results [6, 7]. In addition, target segmentation method based on knowledge and research focus in the near future. Summarized existing successful testing experience, comprehensive decision fusion method of multiple motion detection can greatly improve the accuracy, robustness and precision of motion segmentation [8, 9, 10]. It follows as Fig. 1.

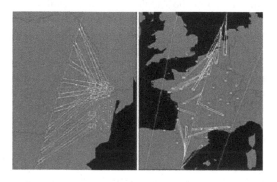

Fig. 1. Pleiades imaging diagram

2 Training Procedure and Structure

Following training procedure is just as Fig. 2. One is input to former system. Another is offline training to stationary neural network.

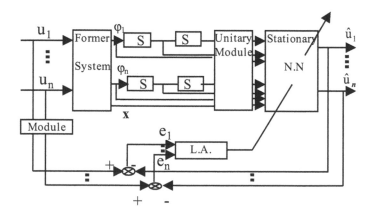

Fig. 2. Sketch map of offline training

2.1   Activate Former System of Ascertain Proper Input

The model can be updated automatically, and automatically converts each pixel classified as background, shadows or moving foreground, in slow-motion case also is able to complete the motion segmentation, and can effectively eliminate the influence of moving objects shadows.

2.2   Accurately Training Data Obtained

Target tracking is based on the cameras collected image sequence analysis, computes a target in the two-dimensional coordinates of image per frame, and depending on the characteristic value, the image associated with the same object in different frames in a sequence, all trajectories of moving objects full, which is in continuous motion video sequences correspond to the target.

3   Input Signals Selection

Current tracking methods can generally be divided into match tracking and motion tracking. Based on match of track algorithm main has regional match track, and profile match track and the features match track, regional match track of basic thought is put advance extraction of movement regional as match template, set a match metric, and in stay match figure as move template search target, put metric take extreme of location judge for best match points, due to match in figure as gray level for, calculation volume is big.

An improved algorithm for region matching pursuit is to put motion into multiple blocks, each block matches, although the amount of calculation is decreased, but cannot solve the inherent defects profiles matching the basic idea is to extract the contour of objects and to track profile properties. Region matching pursuit algorithm by contour matching benefits can be achieved at a very good match, and the complexity is not high, is a focus of current research.

3.1   Nonlinear System

Recent developments two soon based on matching of contour tracking algorithm: active contour tracking algorithm (Snake) and contour model tracking algorithm based on Hausdorff distance. In the military field, video compression, Visual monitoring, and so on to achieve in the background track moving targets in complex situations. Contour matching tracking algorithm track the effect of direct relationship with the initial contour accuracy, how to get very accurate sport target initial profile is to restrict the algorithm universal key.

3.2   MIMO System

In addition, the motion profile update is also very difficult. Same effect this approach in cases of serious disturbance will drop characteristics matching the basic idea is not to track the entire movement as a whole, and track some of these distinctive features, corner or border line. The advantage of this method is that even partial occlusion in the scene, as long as part of the feature is visible, you can still keep track of moving objects. Considerably less than the number of possible matching of feature points cross-correlation algorithm, due to feature carefully selected, when the geometry changes in illumination changes and goals, its matching accuracy can be reduced to a minimum. This method inherits the advantages of high precision and abandon its calculation of the faults and can be fast, reliable and accurate tracking, tracking algorithm based on matching of best value, thus a very wide application. In this algorithm, the feature point selection is very important, it should have on the target's size, location, direction and characteristics of illumination is not sensitive to changes, for example, the gray local edge, corner, isolated points. By contrast, target tracking algorithm based on motion characteristics of complexity is very low, it can generally be divided into associated according to the continuity of the target motion algorithms and motion prediction tracking algorithm. Often are used in a variety of merge tracking algorithm, fewer operations can greatly improve the accuracy of tracking which notable effect on blocking serious occasions, usually by using Kalman filtering to achieve its objectives under the brief disappearance of accurate forecasts or out of the match.

4   Training Sample Selection

4.1   Original Sample Obtaining

It refers to the people's movement pattern analysis and recognition, and natural language to describe. Behavior understanding can simply be considered as temporal and spatial

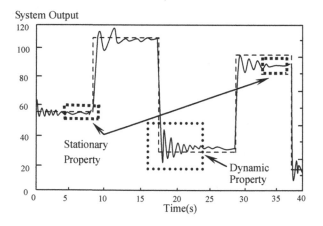

Fig. 3. Dynamic vs. stationary property square wave

changes of data classification, test sequences and calibrated to match the reference sequences representing the typical behavior of. Thus, understanding the key question was how to get reference from the sample sequence and the sequence must be able to handle and matching behavior in the similar patterns of movement in category space and minor feature changes on time scales. It is just like Fig. 3.

4.2 Each Order Derivative Calculation

MIMO is a more sophisticated matching technology for time-varying data, it is a stochastic state machines. MIMO uses involves classification and two stages of training, training phase consists of a specified number of hidden in a hidden Markov model and optimization of State transition and output probability in order to produce the output symbols with a specific movement between matching the observed image features. For each sport category, corresponding to a sequence of MIMO. Matching phase is related to a particular MIMO may correspond to the features of observed probability calculation of test sequences. MIMO in learning skills and continuous data stream than undivided DTW has better advantage, currently is widely used in men's movement pattern matching in the neural network it is also currently interested in matching method for time-varying data. It is just like Fig. 4.

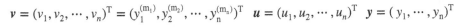

$$\boldsymbol{v} = (v_1, v_2, \cdots, v_n)^{\mathrm{T}} = (y_1^{(m_1)}, y_2^{(m_2)}, \cdots, y_n^{(m_n)})^{\mathrm{T}} \quad \boldsymbol{u} = (u_1, u_2, \cdots, u_n)^{\mathrm{T}} \quad \boldsymbol{y} = (y_1, \cdots, y_n)^{\mathrm{T}}$$

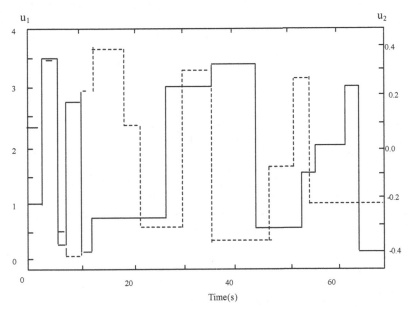

Fig. 4. 2 inputs and 2 outputs system

4.3 Sample Aggregation Obtaining

In recent years, the use of machine learning tools to build statistical models of human behavior research there has been some progress, but behavior is still in its early stages, typical characteristics of the continuous model of matching process is introduced into the body in movement constraints to reduce ambiguity, these conditions are not consistent with the General image, is still understood the difficulties of feature selection and machine learning. Currently state-space for behavior recognition method and template matching method is commonly used in computational cost tradeoff between accuracy and motion recognition, still need to find and develop new technologies in order to improve performance at the same time, and can reduce the complexity.

5 Summary

Proposed classification method based on fuzzy associative memory networks. Static posture as a State each, linked by some probability between these States. Any sequence of movements can be considered for these different transitions between static posture of an ergodic process, joint probability is calculated in the traverse period, the maximum value is selected as criteria for classification. Human motion profile properties are used as the low-level features of learning and recognition learning is the use of MIMO matrix for each category, as fuzzy associative memory networks of knowledge, through the perception of the video stream in real time, final judging types of neural networks, fuzzy associative memory networks of knowledge in different situations also can update at any time. Experimental results show that the algorithms get better recognition results, and have some noise.

References

1. Dai, X., Zhang, K., Zhang, T.: ANN generalized inversion control of turbo-generator governor. IEEE Proc. Gener. Transm. Distrib. **151**(3), 327–333 (2004)
2. Birol, G., Undey, C., Cinar, A.: A modular simulation package for fed-batch fermentation: penicillin production. Comput. Chem. Eng. **26**, 1553–1565 (2002)
3. Dai, X., Zhang, K., Lu, X.: An improved ANN-inversion TCSC controller based on local measurements. IEEE Power Eng. Soc. **4**, 2555–2560 (2003)
4. Wang, W., Dai, X.: An Interactor algorithm for invertibility in general nonlinear system. In: Proceedings of the 5th IEEE World Congress on Intelligent Control and Automation, Hangzhou, China, vol. 1, pp. 59–63, June 2004
5. Nasaroui, O., Gonzdez, F., Dasgupta, D.: The fuzzy artificial immune svstem: motivations, basic concepts and application to clustering and web profiling. Fuzzy Syst. **1**(2), 711–716 (2002)
6. De Castro, L.N., Timmis, J.: Artificial Immune System: A Novel Computational Intelligence Approach. Springer, New York (2012)
7. Jiao, L., Wang, L.: A novel genetic algorithm based on immunity. IEEE Trans. Syst. May Cybern. Part A Syst. Hum. **30**(5), 552 (2010)

8. Wu, X., Chen, Z.: Chaos Study Introductory Remarks. Shanghai Science and technology Publishing House, Shanghai (2012)
9. Jin, G., Zhang, L., Hu, F.-S.: Investigation on space optical system of high F number and high resolution. Opt. Precis. Eng. **15**(2), 155–159 (2007)
10. Ken, F., Sugie, T.: Freedom in coordinate transformation for exact linearization and its application to transient behavior improvement. Automatica **37**(2), 137–144 (2001)

Research and Realization of Crowd Density Estimation Based on Video

Guangdong Zhang and Yan Piao[✉]

College of Electronic and Information Engineering, Changchun University of Science
and Technology, Satellite Road no. 7089, Changchun 130000, Jilin, China
804124051@qq.com

Abstract. Automated estimation of crowd density and early warning for over-crowded situation are essential and valid approaches for public security management. This paper builds a system to detect and monitor the level of congestion aimed at crowded scenes. Obtaining the crowd foreground by Gaussian Mixture Background Modeling to eliminate the bad effect the changing of complex background does to the detecting result. Extracting the textural features of the crowd foreground to avoid the misjudgment issue caused by people covering each other while purely using pixels statistics method to analyze crowd foreground. This paper uses GRNN neural network to train test samples to get the density level classifier. The accuracy of high density for the experimental scene in this paper is up to 92%. Proven the performance is relative good and the system this paper built is practicable and valid.

Keywords: Density estimation · Gaussian mixture background modeling · Textural features · Neural network

1 Introduction

Urban population explosion is a great challenge to the public security management. Public places such as subway station, large mall those can easily occur crowd gathering, further more causing crowded stampede has become the focus and difficulty of public security management. Conventional artificial monitoring needs large personnel investment and exists easily missing alarm problem. Therefore, research of automatic crowd density estimation has been the emphasis of video monitoring and public security management.

In 1995, Davies [1] purposed a pixel-based method for crowd density estimation. This method counts the number of crowd pixels to fit the linear relations between the number of persons in scene and the pixels amount the crowd occupies. In 1998, Marana [2] purposed a new method for crowd density based on image texture feature. This method to some extent is able to solve the difficult in correctly recognizing high density situation but exists the problem of dealing low density situation, the error of estimation is more than 20%.

We purpose a method to distinguish the differences of different crowd density combining pixel information and texture feature. In extracting crowd foreground, we

F. Qiao et al. (eds.), *Recent Developments in Mechatronics and Intelligent Robotics*,
Advances in Intelligent Systems and Computing 691, DOI 10.1007/978-3-319-70990-1_69

adopt Gaussian Mixture Model (GMM) to effectively avoid the error the changing background cause. At last, we employ Generalized Regression Neural Network (GRNN) as density classifier to classify the degree of different crowd situation. Our experiments show that the System we design offers promising results in real world scenes, proven our method valid and workable.

2   Key Algorithms

The realization of crowd density estimation in this paper can be divided as following parts: firstly employ GMM to extract crowd foreground, secondly use HSV color space model to remove foreground shadow, thirdly calculate the proportion of the area of foreground to that of ROI, fourthly extract texture feature with Gray-level Co-occurrence Matrix. In the end, pick training samples to train the Neural Network and obtain the density classifier.

2.1   Extract Crowd Foreground by GMM

Branches swaying, clouds drifting, illumination changing and so on will keep background in change. Given that, we use multi-modal to describe pixels of the scene. Gaussian Mixture Model is an improvement of single Gaussian Model.

For each pixel in frame sequence, we can get an set of its gray value$\{X_1, X_2,....X_n\}$ in time axis.n($1 \leq n \leq K$) Gaussian distribution are used to compose a joint model to describe the distribution of gray values of pixels. At moment t, assume the current observed value of current pixel is X_t, formula (1) gives the probability of the gray value's appearance:

$$P(X_t) = \sum_{i=1}^{k} \omega_{i,t} * \eta(X_t, \mu_{i,t}, \sum i, t) \tag{1}$$

$$\eta(Xt, \mu_{i,t}, \sum i, t) = \frac{1}{2\pi^{n/2} |\sum i, t|^{1/2}} \exp\left\{-\frac{1}{2}(X_t - \mu_{k,t})^T \sum_{k,t}^{-1}(X_t - \mu_{k,t})\right\} \tag{2}$$

In formula (1), $\eta(X_t, \mu_{i,t}, \sum_{i,t})$denotes the probability density function of the number i normal distribution at moment t. $\mu_{i,t}$ denotes the mean value of the normal distribution, $\sum_{i,t}$ the covariance matrix. $\omega_{i,t}$ denotes the weight of the normal distribution in joint distribution, and $\sum \omega_{i,t} = 1$. K is the amount of normal distributions those are used to describe pixels distribution.

To realize GMM to extract crowd foreground, we need following steps:

(1) Model initialization: Use 3–5 normal distribution to describe each pixel's gray value distribution. The mathematical expectation μ_1 of the first normal distribution is given to the first frame's pixels as initial value. Normally, standard deviation σ_1 is given a relative large initial value and all the Gaussian distribution have same initial standard deviation and same weight.

(2) Background model parameter update: for every pixel in one frame, try to match the gray value of it with Gaussian distributions of the joint distribution. If the current gray value of the pixel and the number i($1 < i < K$) meet the in equation $|X_t - \mu_{i,t} - 1|$ $< 2.5\ \sigma_t$, we think they match, the pixel is deemed to be background. Otherwise, continually match the pixel with next distribution. If eventually the pixel doesn't match any one of the joint distribution, it is deemed to be the foreground. In this case, the Gaussian distribution with the smallest value of ω/σ needs to be replaced by a new distribution whose expectation is the current observed value X_t. The other ones remain the same. The weight of each distribution updates as formula (3)

$$\omega_{i,t} = (1 - \alpha)\omega_{i,t-1} + \alpha M_{i,t} \tag{3}$$

In formula (3), α ($0 \leq \alpha \leq 1$) is self-defined learning rate, the value of it is in direct proportion to the background updating rate. For which successfully gets matched, $M_{i,t} = 1$. For the rest, $M_{i,t} = 0$. Expectation μ_t and variance σ^2 update according to formula as below:

$$\mu_t = (1 - \rho)\mu_{t-1} + \rho X_t \tag{4}$$

$$\sigma_t^2 = (1 - \rho)\ \sigma_t^2 + \rho(X_t - \mu_t)(X_t - \mu_t) \tag{5}$$

represents parameter learning rate is expressed as formula

$$\rho = \alpha * \eta(X_t | \mu_t, \sigma_t) \tag{6}$$

The experimental effect of obtaining background by GMM is shown in Fig. 1.

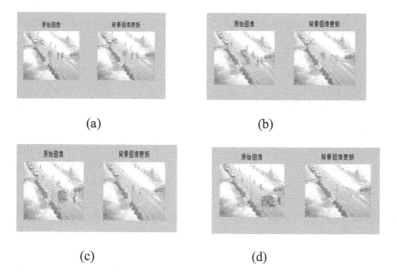

(a)   (b)

(c)   (d)

Fig. 1. The experimental effect of obtaining background by GMM Figure (a)–(d) show the process of obtaining background.

3   Shadow Detection and Removal

The difficulty in distinguishing foreground from its shadow by computer lies in two causes: (1) They are both very much different from background. (2) They both have the same movement feature. There are two ways to detect and remove shadow mainly: the one based on model and the one based on feature. The former use known information about scene, light source and moving objects to create shadow model to distinguish shadow by matching moving object's three-dimensional structure. The later based on feature calculates the value of feature including saturation, hue, geometric features and compares these features of each pixel to those of the corresponding background to detect shadow Fig. 2.

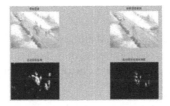

Fig. 2. The experimental effect of removing shadow via HSV color space

One common method of shadow detection is to utilize color feature. HSV color space is one relatively close to perceptual character of human eyes among many others. H represents hue, S represents saturation, V represents value. Formula (7) expresses the transition from RGB color space to HSV color space.

$$H_1 = \arccos\left\{ \frac{[(R-G)+(R-B)/2]}{[(R-G)^2+(R-B)(G-B)^{1/2}]} \right\}$$
$$H = H1\, if (B \le G)$$
$$S = \frac{Max(R,G,B)-Min(R,G,B)}{Max(R,G,B)}$$
$$V = \frac{Max(R,G,B)}{255}$$

(7)

According to research finding, the most primary difference between foreground target and its shadow lies in brightness. The brightness of shadow is clearly darker, the saturation of it is lower, but the hue basically remains the same.

In this paper, we calculate H, S, V of the foreground zone and the corresponding background and use formula (8) to judge whether it's the real foreground or just the shadow of it.

$$sp(x,y) = \begin{cases} 1, & \alpha_s \le \frac{I^V(x,y)}{I^V(x,y)} \le \beta_s^\wedge [I^S(x,y)-B^S(x,y) \le \tau_S]^\wedge |I^H(x,y)-B^H(x,y)| \le \tau_H \\ 0 & \text{other} \end{cases}$$

(8)

$I^H(x, y)$, $I^S(x, y)$, $I^V(x, y)$ respectively represents the H, S, V component value of the input pixel gray value $I(x, y)$. $B^H(x, y)$, $B^S(x, y)$, $B^V(x, y)$ respectively represents that of background pixel. If $SP_{(x, y)} = 1$, the pixel is regarded as shadow, otherwise, it's foreground. Parameter α_S is related to the intensity of shadow. $0 < \alpha_S < \beta_S < 1, \tau_S < 0$.

4 Extract Foreground Texture Feature

Gray-level Co-occurrence Matrix (GLCM) is a two-dimensional correlation matrix, it expresses the probability of the appearance of pixel pairs given certain distance d and certain angle θ. We select suitable d and θ to obtain GLCM of images.

4.1 Feature of GLCM

GLCM is a very valid texture descriptor, it altogether has 14 characteristic quantities [3], but the most commonly used are angular second moment, contrast, entropy and correlation. In this paper, we use contrast and entropy as descriptor to denote crowd foreground texture feature.

(1) Contrast

$$CON = \sum_{i=1}^{g} \sum_{j=1}^{g} \left[(i - j)^2 * P^2(i, j, d, \theta) \right] \tag{9}$$

Contrast expresses the distribution of the element's value of the matrix and the local change in image. It reflects the brightness contrast of one pixel and its neighbourhood [4].

(2) Entropy

$$ENT = - \sum_{i=1}^{g} \sum_{j=1}^{g} p(i, j, d, \theta) * \log 10 p(i, j, d, \theta) \tag{10}$$

Entropy measures the quantity of information and content randomness one image contains. Its value reveals the heterogeneity and complexity [5]. If one image doesn't have texture, entropy equals to zero.

In this paper, we choose parameter $d = 1$, $\theta = 0°, 45°, 90°, 135°$, calculate the average entropy and average contrast of the four directions.

5 Artificial Neural Network

GRNN is a new kind of radial basis function neural network which is first proposed in 1991 by Donad F. Given adequate training samples, GRNN can solve the approximation problem of any smooth function, even with a small number of samples, it is able to achieve relative good prediction. GRNN is capable of processing unstable date with well fault tolerance and robustness.

In this paper, we employ GRNN function in Matlab neural network toolbox to train three degrees of crowd density samples obtain the density classifier.

6   Experiment and Result Analysis

To test the crowd density system we design, we choose a scene to provide experiment samples. Due to the size of region of interest and the pedestrian traffic, we categorize crowd densities as below:

we choose the scene of students coming out of main teaching building after school, we consider the frames with less than 10 pedestrians as low density samples, the ones with more than 10 but less than 25 pedestrians medium density samples, the ones with more than 25 pedestrians high level samples. Altogether, we have 300 frames of each density degree, randomly choose 200 to be training samples, the other 100 frames are left for testing.

We extract three characteristic quantities. The first one is a pixel-based: the proportion of pedestrian foreground's area to ROI's area. The others are texture-based: the average entropy and average contrast of the GLCM of pedestrian foreground in four directions: $0°, 45°, 90°, 135°$. We combine the three quantities as one descriptor of density feature. GRNN receives the input joint-quantities, generates a density classifier, outputs the result of test (output 1 means the frame is judged as low density, output 2 means the frame is judged as medium density, output 3 means the frame is judged as high density).

The effect of crowd density classification is shown in Figs. 3, 4 and 5.

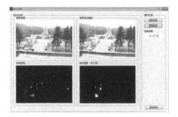

Fig. 3. Effect of low density classification   **Fig. 4.** Effect of medium density classification

Fig. 5. Effect of high density classification

Result date of testing samples is indicated in Table 1:

Table 1. The experimental result date of testing samples

Classification Testing samples	Low density (100 frames)	Medium density (100 frames)	High density (100 frames)
Low density	91	6	0
Medium density	9	86	8
High density	0	8	92
Accuracy	91%	86%	92%

For this scene, the classification accuracy of low density samples is 91%, 9 frames are misjudged as medium density. The classification accuracy of medium density samples is 86%, 6 frames are misjudged as low density, 8 frames are misjudged as high density. The classification accuracy of high density samples is 92%, 8 frames are misjudged as medium density.

The experiment of the experimental scene indicates that for low density and high density samples, the possibility of being misclassified as non-adjacent density degree is extremely low, because low density and high density are distinct from one another a lot in both pixel and texture feature. However, medium density samples have chance to be misclassified as low density and high density. Through analyzing, many of the misjudged are close to the boundary of different density degrees, it's easy to make misjudgment even for human.

The method we proposed combines pixel information and texture information into a joint characteristic quantity to indicate crowd density information. In the experiment, the average accuracy is 89.7% for the experimental scene.

7 Conclusion

In this paper, we employ GMM to extract moving crowd foreground because GMM can well overcome the bad impact on which wind and sunlight changing could generate. We add shadow removal in order to reduce foreground extraction error caused by the shadow. According to the theory that the texture characteristics of different crowd density are different, we adopt GLCM to analyze image texture by calculating characteristic quantities. We choose average entropy and average contrast of four directions to be the texture descriptor, combine with the proportion of crowd foreground's size to the size of ROI size as the image feature of crowd density. We obtain a density classifier by having training sample of different densities sent to GRNN. The experimental effect proves our method valid and workable and to a certain extent novel.

References

1. Maohui T.: The research of crowd density estimation based on video image. Chengtu (2014)
2. Zhang L.: The parametric study background modeling method. Beijing (2011)

3. Yangmin, O., Renhuang, W.: Based on LAB color co-occurrence matrix feature extraction of the distance. J. Guangdong Univ. Technol. **28**(04), 48–50 (2011)
4. Hui W.: Research of wood surface texture pattern recognition method Based on gray level co-occurrence matrix. Harbin (2007)
5. Hang, S.: The crowd flow and density estimation algorithm in video monitoring analysis. Telev. Technol. **33**(11), 100–103 (2009)

Eye Movement Analysis on Driving Behavior for Urban Road Environment

Ziyi Qin[✉]

College of Physics and Engineering, Qufu Normal University, Shandong, China
QNUqinziyi@163.com

Abstract. Under real driving environment, drivers are easily distracted by the external factors, such as vehicles or pedestrians. Overtaking, irritating or even attacking behaviors more often than not occur during driving. Driving behavior is the external manifestation of a driver's psychological and physiological changes. Based on real road driving testing to analyze the driving safety, the paper gets the driving behavior information under real driving environment, and systematically describes the relationship between a driver's eye movement changes and road environment. Research results are capable of providing the basis for the analysis of traffic accidents.

Keywords: Driving behavior · Eye movement analysis · Urban road environment

1 Introduction

In traditional driving environment, a driver mainly gets traffic information from eyes. Driver's visual changes vary from road condition to road condition. In urban areas, traffic accidents are more often than not occurred near intersections of large traffic volume. A driver's perception on the road hazard directly affects driving safety, while the driver's perception and reaction are reflected by changes in eye movement. Therefore, it is very important to design eye change movement tests under different road conditions, to collect the driver's eye movement data, to analyze the relationship between the driver's eye movement change and different road environment, to identify changes of eye movements during driving. Research results provide a theoretical basis for road safety evaluation.

The paper collects transcutaneous electrical eye movement data under actual traffic conditions, such as turning, traffic lights, changing lanes, and avoiding vehicles. Moreover, the paper records specific line condition, analyzes the actual traffic situation in the driver changes, and studies the driver's reactions to actual road traffic information. Driving behavior is often a driver's external manifestation of psychological and physiological changes. Real-time monitoring and analysis supports early detection of operational errors, to avoid accidents.

© Springer International Publishing AG 2018
F. Qiao et al. (eds.), *Recent Developments in Mechatronics and Intelligent Robotics*,
Advances in Intelligent Systems and Computing 691, DOI 10.1007/978-3-319-70990-1_70

2   Related Work

Researches on the driver's visual characteristic have been widely studied in Europe, the United Kingdom, France, and the United States. In 2002, David Crundall, Geoffrey Underwood and Peter Chapman [1] conduct researches on driving test by videoing real traffic scenes. They analyze search features of unskilled and skilled drivers in laboratory within a particular road operator, and design a visual search driver training scheme. The results indicate that training can shorten the driver's gaze duration and enlarge the searching range.

Mark Brackstone and Ben Waterson [2] conduct researches on the driver's eye movement in high speed environment, and analyze the relationship between vehicle speed and the driver's visual characteristics. Results show that as vehicle speed increases, the driver's field of vision becomes narrowed.

In 2010, Mitsubishi's safety car ASV2 is designed with the installed steering wheel mounting cameras to detect eye movements and the driver's fatigue. By using text and voice warning, drivers are informed of driving safety [3].

In 2010, Yong-Fang Li [4] researches the distribution of driver dynamic fixation point. The results show that at the same driving speed, with smaller size of words, the driver's gaze point is not fixed. It takes longer time for the driver to determine flag information. For larger words, driver's visual fixations are relatively concentrated.

Currently, researches mainly focus on generalized driving test on actual roads but fewer researches on microscopic dynamic vision issue or the eye movement parameters specifically. This paper is based on the actual road driving test to research blink frequency, blink duration, visual fixation during the driver's driving process.

3   Field Test Methods

In the real driving test, galvanic skin response testing is conducted for each driver in order to ensure traffic safety. During the experimentation, we collect eye movement data and use Biolab, a behavioral synchronization device to monitor drivers' behaviors. On the computer screen, we can directly observe driver's galvanic skin response curve and the corresponding time stamp.

Galvanic Skin Response
If electric potential on skin exceeds 0.05 V during driving, drivers are in excited state, thus they should not continue driving. If electric potential on skin continuously reduces during driving, drivers' sympathetic activity decreases, thus drivers need to stop and have a rest. Therefore, the paper uses Biolab to get behavior curves and identify whether driving behavior is normal or not. Moreover, this can be used as a driving safety indicator in the monitoring process of the actual road driving.

Test Scheme Design
Drivers are chosen for obtaining eye movement data in field test prior to driving on road. In galvanic skin response test, we test the driver's galvanic skin response values in their calm states. The design process of real driving test scheme is shown in Fig. 1.

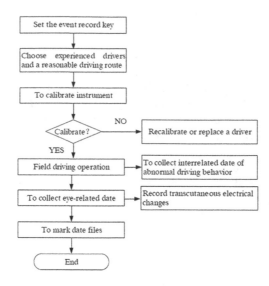

Fig. 1. Design process of driving test

4   Analysis on Visual Characteristic Parameter

4.1   Data Collection

Facelab5.0 eye tracker is installed between the steering wheel and driver's windscreen. Eye tracker should be well fixed to avoid shaking, tracking inaccuracy, and eye movement data deviation during driving. In the process of actual road driving, Facelab5.0 eye tracker fixed in the car.

4.2   Data Analysis

Data collected is put into SPSS software to do variables import analysis [5], and the correlation test, which assumes that the probability of a certain traffic and corresponding eye movement changes at the same time is P. The meaning P in the driving test is as shown in Table 1.

Table 1. The statistical significance of correlation between road conditions and data changes

Correlation P Value	Probability of happen	The connection between the road and data changes	Statistical significance
$P > 0.05$	The probability of happened to occur is more than 5%	The changes between the two groups have no inevitable connection	Changes in the two groups have no significant correlation
$P < 0.05$	The probability of happened to occur is less than 5%	The changes between the two groups have connection	Changes in the two groups have significant correlation
$P < 0.01$	Probability of happened to occur is less than 1%	The changes between the two groups have inevitable connection	Changes in the two groups have very significant correlation

5   Simulation Results

In order to obtain data from real driving process, the paper choose real traffic network of campus in China. Drivers start from and return back to the school gate. During the driving process, drivers will experience different road conditions. When a vehicle starts leaving school gate, it is the deceleration-stop phase. When a driver drives on roads outside the school, we classify roads with low traffic volume and no traffic lights as general roads, and classify roads with with large traffic volume and traffic lights as complicated ones.

The entire driving process takes 28 min and 15 s (1695 s). From starting a vehicle to approaching the school gate, it takes 125 s. From returning to the school gate, it take a total of 1453 s. When a vehicle slows down to a stop, it takes 117 s.

5.1   Galvanic Skin Response

In the entire driving process, the driver's galvanic skin response is maintained at a low value. It indicates that the driver has a driving strong capacity for safe operation in the whole process, and the driver does not have emotional changes, such as tension, anxiety and others.

At the end of the driving process, we collect galvanic skin response data. During data processing, the data is averaged per second to get galvanic skin response values per second. We also plot data curve about galvanic skin response and compare galvanic skin response changes between the time when a driver is in calm state and drives in actual road environment. We also conduct an analysis of the variation range of driver's galvanic skin response. Data curve about galvanic skin response is shown in Fig. 2.

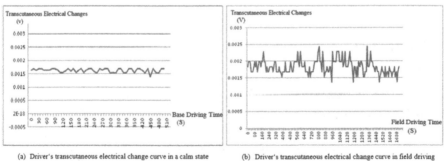

(a) Driver's transcutaneous electrical change curve in a calm state

(b) Driver's transcutaneous electrical change curve in field driving

Fig. 2. Driver's galvanic skin response

When a drive is in calm state, galvanic skin changes of drivers only fluctuate within a narrow range over 0.0015 V. When entering the driving phase where road conditions appear more complicated, galvanic skin changes are between 0.0015 and 0.0025 V. The fluctuation is obviously stronger than that in calm state. In actual driving environment, drivers have a high level of alert due to external stimulation. Psychological changes often result in galvanic skin changes of drivers.

5.2   Driver's Blink Duration

When driving on roads, a driver's blink duration is related to the complexity of the traffic conditions. The paper tests the blink duration when driver drives on base and actual road and analyzes changes of the blink duration.

The paper takes blink duration of base driving as the benchmark, and plots a data curve, and conducts comparative analysis on blink duration changes when driving on roads. Blink duration curve of base driving is shown in Fig. 3.

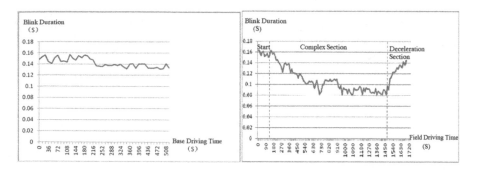

Fig. 3.   Driver's blink duration

In the experiment, driver's blink duration has a change range of less than 0.02 s when driver drives in base, and 0.08 s when driver drives in actual road. At the beginning of actual road driving, blink duration is long. When driving outside the school, blink duration is shorter. Blink duration is below 0. 1 s from 917 to 1552 s time. Blink duration increased from the time when entering the gate to the time when decelerating to stop smoothly.

6   Conclusion

This paper designs and conducts a test of driver's eye movement characteristics when driving on roads. In summary, we can get the following conclusions.

Firstly, during the actual driving test, when drivers are in the driving process, galvanic skin change is larger than that in calm state.

Secondly, during actual driving test, the driver's eye movement change is wider than the base eye movement variations. When drivers are in the base driving test, the blinking frequency and duration curve is steady; when drivers are driving on the actual road, the driver's eye movement change is wider.

Thirdly, driver's blink duration is associated with blinking frequency. With the increasing frequency driver blink, blink duration becomes shorter.

References

1. Underwood, G., Crundall, D.: Peter Chapman selective searching while driving: the role of experience in hazard detection and general surveillance. Ergonomics **45**(1), 1–15 (2002)
2. Brackstone, M., Waterson, B.: Are we looking where we are going an exploratory examination of eye movement in high speed driving. In: Proceedings of the 83rd Transportion Research Board Annual Meeting. Washington (USA): TRB, p. 131 (2004)
3. Srimal, R., Diedrichsen, E.B.R., Curtis, C.E.: Obligatory adaptation of saccade gains. Neurophysiol **99**(3), 1554–1558 (2010)
4. Li, Y.: Research of car driver's dynamic fixation point distribution. Automotive Engineering, Shanghai University of Engineering and Technology, p. 11 (2010)
5. Lin, Z.: Multivariate Analysis-Operation and Application of SPSS. The Peking University Publishing House, Beijing (2007)

Infrared Thermal Wave Image Segmentation Based on 2D Tsallis Entropy and Artificial Immune Algorithm

Jin-Yu Zhang[(✉)]

Xi'an Research Institute of High-Tech,
Xi'an 710025, People's Republic of China
mejyzhng@163.com

Abstract. Infrared image segmentation is widely used in engineering, and it is a very important research topic. In this paper, the two-dimensional minimum Tsallis cross entropy and artificial immune algorithm is organically combined for infrared image segmentation optimization, through clonal selection to enhance optimal threshold search ability, and introduce a high frequency variation and population update to prevent the algorithm into a local optimum degradation. The preliminary test results show that the method is simple and robust, and can obtain good segmentation results.

Keywords: Tsallis entropy · Artificial immune algorithm · Image segmentation · Infrared thermal images processing

1 Introduction

In the infrared image information processing, the segmentation of the image is the separation of the target region from the background, which is an important step in image processing, and has a wide range of applications in engineering. The problem of infrared target segmentation and its inherent particularity is different from the general object segmentation, the difficulty is mainly reflected in: (1) infrared imaging is source imaging, target and boundary in the image are blurred; (2) the target itself has no obvious shape, size, texture and other information can be used; (3) the imaging target area is small, often accompanied by weak signals, and target segmentation under low SNR condition [1].

Image segmentation is an important part of computer vision research field, has been widely studied, and thousands of images segmentation algorithm were proposed, but the existing methods are designed for specific applications, has great pertinence and limitations. And need to carry out in-depth research on infrared thermal wave detection image segmentation and target recognition hand. In recent years, with the development of artificial intelligence technology, more and more new intelligent algorithms have been introduced into the field of image segmentation, which makes the image segmentation more rapid, effective and intelligent [2]. The genetic algorithm is introduced to the maximum inter class variance threshold segmentation method by Zhang [3, 4], which significantly reduces the search time of the best segmentation threshold, making

© Springer International Publishing AG 2018
F. Qiao et al. (eds.), *Recent Developments in Mechatronics and Intelligent Robotics*,
Advances in Intelligent Systems and Computing 691, DOI 10.1007/978-3-319-70990-1_71

the segmentation results more real-time, effective. Based on mathematical morphology and watershed, an image segmentation method is proposed by Zhang [5], realize the defect extraction and segmentation feature of complex thermal wave detection. Many intelligent control methods of Robot Uncalibrated Visual servo system are discussed by Li [6]. The intelligent image recognition method of ARM grasping is studied by Zhang [7].

The artificial immune theory uses the mechanism of biological immune system to build artificial immune system, and uses the theory and method to solve practical problems, which is an intelligent theory extracted from biological system after the neural network and genetic algorithm, is one of the hotspot in the field of artificial intelligence at present. The application of artificial immune system involves many fields such as control, optimization, machine learning, fault diagnosis and robot path planning [8, 9]. In order to solve the problem of low efficiency and poor effective in the infrared thermal image segmentation algorithm, this paper combines the artificial immune algorithm with the threshold segmentation method to achieve the key optimal threshold selection in image segmentation.

2 Two-Dimensional Minimum Tsallis Cross Entropy

(1) Tsallis Cross Entropy

Suppose $p = \{p_i\}$ is a discrete probability distribution, and $0 \leq p_i \leq 1, \sum_{i=1}^{k} p_i = 1$, the Tsallis entropy of the distribution is defined as:

$$S_q = \frac{1 - \sum_{i=1}^{k} (p_i)^q}{q - 1} \tag{1}$$

Where, q is a real number.

Tsallis entropy considers the influence of non additive information in the image, and gives the definition of Tsallis cross entropy. Suppose $P = \{p_i\}, i = 1, \ldots, N$ and $Q = \{q_i\}, i = 1, \ldots, N$ are any two probability distribution, and meet $0 \leq p_i, q_i \leq 1$, $\sum_{i=1}^{k} p_i = \sum_{i=1}^{k} q_i = 1$, then the Tsallis entropy between P and Q as follow:

$$D_q = \sum_{i=1}^{k} p_i \frac{\left(\frac{q_i}{p_i}\right)^{1-q} - 1}{q - 1} + \sum_{i=1}^{k} q_i \frac{\left(\frac{p_i}{q_i}\right)^{1-q} - 1}{q - 1} \tag{2}$$

The cross entropy of Tsallis is not extensive features. When a system is decomposed into two independent subsystems A and B, the total Tsallis cross entropy of the system can be expressed as:

$$D_q(A+B) = D_q(A) + D_q(B) + (1-q)\left(D_q(A) + D_q(B)\right) \tag{3}$$

The Tsallis cross entropy also takes into account the difference in the amount of information between two probability distributions in the sense of Tsallis entropy, which is a generalization of the Kullback distance in the sense of Shannon entropy.

(2) 2D Minimum Tsallis Cross Entropy Threshold Selection

Suppose $f(x, y)$, $1 \leq x \leq M$, $1 \leq y \leq N$, is a $M \times N$ pixel image. Its gray level is L, i.e. $0 \leq f(x, y) \leq L-1$, $g(x, y)$ is the gray mean value of current pixel $f(x, y)$ under 3×3. If r_{ij} is the number, in which $f(x, y) = i$ and $g(x, y) = j$, then the two-dimensional histogram of the image can be expressed as:

$$p_{ij} = \frac{r_{ij}}{M \times N}, i,j = 0, 1, \ldots, L-1 \tag{4}$$

If (s, t) is the threshold vector, the histogram is divided into A, B, C, D four area, as shown in Fig. 1, where A and C represent the background or target class, the regional B and D represent the boundary point or noise point. Due to the small number of boundary points and noise points, the probability is set to 0. The probability of the region A and C representing the target and background is:

$$P_A = \sum_{i=0}^{s} \sum_{j=0}^{t} p_{ij} \tag{5}$$

$$P_C = \sum_{i=s+1}^{L-1} \sum_{j=t+1}^{L-1} p_{ij} \tag{6}$$

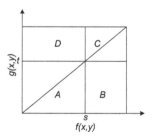

Fig. 1. 2D histogram

The two-dimensional Tsallis cross entropy of A and C in the segmented region is:

$$D_q^A = \sum_{i=0}^{s} \sum_{j=0}^{t} \left(ijp_{ij} \frac{\left(1 - \frac{ij}{\mu_1(s,t)}\right)^q}{q-1} + \mu_1(s,t)p_{ij} \frac{\left(1 - \frac{\mu_1(s,t)}{ij}\right)^q}{q-1} \right) \tag{7}$$

$$D_q^C = \sum_{i=s+1}^{L-1} \sum_{j=s+1}^{L-1} \left(ijp_{ij} \frac{\left(1 - \frac{ij}{\mu_2(s,t)}\right)^q}{q-1} + \mu_2(s,t)p_{ij} \frac{\left(1 - \frac{\mu_2(s,t)}{ij}\right)^q}{q-1} \right) \tag{8}$$

Where, $\mu_1(s, t)$, $\mu_2(s, t)$ are the mean value of target and background.

$$\mu_1(s,t) = \frac{\sum_{i=0}^{s} \sum_{j=0}^{t} ijp_{ij}}{P_A} \tag{9}$$

$$\mu_2(s,t) = \frac{\sum_{i=s+1}^{L-1} \sum_{j=t+1}^{L-1} ijp_{ij}}{P_C} \tag{10}$$

The total two-dimensional Tsallis entropy of the image is:

$$D_q(s,t) = D_q^A + D_q^B + (1-q)D_q^A D_q^B \tag{11}$$

The optimal segmentation threshold selection is to find a threshold vector (s, t), so that Dq (s, t) minimum. The objective realization depends on the optimization of the following artificial immune algorithm.

3 Segmentation Threshold Optimization Based on Artificial Immune Algorithm

The artificial immune algorithm can be used to suppress the degradation in the optimization process with some characteristic information. Variation and random variables can be effectively prevented from falling into local minima, so it is very useful to select the optimal threshold in threshold segmentation. The process is shown in Fig. 2.

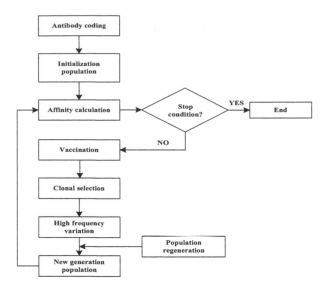

Fig. 2. Optimization process of artificial immune algorithm

(1) Antibody Coding

The gray scale of the L level is represented by the pixel gray value and the field gray value (s, t). For level 256 (2^8) images (s, t) requires a 16 bit binary encoding. The pixel gray value s is arranged in the high 8 bits, and the neighborhood gray value t is arranged in the lower 8 bits. Use x to express:

1	0	1	1	0	0	1	0	0	0	1	0	1	0	0	1

(2) Affinity Calculation

Affinity is a measure of the maturity of antibody. A segmentation threshold pair is an antibody. Two dimensional Tsallis entropy is used as the evaluation index, that is, the optimal threshold value (s, t) is obtained when the threshold value of the two-dimensional Tsallis entropy is the lowest. The specific formula is shown in formula (11).

(3) Vaccination

(1) Select Vaccine

Randomly generated N antibody xi, $i = 1, 2,..., m$, calculated by affinity to select the two most affinity antibodies $R1$, $R2$ as the vaccine.

(2) Vaccine Inoculation

The average affinity of the antibody population was calculated, and the antibody was vaccinated if less than the average affinity. That is, with reference to the vaccine, the vaccine will be vaccinated antibody M on the location of the code value to the corresponding location of the vaccine code value.

Determination of the number of antibody of generation t:

$$Wn = \frac{\alpha \times Ps}{1 + e^{-\beta \times t}}, 0 < \alpha, \beta \leq 1 \tag{12}$$

Where, Ps - size per generation (known);
α, β- setting parameter.
Determination of the number of vaccination:

$$Vn = C \times e^{-\frac{t^2}{2\sigma^2}}, \sigma = \frac{m}{3} \tag{13}$$

Where:
C - binary coded length (known);
m - maximum number of iterations (known).
4) Post Vaccination Operation
R_1 and R_2 were used to obtain M_1 and M_2 after inoculation with M, and the affinity of M_1, M_2 and antibody M was calculated.
(5) Clonal Selection, High Frequency Variation, Population Regeneration
(1) Clonal Selection
According to the size of the affinity cloning

$$n_{ti} = \text{int}\left(\frac{f_{ti}}{\overline{f}_t} + 0.5\right), i = 1, 2, \ldots, Ps \tag{14}$$

In the formula, f_{ti} represent the affinity of the antibody i in generation t, $\overline{f}_t$ is the affinity of the population for generation t. And the largest affinity is sent into the memory.
(2) High Frequency Variation
Mutation rate:

$$P_{ti} = \begin{cases} 0.5 \times \frac{f_{max} - f_{ti}}{f_{max} + f_{ti}} f_{ti} > \overline{f}_t \\ 0.5, f_{ti} \leq \overline{f}_t \end{cases} \tag{15}$$

(3) Population Regeneration
In order to ensure the diversity of the population, some of the antibodies with the least affinity were replaced by random antibodies. Then go to the next generation.

4   Experimental Results

Based on artificial immune threshold segmentation algorithm, two experimental infrared thermal image were processed, the effect of segmentation and three-dimensional display as shown in Fig. 3. It is obvious that the two experimental infrared thermal images are effectively segmented.

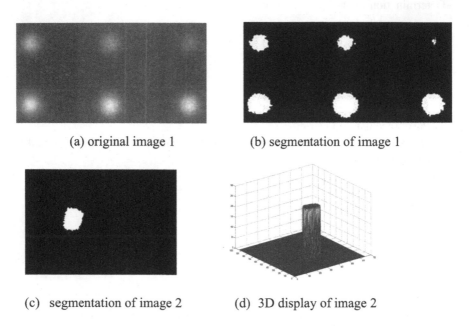

(a) original image 1   (b) segmentation of image 1

(c) segmentation of image 2   (d) 3D display of image 2

Fig. 3. Segmentation results of two experimental images

5   Conclusion

Based on the basic image segmentation method, a segmentation algorithm based on the artificial immune algorithm and two-dimensional minimum Tsallis cross entropy is proposed. Through clonal selection to enhance optimal threshold search ability, introduce the high frequency variation and population update to prevent the algorithm into a local optimum degradation, a good result is achieved.

Acknowledgments. This work is supported by the National Natural Science Foundation of China (Grant No. 51275518).

References

1. Zhang, J.-Y., Zhang, Z.-W., Tian, G., et al.: Infrared Thermal Wave Testing and Images Sequence Processing Technology. National Defence Industry Press, Beijing (2015)
2. Tian, B., Shaikh, M.A., Azimi-Sadjadi, M.R., et al.: A study of cloud classification with neural networks using spectral and textural features. IEEE Trans. Neural. Netw. **1**, 138–151 (1999)
3. Jinyu, Z., Yan, C., Xian-Xiang, H.: IR Thermal image segmentation based on enhanced genetic algorithms and two-dimensional classes square error. In: ICIC 2009, The Second International Conference on Information and Computing Science, 21–22 May 2009, Manchester, England, UK
4. Xu, H., Niu, C., Zhou, Q.: Image segmentation and realization based on enhanced genetic algorithms maximum classes square error, Syst. Simul. Technol Appl. **10**
5. Zhang, W., Cai, F.-H., Bao-Ming, M.A.: Quantitative analysis of infrared thermal image defect based on mathematical morphology. NDT **31**(8), 596–599 (2009)
6. Li, Y., Mao, Z.: Visual servo control based on artificial immune and image direct feedback. J. South China Univ. Technol. (Nat. Sci. Edn.) **37**(5) (2009)
7. Zhang, Y.-B.: The research of manipulator grasping based on image recognition of artificial immune algorithm. Xi'an Scientific University (2009)
8. Zhang, A.-L., Deng, Q.-S.: Research on hybrid artificial immune algorithm. J. North Univ. China **36**(1), 56–60 (2015)
9. Nicholas, W., Pradeep, R., Grep, S., et al.: Artificial immune systems for the detection of credit card fraud: an architecture, prototype and preliminary results. Inf. Syst. J. **22**(1), 53–76 (2012)

Target Image Detection Algorithm for Substation Equipment Based on HOG Feature

Meiting Zhang[1], Weimin Zhang[1], Junyong Hu[1], Yafei Wang[2],
Guoqing Yang[2(✉)], Jian Li[2], and Guangting Shao[2]

[1] State Grid Qinghai Electric Power Maintenance Company, Xi'ning, China
[2] Shandong Luneng Intelligence Technology Co., Ltd., Jinan, China
sdygq2004@163.com

Abstract. A method of substation target detection based on HOG feature is proposed. First of all, the image acquired by the robot is processed by digital image processing technology. Secondly, the HOG feature of the device image is extracted, and the sliding window is used to compare the detected image and the template image. Finally, the features are calculated, and the largest score represents the target position. Experimental results show the algorithm realizes the position detection of the target device in the image and judges whether the target equipment is in the image or not. The algorithm is simple, high accuracy and strong robustness.

Keywords: Substation · Robot · Object detection · HOG feature

1 Introduction

Substation is the core of all levels of power grid, the routine inspection of station equipment is the key technology to ensure the safe operation of power grid [1]. The existing manual inspection mode has the disadvantages such as the large labor intensity, the scattered detection quality, large interference by bad weather etc. Therefore it cannot meet the rapid development of the electric power system. With the rapid development of robot technology, it is possible to carry out the inspection of the equipment on the basis of the mobile robot platform, which combines robot technology with electric power application [2, 3].

Because of the special environment of the substation, how to ensure the precision and accuracy of the robot localization has become a difficult and hot issue in the process of replacing manual inspection [4]. In the implementation of the inspection task, the robot has some errors between the current parking point and the last parking point in the same preset position. There are mechanical errors of rotation and pitch when the cloud terrace carries camera to shoot equipment image. These factors will cause differences between the current shooting angle and the previous shooting angle, and then cause the target device offset, even the absence of the target device. Figure 1 is the normal position of the circular instrument in the image. Figure 2 shows the off tracking of the circular instrument.

© Springer International Publishing AG 2018
F. Qiao et al. (eds.), *Recent Developments in Mechatronics and Intelligent Robotics*,
Advances in Intelligent Systems and Computing 691, DOI 10.1007/978-3-319-70990-1_72

Fig. 1. Normal image

Fig. 2. Deviation image

The above situation brings difficulties to the follow-up target detection and recognition. In order to improve the accuracy of the detection algorithm, it is necessary to detect the target device to determine the position of the target in the image. Therefore, it is of great significance to study a method of target detection.

2 De-noising Method via Filtering

In the process of image formation and transmission, the image quality is often deteriorated by external noise. Therefore, the noise detection is a key step, which provides the basis for the correct classification of pixels in the image. The method of noise detection is to reduce the influence of noise, and various filtering methods can be used in image de-noising.

2.1 Median Filter

Median filter is a nonlinear smoothing technique, which sets the gray value of each pixel to the median value of all the pixels in the neighborhood window. That is the use of an odd number moving window (median window), and the centre value of the moving window is substituted for the middle number that the number of points is sorted according to its numerical value in the window. Median filtering is an optimal filtering method to preserve the boundary and filter the high frequency noise. Especially, it is very effective to deal with the impulsive noise, such as abrupt spikes. The image is not blurred while the noise is being filtered [5].

The window shape and size design of two-dimensional median filter have a great influence on the filtering effect. Different image content and application requirements are often used in different shapes and sizes to achieve satisfactory filtering effect.

2.2 Gaussian Filter

Gaussian filter is a kind of linear smoothing filter, which is suitable for eliminating the noise of the Gaussian and widely used in image processing. Popularly speaking, Gaussian filter is the process of weighted average in the whole image, the value of each pixel is obtained by the weighted average of its own and other pixel values in the neighborhood. The specific operation of the Gaussian filter is as follows: Each pixel in an image is scanned with a template, The weighted average gray value of pixels in the neighborhood of the template is used to replace the value of the central pixel of the template [6].The impulse response function of Gaussian filter is calculated by the following Eq. (1):

$$h(x) = \frac{1}{\alpha\lambda_c}\exp\left[-\pi\left(\frac{x}{\alpha\lambda_c}\right)^2\right] \tag{1}$$

Where, $\alpha = \sqrt{\frac{\ln 2}{\pi}}$, λ_c is the cutoff wavelength of the filter. The original signal can be divided into two parts which include low frequency signal and high frequency signal.

Figures 3 and 4 are the results of median filter and Gauss filter respectively. It can be seen from the figure that the Gaussian filter maintains the integrity of the contour, and the feature matching on target detection has more advantages. Therefore, Gaussian filter is selected to de-noise in this paper.

Fig. 3. Median filter   **Fig. 4.** Gaussian filter

3 HOG Feature

Histogram of oriented gradient (HOG) is a feature descriptor used to perform object detection in computer vision and image processing. It can calculate and count the gradient direction histogram of local image regions to form feature.

3.1 HOG Feature Extraction

In the identification of target equipment, feature extraction is the most important step before classification. The aim is to map the input image to a feature space. HOG is a feature extraction method for gray images. It extracts features of given region of interest in the image by calculating information about gradients [7, 8]. The algorithm works by analyzing the gradient of small cells and blocks, which consist of several cells. The feature space of each cell is obtained by accumulating weighted votes for gradient orientation over each pixel. This means that gradient information of each pixel in each cell will not only vote for its orientation bin, but also for its neighboring orientation bins. These bins of the histograms are combined to get the descriptor of the cell and each descriptor of the cell is catenated to form the descriptor of the block. Histograms are also normalized by L2 norm across blocks. The normalization of the cell's and block's histogram [9] can weaken the dramatic changes between gradient strengths caused by variations of illumination and foreground-background contrast.

3.2   The Implementation Process of HOG Feature

The image to be detected is pre-treated by Gaussian filter, and then the HOG feature extraction is achieved. HOG feature extraction algorithm is as follows:

Step1: image gray;
Step2: gamma normalization;
Step3: calculate the gradient of each pixel in the image (including size and orientation);
Step4: the image is divided into small cells (8 × 8 pixels/cell);
Step5: statistics histogram of each cell, and a descriptor for each cell can be formed.
Step6: several cells will be formed a block, all cell feature descriptors in block are connected in series to obtain the HOG feature descriptor of the block. L2-norm followed by clipping (limiting the maximum values of to 0.2) and renormalizing.
Step7: The HOG feature descriptor of the image can be obtained by combining all the HOG feature descriptors of the block in the image.

3.3   Matching

For two n-dimensional sample points such $a(x_{11}, x_{12}, \cdots, x_{1n})$ and $b(x_{21}, x_{22}, \cdots, x_{2n})$, we can use the concept of angle cosine to measure the degree of similarity between them [10].

$$\cos(\theta) = \frac{a \cdot b}{|a||b|} = \frac{\sum\limits_{k=1}^{n} x_{1k} x_{2k}}{\sqrt{\sum\limits_{k=1}^{n} x_{1k}^2} \sqrt{\sum\limits_{k=1}^{n} x_{2k}^2}} \qquad (2)$$

The sliding window mechanism is used to calculate the similarity of each position. Firstly, the HOG features of the detected image are calculated. Secondly, the similarity between each window feature and template feature is calculated. Finally, select the maximum score position as the final target position.

4   Experimental Results

The experimental data were taken from a substation within one year. There are 1293 positive images that contain the target device, which includes 39 target devices and the average of about 20–60 images per target. There are 731 negative images that do not include the target device. In the experiment, the algorithm process is as follows:

Step1: image pre-processing;
Step2: HOG feature extraction;
Step3: the sliding window is used to compare the detected image feature and the template image feature;
Step4: calculate the score in Step3;
Step5: the position of maximum score is the final result;

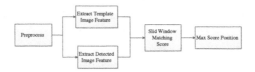

Fig. 5. Algorithmic process

Figures 6 and 7 are the experimental results of a positive sample and a negative sample respectively. Where the red rectangle is the detected position and the green rectangle is the correct position. In the positive sample, the HOG feature scores high, so the red rectangle and the green rectangle are coincident, and the test results are correct. In the negative sample, the HOG feature has a lower score because there is no device target. As shown in Fig. 5, there is only red rectangle in image.

For a binary classification problem, there will appear four cases. If an instance is a positive class and predicted to be positive, it is a true positive. If an instance is a negative class that is predicted to be positive, it is called a false positive. Accordingly, if an instance is a negative class that is predicted to be a negative class, it is called a true negative, and the positive class is predicted to be a negative class, which is a false negative. In the experimental results statistics, this paper uses the quantitative way of the True positive and false positive, as shown in Table 1.

Table 1. Experimental results statistics

FPR (false positive rate)	0.1%	0.01%	0
TPR(true positive rate)	99.14%	97.83%	96.65%

In the experiment, the experimental results are represented by the ROC curve as shown in Fig. 8. Through the comparison of the experimental data, we can find that the algorithm has good performance, high accuracy, and can realize the automatic detection of target equipment. However, it is found that the main difficulties in the process of the algorithm are fuzzy, illumination changes and so on. As shown in Fig. 9, there is a deviation between the detected position and the true position. Where the red rectangle is the detected position and the green rectangle is the correct position.

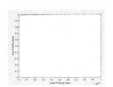

Fig. 6. Positive sample

Fig. 7. Negative sample

Fig. 8. The ROC curve

Fig. 9. Error detection

5   Conclusion

A method of target detection is presented based on HOG features. The method compares the template image HOG feature with the detected image HOG feature, which realizes the automatic detection of the target in the detected image, and ensures the accuracy of the subsequent target analysis. Experimental results show that the method has good robustness and recognition performance, and can solve the problem that the target analysis is not correct due to the deflection of the target of the equipment. So the method can be applied to the intelligent inspection robot in the substation.

Acknowledgments. This work was supported by research and application of indoor combined rail intelligent inspection system based on high altitude non-attended power substation (item serial number: 52282115001A). Thank you for your support of the paper.

References

1. Yang, D.X., Huang, Y.Z., Li, J.G., Li, L., Li, B.D.: Research status review of robots applied in substations for equipment inspection. Shandong. Electr. Power. **42**(1), 30–31 (2015)
2. Zhou, L.H., Zhang, Y.S., Sun, Y., Liang, T., Lu, S.Y.: Development and application of equipment inspection robot for smart substations. Autom. Electr. Power. Sys. **35**(19), 85–86 (2011)
3. Tan, Z.W.: Design of Indoor Substation Inspection Robot System based on VLC, pp. 1–4. North China Electric Power University, Beijing (2016)
4. Wang, K.: Research and Implementation on the Key Technologies and System of Substation Inspection Robot, pp. 3–5. University of Electronic Science and Technology, Sichuan Sheng (2015)
5. Ye, W.Z.: Optimality of the median filtering operator, circuits, systems, and signal processing. CSSP. **30**(6), 1329–1340 (2011)
6. Abhishek, J., Richa, G.: Gaussian filter threshold modulation for filtering flat and texture area of an image. In: 2015 International Conference on Advances in Computer Engineering and Applications, pp. 760–763 (2015)
7. Ren, H.Y., Li, Z.N.: Object detection using edge histogram of oriented gradient. In: 2014 IEEE International Conference on Image Processing, pp. 4057–4061 (2014)

8. Chayeb, A., Ouadah, N., Tobal, Z., Lakrouf, M.: HOG based multi-object detection for urban navigation. In: IEEE International Conference on Intelligent Transportation Systems 17th, 2014, pp. 2962–2967 (2014)
9. Monzo, D., Albiol, A., Sastre, J., Albiol, A.: Precise eye localization using HOG descriptors. Mach. Vis. Appl. **22**(3), 471–480 (2011)
10. Ma, Z.J., Guo, S.F., Li, Y.F.: A noise suppression scheme with EEMD based on angle cosine and fuzzy threshold. Chin. J. Sens. Actuators. **29**(6), 873–875 (2016)

Research on the Method of Finding the Face of Single Mortise-Tenon Work Line Graph

Zhu Tang[✉], Zhiqiang Wang, and Qing Zhu

Beijing University of Technology, Beijing, China
S201525026@emails.bjut.edu.cn

Abstract. The face of the mortise-tenon line graphs provide important information for reconstructing its 3D structure. In this paper, an algorithm is used to identify the face of line graphs. The geometric properties are added to the depth first algorithm to generate the loop set, using backtrack algorithm to find the optimal solution according to the principle that the edges are shared by two faces, which contains solution of special mortise-tenon structure.

Keywords: Face recognition · Backtrack algorithm · Depth first algorithm · Mortise-tenon work

1 Introduction

The mortise-tenon work is a connection mode of convex combination that adopted in two wooden parts. The convexity is called tenon, the combination called mortise. It's the main structure of ancient Chinese architecture, furniture and other wooden instruments, such as the Forbidden City, Tiantan temple and Hanging Temple of Shanxi etc. According to the use of different places, mortise-tenon structure derived thousands of patterns and applied to lots of places. The most common using in ancient furniture is dovetail tenon, such as wood cover, house joint beams etc.

For the question of how to correctly identify the face of the mortise-tenon line graph, this paper applies an algorithm based on the property of each edge being shared by both faces to find the loop combination by recognizing face [1–4] for line graph (including simple, with a curved face and represented by multiple graphs mortise-tenon structure). This algorithm adds the geometric properties to the depth-first algorithm for loop search, and obtains a set of loops that include all the real faces. The most reasonable loop combination is selected from all the subset of the loops that may coexist by applying the backtrack algorithm. Here we only study the structure of the line graph to show the hidden structure. The line graph is defined as a horizontal or approximately horizontal mortise-tenon structure projection, where all the edges and vertices are visible.

© Springer International Publishing AG 2018
F. Qiao et al. (eds.), *Recent Developments in Mechatronics and Intelligent Robotics*,
Advances in Intelligent Systems and Computing 691, DOI 10.1007/978-3-319-70990-1_73

2 Algorithm

2.1 Algorithm Summary

Assume that the mortise and tenon line graph are made up of N graphs $G_1, G_2, ..., G_N$ ($N \geq 1$).

1   generates a loop set SC_i ($1 \leq i \leq N$) from G_i using a depth-first algorithm based on geometric properties.
2   Using the backtrack algorithm to find a subset of the loops from the SC_i.
3   If there are many subsets, using the Property 11 to select the most reasonable combination.
4   If there is a manually added line, merge the eligible faces.
5   $i = i + 1$, go to step 1; until $i = N$, go to step 6.
6   If there is a dashed line in the line graph, look for the real face that is closed by multiple loops.

2.2 Geometric Properties

In the line graph, each edge starts at a vertex, terminates at the other vertex, and is formed by three or more planes crossing. Thus, the degree of each vertex is greater than or equal to 3.

    Property 1: the self-intersecting loop is not a real face in the line graph.

    Property 2: Each edge of a line graph is shared by two different faces.

    Inference 1: A vertex d (v) = 3, then there must be three faces, each of which contains two different edges in the three edges of the vertex.

    Property 3: If there is a string in the loop (as shown in Fig. 1① edge fo), at least one of its two vertices is the intersection of the three edges, then the loop is not the real face.

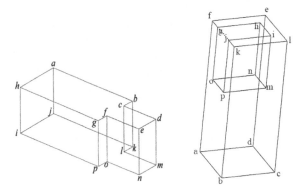

Fig. 1. A pair of open dovetail tenon. ① The tenon line graph. ② The mortise line graph.

    Property 4: If the following two conditions are satisfied, then this loop is not the real face:

(1) The loop has a string, the degree of at least one vertex in the two vertices is 4.
(2) If the string has only one vertex of degree 4, it is not collinear with any other three lines intersecting the same vertex; when the two vertex degrees of the string are 4, for at least one vertex, the string is not collinear with any of the three lines that intersect at this vertex.

Property 5: If the loop has a string completely or partially enclosed in the loop, it is not real face.

Property 6: Let the three vertices a, b, c of a loop contain three edges, and if any of the following conditions are satisfied, this loop is certainly real face:

1) At least two vertex degrees of the loop are 3;
2) $d(a) = 3$, d (b) = 4, the line bc is not collinear with any of line bd and be of the b-point;
3) d (a) = 3, d (b) = d (c) = 4, the line bc is not collinear with any of line bd and be of the b-point or any of line cf and cg of the c-point.

Property 7: The common lines of the two adjacent faces must be collinear in the line graph.

Property 8: If the two adjacent faces have a common line and a common vertex (not one of the two endpoints of the common line), then the line and the vertex must be collinear in the line graph.

Property 9: A loop that contains four lines and does not contain any strings has four vertices a, b, c, d. If any of the following conditions are satisfied, it is proved that the loop is a true face:

1) $d(a) = d (b) = d (c) = d (d) = 3$, ae, cg, bf, dh are four lines connecting a, b, c, d, and at least one pair of lines are not collinear in (ae, cg) (bf, dh);
2) $d(a) = d (d) = d (b) = 3$, d (c) > 3, ae that is a line connecting vertex a and vertex c are not collinear.

Property 10: Let the four edges connected to the vertex v of degree 4 be va, vb, vc, vd. If the loop $C_1 = (a, v, b, ..., a)$ and $C_2 = (c, v, b, ..., c)$ are treated as two real faces, then all loops through va and vc are not real faces.

Property 11: If the two loops have the same virtual line and completely close the line, then the two loops are not real face.

The Property of the application: for Fig. 1①, only 10 loops are real face in all loops. When retrieving a loop with a length of 4, it will get 8 loops that is a real face (Property 9). Remove lines hi, aj, bk, cl, dm, en, fo, gp from the line graph (it has been shared by two faces). In the remaining graphs, the loops (a, b, c, d, e, f, g, h, a) and (i, j, k, l, m, n, o, p, i) must be real face (Property 2). Fig. 1① is relatively simple, and it is easy to find the smallest subset of the loops based on Properties 2 and 9.

2.3   Property-Based Depth-First Algorithm

1. Enter the edges and vertices of the line graph showing the hidden structure.
2. Initialization: Set the 2D array EI to store the edges of the intersection; ECL store collinear edges; ECE store coexistence edges.
3. Search for loops of length 3 in the line graph based on Property 6 to check if it is a real face.
4. Search for loops of length 4 in the line graph based on Property 9 to check if it is a real face.
5. Update ECE if there is a real face. According to Property 7, these edges can't coexist in the subsequent loop.
6. Remove the edges pass through the two faces. Use G to represent the remaining line graph.
7. Stop if there is no edge in G.
8. Select an edge that belongs to the real face from G, if there is no such edge, then choose one. Use l to represent edge and v1, v2 to represent the two vertices of this edge.
9. Find the loop through the edge l in G. According to the ECE check whether there is a clear path (v2, v1, ..., vn) and all vertices are different, if the path P1 exists stored in the Ep, otherwise the path P2 (v1, v2) stored in Ep. Use the depth-first algorithm [5] to start looking for all loop sets that pass P1 or P2 from point v2 according to Ep, EI, ECL, ECE. The search rule: Before adding a new vertex to Ep, review the EI to determine whether the newly formed edge intersects the presence of the current path. If there is, select another vertex; otherwise check ECE, if the new edge and the path of the edge does not coexist, select another vertex; Otherwise, check whether the new vertex leads to a string in the path, if this string makes the loop through this path can't be real face according to Property 3 and 4, select another vertex; otherwise add the new vertex to Ep. When generating a loop, check if there is a closed string in the loop. If present, discard this loop according to Property 5.
10. If there is only one loop in the set Pc and the edge l has been passed twice through the two loops, the ECE is updated according to the Property 7 and the property 10.
11. Remove the edge l from G and use G1 to represent the simplified line graph. Check if there is a vertex with a vertex degree of 1 in G1; if so, remove the edges of these vertices in line graph G1. Repeat this process until all vertex degrees in Gm are at least 2 or no edges exist.
12. Set $G = G_m$, go to step 7.

When using Fig. 1② as a graph input, loop $C_1 =$ (a, b, c, d, a) is a real face (Property 9), update E_{CE}. Another real face that pass edge bc must not pass edge ab and cd, and will be over bk and cl (Property 7). Edge kl and kj, kb are not collinear. We can find only the loop $C_2 =$ (b, k, l, c, b) is the real face in all loop pass edge bc, bk and cl (Property 4). Edge bc must be shared by another loop other than C_1, and then prove that C_2 is real face (Property 2,). Similarly, the loop (a, f, e, d, a) (d, c, l, e, d) is also true. As a result, edge bc, cd, ad, de, cl will be removed from the line graph. When we choose the edge ab we can find a clear path (b, a, f, g) (step 8). According to the updated E_{CE}, we can see that fg, gh and go can coexist, this path will no longer be extended. Two loops (b, a, f, g, o, p, j, k, b) and (b, a, f, g, h, i, j, k, b) can be formed after the path (b, a, f, g) (step 9). There are 224 loops in the Fig. 1②, of which only 10 are true, and the application of the above algorithm will generate 12 loops. Obviously this algorithm is very effective.

2.4 Mortise-Tenon Line Graph

As mentioned above, there are many structures depending on the use of the mortise-tenon work. As shown in Fig. 2, yandaiguo tenon and straight tenon are the most commonly used in the classical furniture. This structure for the appearance and practicality will generally be made with a curved face. The computer has a lot of ambiguity about this, and we have the step of manually adding the desired line, turning the curve into a straight line, using the above algorithm to identify the face, merge the smooth face, and delete the added line. There are four curves in the Fig. 3①, respectively abc, adc, efg, ehg, if the use of four straight lines instead of curved side, there will be overlapping lines do not hold. We add a straight line bf, you can find that there is no change in the number of faces, the cylinder has five faces. Face (a, b, f, e, a) and (c, b, f, g, c) will be merged into one face (a, c, g, e, a), and at this time the cylinder has four faces.

Fig. 2. ① The tenon line graph of yandaiguo tenon. ② The application of yandaiguo tenon and straight tenon in the practice. ③ The tenon line graph of the straight tenon.

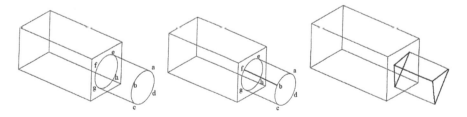

Fig. 3. ① The tenon line graph of the original straight tenon. ② Add line bf in the *middle* picture ③ A straight line instead of a curved edge

It is found that, by the observation and experiment, it is feasible to apply the method based on the straight line obtained by the curve is the infinite approach curve when the following two conditions are satisfied: (1) No two edges are overlapped after the curve becomes a straight line. (2) Loops that are a real face are not self-intersecting.

The mortise and tenon structure has a special matching connection so that there are a large number of line graph as shown in Figs. 2 and 3. When dealing with this line graph, it is necessary to use the algorithm for each geometry to identify the face and obtain all the loops that belong to real faces. Finally, a dashed line is used to identify the true face enclosed in multiple loops. The application of the algorithm in Fig. 4① for face recognition, the two geometrics will get 12 loops (Property 9). According to the two dashed lines kn and ij, it can be seen that loops $C_1 = $ (a, b, c, d, a) and $C_2 = $ (e, f, g, h, e) are located on the same surface, and the real face is enclosed between the loops C_1 and C_2. So there are 11 faces. Similarly, there are four loops that are used to form two real faces that are enclosed in two loops, which leads to a hole in the Fig. 4②.

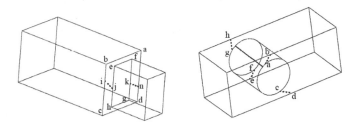

Fig. 4. ① The tenon line graph of dark dowry tenon with *dashed line*. ② The mortise line graph of the straight tenon with *dashed line* (a pair of mortise-tenon structure with Fig. 3)

2.5 Optimal Loop Combination

The construction of this state tree is shown in Fig. 5 when there are only five loops. In addition to the root node, each node of the tree represents a combination of loops. Let SC denote a set of loops obtained from the application of the face recognition algorithm in the line graph, A is a subset, and A ⊂ SC. If the loop in the node causes the edge to pass through more than twice, delete the node; otherwise, it will add a new loop a to the A subset to expand it and then in turn to test. When you find the solution or to the bottom of the branch, you will be returned to the previous layer to continue the search.

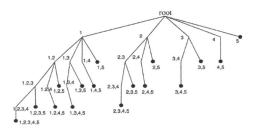

Fig. 5. State tree with only 5 loops

In Fig. 1②, the 8 loops with a length of 4 must be included in the loop combination. For convenience, we assume that it is loop 1, loop (f, e, l, k, j, i, h, g, f) is 2, (a, f, g, o, p, j, k, b, a) is 3, (a, f, g, h, i, j, k, b, a) is 4, (f, e, l, k, j, p, o, g, f) is 5. At this point for the state tree in Fig. 5 we only need to consider the left side of the branch containing loop 1. According to the backtrack algorithm, two solutions A_1 (1, 2, 3) and A_2 (1, 4, 5) can be found. When multiple solutions are obtained, the property 11 can be applied to remove the unreasonable solution (A_2).

Table 1 lists the experimental results of several mortise-tenon line graph mentioned in this paper. The third column is the number of all loops of the graph, and the fourth column is the number of loops obtained using the algorithm based on geometric properties. Numbers can be found that our application of the algorithm is helpful in eliminating some of the loop interference. The fifth column is the solutions of all loop combinations. The most reasonable solution can be selected according to the above. The whole algorithm does not exceed 0.2 s in the process of identifying these simple mortise-tenon line graphs, but when there are too many loops, it will take a lot of time to apply the backtrack algorithm to search. Therefore, it is necessary to further improve the algorithm to deal with the complex mortise-tenon line graph, which will be my future research work.

Table 1. The experimental results

	Mortise and tenon line graph	Total number of loops	Loops based on the property alg.	Solutions based on backtrack alg.	Faces found finally
Figure 1①	Tenon line graph of the open dovetail tenon	322	10	1	10
Figure 1②	Mortise line graph of the open dovetail tenon	224	12	2	10
Figure 4①	The tenon line graph of dark dowry tenon	56	12	1	11
Figure 2③	The tenon line graph of the straight tenon.	42	11	1	9
Figure 4②	The mortise line graph of the straight tenon	42	11	1	8

References

1. Chartrand, G., Oellermann, O.R.: Applied and algorithmic graph theory. McGraw-Hill, New York (1993)
2. Bagali, S., Waggenspack, J.W.N.: A shortest path approach to wireframe to solid model conversion. In: Proceedings of Third Symposium Solid Modeling and Applications, pp. 339–349 (1995)
3. Shpitalni, M., Lipson, H.: Identification of faces in a 2D line drawing projection of a wireframe object. IEEE Trans. Pattern Anal. Mach. Intell. **18**(10), 1000–1012 (1996)
4. Liu, J., Lee, Y.T.: A graph-based method for face identification from a single 2D line drawing. IEEE Trans. Pattern Anal. Mach. Intell. **23**(10), 1106–1119 (2001)
5. Cormen, T.H., Leiserson, C.E.: Introduction to Algorithms. The MIT Press, Cambridge (2005)

The Design and Implementation of a Computer Display Fixed Device Based on "Internet+" Network Education

Lang Pei[(✉)] and Jia Xu

College of Computer Science, Wuhan Qinchuan University, Wuhan, China
11286978@qq.com

Abstract. As the society is becoming superior day-by- day, loads of smart devices are used in different application areas. This is the challenge to the technocrats for forming the intelligent and smart social systems. It requires easy access and fast processing, which is the main focus of any application. In this work, an attempt has been taken into consideration to develop an intelligent e-healthcare system. In e-healthcare system the entities are considered as the patient, the physician, the pathological centre and result as diagnosis, treatment and post care. This paper uses an ANFIS structure for e-healthcare system. Further the ANFIS system is used for disease diagnosis and support to the patient as well as for physicians. For the management of multi-agent system has been satisfied by, using rule based fuzzy parameters. The service can be provided through internet to the patient as well as by the physician. The different situation of patient automatically informs to the doctor similarly the prescription from the doctor for diagnosis can inform to the pathology centre and vice versa. The result of detection communicated to both for desired medicine, monitoring and post care purpose. The performance found to be excellent to satisfy this part of intelligent system.

Keywords: The network education · Displays · Fixtures

1 Introduction

With the continuous development of the Internet era, all kinds of emerging information technology into the classroom, such as class, class, class, whiteboard, baidu it, flip the classroom, learning space and so on, makes the teacher's teaching way and organization form, great changes have taken place in the classroom has found some \"flip\". Internet + education born? Nature is education wisdom, in the face of the Internet are ling lang see everywhere under the support of information technology, teachers how to more reasonable application of these techniques, properly for these technologies to help teachers better teaching fully, solve the problem of the vagaries of teaching, teaching wisdom is particularly important.

In July 2016, the general office of the central committee of the communist party of China, issued by the general office of the state council of the national informatization development strategy guidelines mentioned to promote high quality teaching resources sharing and the equilibrium configuration, and the "Internet+" rapid development in the

© Springer International Publishing AG 2018
F. Qiao et al. (eds.), *Recent Developments in Mechatronics and Intelligent Robotics*,
Advances in Intelligent Systems and Computing 691, DOI 10.1007/978-3-319-70990-1_74

field of education, makes a lot of passion for the Internet users in the daily learning process, through desktop computer monitor at teaching materials need to manually adjust the frustration, and delay time which seriously affect the learning effect, in order to solve this problem, we propose to develop a new type of computer display fixtures, which can solve the user's hands and can improve the learning efficiency of user, so as to solve the "Internet+" in education link due to hardware facilities imperfect impact study the problem of overall effect. This new device also conforms to the concept of "Internet plus" education, and USES this new device to optimize distance learning effect (Table 1).

Table 1. Device structure composition diagram

Serial number	Name	Serial number	Name
1	The base	8	Set screw
2	stud	9	Rotating device
3	Fixed plate	10	Install the tank
4	Rotary joint	11	Adjustable switch
5	Telescopic device	12	Rotary switch
6	Connection plate	13	The control panel
7	Limit hinges	14	Threading hole

2   Design Thought

The basic connotation of "Internet+" network oriented education is based on the Internet and the latest information technology, building intelligent computer assisted learning environment, appropriate to the wisdom of teaching method, using promote learners to become high intelligence and active creativity of the talents, the fusion between the two not only brings the concept of learning style, learning, learning tool innovation, also the personalized service in intelligent network education put forward new demands. And network education, master the use of TV and the Internet and other media teaching mode, it broke through the time and space boundaries, accommodation in the school is different from the traditional teaching mode. Students who use this model are usually amateurs. Because you don't have to go to a specific location, you can have classes anytime. Students can also learn from different channels, such as TV, Internet, tutoring special line, class research society, and face-to-face (correspondence). Network education process without the application of computer, computer monitor effectively fixed need fixing device, the existing fixed device when making adjustment to the display, require users to use both hands to throw, extremely convenient adjustment, need to be improved. And we developed this web-based education computer display fixture, which mainly involves the network education device technology.

3   Basic Structure

This new device to achieve the above purpose, strictly control in every link, the components of each part contains: the base, stud, fixed plate, screw, telescopic device, connecting plate, limit hinge, screw, rotating device, installation groove, telescopic switch, rotary switch, control panel, threading hole; The parts are numbered as follows:

4   System Design Process

4.1   Technical Solutions

This design scheme can effectively solve the present display existing technical shortcomings and the insufficiency, provide users with a simple structure, reasonable design, convenient to use a network education of computer display fixtures, without user's hands difficult operation, adjust the Angle of the display can be realized, the liberation of the user's hands, convenient adjustment, save user time and improve efficiency, more practical.

4.2   Working Principle

Will display through the set screw is installed on the fixed plate, when need to display Angle is adjusted up and down, can through the adjustable switch control telescopic device, so as to control the rotation of the fixed plate, to adjust the Angle of different display; When you need to rotate the monitor, you can control the rotating device by rotating the switch, so that the lever is rotated to adjust the display.

4.3   Concrete Implement Process

First there is a rotating device inside the base, as shown below (Fig. 1):

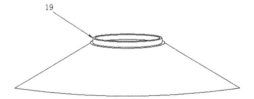

Fig. 1. Shows the structure of the base of this new device

Rotating device of the upper connection have stud, let a good contact with the bottom of the vertical rod and the base Settings, poling a telescopic device by connecting plate, and the telescopic device through the vertical bar at the top of the wall, through the screw parts connected to the fixed plate; The diagram below describes the view of the top of the new device and the bottom structure of the vertical bar (Fig. 2).

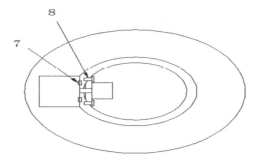

Fig. 2. The top view of this new device.

In this case, the upper edge of the fixed plate is placed on the edge of the installation slot on the upper part of the vertical bar. For the connection plate, the fixed screw is connected to the inner wall of the vertical bar. In the whole design, the telescopic device control circuit through the vertical bar at the bottom of the threading hole connected to the telescopic switch, and the length of the telescopic device of control circuit enough for stud rotating around the required length; In the design, make sure that the control line of the rotating device is connected to the rotary switch. The expansion switch and rotary switch are set on the control panel, and the control panel is fixed to the inner wall of the base. As shown in the diagram below, the link diagram for the spinner is shown (Fig. 3):

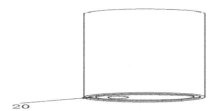

Fig. 3. The bottom structure diagram of the new device neutral bar

Among them: The spinner is composed of fixed pin and rotating plug-in. The outer end of the retractable device has a rotating insert; There are fixed pin holes in the rotating plug and rotating plug-in. The rotating plug is inserted between two rotating plug-ins and is fixed to the pin. The upper part of the base is set up with a positioning edge. The bottom of the vertical bar has a positioning slot with the positioning edge. The location of the location plug is located in the positioning slot to facilitate the connecting limit of the base and vertical bar. The retractable device is an electric telescopic pole; The rotating device described is a positive and reverse rotating motor (Figs. 4 and 5).

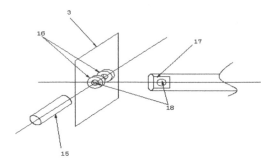

Fig. 4. The connection diagram of the spinoff in this new device

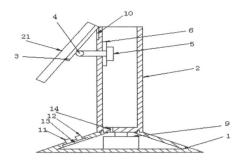

Fig. 5. Location and realization diagram of each component in this new device

4.4 Expected Effect

Display is the entire network in the process of learning a very important configuration, intuitive learners study according to the length of time, sitting position transformation, you need to use to adjust the height of the display and the direction, so, after enabling this new device, will find that when learners study time is a bit long, need to replace sit and watch the learning materials, so, this play is a good chance for the new device, let learners complete liberation hands, convenient adjustment, more important is the practical very strong, simple to use, the cost of production is also very low, very suitable for popularization.

5   Conclusion

Described above, only for this new concept of the design process, rather than the limit, the common technical personnel to this utility model in the field of technical scheme of modified or other equivalent replacement, as long as it is from the spirit and scope of the utility model technical scheme, shall cover the scope of rights requirements of this new device, it is important for learners to bring real convenience and effectiveness.

References

1. Li, D.: Design of the fixed device of the projector and the finite element analysis. Mech. Eng. Autom. **2005**, 40–42 (2013)
2. Cheng, K.: Effective time management for remote learners. Mod. Remote Educ. Res. 2, 56–58 + 72 (2009)
3. Ting, Q., Liu, C., Xiao, N.: The time management of distance learners. Remote Educ. **2004**, 37–41 (2008)
4. Peng, H., Chen, M.: From the patent perspective, the technology of thin-film transistor, liquid crystal display control. Electron. Prod. 7, 92–93 + 91 (2014)
5. Pang, G.: LED display mounting and LED display fixed support (2013)
6. Premlatha, K.R., Geetha, T.V.: Learning content design and learner adaptation for adaptive e-learning environment. Artif. Intell. Rev. **44**(4), 1–23 (2015)
7. Yang, X.: The connotation and characteristics of education in the information age. China Electr. Educ. **01**, 29–34 (2014)

The Motif Detection of Turning Points in Stock Time Series

Zhiyuan Zhou[(⊠)], Wengang Che, Pengyu Quan, and Jing Long

School of Information Engineering and Automation,
Kunming University of Science and Technology,
Kunming 650051, People's Republic of China
15755507846@163.com

Abstract. A stock time series is the description to show the regular patterns of historical data with the changes of time. The method of motif detection is used here to detect the turning points in short-term tendency. In this paper, firstly, the definition of short-term tendency and the important signal turning points are defined here. Secondly the detection of turning points algorithms are also presented. And then the motif detection is proposed to predict the turning points in the short-time tendency. Finally, using the empirical analysis proves the high accuracy by the proposed method.

Keywords: Stock time series · Motif detection · Turning points · Short-term tendency

1 Introduction

The stock market is a significant component of financial market. Obviously, the stock time series is always absorbing by the predictor and is the hot area of all kinds of forecasting methods. A time series is segmented with a data stream segmentation method, the segments are modeled by means of normal distributions with time-dependent means and constant variances, and these models are compared using a divergence measure for probability densities [1]. Mining and studying the data of time series do good to find the potential relationship between those data. With the relationship of data can easily mine the significant information when investing in stock market. For example, lighten up the stock we hold in a suitable time when the stock market is going down, and increasing the stock we hold at the opposite tendency.

To the best of our knowledge, stock markets are influenced by many economical, political and psychological factors, and base on the finding previously unknown patterns in a time series has received much attention in recent years [2], so it is very difficult and meaningful to forecast the future stock prices. However, it is impossible to develop a precise mathematical model that can predict crisp numerical values of the stock prices. This is due to uncertainties involved in these parameters, the uncertain behavior of the factors, and tolerances [3]. The turning points are very important signals for the investor and the predictor, such as the Engulfing pattern, Evening doji star, Hammer, Hanging-man lines, Piercing patterns and so on. Here use the Engulfing pattern as the representative of turning points. When watching the presence of ceiling line and the next is a

© Springer International Publishing AG 2018
F. Qiao et al. (eds.), *Recent Developments in Mechatronics and Intelligent Robotics*,
Advances in Intelligent Systems and Computing 691, DOI 10.1007/978-3-319-70990-1_75

black candle bar in the Candlestick chart, the stock technical analyst will easy to get the knowledge of the opposition of the growth trend and should set the mental stop in order to reduce the losses maximize.

Time series motif is a pattern appearing frequently in a time sequence, useful to forecast the stock temporal tendencies and prices as a reliable part in time series [4, 5]. In many fields, the applications of motif detection can be found, such as, in the application of weather forecasting, flight fares, medicine and stock trade. In this paper, the method of motif detection is proposed to detect the tendency of short-term series. And then take appropriate measures with the combination of the tendency and turning points.

2   The Related Definitions

In order to research the stock time series, the definition below is proposed to help us to mine and to understand the important information.

Definition 1: Stock Time Series: Given a stock time series X and Y, a close price of stock time series is denoted by $X(t_i, \alpha_i)$ where α_i is used to represent close price of a stock time series at the trade day of t_i. And also a open price of stock time series is donated by $Y(t_i, \beta_i)$ where β_i is used to represent open price of a stock time series at the trade day of t_i.

Definition 2: Turning points: Here denote two segment of X and Y by $D_1 = X(t_{i-1}, t_i)$ and $D_2 = Y(t_{i-1}, t_i)$, where i − 1 and i are the starting and ending day indices respectively. If $\alpha_i > \beta_i$, $\alpha_{i-1} > \beta_{i-1}$, $\alpha_i > \beta_{i-1}$ and $\alpha_{i-1} > \beta_i$ with duration of $[t_{i-1}, t_i]$, then the time series, with close price X and open price Y, is considered having upper turning points. If $\alpha_i > \beta_i$, $\alpha_{i-1} > \beta_{i-1}$, $\alpha_i > \beta_{i-1}$ and $\alpha_{i-1} > \beta_i$ with duration of $[t_{i-1}, t_i]$, then the time series, with close price X and open price Y, is considered having lower turning points. Point t_i is defined as a turning points of short-term tendency.

Definition 3: Motif of Turning points: If there are two or more similar segments of stock within stock time series of $[t_{i-1}, t_i]$. This similar segment having the same tuning point pattern is defined as motif of short-term tendency.

3   Turning Points Detection and Algorithms

The methods of detecting the turning points are widely used in the stock time series, which is significant for the investors and institutions. For example, the foundation of the "Golden Cross Point" and "Dead Cross Point" is a common method to detect the turning points. However, it is difficult to insure the accuracy. What's more, the moving average is also lagging behind the current stock time series. So here we applied the motif detection in order to keep pace with the stock and also guarantee the accuracy.

In this paper, the new method, comparing the open price and close price of the adjoining day in stock time series, is a good way to find the turning points, which is well known by stock traders as "turning point" shown in Fig. 1(a) and (b).

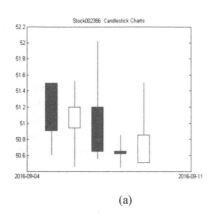
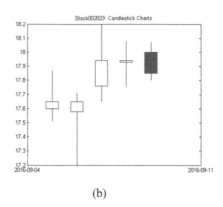

(a)   (b)

Fig. 1.

Here Given the concrete algorithms to detect the tuning points.

Algorithm 1: The Growth Turning points

Step 1: Take a stock time series of the close price $X(t_i, \alpha_i)$ and the open price $Y(t_i, \beta_i)$, Comparing the close price and open price of day i, if $\alpha_i > \beta_i$, go to step 2. Otherwise increase to next day and go to step 1.

Step 2: Comparing the close price and open price of stock at day $i - 1$, if $\alpha_{i-1} > \beta_{i-1}$, then go to step 3. Otherwise increase to next day and go to step 1.

Step 3: Comparing the close price of day$i$ and the open price of day $i - 1$, if $\alpha_i > \beta_{i-1}$, then go to step 4. Otherwise increase to next day and go to step 1.

Step 4: Comparing the open price of day$i$ and close price of day $i - 1$, if $\alpha_{i-1} > \beta_i$, then go to step 5. Otherwise increase to next day and go to step 1.

Step 5: A Growth turning points is detected.

Algorithm 2: The Downward Turning points

A similar algorithm with above one is used where the large than calculation substituted by less than calculation and also the less than calculation is substituted by large than calculation.

4   The Motif of Turning Points Detection

Generally, time series motif is a pattern appearing frequently in a time sequence which is very useful to forecast the stock turning points. In this section, we propose an algorithm to find those motifs of turning points in stock time series.

Algorithm 3: Motif detection of Short-term tendency

Step 1: Take a stock time series of the close price$X(t_i, \alpha_i)$ and the open price of $Y(t_i, \beta_i)$, detect the turning points and remember the time duration.

Step 2: Take another stock time series of the close price$X(t_i, \alpha_i)$ and the open price $Y(t_i, \beta_i)$,detect the turning points and compare if there are similar segment of time series during same time period. If yes, got to step 3, otherwise go to step 2.

Step 3: Remember the number of similar segment detected and then got back to step 2.
Step 4: The motifs of the stock time series are detected.

5   Empirical Analyses

The open and close stock prices of A-shares in Shenzhen among August 2016 are used as the experimental data for short-term turning points detection. First use motif detection method to check out the Engulfing pattern from more than 1500 stocks in A-shares. As

Observed Day	Motif Found	Segment Found	Observed Day	Motif Found	Segment Found
2016-08-03	27	15	2016-08-18	7	6
2016-08-09	72	48	2016-08-24	17	10
2016-08-12	4	3	2016-08-29	7	7
2016-08-16	16	14	2016-08-31	9	3

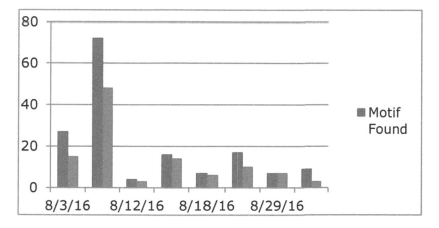

Fig. 2. The observed day, motif found and segment found

then by observe the candlestick charts of all stocks to verify the accuracy of the Engulfing pattern. The result is shown in Fig. 2.

Here easily find that this motif detection of the Engulfing pattern has the high accuracy.

6 Conclusion

In this paper, a motif of turning points is introduced and the algorithms to detect the Engulfing pattern are proposed. An empirical data shows that the proposed algorithm has high accuracy and scalability. Since the turning point detection is based on the motif detection of short-term tendency, those turning points have a lot of information for trend change and are practical for the stock market traders.

References

1. Fuchs, E., Gruber, T., Nitschke, J., Sick, B.: On-line motif detection in time series with SwiftMotif. Pattern Recogn. **42**(11), 3015–3031 (2009)
2. Tang, H., Liao, S.S.: Discovering original motifs with different lengths from time series. Knowl. Based Syst. **21**(7), 666–671 (2008)
3. Atsalakis, Zopounidis, C.: Forecasting turning points in stock market prices by applying a neuro-fuzzy model. Int. J. Eng. Manage. 1(1), 19–28 (2009)
4. Son, N.T., Anh, D.T.: Discovering time series motifs based on multidimensional index and early abandoning, vol. 7653(1), 72–82. Springer, Heidelberg (2012)
5. Jiang, Y.F., Li, C.P., Han, J.Z.: Stock temporal prediction based on time series motifs. Int. Conf. Mach. Learn. **6**, 3550–3555 (2009)
6. Xiaodi, L., Wengang, C., Qingjiang, Z.: Analysis of outliers and public information arrivals using wavelet transform modulus maximum. In: The 2nd IEEE International Conference on Information and Financial Engineering, pp. 176–179 (2010)
7. Castro, N.C., Azevedo, P.J.: Significant motifs in time series. Stat. Anal. Data Mining **5**(1), 35–53 (2012)
8. Lin, Y., Mccool, M.D., Ghorbani, A.A.: Motif and anomaly discovery of time series based on subseries join, vol. 2180(1) (2010)

Research on Video Transmission Technology Based on Wireless Network

Liqiang Liu[✉]

Department of Computer, Hunan Institute of Science and Technology, Yueyang, China
liuliqiang2@163.com

Abstract. With the rapid development of wireless mobile communication technology, the multimedia communication service is increasing rapidly in the mobile environment. Analyzing With the rapid development of wireless communication technology and the continuous improvement of wireless network bandwidth, the multimedia communication service is increasing rapidly in the mobile environment. The communication performance of multimedia transmission in wireless networks with limited resources, the video encoding in the main influencing factors of multimedia transmission and application is studied, through the evaluation of video transmission in wireless network quality of service by using simulation tools, such as data packet time delay, PSNR, image difference etc. Experimental results show that, when the video is transmitted over the wireless network, it is necessary to select the appropriate compress quantization parameters and GOP type according to the network condition, so as to obtain a better reconstructed video quality.

Keywords: Wireless network · Multimedia information · Video coding · Network simulation

1 Introduction

With the rapid development of wireless network and multimedia technology, wireless network business forms have become increasingly rich, the main form of traditional voice will gradually change the new wireless multimedia services for mobile video. The transmission of video stream needs stable bandwidth, and the bandwidth fluctuation will seriously affect the quality of video reconstruction. Therefore, it is necessary to solve the contradiction between wireless channel bandwidth fluctuation and high quality video [1]. Video coding can reduce the amount of data transmission and reduce the energy consumption of communication [2]. The data compression ratio is inversely proportional to the quality of video decoding. The higher the data compression ratio is, the lower the quality of video decoding is. How to realize the compression and transmission of multimedia information efficiently and effectively under the condition that the storage, processing power and energy of single node are severely limited [3]. It is a problem that needs to be solved in order to balance the computational complexity and the amount of communication data.

© Springer International Publishing AG 2018
F. Qiao et al. (eds.), *Recent Developments in Mechatronics and Intelligent Robotics*,
Advances in Intelligent Systems and Computing 691, DOI 10.1007/978-3-319-70990-1_76

2   Wireless Channel Model

Generally speaking, there are two reasons for packet loss in packet radio network, such as congestion loss and wireless loss. The reason for the former is that the amount of data transmission on the network is too large, resulting that in the transmission of network equipment is not timely, so that the queue buffer space is insufficient, and the data packet must be lost. For wireless transmission, it can be divided into two types according to the distribution of data loss. When the distribution of packet loss is quite scattered and average, it belongs to the first type. On the contrary, if the loss occurs in the continuity of the majority, it belongs to the continuous loss. However, the G–E model is used as a continuous packet loss model. In the model, because the data is lost in the way of random dispersion, the average loss probability represents the average loss of data in the transmission process. PGB represents the probability that the transmission channel is going bad, and PBG is just the opposite (Fig. 1). In the steady state, the probability of good and bad

$$\Pi G = \frac{PBG}{PBG + PGB}, \quad \Pi B = \frac{PGB}{PBG + PGB}, \quad P_{avg} = P_G \Pi_G + P_B \Pi_B \tag{1}$$

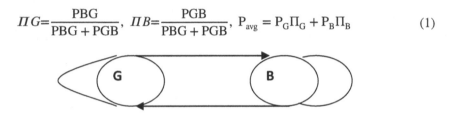

Fig. 1. G–E channel model

3   Video Coding Standard

International Telecommunication Union (ITU-T) has developed a video coding standard including H.261, H.263, H.263+, in the year of 2000 will be the final text through H. 263++.H.26X series standard is a video coding standard for low bit rate video communication, which has a high compression ratio, so it is especially suitable for wireless video transmission. The basic techniques used are DCT transform, motion compensation, quantization, entropy coding, etc. [4].

MPEG-1, MPEG-2 is characterized by the use of digital storage media, high bit rate, they are not suitable for wireless video transmission. People familiar with VCD, DVD is a typical application of MPEG-1, MPEG-2. Subsequently, MPEG organizations have noticed a potential market for low bit rate applications, and began to compete with ITU-T. In the development of MPEG-4, it not only considers the high bit rate applications, but also includes a special low bit rate applications for wireless transmission [5]. The most important feature of MPEG-4 standard is the coding method based on video object [6].

In MPEG encoding, encoding the video stream will be divided into 3 different images respectively: I-frame, P-frame, and B-frame [7]. I-frame is from its own picture data as encoding, which is also not the reference picture, P-frame is a reference to previously encoding I-frame or P-frame and the data itself to do encoding, B-frame is a reference to previously encoding I-frame or P-frame and their own data encoding, the prediction of common relation as shown in Fig. 2 [8].

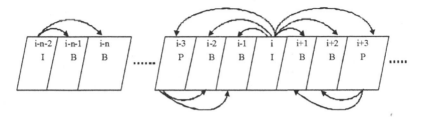

Fig. 2. I, P, B frame relationship

4 Simulation Analysis

The network topology of the simulation experiment is shown in Fig. 3, the connection between the video server and the wireless router is a wired network, and the connection between the Internet and the video receiving computer is 802.11 wireless network. It is assumed that the packet loss occurs in the wireless network, which leads to the poor video quality.

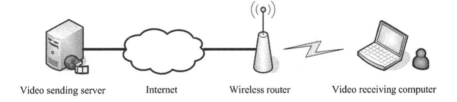

Video sending server   Internet   Wireless router   Video receiving computer

Fig. 3. Network structure

In order to analyze the correlation between image quality and packet loss rate, the quantization parameter is set to 10, the packet length is set to 1024, the length of GOP is set to be 6, 12, 15, and the packet loss rate is set to be 0, 0.02, 0.1. Table 1 is the average value of the PSNR simulation, it can be clearly seen from Table 1. While using the image GOP length is shorter, the mass ratio of image using GOP is long. The quality of the multimedia data transmitted on the network will result in different quality due to

Table 1. Average PSNR value

Packet loss rate	GOP length (6)	GOP length (12)	GOP length (15)
0.0	32.36	32.18	32.09
0.02	29.29	29.09	30.91
0.06	28.65	28.19	28.79
0.1	27.83	26.95	26.13

different compression quantization and network parameters. Figure 4 shows that there is a significant difference in image quality between decoded video images at different PSNR values. Figure 5 shows the packet delay.

(1) Original   (2) PSNR(32.18)   (3) PSNR(26.13)

Fig. 4. Video image

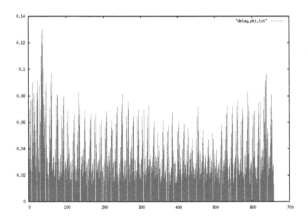

Fig. 5. Packet delay

In order to analyze the relationship between image quality and quantization parameters, the compress quantization parameters are set to be 2, 10, 20, 31. Simulation result the quantization parameter selected value is greater, the compression out of the image quality will be poor, and the choice of quantitative parameters of small will have a better image effect. Because of the unique characteristics of human vision, the imperceptible high frequency signal by using larger quantization parameters, relatively easy to detect low frequency signals with low standards, let video bit stream distribution limited reasonable, so that the quality of the reconstructed video can accept. Figure 6 shows the frame distortion under different Q values.

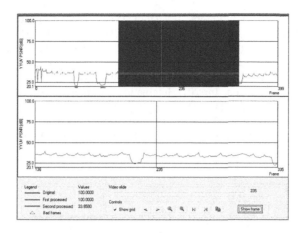

Fig. 6. Frame distortion

5   Conclusion

The error rate of the wireless channel is high, and the encoded video data is relatively small, which is sensitive to the data error, and the single bit error will cause the failure of subsequent data. Reliability and quality assurance of multimedia information transmission is a key factor to be considered in wireless network. Therefore, it is very important to explore the quality of multimedia information transmission in wireless network and analyze the impact of the loss rate of information frame on the multimedia information.

Acknowledgment. This research is supported by Scientific research project of Hunan Provincial Department of Education (16C0721).

References

1. Zhou, L., Yang, Z., Wen, Y., Rodrigues, J.: Distributed wireless video scheduling with delayed control information. IEEE Trans. Circuits Syst. Video Technol. **24**(5), 889–901 (2013)
2. Zhixing, L., Jing, Z., Li, Z.: A distributed JPEG2000 coding algorithm based on mobile agent. In: International Conference on Wireless Communications and Signal Processing, pp. 1–5 (2009)
3. Wang, P., Rui, D., Akyildiz, I.F.: A spatial correlation-based image compression framework for wireless multimedia sensor networks. IEEE Trans. Multimed. **13**(2), 388–401 (2011)
4. Gan, L., Tu, C., Liang, J., Tran, T.D., Ma, K.-K.: Undersampled boundary pre-/postfilters for low bit-rate DCT-based block coders. IEEE Trans. Image Process. **16**(2), 428–442 (2007)
5. Junejo, I., Cao, X., Foroosh, H.: Geometry of a non-overlapping multi-camera network. In: IEEE International Conference on Video and Signal Based Surveillance, p. 43. IEEE Press, Sydney (2006)

6. Wiegand, T., Sullivan, G., Luthra, A.: Draft ITU-T recommendation and final draft international standard of joint video specification (ITU-T Rec. H.264 ISO/IEC 14496-10 AVC). Joint Video Team (JVT) of ISO/IEC MPEG & ITU-T VCEG 8th Meeting document JVT-G050, Geneva, p. 3 (2003)
7. Ling, S., Xie, P., Wu, D., Xu, B.: A multi-interface multi-channel Ad Hoc routing protocol for real time video transmission. In: 2011 International Conference on Mechatronic Science Electric Engineering and Computer (MEC) (2011)
8. Odaka, T.: Personalized MPEG2 video-data transmission system. In: 2008 Third International Conference on Internet and Web Applications and Services, p. 6 (2008)

Real-Time Visuo-Haptic Surgical Simulator for Medical Education – A Review

Yonghang Tai[1,2], Junsheng Shi[1(✉)], Lei Wei[2], Xiaoqiao Huang[1],
Zaiqing Chen[1], and Qiong Li[1]

[1] Institute of Color and Image Vision, Yunnan Normal University, Kunming, China
`shijs@ynnu.edu.cn`
[2] Institute for Intelligent Systems Research and Innovation,
Deakin University, Geelong, Australia

Abstract. Virtual surgery simulations are able to provide reliable and repeatable learning experiences in a safe environment, form acquiring basic skills, to performing full procedures. Yet, a high-fidelity, practical and immersive surgical simulation platform involves multi-disciplinary topics. Such as computer graphics, haptics, numerical calculation, imaging processing as well as mechanics etc. Depending on the detailed simulation, various surgical operations such as puncture, cut, tear, burning and suture may need to be simulated, each comes with very specific requirement on medical equipment and micro skills. In this paper, we review a number of previous simulators of haptic-enabled medical education in different surgical operations, and identify several techniques that may improve the effectiveness of the pipelines in both visual and haptic rendering. We believe that virtual surgery simulation has enormous potential in surgical training, education and planning fields of medical advancement, and we endeavor to push the boundaries in this field through this review.

Keywords: Virtual surgery · Medical education · Visuo-hapitc · Surgery

1 Introduction

Conventional medical interventions aim to address the problems of educating and training residents: these procedures usually take extended time to learn so as to complications and risks in relation to the actual surgical procedures [1]. Virtual surgery simulations are able to provide reliable and repeatable learning experiences in a safe environment, thus allowing the trainees to gradually improve their skills and eventually capable of handling real-life procedures. Yet, this emerging simulation is an extremely complicated multi-disciplinary research involved computer graphics, haptics, numerical calculation, medical imaging as well as mechanics etc. [2]. Domain knowledge associated with different surgical operations such as puncture, cut, tear, burning, suture are keys to immersive training but at the same time extremely difficult to represent [2]. Nevertheless, it is believed virtual surgery simulations have enormous potential in the surgical education and planning applications [4–6].

© Springer International Publishing AG 2018
F. Qiao et al. (eds.), *Recent Developments in Mechatronics and Intelligent Robotics*,
Advances in Intelligent Systems and Computing 691, DOI 10.1007/978-3-319-70990-1_77

As demonstrated in Fig. 1, both visual and haptic pipelines in virtual surgical simulation are required to be rendered in real-time during the system operation. When collisions are detected between the tip of the surgical instrument and the tissue surface, the contact point (line or surface) needs to be immediately update with the position information to the physical-based model. The physical-based model based on its own mechanical property, will calculate the force magnitude and direction between the surgical instrument and the virtual tissue, while physical-based soft tissue calculates the displacement of deformation in real time by the stress point by the external force. According to the calculated displacement and direction of each point, the node position is constantly updated and the virtualization pipeline is constantly refreshed on the screen. It can be observed that in a very short period of time, the entire surgical system needs to accomplish all of the above tasks, which is a huge challenge for the real-time requirement.

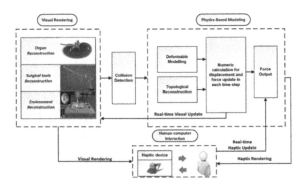

Fig. 1. Visuo-haptic pipeline in virtual surgery simulation

2   Literature Survey

Combining realistic visualization with haptic feedback, provides repetitive, hygienic and immersive simulation experiences for various surgical procedure training. This also increases the effectiveness of surgical skill training and reduces the risks associated with complicated procedures. There have been few previous works on this in the literatures. In this paper, we only focus on the medical simulators with haptics feedback. One of the most common surgeries is laparoscopy Halic [3] conducted an inexpensive virtual reality laparoscopic cholecystectomy surgical simulator, then Gaudina et al. and De Paolis [4] utilized the same haptic device Omni and based on java scripts application. Most of laparoscopic simulator employed dual hands operation and without acceleration of GPU. Endoscopy could examine the interior lesions of a human visceras and cavities of body facilitated. Delorme et al. [5] developed the system named NeuroTouch for cranial microsurgery training with endoscope, Jiang et al. [6] integrated another endoscopic simulator for third ventriculostomy, and Punak et al. [7] utilized a 4-DoF haptic devices to simulate the three steps of develop an endoscopic endonasal surgery. Apart from the minimally invasive surgery simulator, Dental training is another popular medical simulation, Tse et al. [8] made the first haptic dental procedures simulator called hapTEL which allows dental drilling and removing. Another oral implantation simulator was developed by Chen [9] named "Computer Assisted Preoperative Planning for Oral Implant Surgery" (CAPPOIS) can set the medical environment and store the data of trainees. A recent research made by Dangxiao Wang [10] extended this technique by employing dual-manual interaction to facilitate practical implementations in dental simulation. Arthroscopy [11] is another significant surgery which has been applicated in virtual simulations which Heng et al. developed the first arthroscopic surgical simulator by OpenGL [12]. After He designed a bone drilling simulator for fracture surgery [13] and Cecil [14] created a haptic-enable trainer named "Less Invasive Stabilization System" (LISS). Most of the arthroscopy simulator were focus on bone drilling training

Table 1. Comparison of existing simulator in various virtual surgeries

	Time	Authors	Devices	Handle	DOF	Validation	Details
Laparoscopy	2010	Halic and De	Omni	Dual	6/3	No	Gastric banding task
	2012	Gaudina et al.	Omni	Dual	6/3	No	Laparoscopic surgery
	2012	De Paolis	Falcon & Omni	Dual	6/3	No	Laparoscopic suture
	2013	Hernansanz et al.	Desktop	Single	6/3	No	Smoke & Bleeding
Endoscopy	2011	Punak et al.	Omni	Dual	6/3	No	Close a wound
	2012	Delorme et al.	Freedom 6s	Dual	6/3	No	Tumor-debulking & Aspirator
	2013	Jiang et al.	Omni	Dual	6/3	No	Ventriculostomy
Dental	2010	Tse et al.	Falcon	Single	3/3	No	Drill
	2012	Wang et al.	Desktop	Single	6/3	No	Check and Tongue
	2012	Chen et al.	Omega 6	Single	4/3	No	Drill
	2012	Kosuki and Okada	Omni	Single	6/3	No	Touch and Drill
Orthopedics	2004	Heng et al.	Own device	Dual	4/3	No	Knee Arthroscopy
	2013	Chen and He	Premum	Single	6/6	No	Driller
	2013	Cecil et al.	Premum	Single	6/6	No	LISS surgery
Biopsy	2011	Ni et al.	Omni & Premium	Single	6/3	No	Needle
	2013	Selmi et al.	Omni	Single	6/3	No	Ultrasonic & Needle
Liver	2010	Yi et al.	Omega 3	Single	3/3	No	Cutting
Optometry	2012	Wei et al.	Omni	Single	6/3	Yes	Object Remove & Injection
Suturing	2010	Jia and Pan	Desktop	Single	6/3	No	Needle
	2010	Payandeh	Omni	Dual	6/3	No	Needle
	2012	Ricardez et al.	Omni	Dual	6/3	No	Needle
	2014	Choi et al.	Omni	Dual	6/3	No	Needle
Palpation	2011	Coles et al.	Falcon & Omni	Dual	6/3	No	Finger
	2012	Ullrich and Kuhlen	Omni	Dual	6/3	No	Hand/Needle
	2014	Li et al.	Omni	Single	6/3	Yes	Tumor

with force feedback. Biopsy, which is a puncture surgery for sampling of cells or tissues for examination, in this area Ni [15] developed a virtual reality simulator of "Ultrasound-Guided" (UG) liver biopsy simulating, Selmi [16] conducted a complete studying environment provides didactic teaching for numbers of exercises of biopsy application. Yi et al. adapted and calibrated geometry and size of soft tissues by stewing special spots on the surface [17]. In optometry, Wei [18] proposed an immersive configurable and extendable surgical station that implements surgical simulation with and Augmented Reality (AR) techniques. Additionally, this simulator employed the head-mounted display (HMD) to rendering the visual simulation, and AR techniques (by cameras) to mimic the slit lamp orientation during this simulation [18]. For some specific surgical operation, needle puncture and suturing simulator has been developed in many study, Jia [19] and Payandeh [20] are conducted a needle suturing simulator in 2010, after that Ricardez et al. [21] focused on create an external suture surgical environment named SutureHap. Palpation is surgeons analyses the force feedback of tissues and viscera to detect anomalies on body surface via their fingers. Latest development is Li [22] created a tumor location simulator facilitated by deformable tissue probing data. Before this, another palpation simulator is proposed by Ullrich and Kuhlen [23]. This system implements needle puncture applications, which is a normally procedure palpation. PalpSim is a simulator created by Coles et al. [24] in 2011, it was conducted to replace real surgical scenarios based on AR with three haptic devices. Table 1 shows the comparison of existing simulators mentioned above in various virtual surgeries.

3   Problem Identification

As virtual surgical simulations are entirely in the virtual world, each single step mentioned above is computationally demanding in real-time. Balance between accuracy and speed will be the principal challenge for every simulator design. Apart from that, haptic rendering algorithms and mechanical behaviors of various deformable organs are often much different. Additionally, different internal structures are another computationally challenge to implement, physiology relevant simulations during surgery procedure (such as breathing, heartbeat, blood flow, etc.) should also be reproduced faithfully. we followed the trend of haptic-enabled and physics-based volumetric models capable of describing with great accuracy in mechanical behavior of anatomical structures and the major challenges based on this specific research are as follows.

(1) **Visualization rendering**, visual rendered in terms of a number of visible features that reconstructed in simulated environments, not only the basic aspects such as colors, brightness, shadows, reflection as transparency etc., but the special human organs and surgical tools need to be rendering.

(2) **Haptics-Collision detection**, primary challenge of haptics rendering is collision detection algorithms, which is the contact checking among virtual objects during the surgical simulation. Without this technology, surgical tools could go through organs and other obstacles during the simulation. The collision detection algorithm should be updated in all intersections among contact objects in each time step.

(3) **Haptics-Force modelling**, after collision detection between two objects has been detected, a variety of responses to this contact may be formulated. The two solids may rebound away from each other, slide (with or without friction), or settle into relative static contact [48]. This step must also be carried out at each time step.

(4) **Deformable tissue dynamic modelling**, the simulator must also consider the physics involved in the deformation of anatomical structures. Indeed, when a pressure is exerted on a given organ, the latter must deform and react according to its mechanical properties. The force feedback must take place at a very high frequency (often between 500 and 1000 Hz) for a smooth sense of touch.

4   Discussion

As mentioned above, visuo-haptic modelling the deformation of anatomical structures in real-time is the major challenge in surgical simulation. In this paper, we followed the trend of physics-based volumetric models capable of describing with great accuracy in mechanical behavior of anatomical structures. What more, we strived to identify a multiply efficient visuo-haptic framework to model various kinds of surgical simulation in real-time. Such advanced models will not only permit to increase the realism of future training systems, but will also act as a bridge toward the development of patient-specific preoperative planning as well as VR/AR tools for the operating room. Here are the details we discussed:

(1) **Patient-specific demo reconstruction**, individual medical planning or per-operative guidance requires the models with patient-specific biomechanical data, however rarely studies utilized patient-specific data, for example, Computed Tomography (CT) or Magnetic Resonance Imaging (MRI), to reconstructed tissue deformation, tool-tissue interaction and tumour locations.

(2) **Visuo-haptic Multi-function framework,** comparison of existing simulator in virtual surgery in Table 1 demonstrated that most of the existing virtual surgery are focused on only one kind surgery, or one kinds of operation such as puncture or drilling. Through this review, we prospect to propose a uniform surgical simulator which not only includes various kinds of surgical motion, but composite of multiple surgery models.

(3) **Modelling of volumetric deformation**, we will conduct the biomechanical experiments to obtain the original data to set tissue deformation and the tool-tissue interaction force model. Inspired by continuum mechanics and CG animation, physics-based simulation approaches will combine FEM and PBD, leading to an effective, robust, yet precise simulate algorithm fits for multiple non-linear viscoelasticity tissue deformation situations.

(4) **GPU implementation**, GPU was only utilized to accelerate certain parts of the graphics pipeline in recent years, accurate non-linear and viscoelastic FEM procedure will develop by CUDA-based re-implementation within our proposed framework to accelerate the numerical computation.

5 Conclusion and Future Work

In this paper, we review plenty of existing simulators of haptic-enabled medical education in different surgical operations, we then identify the mainly challenges and advanced models within the simulation and training procedures which can be improved through the recent advances in visualization rendering, collision detection, haptics-force modelling and deformable tissue dynamic modelling. In the future, we plan to work on the implementation of these proposed improvements and aim for a more effective and endeavor to push the boundaries in this field.

References

1. Vanlehn, K.: The behavior of tutoring systems. Int. J. Artif. Intell. Educ. **16**(3), 227–265 (2012)
2. Gazibara, T., Marusic, V., Maric, G.: Introducing E-learning in epidemiology course for undergraduate medical students at the Faculty of Medicine, University of Belgrade: a pilot study. J. Med. Syst. **40**(3), 1–12 (2015)
3. Halic, T., De, S.: Lightweight bleeding and smoke effect for surgical simulators. In: IEEE Virtual Reality Conference (VR), pp. 271–272 (2010)
4. De Paolis, L.T.: Serious game for laparoscopic suturing training. In: IEEE Sixth International Conference on Complex, Intelligent, and Software Intensive Systems (CISIS), pp. 481–485 (2012), pp. 271–272 (2010)
5. Delorme, S., Laroche, D., DiRaddo, R., Del Maestro, R.F.: NeuroTouch: a physics-based virtual simulator for cranial microneurosurgery training. Neurosurgery **71**, 32–42 (2012)
6. Jiang, D., Hovdebo, J., Cabral, A., Mora, V., Delorme, S.: Endoscopic third ventriculostomy on a microneurosurgery simulator. Simulation **89**(12), 1442–1449 (2013). Transactions of The Society for Modeling and Simulation International
7. Punak, S., Kurenov, S., Cance, W.: Virtual interrupted suturing exercise with the Endo stitch suturing device. In: Advances in visual computing, pp. 55–63. Springer Heidelberg (2011)
8. Tse, B., Harwin, W., Barrow, A., Quinn, B., Diego, J.S., Cox, M.: Design and development of a haptic dental training system - hapTEL. In: Haptics: Generating and Perceiving Tangible Sensations. Lecture Notes in Computer Science, vol. 6192, pp. 101–108. Springer, Heidelberg (2010)
9. Chen, X., Lin, Y., Wang, C., Shen, G., Wang, X.: A virtual training system using a force feedback haptic device for oral implantology. In: Transactions on Edutainment VIII, pp. 232–240. Springer, Berlin (2012)
10. Wang, D., Zhang, Y., Hou, J., Wang, Y., Lv, P., Chen, Y., Zhao, H.: Dental: a haptic-based dental simulator and its preliminary user evaluation. IEEE Trans. Haptic **5**(4), 332–343 (2012)
11. Heng, P.-A., Cheng, C.-Y., Wong, T.-T., Xu, Y., Chui, Y.P., Chan, K.-M., Tso, S.-K.: A virtual-reality training system for knee arthroscopic surgery. IEEE Trans. Inf. Technol. Biomed. **8**(2), 217–227 (2004)
12. Crespo, L.M., Reinkensmeyer, D.J.: Effect of robotic guidance on motor learning of a timing task (2008)
13. Powell, D., O'Malley, M.K.: Efficacy of shared-control guidance paradigms for robot-mediated training. In: IEEE World Haptics Conference, pp. 427–432 (2011)

14. Cecil, J., Ramanathan, P., Rahneshin, V., Prakash, A., Pirela-Cruz, M.: Collaborative virtual environments for orthopedic surgery. In: IEEE International Conference on Automation Science and Engineering (CASE), pp. 133–137 (2013)
15. Ni, D., Chan, W.-Y., Qin, J., Chui, Y.-P., Qu, I., Ho, S., Heng, P.A.: A virtual reality simulator for ultrasound-guided biopsy training. IEEE Comput. Graph. Appl. **31**(2), 36–48 (2011)
16. Selmi, S.-Y., Fiard, G., Promayon, E., Vadcard, L., Troccaz, J.: A virtual reality simulator combining a learning environment and clinical case database for image-guided prostate biopsy. In: IEEE 26th International Symposium on Computer-Based Medical Systems (CBMS), pp. 179–184 (2013)
17. Yi, N., Xiao-jun, G., Xiao-ru, L., Xiang-feng, X., Wanjun, M.: The implementation of haptic interaction in virtual surgery. In: International Conference on Electrical and Control Engineering (ICECE), pp. 2351–2354 (2010)
18. Wei, L., Najdovski, Z., Abdelrahman, W., Nahavandi, S., Weisinger, H.: Augmented optometry training simulator with multi-point haptics. In: IEEE International Conference on Systems, Man, and Cybernetics (SMC), pp. 2991–2997 (2012)
19. Jia, S., Pan, Z.: A preliminary study of suture simulation in virtual surgery. In: International Conference on Audio Language and Image Processing (ICALIP), pp. 1340–1345 (2010)
20. Payandeh, S., Shi, F.: Interactive multi-modal suturing. Virtual Reality **14**(4), 241–253 (2010)
21. Ricardez, E., Noguez, J., Neri, L., Munoz-Gomez, L., Escobar-Castillejos, D.: SutureHap: A suture simulator with haptic feedback. In: Workshop on Virtual Reality Interaction and Physical Simulation VRIPHYS, pp. 79–86 (2014)
22. Li, Y., Wang, C., Shen, G., Wang, X.: A virtual training system using a force feedback haptic device for oral implantology. In: Transactions on Edutainment VIII, pp. 232–240. Springer, Berlin (2012)
23. Ullrich, S., Kuhlen, T.: Haptic palpation for medical simulation in virtual environments. IEEE Trans. Vis. Comput. Graph. **18**(4), 617–625 (2012)
24. Coles, T.R., Meglan, D., John, N.W.: The role of haptics in medical training simulators: a survey of the state of the art. IEEE Trans. Haptic **4**(1), 51–66 (2011)

Development of NSCLC Precise Puncture Prototype Based on CT Images Using Augmented Reality Navigation

Zhibao Qin[1], Yonghang Tai[1,2], Junsheng Shi[1(✉)], Lei Wei[2],
Zaiqing Chen[1], Qiong Li[1], Minghui Xiao[3], and Jie Shen[3]

[1] Institute of Color and Image Vision, Yunnan Normal University, Kunming, China
shijs@ynnu.edu.cn
[2] Institute for Intelligent Systems Research and Innovation,
Deakin University, Geelong, Australia
[3] Yunnan First People's Hospital, Kunming, China

Abstract. According to the gray value in CT image sequence of NSCLC, the visualization of CT images can facilitated surgeries to analyze and judge detailed characteristics inside the tumor. Through the color distribution of the heat map, it can identify a representative position for tumor puncture biopsy. CT images sequence is reconstructed three-dimensional model, including the skin, bone, lungs and tumor, the reconstructed model and the patient's real body are registered. The four-dimensional heat map of the tumor as a reference to determine the space location of the puncture. In this paper, we conducted reconstructions of a four-dimensional heat map of region of interest and designed the best puncture path. The use of AR navigation technology to guide the puncture biopsy, which achieved the purpose of precise puncture, and facilitate the doctor to take a sample to do pathology analysis. Validations demonstrated our precise puncture system based on AR navigation and four-dimensional heat map reconstruction perform a greatly improved the accuracy of tumor puncture and the diagnoses rate of tumor biopsy

Keywords: NSCLC · Heat map · Reconstruction · Precise puncture · AR

1 Introduction

The individual diagnosis and treatment research of lung cancer turned into bottleneck stage, which means creative ideas and disciplines need to be introduced. At the present stage, the molecular medicine and the individualized treatment plan are to detect the pathological characteristics of the tumor through genomics and proteomics. The greatest dependence of this approach is necessary to obtain some of the tumor tissue as a test specimen by endoscope clamp, biopsy or surgical resection. Medical imaging plays an important role in the detection, diagnosis and treatment of cancer. Studies have shown that the phenotype characteristics of imaging has a strong correlation with the phenotypic characteristics of the tumor [1]. Radiomics consider that through in-depth study and analysis of the imaging features, we can get more phenotypic features and hidden information in the image, and obtain some imaging features of genomics and proteomics [2–4].

© Springer International Publishing AG 2018
F. Qiao et al. (eds.), *Recent Developments in Mechatronics and Intelligent Robotics*,
Advances in Intelligent Systems and Computing 691, DOI 10.1007/978-3-319-70990-1_78

Because of the heterogeneity of space and time in the process of tumor growth, a small number of tissue samples obtained at any time and randomly can't represent the characteristics of the whole tumor tissue and reduce the diagnostic rate of pathological examinations. And the opacity of human tissues and organs, doctors and general instruments is difficult to accurately determine the spatial location of the tumor and puncture points, the selected puncture point is not representative. Therefore, the urgent need to explore the representative puncture point for puncture biopsy, precise positioning technology, research and development of innovative integrated precision biopsy system.

2   Related Works

Haaga et al. first reported CT-guided percutaneous biopsy. Lughezzani G et al. used transrectal ultrasound guided puncture biopsy to improve the positive findings of biopsy, and found that with the increase in the number of puncture needle, no significant increase in complications [5]. Solomon et al. completed the first clinical trial of pulmonary bronchoscopy with 3D CT and magnetic navigation in 2000 [6]. Laurent F et al. compared the accuracy and complication rate of two different CT-guided transthoracic needle biopsy techniques: fine needle aspiration and automated biopsy device [7]. Xu Jing et al. describe a three-dimensional ultrasound navigation system for liver cancer surgery to track puncture needle and real-time display slice images [8]. SONN et al. studies found that the diagnostic rate of biopsy with MRI-TRUS real-time fusion system is 3 times than the diagnostic rate of biopsy with traditional puncture system [9]. Grimson first described an image-guided neurosurgery system based on AR navigation techniques to improve the robustness and accuracy of surgery [10]. Most of the above studies are based on the entire lesion to operate, the choice of puncture position is not representative, affecting the accuracy of puncture and the diagnosis rate of pathology. We present a new method to determine the precise location of the lesion.

3   Implement Detail

In this paper, the most common non-small cell carcinoma of lung cancer is the research object, the precise puncture of lung tumors is the target of research. Do a quantitative analysis for density characteristic of tumor in imaging and construct four-dimensional heat map model for the tumor, combining with the graphics registration and augmented reality technology to build the precise puncture system of tumors. The specific flowchart is shown in Fig. 1.

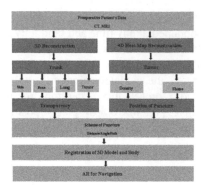

Fig. 1. The flow chart of precise puncture

3.1   Image Segmentation of Lung Tumor

The segmentation of lung tumors using the method of semi-automatic and threshold segmentation. By observing the gray value of CT images and the regional distribution of each organ tissue. In order to reduce the amount of data processing method, using semi-automatic segmentation to roughly extract the lung tumor area complete, and do a further detail by threshold segmentation for the tumor to remove the adjacent tissue and vascular, and then we will get more accurate region of the lung tumor.

3.2   3D Reconstruction Models of Patients

A series of two-dimensional human CT images are reconstructed into three-dimensional model, showing a more intuitive effect on doctors, and it is convenient to observe and analyze the geometric features such as size and spatial position of each organization. At the same time, the three-dimensional model can accurately determine the location of the tumor, to provide help for the puncture path planning. In this paper, the 3D surface reconstruction method is used to reconstruct the skin, bones, lungs and tumors of the patients, and adjust the appropriate transparency. The doctor can see the model of human internal tumor and other organization clearly from the outside.

3.3   4D Heat Map of Lung Tumor

The four-dimensional visualization model of tumor's density can help doctors to more intuitively observe the density distribution of the whole tumor, including the size and gradient of density in different locations and density. We can roughly determine the necrotic area of the tumor, reduce the incidence of false puncture, thereby improving the success rate of puncture and the diagnosis rate of pathological examination. In this paper, we employed Matlab as the toolbox to reconstruct the 4D heat map model for the region of tumor base on volume reconstruction, and add the appropriate transparency to the model. Because the entire tumor has the same transparency, the color information of a region within the tumor is visualized well. We show a series of opaque slices with heat map on the transparent model by Matlab program to solve the problem. We have an analysis for the 4D heat map of the tumor, because the region with bigger color change gradient is more representative, likely associated with tumor heterogeneity greatly. Studies have shown that the depth of puncture is positively correlated with the incidence of complications of pneumothorax and pulmonary hemorrhage. So we chose the tumor position with closer to the puncture point of skin and bigger color change gradient as puncture point.

3.4 Puncture Navigation Base on Argument Reality

Augmented reality (AR) is the technology that combines computer-generated virtual objects or scenes into real scenes, presenting the scene to the user through the display of the device, improving the observer's cognitive ability and interactive experience in the real world [11]. We have imported 3D body models into Unity3D, which control the spatial transformation of the models, such as rotation, motion and zoom, and set the color and transparency of the models. Calling the AR camera presents the patient in the real environment on the screen, adjusting the size and position of the model in the Unity3D scene, and making it have the same size with the real patient on the screen, and then register them. The registration results can be presented to the doctor the more intuitive distribution of the organs and tissues within the patient, convenient for doctors to design puncture paths, avoiding blood vessels and other important tissues. Unity 3D regard the center of patient model as origin of coordinate, and the spatial coordinates of the tumor's puncture point and the skin's entry point in the model can be displayed so that the relative position of the two can be calculated. Then we will be able to design puncture path planning, including entry points, puncture direction, puncture depth and puncture points.

4 Results

We obtained CT image data from patient A and patient B who had non-small cell lung cancer from Yunnan First People's Hospital in Kunming. The skin, bones, lungs and tumors of the CT image sequences of group A and group B were reconstructed, and the transparency was adjusted.

We selected CT image with lung tumor from group A and group B, group A have 15 slices, and group B have 20 slices. After the semi-automatic segmentation of the tumors, the result is shown in Fig. 2.

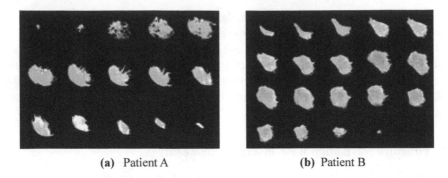

(a) Patient A   (b) Patient B

Fig. 2. The semi-automatic segmentation of tumors of patient A and patient B

Using Matlab software to do the threshold segmentation for the tumor region which after semi-automatic segmentation, and reconstruct the tumor to 4D heat map model, the effects are shown in Fig. 3. We selected point P and point Q of patient A and patient B as the puncture points, and found out the corresponding puncture point P' and point Q' in the corresponding 3D tumor models. Through the distribution of organs and tissues in 3D models, planning an optimal puncture path to avoid the important organs and tissues. Selecting the M, N points as skin entry point of patients A, B. Connect the P' and M, Q' and N with the needle model in unity3D.

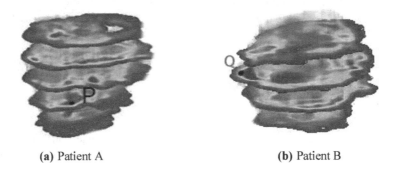

(a) Patient A   (b) Patient B

Fig. 3. 4D heat map of patient A and patient B

We utilized 16G puncture biopsy needle to puncture the tumor tissue for biopsy. After doing registration on patient's 3D models and real patient, using the needle model to guide the puncture needle puncture. The results of 3D body models and precise puncture navigation are shown in Fig. 4.

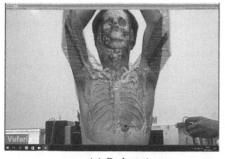
(a) Patient A

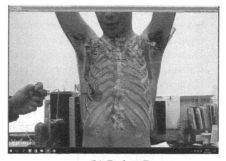
(b) Patient B

Fig. 4. AR navigation puncture of patient A and patient B

5   Conclusion and Future Works

In this paper, we reconstruct 3D model of the patient and 4D heat map model of the tumor, determine the representative puncture point in the tumor, design the puncture path planning based on image registration and AR navigation technology, and realize the precise puncture of non-small cell lung cancer. This is the first time to propose a precise puncture system based on 4D heat map reconstruction and AR navigation technology. At present, we can extract the density details by radiomics of NSCLC, to compare the result of pathological examination obtained by radiomics features and tumor heterogeneity, and found a high relationships between them. The future works will focus on the distribution of vessels and nerve during the puncture surgery, there will be a small dislocation between the actual patient and pre-reconstruction model, which will reduce the success rate of puncture and the diagnosis rate of pathology examination, and increase the risk of puncture. Other future works are to add a three-dimensional model of blood vessels, consider the breathing movement of the lungs when puncture, and create a 5D CT model of lung tumor to provide more accurate puncture results for precise puncture.

References

1. Ho, L.S., Nariya, C., Ja, K.S., et al.: Correlation between high resolution dynamic MR features and prognostic factors in breast cancer. Korean J. Radiol. **9**(1), 10–17 (2008)
2. Panpan, H., Jiazhou, W., et al.: Radiomics in precision radiotherapy. Chin. J. Radiat. Oncol. **26**(1), 103–106 (2017)
3. Lambin, P., Riosvelazquez, E., et al.: Radiomics: extracting more information from medical images using advanced feature analysis. Eur. J. Cancer **48**(4), 441–446 (2012)
4. Guinney, J., Dienstmann, R., Wang, X., et al.: The consensus molecular subtypes of colorectal cancer. Nat. Med. **21**(11), 1350 (2015)
5. Lughezzani, G., Sun, M., Budäus, L., et al.: Effect of the number of biopsy cores on prostate cancer detection and staging. Future Oncol. **6**(3), 381–390 (2010)

6. Solomon, S.B., White Jr., P.M.D., Wiener, C.M., et al.: Three-dimensional CT-guided bronchoscopy with a real-time electromagnetic position sensor: a comparison of two image registration methods. Chest **118**(6), 1783–1787 (2001)
7. Laurent, F., Latrabe, V., Vergier, B., et al.: Percutaneous CT-guided biopsy of the lung: comparison between aspiration and automated cutting needles using a coaxial technique. Cardiovasc. Intervent. Radiol. **23**(4), 266–272 (2000)
8. Jing, X., Xiangdong, Y., et al.: Intra-operation 3D ultrasound navigation system in intervention therapy for liver cancer. Chin. J. Biomed. Eng. **26**(5), 719–723 (2007)
9. Sonn, G.A., Natarajan, S., Margolis, D.J., et al.: Targeted biopsy in the detection of prostate cancer using an office-based MR-US fusion device. J. Urol. **189**(1), 86 (2013)
10. Grimson, E., Leventon, M., et al.: Clinical experience with a high precision image-guided neurosurgery system. In: Medical Image Computing and Computer-Assisted Intervention, MICCAI 1998, pp. 63–73 (1998)
11. Azuma, R., Baillot, Y., Behringer, R., et al.: Recent advances in augmented reality. IEEE Comput. Graph. Appl. **21**(6), 34–47 (2001)

A Binocular Vision Motion Capture System Based on Passive Markers for Rehabilitation

Zhang Mei[1], Lin Lin[2(✉)], Yang Lei[1], Liang Lan[2], and Fang Xianglin[2]

[1] School of Electrical Engineering and Intelligentization, Dongguan University of Technology, Dongguan, China
[2] School of Information Engineering, Guangdong Medical University, Dongguan, China
lynwindsent@163.com

Abstract. A new type of posture tracking for rehabilitation is introduced to deal with the standard digital video image and real-time tracking the object with computer vision method. Impassive infrared marker is introduced to help spatial localization and angle measurement. The tracking accuracy of location is <1 mm and the angle measurement accuracy is <0.05 rad that can meet requirements of rehabilitation exercise instruction. The instrument is developed from commercial binocular camera which has very low price and high reliability.

Keywords: Posture tracking · Rehabilitation · Infrared marker

1 Introduction

With the improvement of public health construction, as well as increasing social demand of the rehabilitation of the physical and mental disabilities, the rehabilitation medical engineering is becoming increasingly important. Research shows that most people can restore the loss of function in whole or in part by equipment of rehabilitation instruments and improve the quality of life. In recent years, development of rehabilitation system has become a research hotspots, especially in the field of computer vision and clinical medicine. Based on the human motion tracking system in the treatment of patients, doctors can track the patient's movement through the network and perform identification and correction of the wrong action as well as rehabilitation training to help patients restore limb function. Human motion tracking is one of the key technologies of remote rehabilitation system. It is of great theoretical significance and potential application value to carry out the research of remote rehabilitation system based on human motion tracking. Optical navigation and tracking systems often use infrared laser emitting diode as the marker and only the infrared rays may be detected by the image sensor [1–4]. This method needs power supply unit by wire, which limits the normal range of motion. Battery will also increase the weight and volume of the equipment with reduction of the endurance. On the other hand, it is very difficult to distinguish all markers so that the light markers must shine in a certain order when the object is in a moving state [5–7]. Identification of the object at different time and location might cause the measurement error of the attitude and the moving speed of the measured object with the lower the

© Springer International Publishing AG 2018
F. Qiao et al. (eds.), *Recent Developments in Mechatronics and Intelligent Robotics*,
Advances in Intelligent Systems and Computing 691, DOI 10.1007/978-3-319-70990-1_79

sampling frequency, the greater the error. Impassive optical maker, little ball or patterned plate, can be installed on the body to improve optical positioning and tracking system [8–10]. These spherical or discoid the surface of the marker is composed of a layer of spherical particles optical coating, the camera can detect the infrared light reflected back to the camera lens, so that the marker in the infrared shadow shown as a bright circular spot in the image. This method avoids the disadvantages of using wired power supply to the identifier with convenient and reliability. On the other hand, markers need to have a special coating to help optical sensor capture the target image into the host computer for processing. In this paper, we provide a posture recognition system with impassive marker based on commercial KINECT 2.0 equipment. User can also carry out the second development according to the positioning requirements measurement space, while tracking the target.

2 Methods

The human limbs mounted by infrared reflector ball can easily through the image domain algorithm from the background identification. By marking the external shape of the limb structure without aliasing, one can directly determine which feature points belonging to the same part [11]. In order to ensure in arbitrary pose in intelligent visual identification by feature points and unambiguous discrimination, the relative positions of markers have significant difference. The center of marker can be calculated according to its geometry feather, which can increase the accuracy to millimeters.

First of all, the world coordinate system of observer is defined as reference identification unit. Taking the common edge of the two-orthogonal plane of the reference identification unit as the X axis of the world coordinate system, the left side of the public side is the Y axis and the vertical axis side as Z. The intersection of the three axes is the origin of the world coordinate system represented as (x_i, y_i, z_i). According to the pinhole camera model, the two-dimensional projection of the characteristic line in the camera imaging plane can be expressed as

$$[a, b, c] \begin{bmatrix} x_i \\ y_i \\ 1 \end{bmatrix} = 0 \tag{1}$$

When there is an angle between the world coordinate system and the camera coordinate, we need round the world coordinates of the X, Y and Z axis with rotation angle α, β and γ respectively. The rotation matrix can be written in the following form.

$$R = R_x R_y R_z = \begin{bmatrix} 1 & 0 & 0 \\ 0 & a_x & b_x \\ 0 & -b_x & a_x \end{bmatrix} \begin{bmatrix} a_y & 0 & -b_y \\ 0 & 1 & 0 \\ b_y & 0 & a_y \end{bmatrix} \begin{bmatrix} a_z & b_z & 0 \\ -b_z & a_z & 0 \\ 0 & 0 & 0 \end{bmatrix} \tag{2}$$

According to the rotation matrix, the rotation angle can be obtained to adjust the patient's posture and action.

3   Results

The capture system is based on commercial Kinect 2.0 of Microsoft Co. Lt. with i7-2.5G CPU, 1980*1080 pixels imaging capability and cognitional ability of 25 skeletal points per person. The instrument has an infrared light source, a color camera, an infrared camera and four microphones. These sensors are used to obtain RGB, depth and audio data. The system determinate z coordinate by speckle patterns which are unique for every specific distance from the x-y plane of camera.

The Camshift algorithm is introduced to calculate the probability lines and the position of the window centroid. Tracking algorithms based on the motion of target is programmed in Visual Studio 2010 development environment with combination of software development kit from Microsoft Corp. A single human motion tracking experiment results show that the algorithm is suitable for human motion in the range of Kinect sensor's field of vision. The algorithm uses the depth information tracking with the background being eliminated. The system has strong robustness and will not be affected by illumination and background.

We took two postures, A-Pose and T-Pose, of the volunteer to record the Coordinate information of her 20 joints. The centroid of each joint in x-y plane was also calculated by Camshift algorithm based on color and depth images of the body. The results of Kinect with or without markers and calculation of algorithm are compared in Fig. 1.

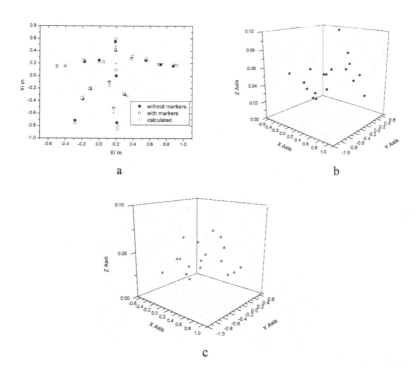

Fig. 1. Results of imaging and calculation of algorithm based on Kinect 2.0 without marker

The results of Camshift algorithm were verified by the data from software development kit (SDK) 2.0 with root-mean-square error remaining less than 8%.

The mean error of measurement with infrared markers is obviously decreased as shown in Fig. 2. We also took another ten more times measurement to prove the robustness of the method. All the tests showed good agreement with the data of skeleton graph generated by Kinect SDK. Next, we selected the rehabilitation training on both sides of the body with eight joints to obtain the data volatility and quantitative explanation. The purpose is to rationalize the training range of the human body to determine the range of acceptable data volatility. The data distribution of eight joint points is shown in Fig. 2 with standard error distribution. The data was measured by an older version system with higher error of 10%.

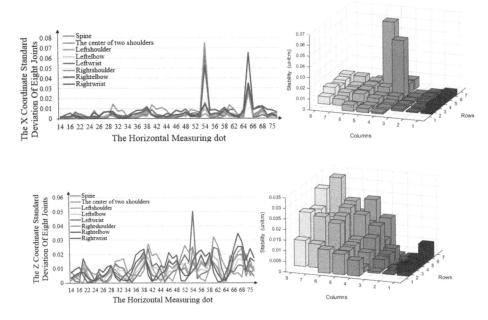

Fig. 2. The spatial stability distribution line of measuring coordinate X, Z value

4   Summary

In this paper, an optical positioning and tracking technology uses the infrared light spot as a marker to locate the image, and the center of the marker is determined by the light intensity image of the infrared spot. The markers are mounted onto the patient to provide more accurate location information. The system is built up based on a commercial binocular vision motion capture instrument which is only 1/100 cost of general infrared binocular tracking system. The tracking accuracy of location is 1 mm and the angle measurement accuracy is less than 0.05 rad that can meet requirements of rehabilitation exercise instruction.

Acknowledgement. The authors acknowledge the support given by Natural Science Foundation of Guangdong Province (2014A030310258), Educational Commission of Guangdong Province (YQ201402) and Guangdong science and technology plan project (2015A020214024).

References

1. Jongwook, K., Dongjun, S., Dongseok, J.: A study on the control method of 3-dimensional space application using kinect system. Int. J. Comput. Sci. Netw. Secur. **11**(3), 55–59 (2011)
2. Kuehn, T.: The Kinect sensor platform. Adv. Med. Technol. **40**(9), 2192–2198 (2011)
3. Khoshelham, K.: Accuracy and resolution of Kinect depth data for indoor mapping application. Sensors **12**(2), 1437–1454 (2012)
4. Papadopoulos, G.T., Axenopoulos, A., Daras, P.: Real-time skeleton-tracking-based human action recognition using kinect data. Lecture Notes in Computer Science, pp. 473-483. Springer, New York (2014)
5. Corporation, H.P.: A review on technical and clinical impact of microsoft kinect on physical therapy and rehabilitation. J. Med. Eng. **2014**, 846514 (2014)
6. Arai, K., Asmara, R.A.: Human gait gender classification in spatial and temporal reasoning. Int. J. Adv. Res. Artif. Intell. **1**(2), 1–6 (2012)
7. Moore, B.A., Pita, D., Tyrrell, L.P.: Vision in avian emberizid foragers: maximizing both binocular vision and fronto-lateral visual acuity. J. Exp. Biol. **218**(9), 1347–1358 (2015)
8. Gonzalez, D.A., Niechwiej-Szwedo, E.: The effects of monocular viewing on hand-eye coordination during sequential grasping and placing movements. Vis. Res. **128**, 30 (2016)
9. Park, S.W., Park, H.S., Kim, J.H.: 3D displacement measurement model for health monitoring of structures using a motion capture system. Measurement **59**, 352–362 (2015)
10. Bujang, K., Nazri, A.F.A., Azam, A.F.A.: Development of a motion capture system using Kinect. Eur. J. Ophthalmol. **76**(11), 531–540 (2015)
11. Jia, Z., Yang, J., Liu, W.: Improved camera calibration method based on perpendicularity compensation for binocular stereo vision measurement system. Opt. Express **23**(12), 15205–15223 (2015)

Highlight Removal of Workshop Floor Images

Sibo Quan$^{(\boxtimes)}$ and Weiguang Li

School of Mechanical and Automotive Engineering,
South China University of Technology, Guangzhou 510640, China
quan.sibo@mail.scut.edu.cn

Abstract. Manufacturing workshops floors are usually paved with ceramic tiles or painted with resin, which brings about strong specular reflection under sunlight or fluorescent lighting and is a serious disturbance for many actions based on machine vision, thus needs to be removed. However, prior arts focused mainly on weak specular reflection. To address the issues, we put forward a gradual chromaticity concept, separate weak specular reflection first, then reduce critical chromaticity value step by step and separate stronger specular reflection iteratively. Finally, we propose an inpainting approach for removing remainder highlight which still uses chromaticity constraint and operates in single pixel. The implement with the proposed method shows a quite improved result.

Keywords: Workshop · Highlight · Gradual chromaticity · Specularity removal

1 Introduction

Though specular highlight information may be useful in some applications such as illumination chromaticity estimation and shape-from-shading reconstruction [1], it is an adverse influence for a lot of vision applications, especially those algorithms related to feature extraction and matching [2], because the intrinsic characteristic of non-Lambertian surface is disguised by highlight. Therefore, highlight removal is an important task.

Highlight removal studies have main two paths including hardware and software based methods. The hardware-based method means improving lighting condition to capture information about the object's inherent properties [3], which requires polarizing filter setups or diffuse plate. The software-based path is more attractive, particularly those approaches using single image to remove specularity. Influentially, Tan et al. [4, 5] put forward a specular-free(SF) image concept based on chromaticity information and gave a specular-to-diffuse mechanism algorithms; afterwards other researchers also developed some algorithms improving the efficiency and accuracy of specular separation, such as modified SF, bilateral filtering [6, 7]. Recent further researches such as brightness ratio method [8], Zeta-image method etc. [9], were still substantially chromaticity-intensity-based.

Manufacturing workshops floors are usually paved with ceramic tiles or painted with resin, which brings about strong specular reflection under sunlight or fluorescent lighting. For many actions such as visual-guiding robot operation, 3D reconstruction and automated guided vehicle navigation and so on, the highlight is a serious

© Springer International Publishing AG 2018
F. Qiao et al. (eds.), *Recent Developments in Mechatronics and Intelligent Robotics*,
Advances in Intelligent Systems and Computing 691, DOI 10.1007/978-3-319-70990-1_80

disturbance thus needs to be removed. However, how to remove highlight in manufacturing scene image there is a severe lack of application-oriented research. To address the issues, we propose an approach based on single pixel operation to remove weaker and stronger highlight which does not need specular-free image as reference, and avoids excessive intensity decrease.

2   The Limitations of Previous Works

Here we briefly review specular-to-diffuse mechanism [4, 5] which was based on Shafer's dichromatic model: the reflectance of non-Lambertian surface is a linear combination of diffuse and specular component [10]:

$$I_c(x) = I_c^D(x) + I_c^S(x) \tag{1}$$

where $I_c(x) = [I_r(x), I_g(x), I_b(x)]^T$ is the RGB color vector at pixel x, c expresses either channel of R, G, and B, and $I_c^D(x)$ and $I_c^S(x)$ are the diffuse reflection component and specular reflection component, respectively.

Define normalized maximum image chromaticity as:

$$\sigma'_{max}(x) = \frac{I'_{max}(x)}{\Sigma I'_c(x)} \tag{2}$$

where $I'_c(x)$ is the normalized R,G,B color vector. In order to simplify expression we substitute $\sigma'_{max}(x)$ and $I'_{max}(x)$ with $\tilde{\sigma}'$ and $\tilde{I}'$ respectively in the following.

Similarly, there are diffuse chromaticity and specular reflection chromaticity:

$$\Lambda'_c(x) = \frac{I_c'^D(x)}{\Sigma I_c'^D(x)}, \ \Gamma'_c(x) = \frac{I_c'^S(x)}{\Sigma I_c'^S(x)} \tag{3}$$

For computer synthetic images, under white illumination (it is achromatic, $I_r = I_g = I_b$, $\Gamma = 1/3$), a pixel including specular component will have varied $\tilde{\sigma}'(x)$, but a full diffuse pixel has $\tilde{\sigma}'(x) = \Lambda'(x)$, and all full diffuse pixels in a color clustering have a constant value of Λ'. For a specular pixel x_1 and a diffuse pixel x_2 with exactly same diffuse chromaticity Λ', there is $\tilde{\sigma}'(x_2) > \tilde{\sigma}'(x_1)$; by subtracting intensity in three color channels of x_1 with a small scalar number iteratively, then $\tilde{\sigma}'(x_1)$ will increase till to $\tilde{\sigma}'(x_2)$, i.e. equal to Λ', then x_1 becomes a diffuse pixel. From $\tilde{\sigma}'(x_1)$ to $\tilde{\sigma}'(x_2)$ the process has specular-to-diffuse mechanism formula:

$$m_d(x_1) = \frac{\tilde{I}'(x_1)[3\tilde{\sigma}'(x_1) - 1]}{\tilde{\sigma}'(x_1)[3\Lambda'(x_1) - 1]} \tag{4}$$

$$I'^D(x_1) = m_d(x_1)\Lambda' \tag{5}$$

where $m_d(x)$ is the diffuse coefficients which corresponds to the surface geometry.

This specular-to-diffuse mechanism can only separate weak specular reflection for real image. When there are some low saturation pixels or stronger specular reflection, it will fail to work. If the values of I'_r, I'_g and I'_b are close to each other, we may see $\tilde{\sigma}'$ $\approx 1/3$ in Eq. (2) then $m_d \approx 0$ in Eq. (4) and $I'^D \approx 0$ in Eq. (5). So Tan et al. [11] directly removed achromatic pixels by using simple thresholding in maximum chromaticity. Most of given case images in previous papers were well-finding, otherwise, the processed images by previous methods showed a lump of black corresponding original strong highlight region, like the case "workshop" seen in Fig. 1.

| (a) | (b) | (c) | (d) |

Fig. 1. The result of highlight removal for workshop scene. (a) Original image; (b) Tan's method; (c) Shen's method; (d) Proposed method

3   Proposed Method

Weak specular reflection barely affects the identification of object status, but real-world images are often affected by quite strong highlight, which has a large impact on machine vision for object recognition. Motivated by the above analysis of existing problems, we propose an application-oriented approach to remove weaker or stronger highlights. We assume that the images are photoed under white illumination and that the object in the image has one dominant hue.

3.1   Highlight Definition and Separability Judgement

In theory, any deviation from ground truth image implies anamorphose existence. If a pixel's intensity is larger than that of ground truth image then the pixel has specular component. For a computer synthetic image with white illumination, pixels have specular components in varying degree can become full diffuse through decreasing identical intensity in R, G, B channels. However, for a real image the situation is quite different. Because of ununiform illumination and sensor noise, also maybe different reflection properties of object surface on R, G, B, the highlight may not be the identical in three channels, in particular under the illumination without uniformity calibration. So what is highlight and which extent highlight can be removed is still an open problem discussed hardly.

When a pixel has weak specular reflection, since its hue is close to that of diffuse pixel, the difference between $\tilde{\sigma}'$ and Λ' comes from its saturation difference. Through decreasing its intensity in three channels synchronously we can adjust its saturations while retain its hue then elevate its $\tilde{\sigma}'$ value up to Λ' then we remove the weak specularity. However, when a pixel has strong specularity, there is a biggish difference between its $\tilde{\sigma}'$ and Λ' which comes from not only saturation but also obvious hue difference. By decreasing identical intensity in R, G, B channels synchronously, even though specular pixels attain the required diffuse chromaticity finally, because its intensity needs to be lowered too much, even down to 0 or negative, so they look a lump of black (like the Fig. 1).

Color is expressed by intensity and chromaticity conjointly, and chromaticity is expressed by hue and saturation conjointly. For a certain hue, an observer may feel different color due to varied saturation. Obviously, in a real image the more concentrated its hue distribution, and the higher the saturation, the better the specular component separability is. Simply, when a specular pixel's $I'_{max} - I'_{min}$ value is much less than that of diffuse pixel, its specularity is difficult to be removed. Depending on one's acceptability on image intensity dispersibility, one may point a permissive diffuse intensity lower limit $\tilde{I}'_L$; when specular pixel's intensity is decreased down to $\tilde{I}'_L$ or $I'_{min} = 0$ but its $\tilde{\sigma}'$ does not yet arrive at Λ', we judge the pixel's specular component can't be separated normally.

3.2   Highlight Removal Based on Gradual Chromaticity

From the application-oriented goal we put forward a gradual chromaticity concept and a practical method, which involves a single-pixel operation and neither needs specular-free image nor specular pixel identification, just is a straightforward calculation. According to $\tilde{\sigma}'$ histogram we choose a $\tilde{\sigma}'$ value having larger probability as Λ' first, then decide upon an acceptable chromaticity deviation, say, 10% or 15% for determining the minimum permissive Λ'_L, which means chromaticity may gradually decrease to $\Lambda'_L = 90\%\Lambda'$ or $85\%\Lambda'$; correspondingly, limit $\tilde{I}'_L$ (preventing the pixel intensity from excessive decrease to avoid the "a lump of black" phenomenon). For a specular pixel x with $\tilde{I}'(x)$, $I'_{med}(x)$ and $I'_{min}(x)$, after removing specular reflection we denote the new three channel intensities as $\tilde{I}'^D(x)$, $I'^D_{max}(x)$ and $I'^D_{min}(x)$, respectively. Based on white illumination ($I'_r = I'_g = I'_b$), the minuend:

$$\Delta I'(x) = I'_{min}(x) - I'^D_{min}(x) = I'_{med}(x) - I'^D_{med}(x) = \tilde{I}'(x) - I'^D(x) \qquad (6)$$

After specular component removal the pixel's chromaticity is:

$$\Lambda' = \frac{\tilde{I}'^D(x)}{\tilde{I}'^D(x) + I'^D_{med}(x) + I'^D_{min}(x)} \qquad (7)$$

Combining Eqs. (6) and (7), we obtain:

$$\Delta I'(x) = \frac{\Lambda'_1 [\tilde{I}'(x) + I'_{med}(x) + I'_{min}(x)] - \tilde{I}'(x)}{3\Lambda'_1 - 1} \qquad (8)$$

$\Delta I'(x)$ is the specular component we search. Then

$$
\begin{cases}
\tilde{I}'^{D}(x) = \tilde{I}'(x) - \Delta I'(x) \\
I'^{D}_{med}(x) = I'_{med}(x) - \Delta I'(x) \\
I'^{D}_{min}(x) = I'_{min}(x) - \Delta I'(x).
\end{cases}
\tag{9}
$$

Equation (9) simply contains the diffuse components of the pixel while specular component is removed. If $I'_{min}(x) < \Delta I'(x)$, the pixel has insufficient original intensity to obtain pointed diffuse chromaticity Λ', or if the intensity of the specular pixel is reduced to $I'(x) = \tilde{I}'_L$ but its $\tilde{\sigma}'$ does not yet reach Λ', we judge it impossible to separate the specular component using the formula (9), so we retain its original intensity for next processing.

Using the formula (9) we will successfully process those pixels with weak specular reflection; then reducing the diffuse chromaticity critical value from $\Lambda'_1, \Lambda'_2 ...$ to Λ'_L step by step, by again running Eqs. (6)–(9) we will iteratively separate stronger specular reflection.

3.3   Inpainting

Although above gradual chromaticity method enables continuous color and intensity transition, when the specular reflection is very strong such that it approximates achromaticity, in which case inpainting has to be used for residual strong highlight. Due to the fact that there is no generally suitable method capable of automatically detecting or representing the damaged region [12], existing approaches all require the region that needs to be restored to be manually selected by the user. Now our main goal is structure-based color recovery for those pixels contaminated by quite strong highlight, and we still take chromaticity as constraint to perform hierarchic inpainting and operate on single pixel, which differs from those methods using neighboring pixel information. We classify all remaining highlighted pixels into n grades according to their saturation values and assign n diffuse chromaticity values from Λ' to Λ'_n to them, each saturation grade corresponds with a special diffuse chromaticity; and then, we replace the intensities of the highlighted pixels in each grade with the intensity of corresponding critical diffuse point.

4   Implement

We take "workshop" as the example. The image is captured by a common digital camera under incandescent light bulb illumination. As seen in Fig. 1(a), the green paint floor has quite strong specular reflection which will be a severe disturbance for AGV autonomous run. We take a higher $\tilde{\sigma}'$ value as the Λ' and set $\Lambda'_L = 0.9\Lambda'$ to iteratively separate the image specular reflection and make hierarchic inpainting. Compared with the highlight removal using previous identical chromaticity method, the processed result of proposed approach is quite improved.

5   Conclusions

The effect of highlight on real-world images and the separability of specular reflection are clearly explained and evaluated. With a new model and through straightforward calculation, the proposed application-oriented method can removes weak and stronger highlights of real-world images and avoid excessive intensity reduction. The proposed gradual chromaticity concept provides a wider path to separate specular reflection and let the remained highlight pixels needing inpainting be least, whilst an acceptable color deviations is retained. Experiment testifies the proposed approach achieve a better result for highlight removal, this is quite useful for solving the problem of specular reflection disturbance in manufacturing scene.

Acknowledgements. This research was supported by Innovation Project of Guangdong, China under grant 2013B011301026.

References

1. An, D., Suo, J., Wang, H.: Illumination estimation from specular highlight in a multi-spectral image. Opt. Express **23**(13), 17008–17023 (2015)
2. Zhang, H., Han, S., Liu, S., et al.: 3D shape reconstruction of large specular surface. Appl. Optics **51**, 7616–7625 (2012)
3. Missael, G.N., Ignacio, E., Christopher, E.: Surface normal reconstruction using circularly polarized light. Opt. Express **23**(11), 14391–14406 (2015)
4. Tan, R., Ikeuchi, K.: Separating reflection components of textured surfaces using a single image. IEEE T. Pattern Anal **27**(2), 178–193 (2005)
5. Tan, R.T., Nishino, K., Ikeuchi, K.: Separating reflection components based on chromaticity and noise analysis. IEEE T. Pattern Anal **26**, 1373–1379 (2004)
6. Yoon, K., Choi, Y.: Fast separation of reflection components using a specularity-invariant image representation. In: Proceedings of IEEE Conference on Image Processing, pp. 973–976 (2006)
7. Jung, C., Jiao, L., Qi, H., Yang, Y.: Specularity removal based on reflection component separation and joint bilateral filtering. Opt. Eng. **51**, 107005 (2012)
8. Shen, H., Zheng, Z.: Real-time highlight removal using intensity ratio. Appl. Optics **52**, 4483–4493 (2013)
9. Je, C., Park, H.M.: BREN: body reflection essence-neuter model for separation of reflection components. Opt. Lett. **40**, 1940–1943 (2015)
10. Shafer, S.A.: Using color to separate reflection components. Color Res. Appl. **10**(4), 210–218 (1985)
11. Tan, R., Nishino, K., Ikeuchi, K.: Color constancy through inverse-intensity chromaticity space. Opt. Soc. Am. A **21**(3), 321–334 (2004)
12. Lu, Z., et al.: A novel exemplar-based image completion scheme with adaptive TV-constraint. In: Proceedings of the IEEE 4th International Conference on Genetic and Evolutionary Computing, pp. 94–97 (2010)

Hyperspectral Image Feature Extraction Based on Multilinear Sparse Component Analysis

Dong Guangjun, Zhou Yawen$^{(\boxtimes)}$, Shao Lei, and Yao Qiangqiang

Information Engineering University, Zhengzhou, Henan, China
yawen236236@163.com

Abstract. Feature extraction generates a low-dimensional representation of high-dimensional sample data, which retaining most of the information regarding the spatial feature and the spectral feature. A new hyperspectral image tensor feature extraction method based on Multilinear Sparse Principal Component Analysis is proposed in this paper. Experimental results indicate that the proposed method can maintain the spatial-spectral information and discriminating information, which has better classification accuracy than other algorithms when it is applied to the classification images, and the overall classification accuracies reach 96.36 and 95.00%, respectively.

Keywords: Feature extraction · Hyperspectral imagery · Multilinear sparse principal component analysis · Classification

1 Introduction

The technology of hyperspectral remote sensing is a milestone in modern remote sensing and has been widely used in agricultural production, target recognition, military reconnaissance, urban planning and other fields [1, 2]. With the development of hyperspectral imagery, spatial structure is considered as complementary information of spectral features in the hyperspectral image dimensionality reduction. It is now commonly accepted that using the tensor feature accepts significant advantages in dimension reduction.

Typical tensor objects in pattern recognition or machine vision applications are usually specified high-dimensional tensor spaces. Complexity is a character of the high-dimensional data, which has amount of calculation and problem of small-samples. However, since the tensor object has a strong correlation with the surrounding object, it can be assumed that the tensor space is highly constrained in the low-dimensional manifold space. Its advantage is the ability to directly to the three-dimensional tensor form of projection data to low dimensional tensor subspace without changing the spatial structure of the image data. On this basis, improve the calculation speed, processing efficiency and solve the problem of high dimension and small sample.

Although tensor-based feature extraction can be used to solve the above problems of hyperspectral image dimensionality reduction, it can obtain better classification effect. However, it is reasonable to consider the sparse structure of the data due to the variety of complex objects in the high spectral image and the irregular distribution of samples. Therefore, this paper introduces sparse representation into MPCA to reduce

© Springer International Publishing AG 2018
F. Qiao et al. (eds.), *Recent Developments in Mechatronics and Intelligent Robotics*,
Advances in Intelligent Systems and Computing 691, DOI 10.1007/978-3-319-70990-1_81

the dimensionality of hyperspectral remote sensing image, and aim at obtaining more accurate classification using the advantages of sparse and tensor learning.

The paper is structured as follows. In Sect. 2, a survey on related work is discussed. Section 3, explains the proposed work where the processes involved in Multilinear Sparse Principal Component Analysis (MSPCA) based feature extraction and classification are discussed. Results are discussed in Sect. 4 followed by conclusions and future work to be carried out in Sect. 5.

2 Literature Survey

Recently, scholars have proposed a variety of tensor-based feature extraction methods for hyperspectral imagery. The following introduction is divided into high-order tensor analysis and multilinear subspace related to tensor object.

First, relevant researches on high order tensor are introduced. Lathauwer LD *et al.* [3] extend Singular Value Decomposition (SVD) to High Order Singular Value Decomposition (HOSVD) for higher-order tensors. And based on the idea of HOSVD, they proposed multilinear generalization of multi-symmetric tensor with eigenvalue decomposition by contrasting the tensor symmetry. In the same year, Lathauwer LD *et al.* [4] had research alternating Least-Squares on rank-$(R_1, R_2, \cdots, R_N)$, approximation of the higher order tensor. Lee CS *et al.* [5] used HOSVD to simulate the multilinear generation model for subsequent analysis. In [5–7], the tensor data is projected in the original coordinates without the center of the data. In the attempt of the feature problem solution, the feature decomposition is affected by the mean of the data set in each mode.

Followed by the study of multilinear subspace of tensor object feature extraction and classification. Lu H et al. [8] proposed Multilinear Principal Component Analysis (MPCA) to obtain a multi-linear transformation (projection) set. D. Xu *et al.* [9] encoded an image object into a second or higher order tensor to derive a representative subspace, and proposed Concurrent Subspaces Analysis (CSA). Both Multilinear Discriminant Analysis (MLDA) [10] and Discriminant Analysis with Tensor Representation (DATER) [11] use an iterative algorithm similar to ALS to maximize the tensor criterion. However, the variables of MLDA can't converge and are affected by parameters. Since the subspace dimension of the tensor object may be very high, the exhaustive method determines that the parameter is not feasible. Thus, the algorithm of Yan S *et al.* [10, 11] can not determine the subspace dimension comprehensively and systematically.

3 MSPCA for Hyperspectral Image Feature Extraction

Assume that the training samples are represented as the nth-order tensor $\{X_i \in R^{m_1 \times m_2 \times \cdots \times m_n}, i = 1, 2, \ldots, N\}$, where N denotes the total number of the training samples. According to the relationship between multilinear regression problems in any given mode and the target function of MPCA, the proposed method imposes a lasso penalty on the normalized regression representation of the MPCA regression criteria to obtain the multilinear sparse principal vectors. MSPCA can be stated as follows:

$$\min_{B_k, U_k} \sum_i \left\| X_i^k - B_k U_k^T X_i^k \right\|_F^2 + \alpha_k \| U_k \|_F^2 + \sum_j^{d_k} \beta_{k,j} | u_k^j | (\forall k) \quad s.t.\, B_k^T B_k = I_k \quad (1)$$

where $\beta_{k,j} > 0$.

The steps of MSPCA are as follows:

Step 1: Center the training samples.

Step 2: Initialize U_k^1, $B_k^1 |_{k=1}^n$ as arbitrary columnly-orthogonal matrices.

Step 3: For $t = 1:T_{\max}$, $k = 1:n$, compute X_i^k and perform the mode-k flattening of the nth-order tensor to matrices, where $X_i^k = X_i \times {}_1 U_1^{tT} \cdots \times {}_{k-1} U_{k-1}^{tT} \times {}_{k+1} U_{k+1}^{tT} \cdots \times {}_n U_n^{tT}$.

Step 4: Repeat the Elastic Net and SVD until $\widehat{U}_k$ converges.

Step 5: Out the multilinear sparse subspaces.

MSPCA solution is as follows:

According to (1), we have

$$\sum_i \left\| X_i^k - B_k U_k^T X_i^k \right\|_F^2 + \alpha \| U_k \|_F^2 + \sum_j^{d_k} \beta_{k,j} | u_k^j |$$
$$= tr\left(\sum_i X_i^k X_i^{k^T} \right) + \sum_j \left(u_k^{jT} \left[\left(\sum_i X_i^k X_i^{k^T} \right) + \alpha \right] u_k^j - 2 b_k^{jT} \left(\sum_i X_i^k X_i^{k^T} \right) u_k^j + \beta_{k,j} | u_k^j | \right) \quad (2)$$
$$= tr(S_T^k) + \sum_j \left(u_k^{jT} \left(S_T^k + \alpha \right) u_k^j - 2 b_k^{jT} S_T^k u_k^j + \beta_{k,j} | u_k^j | \right)$$

if B_k is given, then the optimal sparse solutions can be converted to the following formula.

$$u_k^{jT} \left(S_T^k + \alpha \right) u_k^j - 2 b_k^{jT} S_T^k u_k^j + \beta_{k,j} | u_k^j | \quad j = 1, \ldots, m_k' \quad (3)$$

On the other hand, we also have

$$\sum_i \left\| X_i^k - B_k U_k^T X_i^k \right\|_F^2 + \alpha \| U_k \|_F^2 + \sum_j^{d_k} \beta_{k,j} | u_k^j |$$
$$= tr\left(\sum_i X_i^k X_i^{k^T} \right) - 2 tr\left[B_k^T \left(\sum_i X_i^k X_i^{k^T} \right) U_k \right] + tr\left[U_k^T \left(\sum_i X_i^k X_i^{k^T} + \alpha I_k \right) U_k \right] + \sum_j \beta_{k,j} | u_k^j |$$
$$\qquad (4)$$

when U_k is known and fixed, the problem of minimizing (4) becomes the following maximization problem:

$$\max tr\left[B_k^T \left(\sum_i X_i^k X_i^{k^T} \right) U_k \right] \quad s.t.\, B_k^T B_k = I_k \quad (5)$$

According to Theorem 4 in [5], the optimal solution of the above problem is:

$$\boldsymbol{B}_k^* = \bar{\boldsymbol{U}}\bar{\boldsymbol{V}}^T \tag{6}$$

Where $\bar{\boldsymbol{U}}$ and $\bar{\boldsymbol{V}}$ are the SVD of $\left(\sum_i X_i^k X_i^{k^T}\right)U_k$, i.e. $\left(\sum_i X_i^k X_i^{k^T}\right)U_k = \bar{\boldsymbol{U}}\bar{\boldsymbol{D}}\bar{\boldsymbol{V}}$.

4 Experimental Results and Analysis

4.1 Datasets

In order to verify the correctness and effectiveness of the proposed method, Jiaxing dataset and Cuprite dataset were selected to carry out the experiments.

(1) Jiaxing dataset was collected by the AISA sensor over the Jiaxing, Zhe Jiang, China. The image size in pixels is 525×356. The number of data channels in the acquired image is 126 (with spectral range from 0.38 to 2.5). Figure 1(a) shows the band combination (95, 64 and 33) of the image, while Fig. 1(b) shows six reference classes of interest.

 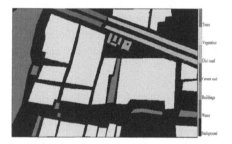

(a) False color image   (b) Ground truth map

Fig. 1. Jiaxing hyperspectral image

(2) Cuprite dataset was collected by AVIRIS sensor over Nevada, America. It consists of 204 spectral bands with 500×320 pixels and spectral resolution of 10 nm. Figure 2(a) shows the band combination (178, 111 and 33) of the image, while Fig. 2(b) shows eight reference classes of interest.

4.2 Experimental Settings

To evaluate the validity of proposed MSPCA method, we compare the proposed method with PCA, MPCA and SPCA. In supervised feature extraction classification methods, fifty samples from each category are selected as the training set and the remaining ones as the testing set. In experiments, the paper set the extracted dimension as $C-1$, and C is the number of the classes. This paper classifies the result of feature

(a) False color image

(b) Ground truth map

Fig. 2. Cuprite hyperspectral image

extraction methods using SVM classifier and regards OA, AA and Kappa coefficients as the classification and evaluation indexes.

4.3 Jiaxing Data Experimental Results and Analysis

For performance, three quantitative indexed are used: overall accuracy (OA), average accuracy (AA) and kappa coefficient (Kappa).

As shown in Table 1, the methods with MSPCA achieve better performance compared with other methods. The overall classification accuracy, average classification accuracy and Kappa coefficient reach highest in our proposed method. Especially for the water, the accuracy can reach 85.81%.

Table 1. Classification accuracy of different class data with different methods on Jiaxing data

#	Class	Train samples	Test samples	Original	PCA	SPCA	MPCA	MSPCA
1	Trees	50	4596	99.72	**99.99**	**99.99**	**99.99**	**99.99**
2	Vegetation	50	72888	**92.26**	88.98	87.07	88.80	90.26
3	Dirt road	50	6052	94.33	98.03	98.19	**98.66**	98.38
4	Cement road	50	2538	98.96	99.62	99.62	99.62	**99.60**
5	Buildings	50	1098	91.34	94.25	95.80	95.96	**96.07**
6	Water	50	14414	64.16	79.61	74.98	82.33	**85.81**
	OA			91.84	94.76	95.65	96.12	**96.36**
	AA			90.13	93.41	92.61	94.23	**95.02**
	Kappa			83.52	89.12	90.78	91.81	**92.33**

The classification maps in the Jiaxing image are illustrated in Fig. 3. It can be seen that the MSPCA method has less category noise in the classification diagram, and the spatial continuity is effectively enhanced.

(a) Ground   (a) Original   (a) PCA   (a) SPCA   (a) MPCA   (a) MSPCA

Fig. 3. Classification maps with different methods on Jiaxing data set

4.4 Cuprite Data Experimental Results and Analysis

In Cuprite dataset experiment, this paper use the same parameter setting method as the Jiaxing data, the classification results are shown in Table 2 and the classification maps are illustrated in Fig. 4. From Table 2, we can see that the overall accuracy, average accuracy and Kappa coefficient in our proposed method are highest, and most of the classes can reach highest classification accuracy in our method. Especially for the Tuff, the classification accuracy can reach 93.33%.

Table 2. Classification accuracy of different class data with different methods on Cuprite data

#	Class	Train samples	Test samples	Original	PCA	SPCA	MPCA	MSPCA
1	Illite	50	552	**96.43**	80.98	80.98	81.16	81.70
2	Dickite	50	39	48.38	**94.87**	**94.87**	**94.87**	**94.87**
3	Tuff	50	240	68.47	84.58	87.92	88.75	**93.33**
4	Dry lake	50	235	86.93	95.32	94.47	**95.74**	**95.74**
5	Quartz	50	983	95.48	99.70	99.70	**100**	**100**
6	Kaolinite	50	723	**99.30**	97.36	97.25	97.25	98.17
7	Buddingtonite	50	345	89.49	**99.42**	98.84	99.71	**99.42**
8	Alumstone	50	712	**99.81**	97.33	97.47	97.33	96.77
	OA			93.14	94.09	94.21	94.47	**95.00**
	AA			85.54	93.69	93.94	94.35	**95.00**
	Kappa			91.70	92.83	92.97	93.29	**93.93**

The classification maps in the Jiaxing image are illustrated in Fig. 4. The results also confirm the validity and determination of the proposed algorithm.

(a) Ground   (a) Original   (a) PCA   (a) SPCA   (a) MPCA   (a) MSPCA

Fig. 4. Classification maps with different methods on Cuprite data set

5   Conclusion

This paper analyzes the current tensor-based feature extraction methods and proposes MSPCA algorithm for hyperspectral imagery. Jiaxing and Cuprite experiments indicate that the proposed method effectively improves ground object classification accuracy of hyperspectral image and the classification maps have better spatial continuity.

References

1. Bioucas-Dias, J.M., et al.: Hyperspectral remote sensing data analysis and future challenges. IEEE Geosci. Remote Sens. Mag. **1**(2), 6–36 (2013)
2. Tong, Q., Xue, Y., et al.: Progress in hyperspectral remote sensing science and technology in China over the past three decades. IEEE J. Sel. Top. Appl. Earth Obs. Remote Sens. **7**(1), 70–91 (2014)
3. Lathauwer, L.D., De Moor, B., Vandewalle, J.: A multilinear singular value decomposition. SIAM J. Matrix Anal. Appl. **21**(4), 1253–1278 (2000)
4. Lathauwer, L.D., De Moor, B., Vandewalle, J.: On the best rank-1 and rank-$(R_1, R_2,..., R_N)$ approximation of higher-order tensors. SIAM J. Matrix Anal. Appl. **21**(4), 1324–1342 (2000)
5. Lee, C.S., Elgammal, A.: Towards scalable view-invariant gait recognition: multilinear analysis for gait. Springer, Berlin (2005)
6. Lathauwer, L.D., Vandewalle, J.: Dimensionality reduction in higher-order signal processing and rank-$(R_1, R_2,..., R_N)$ reduction in multilinear algebra. Linear Algebra Appl. **391**(1), 31–55 (2004)
7. Vasilescu, M., Terzopoulos, D.: Multilinear analysis of image ensembles: Tensorfaces. In: Computer Vision—ECCV 2002, pp. 447–460 (2002)
8. Lu, H., Plataniotis, K.N., Venetsanopoulos, A.N.: MPCA: multilinear principal component analysis of tensor objects. IEEE Trans. Neural Netw. **19**(1), 18–39 (2008)
9. Xu, D., Yan, S., Zhang, L., et al.: Concurrent subspaces analysis. In: 2005 IEEE Computer Society Conference on Computer Vision and Pattern Recognition (CVPR 2005), vol. 2, pp. 203–208 (2005)
10. Yan, S., Xu, D., Yang, Q., et al.: Discriminant analysis with tensor representation. In: 2005 IEEE Computer Society Conference on Computer Vision and Pattern Recognition (CVPR 2005), vol. 1, pp. 526–532 (2005)
11. Yan, S., Xu, D., Yang, Q., et al.: Multilinear discriminant analysis for face recognition. IEEE Trans. Image Process. **16**(1), 212–220 (2007)

Solder Paste on the Quality of the Welding and 3D Detection Method

Junjie Lv[(✉)]

School of Electronic Information Engineering, Wuhan Polytechnic College,
Wuhan, Hubei, People's Republic of China
593316640@qq.com

Abstract. SPI (Solder Paste Inspection) refers to the solder paste detection system, the main function is to detect the quality of solder paste printing, including volume, area, height, XY offset, shape, bridge and so on. How to quickly and accurately detect very small solder paste, the general use of PMP (Chinese translation for the phase modulation profile measurement technology) and Laser (Chinese translation for the laser triangulation technology) detection principle.

Keywords: SPI · Laser · Solder paste detection · Phase modulation

1 Laser Triangulation Technology

With the electronics industry toward the miniaturization, intelligent, integrated direction of development, smaller components and denser IC pin spacing on the solder paste printing requirements to improve the solder paste printing in SMT printing production has become a key process, Affecting the quality of welding products. Solder paste is a uniform solder alloy powder and the stability of the mixture of uniform flux according to certain proportion of the paste body. When welding assembly components can make the surface of lead or endpoint and shapes on the PCB solder alloy connection. Solder paste in the metal content increases, the solder paste viscosity increase, can effectively resist the force created by the preheating process of vaporization, the increase of the content of metal to metal powder are close, in the melt when combined with without being blown, the increase of metal content may also reduce the solder paste printing after collapsing. The amount of flux in solder paste if too much, can cause local cave solder paste, result in tin beads is easy to produce. Active hours solder paste, flux, flux oxidation ability is reduced, it will make the solder ball easy to produce. (flux activity levels into R inactive, RMA moderate activity levels, RA, fully active, RSA super activity). In the solder paste, the higher the degree of metal oxide in welding metal powder combined with the resistance, the greater the solder paste and solder and components between the less invasion, thus reduce solderability. Metal powder particle size is smaller, the greater the total surface area of solder paste, so that a higher degree of oxidation of fine powder, lead to solder ball phenomenon. Most of the solder paste for short periods of time can endure 26.6 DHS C is the highest temperature, temperature can lead to high flux and solder paste from ontology, and transform the liquidity of the

© Springer International Publishing AG 2018
F. Qiao et al. (eds.), *Recent Developments in Mechatronics and Intelligent Robotics*,
Advances in Intelligent Systems and Computing 691, DOI 10.1007/978-3-319-70990-1_82

solder paste, further lead to bad printing. Generally do not recommend freezing solder paste, because can lead to precipitation of catalysts, reduce the welding performance. All the solder paste will absorb moisture, so to avoid in high humidity. If inhaled too much moisture, tin in use will cause explosion, residue, tin ball, in the reflow soldering components will shift, and poor welding, etc. According to the above situation, in the process of teaching should make students clear, training workshop of solder paste should be stored in an airtight form in constant temperature, and humidity of the air-conditioner, save the temperature of 0, 10 DHS C (in such conditions can be kept up to 6 months). If the high temperature alloy powder in the solder paste and flux react will reduce viscosity activity affect performance; Resin will produce crystallization in the low temperature solder, solder paste is form bad; If a short period of time the solder paste repeatedly appear different in different environment temperature change can make the solder paste, flux performance change and influence welding quality. Solder paste, solder paste storage environment influence on process and use requirements, solder paste and control decision-making aspects in this paper, the defects caused by the importance of solder paste in SMT production printing, the solder paste is in use and welding problems arising from the targeted control decisions are put forward.

Using the detection light source for the laser, the laser beam in different height plane distortion, the detection head in a certain direction of continuous movement, the camera according to set the time interval to take pictures, so as to obtain a set of laser distortion data, and then calculate the test results (Fig. 1).

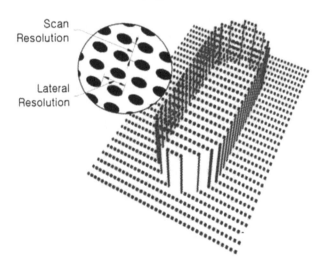

Fig. 1. Laser beam in different height plane distortion

Advantages: Faster detection
Disadvantages:

(1) Laser resolution is low, generally only 10–20 um level.
(2) Single sampling, low repeatability accuracy.

(3) Sampling in the movement, the external shock and transmission vibration on the detection of a greater impact.

(4) Laser monochromatic light on the PCB board color is weak.

2   PMP Phase Modulation Contour Measurement Technology

(1) Use a white light source to measure the solder paste by changing the phase of the structural grating (Fig. 2)

Fig. 2. Solder paste by changing the phase of the structural grating

(2) The height value of the high precision is obtained by using the gray change of the structure grating (Fig. 3)

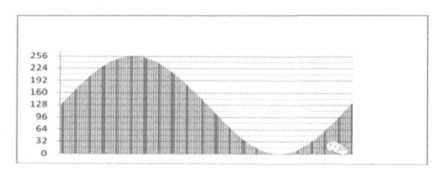

Fig. 3. The height value of the high precision

(3) With the phase change, for each solder paste 8 times to ensure that the detection of high repeatability accuracy (Fig. 4)

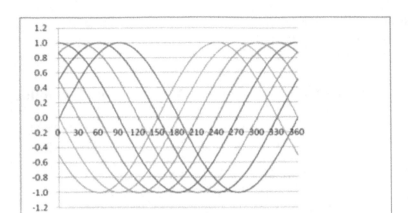

Fig. 4. The solder paste 8 times to ensure

(4) PMP technology is divided into FOV walk-off and Scan scanning two detection methods

   When the test is carried out, the motion is not sampled and does not move during sampling. Minimizing the impact of vibration on the detection.
   Advantages:

(1) PMP principle detection resolution is high, 0.37um.
(2) Stable multi-sampling, detection of high repeatability accuracy.
(3) The color of the PCB is not critical.

   Disadvantages: the speed is relatively slow.
   The phase change of the structural grating is formed by the continuous motion of the detecting head. Sampling while moving.
   Advantages:

(1) PMP principle detection resolution is high, 0.37um.
(2) The color of the PCB is not critical.
(3) Multiple sampling, detection repeatability than laser-type equipment.

   Disadvantages: The impact of external shocks, detection repeatability is low.

3 The Structure of Programmable Grating (PSLM)

Programmable grating structure (PSLM): implements the software control of the grating structure movement, avoid the traditional piezoelectric ceramics (PZT) motor drive the necessary machinery, glass Moore grain grating reduce mechanical wear and customer maintenance cost.

Using advanced phase contour modulation measurement technique (PMP), 8 bits of gray-scale resolution, 0.37 micron detection resolution, compared with the laser measuring accuracy increased two orders of magnitude, greatly improving the equipment of detection ability and scope of application.

The influencing factors on the quality of the solder paste welding, solder paste viscosity and quality with time, environment temperature, humidity, environmental health and change; Therefore, the establishment of a complete set of printing process control file is very necessary. To ensure the quality of SMT products, it is necessary to analyze the key factors in the production of each link of the research, work out effective control method, as a key process of solder paste printing is Paramount, only to develop a suitable parameters, and master the law between them, to get the high quality of the solder paste printing quality.

References

1. Mallik, S., Schmidt, M., et al.: Influence of solder paste components on rheological behavior. In: 2nd Electronics Systemintegration Technology Conference (IEEE), pp. 1135–1140 (2008)
2. Wanda, H., Ningcheng, L.: Solder beading in SMT challenge. SITE Magazine (1987)
3. Ningcheng, L.: Optimizing reflow profile via defect mechanism analysis. IPC Printed Circuits Expo (1998)
4. Al-Sakran, H.O.: Framework architecture for improving healthcare information systems using agent technology. Int. J. Manag. Inf. Technol. (IJMIT) 7(1), 17 (2015)
5. Devi, C.S., Ramani, G.G., Pandian, J.A.: Intelligent E-healthcare management system in medicinal science. Int. J. PharmTech Res. 6(6), 1838–1845 (2014)
6. Wang, Y.L., Li, G.Z., Xu, S.W., Liu, G.P., Wang, Y.Q.: Symptom selection of inquiry diagnosis data for coronary heart disease in traditional Chinese medicine by using social network techniques. In: IEEE International Conference on Bioinformatics and Biomedicine Workshop (2010)

Author Index

Printed in the United States
By Bookmasters